AF817591

Mass Spectrometry in Biomedical Research

Mass Spectrometry in Biomedical Research

Edited by
SIMON J. GASKELL
Tenovus Institute for Cancer Research
University of Wales College of Medicine
Cardiff, UK

(Present address:
Baylor College of Medicine
Houston, Texas, U.S.A.)

JOHN WILEY & SONS
Chichester · New York · Brisbane · Toronto · Singapore

Copyright © 1986 by John Wiley & Sons Ltd.

All rights reserved.

No part of this book may be reproduced by any means, or transmitted, or translated into a machine language without the written permission of the publisher

Library of Congress Cataloging-in-Publication Data:

Mass spectrometry in biomedical research.

Includes index.
1. Mass spectrometry. 2. Biological chemistry—Technique. 3. Biomolecules—Analysis. I. Gaskell, S. J.
[DNLM: 1. Spectrum Analysis, Mass. QC 454.M3 M4143]
QP519.9.M3M36 1986 574.19′285 86–5601

ISBN 0 471 91045 7

British Library Cataloguing in Publication Data:

Mass spectrometry in biomedical research.
1. Mass spectrometry 2. Biological chemistry—Technique
I. Gaskell, S. J.
574.19′285 QP519.9.M3

ISBN 0 471 91045 7

Typeset by Bath Typesetting Ltd.
Printed and bound in Great Britain by Biddles.

CONTENTS

PART III TRACE ANALYSES

CONTRIBUTORS

Michael P. Barbalas — *Department of Drug Metabolism, Hoffmann-La Roche Inc., Nutley, NJ 07110, U.S.A.*

Olof Beck — *Pasarow Analytical Neurochemistry Facility, Nancy Pritzker Laboratory of Behavioral Neurochemistry, Stanford University School of Medicine, Stanford, CA 94305, U.S.A. (Present address: Department of Toxicology, Karolinska Institute, Stockholm S-104-01, Sweden).*

Robert S. Bordoli — *VG Analytical Ltd., Floats Road, Wythenshawe, Greater Manchester M23 9LE, U.K.*

Harry O. Brotherton — *Chemistry Department, University of Florida, Gainesville, FL 32611, U.S.A. (Present address: Chemistry Department, Northeast Louisiana University, Monroe, LA 71209, U.S.A.)*

Richard M. Caprioli — *Department of Biochemistry and Molecular Biology and the Analytical Chemistry Center, The University of Texas Medical School at Houston, Houston, TX 77030, U.S.A.*

Mauro E. Castro — *Department of Chemistry, Texas A & M University, College Station, TX 77843, U.S.A.*

Catherine E. Costello — *Mass Spectrometry Facility, Department of Chemistry, Massachusetts Institute of Technology, Cambridge, MA 02139, U.S.A.*

Thomas Covey — *Drug Testing and Toxicology, New York State College of Veterinary Medicine, Cornell University, 925 Warren Drive, Ithaca, NY 14850, U.S.A.*

Anne Dell — *Department of Biochemistry, Imperial College, London, SW7 2AZ, U.K.*

Peter J. Derrick — *School of Chemistry, University of New South Wales, P.O. Box 1, Kensington, New South Wales, Australia.*

Dominic M. Desiderio — *Departments of Neurology and Biochemistry and The Charles B. Stout Neuroscience Mass Spectrometry Laboratory, University of Tennessee Center for the Health Sciences, Memphis, TN 38163, U.S.A.*

Kym F. Faull — *Pasarow Analytical Neurochemistry Facility, Nancy Pritzker Laboratory of Behavioral Neurochemistry, Stanford University School of Medicine, Stanford, CA 94305, U.S.A.*

Genevieve H. Fridland	*The Charles B. Stout Neuroscience Mass Spectrometry Laboratory, University of Tennessee Center for the Health Sciences, Memphis, TN 38163, U.S.A.*
William A. Garland	*Department of Drug Metabolism, Hoffmann-La Roche Inc., Nutley, NJ 07110, U.S.A.*
Simon J. Gaskell	*Tenovus Institute for Cancer Research, University of Wales College of Medicine, Heath, Cardiff, CF4 4XX, U.K.*
Valerie J. Gould	*Tenovus Institute for Cancer Research, University of Wales College of Medicine, Heath, Cardiff, CF4 4XX, U.K.*
Brian N. Green	*VG Analytical Ltd., Floats Road, Wythenshawe, Greater Manchester, M23 9LE, U.K.*
Lutz Grotjahn	*Gesellschaft für Biotechnologische Forschung mbH, Mascheroder Weg 1, D-3300 Braunschweig, Federal Republic of Germany.*
Timothy W. Harper	*Department of Pharmacology, University of Colorado Health Sciences Center, 4200 East 9th Avenue, Denver, CO 80262, U.S.A.*
David J. Harvey	*Department of Pharmacology, University of Oxford, South Parks Road, Oxford, OX1 3QT, U.K.*
Jack Henion	*Drug Testing and Toxicology, New York State College of Veterinary Medicine, Cornell University, 925 Warren Drive, Ithaca, NY 14850, U.S.A.*
Masataka Ishibashi	*Research Laboratories of Pharmaceutical Group, Nippon Kayaku Co., 3–31 Shimo, Kita-Ku, Tokyo 115, Japan.*
Ian Jardine	*Department of Pharmacology, Mayo Clinic, Rochester, MN 55905, U.S.A.*
Jodie V. Johnson	*Chemistry Department, University of Florida, Gainesville, FL 32611, U.S.A.*
Maw Rong Lee	*Chemistry Department, University of Florida, Gainesville, FL 32611, U.S.A.*
Mike S. Lee	*Chemistry Department, University of Florida, Gainesville, FL 32611, U.S.A.*
Heather M. Leith	*Tenovus Institute for Cancer Research, University of Wales College of Medicine, Heath, Cardiff, CF4 4XX, U.K.*
James A. McCloskey	*Department of Medicinal Chemistry, College of Pharmacy, University of Utah, Salt Lake City, UT 84112, U.S.A.*
David S. Millington	*Division of Genetics and Metabolism, Department of Pediatrics, Box 3028, Duke University Medical Center, Durham, NC 27710, U.S.A.*
Hiroshi Miyazaki	*Corporate Planning Office, Nippon Kayaku Co., 1-11-12 Fujimi, Chiyoda-ku, Tokyo 102, Japan.*

Robert C. Murphy	*Department of Pharmacology, University of Colorado Health Sciences Center, 4200 East Ninth Avenue, Denver, CO 80262, U.S.A.*
Maria Panico	*Department of Biochemistry, Imperial College, London, SW7 2AZ, U.K.*
Vernon Reinhold	*Department of Biological Chemistry, Harvard Medical School, and Department of Nutrition, Harvard School of Public Health, Boston, MA 02115, U.S.A.*
Peter Roepstorff	*Department of Molecular Biology, Odense University, DK-5230 Odense M, Denmark.*
David H. Russell	*Department of Chemistry, Texas A & M University, College Station, TX 77843, U.S.A.*
Margaret M. Sheil	*School of Chemistry, University of New South Wales, P.O. Box 1, Kensington, New South Wales, Australia.*
Kenneth M. Straub	*Department of Drug Metabolism, Smith Kline and French Laboratories, Philadelphia, PA 19101, U.S.A.*
Bo Sundqvist	*Tandem Accelerator Laboratory, University of Uppsala, Box 533, S-75121 Uppsala, Sweden.*
Keiko Watanabe	*Research Laboratories of Pharmaceutical Group, Nippon Kayaku Co., 3–31 Shimo, Kita-ku, Tokyo 115, Japan.*
Joanne O. Whitney	*Department of Pharmaceutical Chemistry and Liver Center, University of California, San Francisco, CA 94143, U.S.A.*
Kouwa Yamashita	*Research Laboratories of Pharmaceutical Group, Nippon Kayaku Co., 3–31 Shimo, Kita-ku, Tokyo 115, Japan.*
Richard A. Yost	*Chemistry Department, University of Florida, Gainesville, FL 32611, U.S.A.*

EDITOR'S PREFACE

It may be argued that in the mid-1970's the subject of mass spectrometry had reached something of a plateau. Important applications were being made in the biomedical field, and others, but the range of applicability of mass spectrometry was considered to be quite rigidly circumscribed. Volatile compounds of (in biochemical terms) low molecular weight could be handled with some ease but many important analytes of biochemical interest were excluded on grounds of polarity, lability or molecular weight. The developments of recent years in ionization techniques and methods for the analysis of compounds of higher molecular weight have dramatically changed this position. The impact on biomedical research has been such that the advancement of mass spectrometric techniques and of certain areas of biomedical research are now inextricably linked and the demands of the biochemist constitute the principal driving force in the development of mass spectrometry.

Such is the justification for the present book, which seeks to bring together contributions from innovators in mass spectrometry whose major motivation is the desire to solve problems in biomedical research. Contributors were asked to summarize their own work in a particular area and place it in the context of work performed by others. The authors were not specifically asked to review the area concerned, though several have chosen to produce useful surveys. Some variation in style is, of course, evident between the various chapters but a sufficient uniformity of presentation has been imposed to facilitate cross-referencing between chapters.

The chapters are divided, somewhat artificially, into three Parts. The first two—'Analyses of labile and polar compounds' and 'Analyses at high mass'—reflect the major developments noted above. The third Part—'Trace analyses'—concerns an area where progress has been steady, rather than a result of dramatic innovation, but which has nevertheless made a major contribution to biomedical research.

The divisions between the Parts are necessarily blurred; several of the chapters would fit equally comfortably in two of the Parts but their location reflects the principal emphasis. The 'Overview' chapters have been incorporated to highlight overlaps between the subject matter of the separate Parts, to provide important background material and to fill any gaps in the coverage provided by the contributed chapters. It may be obvious that the 'Overview' chapters—and the overall balance of the book—reflect the Editor's personal perception of the state of the subject. Nevertheless, it is hoped that this volume, *in toto*, will give an impression of what may currently be achieved in biomedical mass spectrometry, a sense of the areas of present excitement and a feeling for those directions where future developments may be most important.

ACKNOWLEDGEMENTS

It is an unfortunate but logical fact that scientists whose contributions to a book such as this are most to be welcomed are also those with least time to devote to such work. I am therefore particularly grateful that virtually all the invitations to contribute were accepted and that, in the main, the various deadlines were met. I thank all the authors for their contributions.

Part of the editorial work for this volume was carried out during a year spent in the Laboratory of Molecular Biophysics, National Institute of Environmental Sciences (U.S.A.). I am grateful to Dr. Colin Chignell, Laboratory Chief of Molecular Biophysics, for that opportunity. My principal debt with respect to my work, however, is to the Tenovus Institute for Cancer Research, its Director, Professor Keith Griffiths, and to colleagues and friends during my 9-year association with the Institute. Much generous funding was provided by Tenovus, a remarkable organization which has supported, and continues to support, innovative research in basic and clinical sciences related to cancer research.

Finally, I am grateful to my wife and family for their support during this and other projects which must have seemed less important to them than they did to me.

Simon J. Gaskell

COMMON ABBREVIATIONS

APCI	: atmosphere pressure chemical ionization
B	: magnetic sector *or* magnetic field strength
CA	: collisional activation
CAD	: collisionally-activated decomposition
CI	: chemical ionization
CID	: collision-induced decomposition
DCI	: direct/desorption chemical ionization
DLI	: direct liquid introduction
E	: electric sector or electric sector voltage
EC	: electron capture
EI	: electron impact, electron ionization
FAB	: fast atom bombardment
FD	: field desorption
FT–ICR	: Fourier transform-ion cyclotron resonance
GC	: gas chromatography
GC–MS	: gas chromatography–mass spectrometry
HPLC	: high pressure/performance liquid chromatography
LC–MS	: liquid chromatography–mass spectrometry
MIKES	: mass-analysed ion kinetic energy spectrometry
MS	: mass spectrometry
MS/MS	: tandem mass spectrometry
m/z	: mass/charge ratio
PD	: plasma desorption
Q	: quadrupole
SIM	: selected ion monitoring
SIMS	: secondary ion mass spectrometry
SRM	: selected reaction monitoring
TMS	: trimethylsilyl
TOF	: time-of-flight
TSP	: thermospray

PART I

ANALYSES OF LABILE AND POLAR COMPOUNDS

Mass Spectrometry in Biomedical Research
Edited by S. J. Gaskell
© 1986 John Wiley & Sons Ltd.

Chapter 1

OVERVIEW TO PART I

Simon J. Gaskell

1.1. GENERAL INTRODUCTION

The extremely rapid pace of development of mass spectrometry and related techniques provides a constant challenge to researchers in both the biomedical and physical sciences who wish to keep abreast of the most recent advances. Pertinent review papers and books are cited throughout this volume. As general reviews, the Specialist Periodical Reports produced by the U.K. Royal Society of Chemistry (Rose, 1986, and earlier volumes) are useful, while the impressive biennial reviews by Burlingame and colleagues (Burlingame *et al.*, 1986) have tended to be more critical. The role of a book such as this is to focus attention on particular areas of rapid development and to provide an insight into the problems and potentials, as seen through the eyes of those active in the field.

The general areas of most important recent development have been: (i) analyses of labile and polar compounds; (ii) analyses at high mass; and (iii) trace analyses. The work of individual researchers does, of course, frequently contribute to, and is dependent on, progress in more than one of these areas but the division of this book into these categories has been made to emphasize the general trends. The Overview chapters are designed to highlight the major developments, provide appropriate background material and indicate overlap between the Parts. Table 1 lists the topics covered in each Overview chapter.

This chapter is concerned principally with ionization techniques and their associated sample introduction methods. The contributed chapters in this Part (and, indeed, throughout the book) clearly indicate the major impact that the newer ionization methods have had on a wide range of biomedical problems. The chapters also indicate, however, the continuing value of, for example, electron impact mass spectrometry, particularly in problems of structure elucidation (Chapters 2 and 6). One of the major impacts of techniques such as fast atom bombardment, of course, has been the ability to generate gas phase ions from relatively high molecular weight compounds. This has in turn created new demands for the analysis of high mass ions and this topic is covered specifically in Part II.

TABLE 1. Coverage of the overview chapters.

Part	Overview chapter	Topics covered
I. Analyses of labile and polar compounds	1	Ionization methods and associated sample introduction techniques Liquid chromatography–mass spectrometry
II. Analyses at high mass	9	Mass analyzers; tandem mass spectrometry
III. Trace analyses	18	Sensitivity and selectivity of detection; quantitative procedures

1.2. IONIZATION TECHNIQUES

The major extension of the applicability of mass spectrometry in biomedical research has in large measure been attributable to the development of new ionization techniques and their associated sample handling methods. Table 2 indicates the major techniques applied to the analysis of polar, labile and involatile compounds.

The field desorption technique provided some of the earliest breakthroughs in the analysis of polar and labile compounds; it is discussed below (Section 1.2.1.). In direct or desorption chemical ionization (DCI) methods the sample probe is introduced into the reagent gas plasma. Evaporation processes are favoured over decomposition processes either by rapid heating or by the choice of a probe tip material (e.g. Teflon) which reduces the interaction between the sample and the probe. The major ionization process is probably gas-phase CI. (An analogous extension to the applicability of EI techniques is observed when the sample probe is placed in close proximity to the electron beam). DCI is discussed in more detail by Reinhold in Chapter 11 (Section 11.2.2.1).

Biochemically significant analyses have also been performed with laser desorption but the major development in the analysis of polar and labile compounds has been the use of secondary ion mass spectrometry with a liquid matrix; this is discussed in detail below (Section 1.2.2). Californium plasma desorption has been applied principally in conjunction with time-of-flight mass spectrometry for the analysis of high molecular weight compounds; the technique is described by Roepstorff and Sundqvist in Chapter 15 and the implications for high mass analyzers are discussed in Chapters 9 and 15.

The thermospray ionization technique resulted from development work in combined liquid chromatography–mass spectrometry, which is discussed in Section 1.3.

TABLE 2. Principal ionization techniques applicable to the analysis of polar, labile and involatile compounds.

TECHNIQUE	COMPATIBLE INSTRUMENTATION*	SEE CHAPTERS	PERTINENT REVIEWS
Field desorption (FD)	Sector	3,14,24	Beckey (1977), Schulten (1979), Wood (1982)
Direct/desorption chemical ionization (DCI)	Sector, quadrupole	11	Cotter (1980)
Laser desorption (LD)	Sector, FT-ICR	17	Kistemaker *et al.* (1981)
Secondary ion MS (SIMS)	TOF, FT-ICR sector, quadrupole	17	Benninghoven *et al.* (1985), Day *et al.* (1980)
Fast atom bombardment (FAB)/liquid SIMS	Sector, quadrupole FT-ICR†	2–14,24	Barber *et al.* (1982)
^{252}Cf-plasma desorption (PD)	TOF FT-ICR	15,16	Macfarlane (1981, 1983)
Thermospray (TSP)	Quadrupole, sector	6,7,16	Vestal (1983a, 1984)

* Sector: magnetic instrument, usually double-focusing; FT-ICR: Fourier transform-ion cyclotron resonance; TOF: time-of-flight.

† Using a dual cell or tandem quadrupole-ICR design.

1.2.1. FIELD DESORPTION

Field desorption mass spectrometry was developed by Beckey and co-workers (Beckey, 1969) as an outgrowth of field ionization of gaseous samples. The principles of field ionization/desorption are complex (Beckey, 1977) but the production of ions may be considered to result from the distortion of potential curves for the interaction of analyte molecules with a solid metal surface in the presence of an intense applied electric field. There has been extensive discussion (Burlingame *et al.*, 1980) concerning the relative importance of the field ionization and other contributory mechanisms, such as thermally-assisted processes (Cotter and Yergey, 1981, and references therein). While the situation is mechanistically complex, there can be little doubt that the applied field facilitates charge separation. In other procedures of ion production where the *applied* field is weak or absent (notably thermospray ionization; Section 1.3), locally intense electric fields may be generated and play a significant role in gaseous ion production.

Substantial contributions to biomedical MS have been made using FD; Schulten (1979) has reviewed many of these studies. More widespread application to the analysis of biological samples, however, has been restricted, in part, by the dependence of spectral appearance on analytical conditions (as in the studies of a steroid glucuronide by Adlercreutz *et al.*, 1974). A more important discouragement to further use, however, has been the experimental difficulty of the technique, limiting its most productive use to a relatively small number of expert practitioners (compared to the later FAB/SIMS procedures). In certain instances, however, there is a continuing analytical utility for FD, arising principally from the lack of 'chemical' noise which is achieved in the absence of a liquid matrix.

1.2.2. FAST ATOM BOMBARDMENT

Fast atom bombardment (FAB) is properly considered to be a variant of secondary ion mass spectrometry (SIMS). Much of the groundwork for organic SIMS was laid by Benninghoven and co-workers (Benninghoven and Sichtermann, 1978) who recognized the importance of establishing appropriate conditions for 'static' SIMS, in which sample damage is minimal. FAB was originally conceived by Barber and co-workers (Barber *et al.*, 1981a) as a means of avoiding surface charging of the sample and the difficulty of focusing a primary ion beam in the presence of the high source potential of a double-focusing mass spectrometer. The enormous impact of FAB MS, however, has resulted from the innovation, also by Barber and co-workers (1981b), of sample introduction in a liquid matrix, such as glycerol. Perhaps ironically, it has subsequently been shown by Aberth *et al.*, (1982), and by others, that equivalent results to argon FAB may be obtained using a primary beam of Cs^+ ions. Thus the charge state of the primary beam does not appear to be important to secondary ion production, though its momentum certainly is. The principal advantage of the use of the liquid matrix is the prolongation of the ion signal, presumably due to a replenishment of the surface population of analyte. This is achieved at the cost of an increased background signal attributable to the liquid matrix itself.

The mechanisms involved in production of gas phase ions in FAB/SIMS are still the subject of debate. Direct sputtering of 'preformed' ions certainly takes place, but the observation of ions from species not ionic in the condensed phase (such as hydrocarbons) suggests the importance of ion/molecule reactions in the selvedge (the high pressure region adjacent to the target surface) (Day *et al.*, 1980). Despite the incomplete understanding of mechanisms, several important implications for biochemical analyses are clear.

(i) Under certain conditions, the population of gaseous ions produced may accurately reflect the composition of the sample solution (Chapter 4, Section 4.4). Nevertheless, the relative abundances of the ions formed cannot be assumed to be the same as the relative concentrations of the components of the sample mixture. For quantitative analyses, this emphasizes that, in general, the use of isotopically-labelled internal standards is highly desirable.

(ii) The chemistry of the condensed phase, both before and after exposure to the primary atom beam, may significantly affect the observation of gaseous ions. Thus, reactions between the components of the sample solution (Chapter 3, Section 3.4) and radiation damage of the sample (Field, 1982) may occur. Under appropriate circumstances, well-understood condensed phase effects may assist in sample characterization (Chapter 4, Section 4.5).

(iii) Since it is the surface of the liquid matrix that is sampled in the FAB process (Beckner and Caprioli, 1984), modification of the chemistry of the analyte to increase surface concentration will also increase gaseous ion yield.

(iv) Greatest sensitivities of detection are generally observed for analytes ionized in the liquid matrix. Frequently, therefore, adjustment of pH leads to improved secondary ion yield. A number of workers have developed derivatization techniques to convert neutral analytes to ionic products. DiDonato and Busch (1985), for example, reacted ketosteroids with Girard's reagents to yield cationic hydrazone derivatives which gave FAB sensitivities substantially higher than the parent steroids. In Chapter 3, Costello discusses sample preparation procedures designed to maximize the effectiveness of FAB MS procedures.

1.3. LIQUID CHROMATOGRAPHY–MASS SPECTROMETRY (LC–MS)

The combination of the full capabilities of high performance liquid chromatography and mass spectrometry in a tandem technique is extremely attractive for the characterization of polar and labile compounds, particularly where complex mixture analysis is required. The pressure, phase and flow incompatibility of the two techniques, however, has ensured that development of effective coupling methods has been slow. The three principal approaches—transport (moving belt), direct liquid introduction and thermospray—have been reviewed and compared (Desiderio and Fridland, 1984; Games, 1986) and only brief discussion is included here.

(i) Moving belt interface. The design objectives for this interface included the retention of the capability for electron impact MS, necessitating the removal of the LC solvent. This is achieved by evaporation from the moving belt following deposition of the LC column eluent. The less volatile solutes are evaporated by flash heating after introduction, on the moving belt, into the MS ion source. In

EI and CI applications, the requirement for volatilization of the analytes limits the suitability of the interface for the analysis of involatile and labile compounds. The belt, however, may be used as the sample support for desorption ionization, and encouraging results for on-line LC-FAB MS have been reported (Stroh *et al.*, 1985; see also Chapter 11, Section 11.2.2.3).

(ii) Direct liquid introduction (DLI). The direct introduction of LC eluent into the ion source of an MS, with suitably enhanced pumping capacity, is an attractively simple arrangement. Chemical ionization of the solutes occurs, with ions derived from the LC solvent by electron impact serving as reagent ions. The composition of the reagent plasma so formed may not be ideal for efficient ionization of the solutes. The requirement for volatilization of the analyte generally remains, although the detection of gas-phase ions, derived from analytes ionized in solution, has been made in the absence of an electron beam (Arpino and Beaugrand, 1984). This indicates direct sampling of condensed phase ions and suggests mechanistic similarities to the thermospray interface discussed below. DLI interfaces generally accommodate a minor proportion of the eluent of a conventional LC column, but the total eluent may be introduced if microbore columns are used or if the mass spectrometer is equipped with an atmospheric pressure CI source (Chapter 26). The direct coupling of supercritical fluid chromatography and MS (Smith and Udseth, 1983) is likely to represent an area of future development.

(iii) Thermospray (TSP). The thermospray interface developed by Vestal and co-workers (Blakley and Vestal, 1983, and references there cited) arguably represents the most versatile LC–MS coupling technique yet developed. Originally conceived as an efficient procedure for controlled vaporization of relatively high flow rates (1–2 ml/min) of LC eluent, the technique has led to the recognition of new ionization processes of particular significance to the analysis of highly polar compounds. Thermospray vaporization is achieved in a controlled fashion by passage of LC eluent through a heated capillary. When the liquid contains a substantial concentration of dissolved ions, the superheated mist so formed includes a proportion of charged (both positive and negative) droplets. Further evaporation of the droplets, in the TSP ion source, generates locally intense electric fields which facilitate direct evaporation of ions. TSP LC–MS experiments generally employ volatile buffers, such as ammonium acetate, as components of the LC liquid phase, so that gaseous NH_4^+ and CH_3COO^- ions are generated in the TSP source by the mechanism described above. Production of gas phase ions from analytes which are also ionized in solution occurs by the same mechanism. Analytes not ionized in solution evaporate as neutrals but may then be ionized via ion/molecule reactions with the plasma generated from the buffer constituents. Thus, ionization of neutrals is a classical CI process and recent work has indicated that ion formation is predictable on the basis of known gas-phase ion chemistry (Bursey *et al.*, 1985; Parker *et al.*, 1986). The yield of ions from neutral analytes may be increased by incorporation into the ion source of a heated filament to produce an electron beam. Further details of the TSP interface and the mechanisms of ionization involved are to be found in publications by Vestal (Vestal, 1983b, 1984; Gartiez and Vestal, 1985). The application of TSP LC–MS to the analysis of nucleosides is described by McCloskey in Chapter 6 (Section 6.2.1).

REFERENCES

Aberth, W., Straub, K. M., and Burlingame, A. L. (1982). Secondary ion mass spectrometry with cesium ion primary beam and liquid target matrix for analysis of bioorganic compounds, *Anal. Chem.*, **54**, 2029–2034.

Adlercreutz, H., Soltmann, B., and Tikkanen, M. J. (1974). Field desorption mass spectrometry in the analysis of a steroid conjugate, estriol-16α-glucuronide, *J. Steroid Biochem.*, **5**, 163–166.

Arpino, P. J., and Beaugrand, C. (1984). Fast injection of liquid solutions into a quadrupole mass spectrometer using short fused silica columns, in *Proc. 32nd Ann. Conf. Mass Spectrometry and Allied Topics, San Antonio,* American Society for Mass Spectrometry, pp. 199–200.

Barber, M., Bordoli, R. S., Sedgwick, R. D., and Tyler, A. N. (1981a). Fast atom bombardment of solids (FAB): A new ion source for mass spectrometry, *J. Chem. Soc. Chem. Commun.*, **1981**, 325–327.

Barber, M., Bordoli, R. S., Garner, G. V., Gordon, D. B., Sedgwick, R. D., Tetler, L. W., and Tyler, A. N. (1981b). Fast-atom bombardment mass spectra of enkephalins, *Biochem. J.*, **197**, 401–404.

Barber, M., Bordoli, R. S., Elliott, G. J., Sedgwick, R. D., and Tyler, A. N. (1982). Fast atom bombardment mass spectrometry, *Anal. Chem.*, **54**, 645A–657A.

Beckey, H. D. (1969). Field desorption mass spectrometry: a technique for the study of thermally unstable substances of low volatility, *Int. J. Mass Spectrom. Ion Phys.*, **2**, 500–503.

Beckey, H. D. (1977). *Principles of Field Ionization and Field Desorption Mass Spectrometry*, Pergamon, Oxford.

Beckner, C. F., and Caprioli, R. M. (1984). Quantitative aspects of fast-atom bombardment mass spectrometry, *Biomed. Mass Spectrom.*, **11**, 60–62.

Benninghoven, A., and Sichtermann, W. K. (1978). Detection, identification and structural investigation of biologically important compounds by secondary ion mass spectrometry, *Anal. Chem.*, **50**, 1180–1184.

Benninghoven, A., Rudenauer, F., and Werner, H. D. (1985). *Secondary Ion Mass Spectrometry*, Wiley, New York.

Blakley, C. R., and Vestal, M. L. (1983). Thermospray interface for liquid chromatography/mass spectrometry, *Anal. Chem.*, **55**, 750–754.

Burlingame, A. L., Baillie, T. A., Derrick, P. J., and Chizhov, O. S. (1980). Mass spectrometry, *Anal. Chem.*, **52**, 214R–258R.

Burlingame, A. L., Baillie, T. A., and Derrick, P. J. (1986). Mass spectrometry, *Anal. Chem.*, **58**, 165R–211R.

Bursey, M. M., Parker, C. E., Smith, R. W., and Gaskell, S. J. (1985). Gas-phase ionization of selected neutral analytes during thermospray liquid chromatography/mass spectrometry, *Anal. Chem.*, **57**, 2597–2599.

Cotter, R. J. (1980). Mass spectrometry of nonvolatile compounds. Desorption from extended probes, *Anal. Chem.*, **52**, 1589A–1606A.
Cotter, R. J. and Yergey, A. L. (1981). Thermally produced ions in desorption mass spectrometry, *Anal. Chem.*, **53**, 1306–1307.
Day, R. J., Unger, S. E., and Cooks, R. G. (1980). Molecular secondary ion mass spectrometry, *Anal. Chem.*, **52**, 557A–572A.
Desiderio, D. M., and Fridland, G. H. (1984). A review of combined liquid chromatography and mass spectrometry, *J. Liquid Chromatog.*, **7**, 317–351.
DiDonato, G. C., and Busch, K. L. (1985). Derivatization of ketosteroids for fast atom bombardment mass spectrometry, *Biomed. Mass Spectrom.*, **12**, 364–366.
Field, F. H. (1982). Fast atom bombardment study of glycerol: mass spectra and radiation chemistry, *J. Phys. Chem.*, **86**, 5115–5123.
Games, D. E. (1986). The combination of liquid chromatography and mass spectrometry, in *Advances in Mass Spectrometry, 1985* (Ed. J. F. J. Todd), pp. 323–342, Wiley, Chichester, in press.
Gartiez, D. A., and Vestal, M. L. (1985). Thermospray LC/MS interface: principles and applications, *LC Mag.*, **3**, 334–346.
Kistemaker, P. G., van der Peyl, G. J. Q., and Haverkamp, J. (1981). Laser desorption mass spectrometry, in *Soft Ionization Biological Mass Spectrometry* (Ed. H. R. Morris), pp. 120–136, Heyden, London.
Macfarlane, R. D. (1981). ^{252}Californium plasma desorption mass spectrometry (^{252}Cf-PDMS), in *Soft Ionization Biological Mass Spectrometry* (Ed. H. R. Morris), pp. 110–119, Heyden, London.
Macfarlane, R. D. (1983). Californium-252 plasma desorption mass spectrometry. Large molecules, software and the essence of time, *Anal. Chem.*, **55**, 1247A–1264A.
Parker, C. E., Smith, R. W., Gaskell, S. J., and Bursey, M. M. (1986). $[M + A]^-$ or $[M - 1]^-$? How ion formation depends on the ionic additive in thermospray liquid chromatography/negative ion mass spectrometry, *Anal. Chem.*, in press.
Rose, M. E. (1986). Senior Reporter, *Mass Spectrometry, Specialist Periodical Report*, vol. 8, Royal Society of Chemistry, London.
Schulten, H.-R. (1979). Biochemical, medical and environmental applications of field ionization and field desorption mass spectrometry. *Int. J. Mass Spectrom. Ion Phys.*, **32**, 97–284.
Smith, R. D., and Udseth, H. R. (1983). Mass spectrometry with direct supercritical fluid injection, *Anal. Chem.*, **55**, 2266–2272.
Stroh, J. G., Cook, J. C., Milberg, R. M., Brayton, L., Kihara, T., Huang, Z., Rinehart, K. L., and Lewis, I. A. S. (1985). On-line liquid chromatography/fast atom bombardment mass spectrometry, *Anal. Chem.*, **57**, 985–991.
Vestal. M. L. (1983a). Ionization techniques for nonvolatile molecules, *Mass Spectrom. Rev.*, **4**, 447–480.
Vestal, M. L. (1983b). Studies of ionization mechanisms involved in thermospray LC-MS, *Int. J. Mass Spectrom. Ion Phys.*, **46**, 193–196.
Vestal, M. L. (1984). High-performance liquid chromatography–mass spectrometry, *Science*, **226**, 275–281.
Wood, G. W. (1982). Field desorption mass spectrometry: applications, *Mass Spectrom. Rev.*, **1**, 63–102.

Mass Spectrometry in Biomedical Research
Edited by S. J. Gaskell
© 1986 John Wiley & Sons Ltd.

Chapter 2

MASS SPECTROMETRY OF LEUKOTRIENES

Robert C. Murphy
and
Timothy W. Harper

2.1. INTRODUCTION

Numerous studies of the metabolism of arachidonic acid by mammalian cells have revealed several complex cascade mechanisms whereby this polyunsaturated fatty acid is first liberated from phospholipids as the free carboxylic acid only to be subject to oxidative conversion into complex products. Mass spectrometry has played a central role in the elucidation of the structures of these metabolites, at first with the description of the cyclooxygenase products, prostaglandins, prostacyclin and thromboxane (Samuelsson *et al.*, 1978), then later with the lipoxygenase products, 5-hydroxyeicosatetraenoic acid [5-HETE] (Borgeat *et al.*, 1976) and 12-hydroxyeicosatetraenoic acid [12-HETE] (Hamberg and Samuelsson, 1974). More recently, the leukotrienes, also members of the 5-lipoxygenase cascade, were discovered and characterized as 5(S), 12(R)-dihydroxyeicosatetraenoic acid (LTB_4) (Borgeat and Samuelsson, 1979), 5-hydroxy-6-glutathionyl-eicosatetraenoic acid (LTC_4) (Murphy *et al.*, 1979), 5-hydroxy-6-cysteinylglycyl-eicosatetraenoic acid (LTD_4) (Morris *et al.*, 1980) and 5-hydroxy-6-cysteinyl-eicosatetraenoic acid (LTE_4) (Lewis *et al.*, 1980). The mass spectrometry of the leukotrienes will be the topic of this chapter.

The 5-lipoxygenase metabolites mentioned above are formed by a pathway involving multiple enzymatic steps shown in Figure 1. These compounds also possess important biological properties, a few of which are listed in Table 1. It is the expression of these properties *in vivo* which justifies the use of the term 'target cell-derived' lipid mediators to describe their role. These molecules are not stored preformed within cells, nor do they enter the circulation to affect

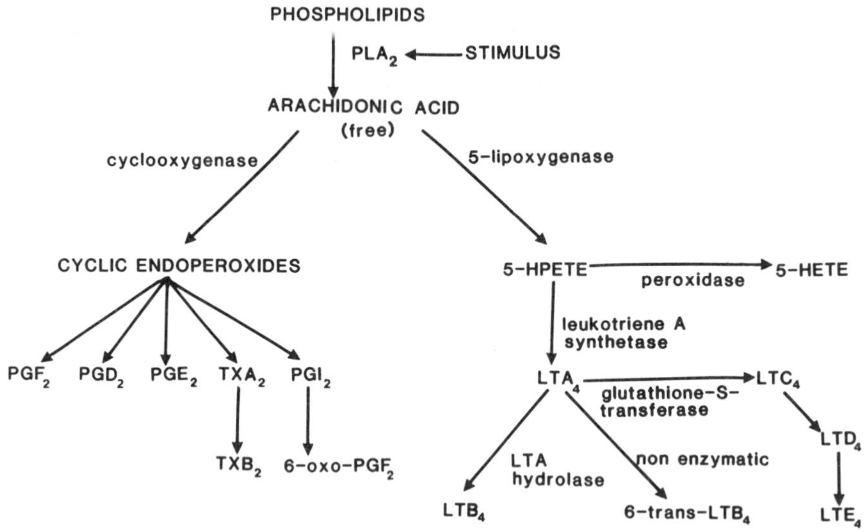

Figure 1: Arachidonic acid cascade leading to the formation of prostaglandins (PG), thromboxanes (TX), leukotrienes (LT), 5-hydroperoxyeicosatetraenoic acid (5-HPETE) and 5-hydroxyeicosatetraenoic acid (5-HETE).

TABLE 1 Biological properties of 5-lipoxygenase metabolites of arachidonic acid

Metabolite	Property
5-HETE	Causes chemotaxis of human neutrophils
LTB_4	Causes chemokinesis and chemotaxis of human neutrophils at low concentrations
$LTC_4(D_4/E_4)$	Slow reacting substance of anaphylaxis Smooth muscle contraction (bronchial smooth muscle) Increase in vascular permeability

distal organs; rather they are synthesized by certain cells in response to specific stimuli. It is thought that nearby cells are then affected by the leukotriene through intercellular chemical communication. Mass spectrometry plays a continuing role in the detection and quantification of leukotriene production, providing a tool for investigators interested in unravelling the aetiology of certain pathophysiological disorders as well as facilitating the study of pharmacological control of leukotriene biosynthesis. One unanticipated observation with these molecules germane to the topic of this chapter was that the mass spectra of the unsaturated hydroxy-acid derivatives have abundant ions which appear to be vinyl bond cleavage reactions. It is through a detailed study of the mass spectrometry of the 5-lipoxygenase metabolites that an appreciation of the unique gas phase ion chemistry of these molecules was obtained quite apart from their unique biological properties (Middleditch and Desiderio, 1973).

2.2. 5-HYDROXYEICOSATETRAENOIC ACID (5-HETE)

The mass spectrum of the methyl ester (ME), trimethylsilyl ether (TMS) derivative of 5-HETE is presented in Figure 2. The molecular ion is present but not particularly abundant, which is typical of such fatty acid derivatives. As

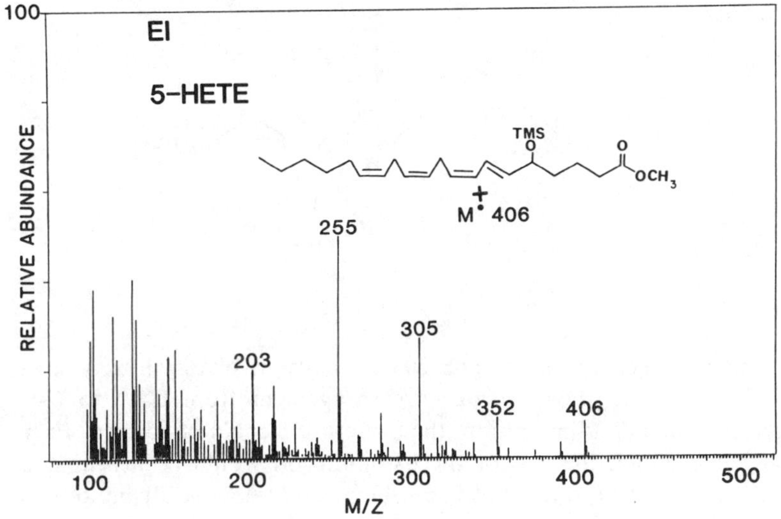

Figure 2: Electron impact mass spectrum (70 eV) of the methyl ester, trimethylsilyl ether derivative of 5-hydroxyeicosatetraenoic acid (5-HETE). Ion abundances are expressed relative to m/z 73 (100%).

with methyl ester, TMS ether derivatives of other hydroxy fatty acids, typical ions at (M-15), (M-31) and (M-90), (m/z 391, 375 and 316, respectively), are observed; the most abundant ions, however, occur at m/z 305, 255 and 203. M/z 305 can be readily rationalized as cleavage of the C-4,5 bond with charge retention on the resultant oxonium ion. Yet, the other abundant ions formally arise from cleavage of vinyl carbon–carbon bonds as suggested in Structure **1**.

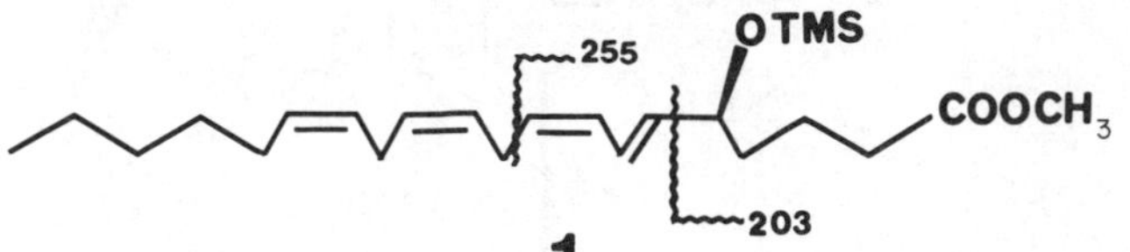

1

We have found this unexpected behaviour to be a rather characteristic property of the lipoxygenase metabolites of both arachidonic and docosahexaenoic acids. The discussion concerning the mass spectrometry of LTB_4, which follows, summarizes detailed studies of these fragmentation processes, which we feel involve rearrangement prior to carbon–carbon cleavage.

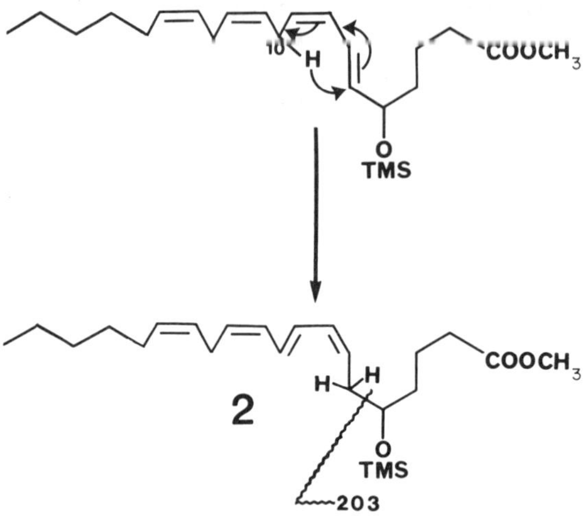

2

As one example, the ion at *m/z* 203 formally involves vinylic cleavage of C-5,6. Yet, with a prior hydrogen rearrangement from C-10 to C-6, and concerted double bond migration, the Structure **2** could be formed. From this rearranged product, the loss of the oxonium ion would be favoured by the stability of the radical formed at C-6, which would now be allylic to the newly formed conjugated triene. This occurrence of formal vinyl cleavages is not typically observed for the prostaglandins (Middleditch and Desiderio, 1973); however, the natural prostaglandins do not contain conjugated dienes within their carbon skeletons. Nonetheless, similar proton rearrangements and formation of conjugated polyenes as in Structure **3** can be readily rationalized for the other major ions in the electron impact mass spectrum of 5-HETE (TMS, ME).

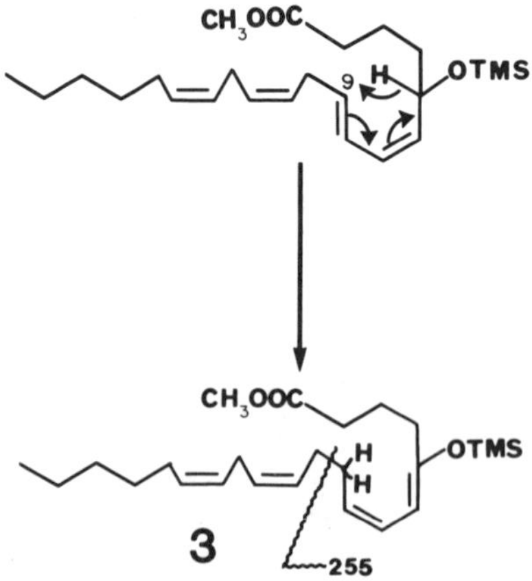

3

2.3. LEUKOTRIENE B_4 AND 6-TRANS-LTB_4 ISOMERS

The electron impact mass spectra of the methyl ester, trimethylsilyl ether derivatives of LTB_4 and the 5(S),12(R)- and 5(S),12(S)-6-trans LTB_4 isomers

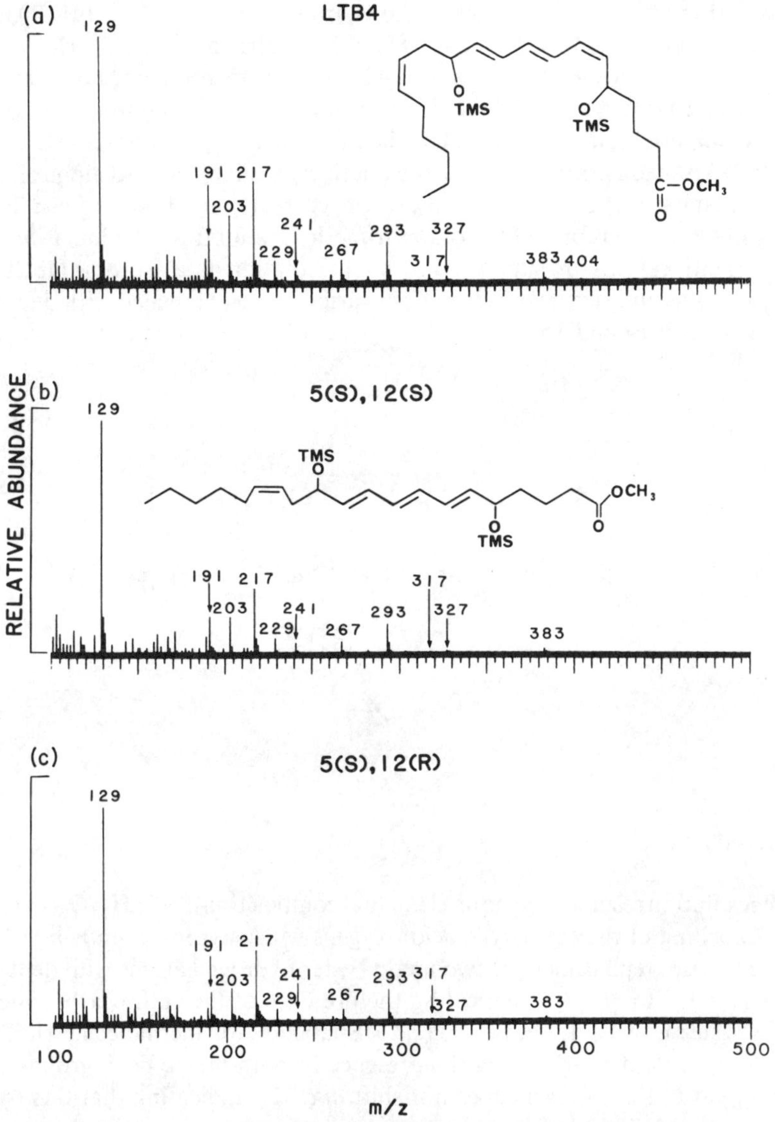

Figure 3: Electron impact mass spectrum (70 eV) of the methyl ester, trimethylsilyl ether derivatives of leukotriene B_4 (a), 5(S),12(S)-6-trans-LTB_4 (b) and 5(S),12(R)-6-trans-LTB_4 (c).

were first reported by Borgeat and Samuelsson (1979). Although these workers reported the tentative identification of the major ions, a thorough study of the fragmentation pathways by high resolution and stable isotope labelling techniques was only recently performed (Harper, 1985). Figure 3a is the mass spectrum of the LTB_4 derivative; it is essentially identical to those of the 6-trans isomers (Figure 3b,c). Characteristic ions are found at *m/z* 129, 191, 203, 217, 229, 241, 267, 293, 317, 327 and 383. With sufficient material, (M-90) and (M-15) ions have been observed at *m/z* 404 and 479, respectively; however, these ions are often lost in the instrument background when dealing with biologically-derived material. In no instance has the molecular ion been observed. The ion at *m/z* 383 is readily rationalized as the relatively straightforward alpha cleavage ion formed by cleavage of the 12,13 carbon–carbon bond. Likewise *m/z* 293 is a straightforward ion formed by loss of trimethylsilanol from the ion at *m/z* 383. These identifications are supported by high resolution data, isotopic labelling of oxygen atoms and results obtained with various species labelled with deuterium at C-8, 10, 12, 14 and 15.

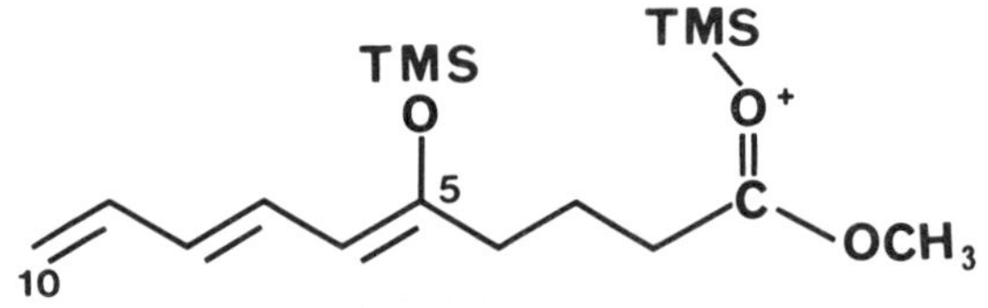

m/z 327

4

2.3.1. *m/z* 327

High resolution results indicate an elemental composition of $C_{16}H_{31}O_3Si_2$ for *m/z* 327. Labelling of the carboxylic acid oxygens with oxygen-18 shifts *m/z* 327 to *m/z* 331. Also, replacement of two methyl ester hydrogen atoms with deuterium shifts *m/z* 327 to *m/z* 329, indicating the presence of the entire carboxylic acid methyl ester moiety. Likewise, [$^2H_9$]TMS-labelling of the molecule shifts *m/z* 327 to *m/z* 345, demonstrating the presence of both intact TMS groups. [^{18}O] labelling at C-12, however, does not shift *m/z* 327, indicating that this oxygen atom is absent. Since high resolution data require the presence of three oxygen atoms, the oxygen from C-5 must be present as well as the carboxylic acid oxygens. High resolution data also require the presence of a C_8H_{10} portion of the hydrocarbon backbone. These facts suggest that the *m/z* 327 ion is a rearrangement ion formed by a pathway involving migration of the TMS group from the oxygen at C-12 to another point along the molecule, most likely to the carbonyl oxygen of the carboxylic acid moiety, with loss of the hydrocarbon backbone from C-10 to C-20. Rearrangement ions involving similar migrations of TMS groups to carboxylic acid TMS esters have been well recognized (Draffan *et al.*, 1968; Capella and Zorzut, 1968). A possible rearrangement pathway leading to formation of an ion with the required structure could arise by initial ionization of the C-10 double bond, followed by a radical site initiated transfer of the hydrogen atom from C-5 to C-10. This is a rather activated position for the hydrogen atom due to its position allylic to a conjugated diene. The resulting radical would be stabilized by resonance delocalization of the radical site over carbons 5 to 9. Charge-induced electrocyclization between carbons 11 and 15 with cleavage of the 12,13 carbon–carbon bond and transfer of the TMS group from the oxygen at C-12 to the carbonyl oxygen of the carboxylic acid moiety follows. Radical site initiated double bond migration and cleavage of the 10,11 carbon–carbon bond produces the ion at *m/z* 327 (Structure **4**).

2.3.2. *m/z* 317

From the observed shifts in *m/z* 317 in the mass spectra of the various isotopically-labelled derivatives, *m/z* 317 was deduced to contain both intact TMS groups, the oxygen from C-12 and the entire carboxylic acid methyl ester portion. High resolution results yield an elemental composition of $C_{14}H_{29}O_4Si_2$, indicating that only a C_7H_8 fragment of the eicosanoid backbone is present. These results imply that *m/z* 317 must be formed by a second rearrangement pathway, most likely involving migration of the OTMS group from C-12 to C-7 after ionization of the triene and a proton rearrangement, to yield Structure **5**.

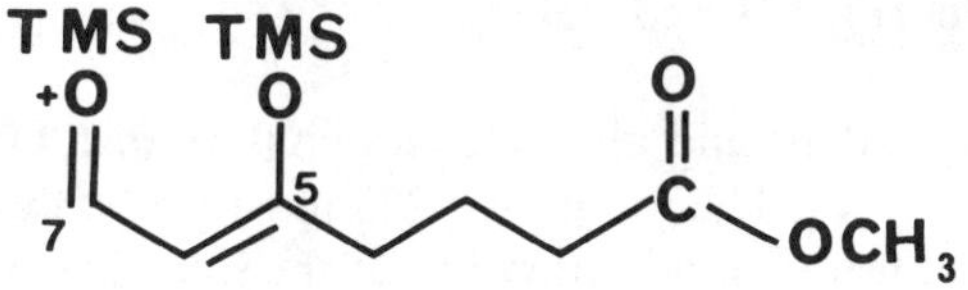

m/z 317

5

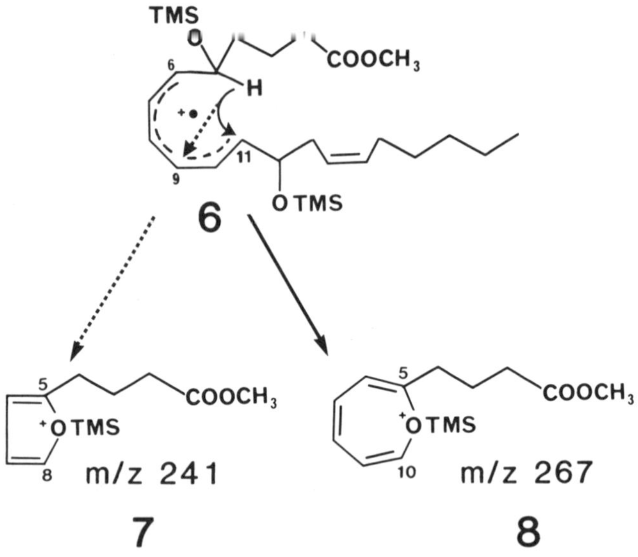

2.3.3. *m/z* 267 and 241

All the stable isotope labelling and high resolution data are consistent with cleavage of the double bonds at positions 8-9 and 10-11, respectively, with proton losses. Since cleavages of double bonds would be expected to be unlikely events, these relatively abundant ions are thought to arise via rearrangements involving the molecular ion, which is produced by removal of an electron from one of the triene molecular orbitals (**6**). From such a delocalized radical cation, a single proton rearrangement, cleavage of the adjacent C–C bond with formation of a new C–O bond with the C-5 TMS ether would yield the stable 5- and 7-membered cyclic ions (**7**) and (**8**).

2.3.4. *m/z* 203 and 217, 191 and 129

These relatively abundant and characteristic ions from LTB_4 also arise from rearrangement processes. The ion at *m/z* 203 most probably originates via the same process which leads to the formation of the same ion from 5-HETE ME, TMS discussed above. *M/z* 217 is commonly observed in the mass spectra of TMS derivatives of sugars and dicarboxylic acids (Sherman *et al.*, 1970) but is not present in the mass spectra of the methyl ester, trimethylsilyl ether derivatives of the 5,6-dihydroxyeicosatetraenoic acids (Borgeat and Samuelsson, 1979). While the structure of *m/z* 217 has been previously proposed to be as

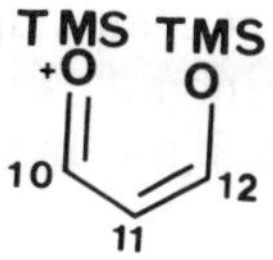

m/z 217

9

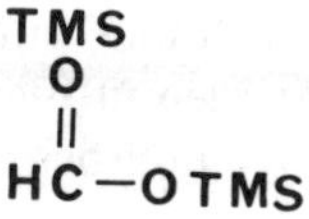

m/z 191

10

Structure **9**, the ion is formed by a rather complex pathway involving migration of the OTMS group from C-5 to C-10 followed by cleavage of the 9-10 and 12-13 C–C bonds. The ions at *m/z* 191 (**10**) and 129 have multiple origins and are prominent ions in the mass spectra of TMS derivatives of various molecules including sugars (Sherman *et al.*, 1970), steroids (Eneroth *et al.*, 1964), dicarboxylic acids (Draffan *et al.*, 1968) and hydroxy acids (Capella *et al.*, 1968).

The most striking feature of the electron impact mass spectra of the TMS ether, methyl ester derivative of the 5-lipoxygenase hydroxy acid metabolites is the preponderance of rearrangements of the molecular ion. These rearrangements result in the production of ions, which formally appear to involve vinyl cleavages and which account for most of the mass spectral ion current. The conjugated diene in 5-HETE and the conjugated triene interrupting the two TMS ethers in LTB_4 no doubt serve as sites for initial ionization from which a host of energetically favourable rearrangements of protons as well as oxygen atoms takes place. The extent to which this behaviour predominates is unique to

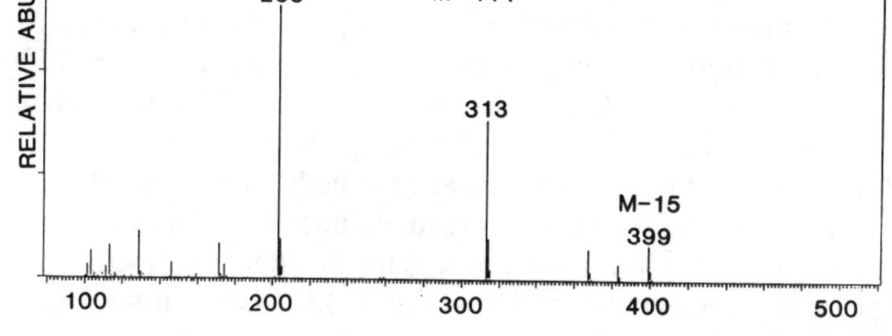

Figure 4: Electron impact mass spectrum (70 eV) of the methyl ester, trimethylsilyl ether derivatives of 5-hydroxyeicosanoic acid obtained from hydrogenation of 5-HETE. Ion abundances are expressed relative to *m/z* 73 (100%).

this case, and is not found with derivatives of other compounds, including the prostaglandins. As seen in Figures 4 and 5, saturation of the double bonds in 5-HETE and LTB_4 abolishes these extensive rearrangement processes.

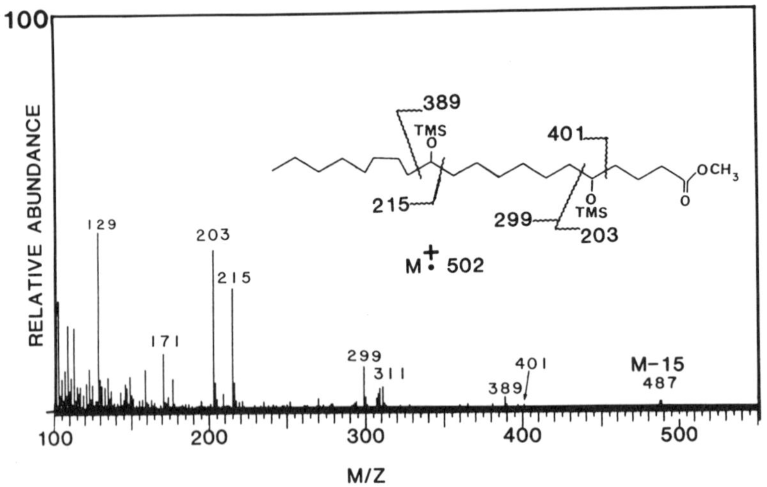

Figure 5: Electron impact mass spectrum (70 eV) of the methyl ester, trimethylsilyl ether derivative of 5,12-dihydroxyeicosanoic acid obtained from hydrogenation of LTB_4. Ion abundances are expressed relative to *m/z* 73 (100%).

2.4. CHEMICAL IONIZATION

Despite the widespread use of electron impact ionization for the structure elucidation of the various monohydroxy and dihydroxyeicosanoids, this technique has drawbacks due to extensive rearrangement of the molecule as we have observed above. This tends to obscure the structural details that mass spectrometry could otherwise provide for these molecules. Chemical ionization mass spectrometry has enjoyed widespread use in more recent years, primarily due to the fact that chemical ionization is a much less energetic process than electron impact ionization (Harrison, 1983). This being the case, it is not too surprising that 5-HETE and LTB_4, derivatized as the methyl ester trimethylsilyl ether, behave in a much more straightforward manner in terms of fragmentation under chemical ionization conditions. The positive ion methane chemical ionization mass spectrum obtained from the 5(S),12(R)-6-trans-LTB_4 derivative is shown in Figure 6. As with electron impact, the mass spectra of LTB_4 and its 6-trans-isomers are very similar. Perhaps the most striking feature of the chemical ionization mass spectra of these molecules is the presence of rather abundant high mass ions, which is in sharp contrast to the paucity of such ions

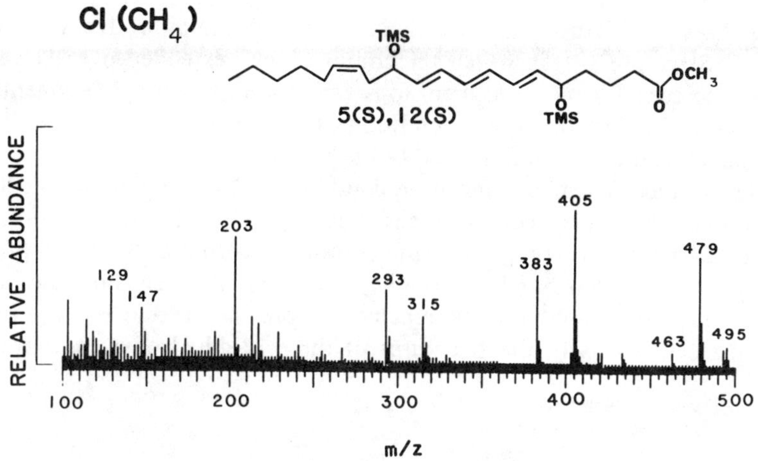

Figure 6: Methane chemical ionization mass spectrum of the methyl ester, trimethylsilyl ether derivative of 5(S),12(S)-6-trans-LTB_4.

under electron impact conditions. $(M-H)^+$ and $(M+H)^+$, respectively, are observed at *m/z* 493 and 495. In addition, (M-15) and (M-31) at *m/z* 479 and 463 are readily observed and arise from the loss of methyl and methoxy groups, respectively.

Simple cleavage and neutral loss fragments which are observed in electron impact mass spectra at *m/z* 383 (Structure **11**) and 293 (**12**) are also present under chemical ionization conditions, but in much greater abundance. Also, the loss of trimethylsilanol from $(M+H)^+$, most probably with charge retention on the 5-carbon atom, results in the appearance of a new ion, *m/z* 405 (Structure

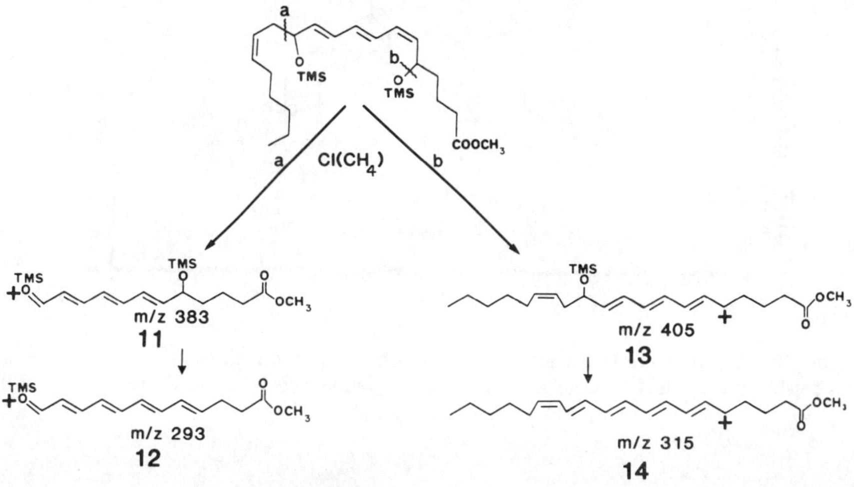

13) as shown in Figure 6. Further loss of trimethylsilanol produces the ion at *m/z* 315 (Structure **14**), in which the conjugation is extended over 11 carbon atoms. The complex rearrangement ions which were the major fragmentation pathways under electron impact conditions, however, are of much less importance under chemical ionization conditions. Nonetheless, a major ion at *m/z* 203 is observed under chemical ionization conditions. This ion is undoubtedly the same ion as that observed under electron impact conditions; its formation involves a complex rearrangement and proton shift from C-10 to C-6 prior to cleavage of the C-5,6 bond. Thus, the behaviour of these molecules under chemical ionization conditions is much more predictable and is more readily rationalized in terms of the structures of the 5,12-dihydroxyeicosatetraenoic acids.

2.5. ELECTRON CAPTURE NEGATIVE CHEMICAL IONIZATION

As shown in Figure 7, the pentafluorobenzyl ester trimethylsilyl ether derivative of LTB_4 yields a very simple negative ion mass spectrum after capturing a thermal electron. With this relatively gentle ionization technique, the base peak is the carboxylate anion at *m/z* 479, which arises from the loss of the pentafluorobenzyl group (Strife and Murphy, 1984a). The behaviour of the LTB_4 derivative is characteristic of all carboxylic acid esters derivatized in this

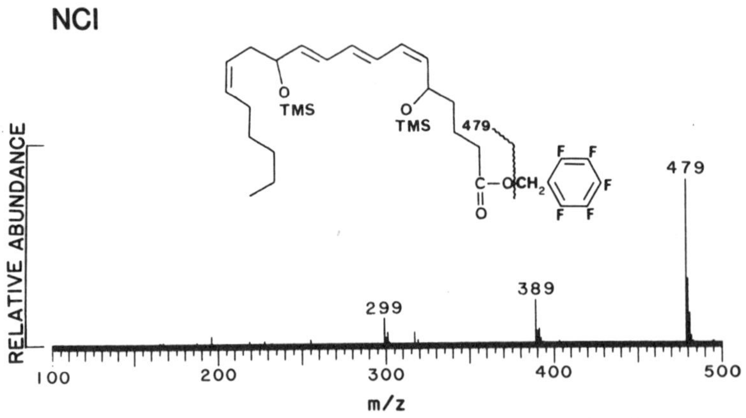

Figure 7: Electron capture chemical ionization mass spectrum (methane as mediating gas to thermalize electrons) of the methyl ester, trimethylsilyl ether derivative of LTB_4 (negative ions).

manner (Strife and Murphy, 1984b). Negative ion chemical ionization using pentafluorobenzyl ester derivatives is a recent development in the mass spectral analysis of both leukotrienes and prostaglandins which produces enhanced ion currents useful for quantitative analysis (Blair *et al.*, 1983). The other ions in the mass spectrum, *m/z* 389 and *m/z* 299, are probably formed by subsequent neutral losses of one and two molecules of trimethylsilanol, respectively, from *m/z* 479. No molecular ion (molecular weight = 606) has ever been observed under negative ion chemical ionization conditions for the pentafluorobenzyl ester trimethylsilyl ether derivative of LTB_4. We had previously postulated that an explanation for this rather large ion current carried by the perhaps resonance stabilized carboxylate anion is due in part to the lack of stability of the potential tropylium anion or benzyl anion. In contrast to the aromatic tropylium cation, the tropylium anion would not fit the $(4n + 2)$ π-electron rule and, as such, would be an antiaromatic molecule (Strife and Murphy, 1984b). Thus, since benzyl anions are not stabilized, there is no driving force for their formation and the electron is captured by the resonance stabilized carboxylate group. Whatever the exact process which dominates this fragmentation behaviour, the structurally-significant carboxylate anions are produced in high yield with little subsequent fragmentation. This technique, therefore, provides an excellent way to establish the molecular weight of molecules which are monocarboxylic acids in a very sensitive and specific manner.

Yet, the simplicity of the electron capture negative chemical ionization mass spectrum, which is its major advantage, is also its major disadvantage. Unlike the positive ion chemical ionization experiment, where simple cleavage ions containing structural information are observed, the fragmentation observed for LTB_4 under negative ion conditions consists of the relatively uninformative losses of trimethylsilanol. Consequently, electron capture chemical ionization, although providing useful molecular weight information and a potentially powerful technique for quantitative analysis of LTB_4 and related compounds, is inferior to positive ion chemical ionization and electron impact for providing other structural information. Therefore, when a sensitive quantification assay is required, such as the measurement of LTB_4 in biological fluids, electron capture chemical ionization is the clear method of choice. For unambiguous characterization of a dihydroxyleukotriene, however, positive ion chemical ionization and/or electron impact is preferred.

Another advantage of the pentafluorobenzyl ester derivative is the ease with which these derivatives can be formed at low levels. In biological extracts, picomole and subpicomole amounts of LTB_4 can be derivatized and detected, as illustrated by the detection of LTB_4 in the lungs of infants with pulmonary bronchodysplasia (Westcott *et al.*, 1986). These infants, because of their severe pulmonary disease, are supported by respirators and during the maintenance of the airways with mechanically-assisted respiration, periodic lavage of the lungs takes place. Analysis of the lung fluid was carried out by mass spectrometry following extraction by high pressure liquid chromatography (HPLC). The HPLC effluent in the region where LTB_4 was suspected to elute was derivatized to the pentafluorobenzyl ester, TMS ether and analyzed by GC–MS. A substance yielding negative ions at *m/z* 479 eluted precisely at the capillary retention time of LTB_4.

2.6. OXYGEN-18 LABELLED 5-LIPOXYGENASE PRODUCTS

In 1976, we began investigating the potential use of [^{18}O] labelled molecules to serve as internal standards for quantitative analysis by gas chromatography–mass spectrometry (Zirrolli and Murphy, 1976). Significant advantages arise from the use of stable isotope-labelled internal standards in the analysis of eicosanoids (Green *et al.*, 1978). These advantages arise primarily from the identical chemical behaviour of the labelled and unlabelled compounds and, therefore, identical gas chromatographic and mass spectrometric behaviour. However, the use of these substances has been limited, primarily due to the limited availability of deuterium-labelled substances, which have to be synthesized, usually by very tedious processes. Since all of the eicosanoids have a carboxyl moiety in their structure, originating from the carboxyl group of arachidonic acid, we have more recently been considering the use of carboxyl [^{18}O] labelled internal standards. The advantage of such substances are several-fold. First of all, there is a lack of a theoretical isotope effect. Secondly, they are rather easily synthesized by choosing one of the three general methods for the synthesis of carboxyl [^{18}O] labelled internal standards we have developed. These syntheses are based upon: (i) strong acid-catalysed exchange (Zirrolli and Murphy, 1976); (ii) hydrolysis of methyl esters in [^{18}O] base followed by remethylation with diazomethane and resaponification (Murphy and Clay, 1982); and (iii) enzyme-mediated processes (Westcott *et al.*, 1985). Thus, we have been able to make [^{18}O] labelled species for nearly all of the eicosanoids and 5-lipoxygenase products. Unfortunately, there is one significant disadvantage—the potential for loss of label through back exchange during the work-up procedure following addition of the internal standard to the biological fluid. Therefore, a great deal of effort must be made to eliminate the possibility that back exchange does occur. [^{18}O] labelled species have been used in numerous applications including the measurement of 5-HETE and the study of the damage of perfused rat lung (Burghuber *et al.*, 1985) and also the measurement of LTB_4 in lung lavage fluid of infants with pulmonary bronchodysplasia (Westcott *et al.*, 1986). One significant advantage of the carboxyl [^{18}O] labelled species is related to the process of electron capture chemical ionization. The stabilized carboxylate anion, which carries large ion currents, contains both oxygen-18 atoms; therefore, the carboxylate anion derived from the stable isotope-labelled molecule is advanced four mass units above that of the endogenous substrate which is being measured by this internal standard.

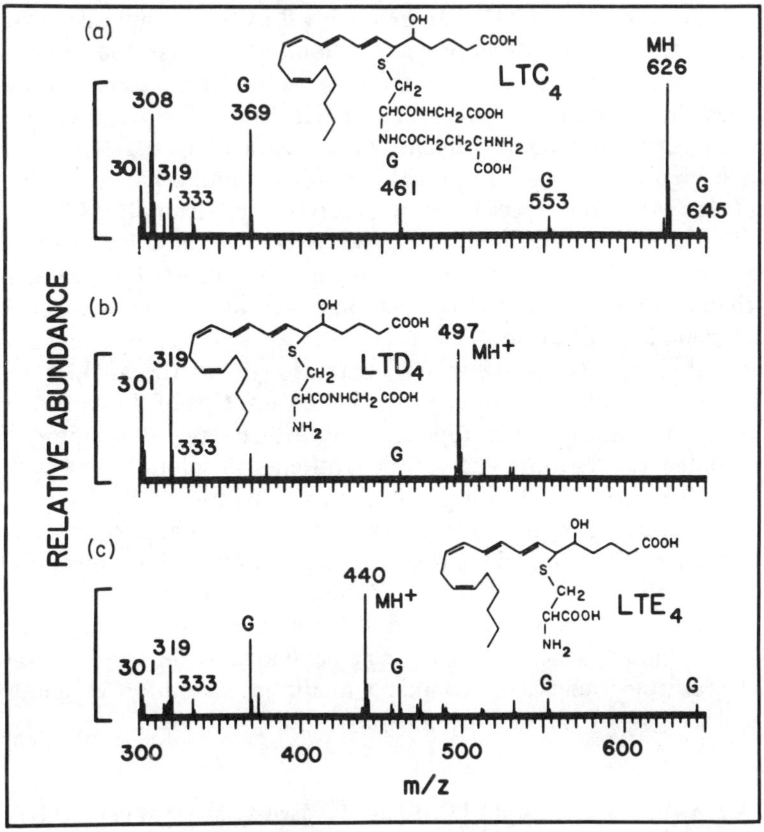

Figure 8: Fast atom bombardment mass spectra of LTC_4(a), LTD_4 (b) and LTE_4 (c). Glycerol was used as matrix and ions from the matrix are labelled (G). Xenon used as particle source, 6 keV.

2.7. FAST ATOM BOMBARDMENT

Up to now, we have considered the mass spectra of those 5-lipoxygenase products which can be made amenable to analysis by gas chromatography–mass spectrometry via conversion to volatile derivatives. Yet, a major branch of the 5-lipoxygenase cascade contains the sulphidopeptide leukotrienes (LTC_4, D_4 and E_4), which all have a thioether bond at C-6, a hydroxyl at C-5 and four centres of unsaturation along the former arachidonic acid backbone. We have found these molecules to be quite resistant to the attempts to analyze them by conventional mass spectrometric techniques, including direct probe following derivatization reactions (Mathews *et al.*, 1981). Morris *et al.* (1980) have reported, however, that LTD_4 can be analysed as the methyl ester, *N*-acetyl, TMS ether derivative by direct probe insertion. Nonetheless, the advent of fast atom bombardment has permitted the observation of molecular ions arising from these molecules using glycerol as well as other matrices (Murphy *et al.*, 1982). The fast atom bombardment mass spectra of LTC_4, D_4 and E_4 are shown in Figure 8. Using 6–7 keV xenon particles, quite abundant positive ion currents for $(M + H)^+$ species can be observed for 1–2 nmol of LTC_4, LTD_4 and LTE_4 at *m/z* 626, 497 and 440, respectively. Typically, fragment ions are observed at 301, 319 and 333, and are thought to originate from the arachidonic acid chain, primarily due to the fact that they are common to all of the sulphidopeptide leukotrienes.

This technique has been a major breakthrough for the analysis of these species, which are difficult to study by other means. One of the features of this class of leukotrienes is the rapidity with which they decompose. Rapid degradation has been observed with leukotrienes in solution, and with both synthetically prepared and biologically-derived leukotrienes. We have used fast atom bombardment mass spectrometry extensively to characterize numerous degradation products of LTC_4 following separation by high pressure liquid chromatography, as shown in Table 2. From the shift of the $(M + H)^+$ ions, the shift of fragment ions, which arise from the arachidonic acid chain, and UV absorption characteristics, we have been able to identify the major degradation products as compounds corresponding to the incorporation of one or two atoms of oxygen (Westcott *et al.*, 1984).

TABLE 2 Autodecomposition of LTC_4 (50 μg). Products separated by gradient HPLC and analyzed by UV spectroscopy and FAB-mass spectrometry.

Fraction	UV	FAB $(M + H)^+$	FAB Peptide Ion	Assignment
21	~230	660	308	$LTC_4 + H_2O_2$
22	~230	660	308	$LTC_4 + H_2O_2$
26,27	307(t)*	660,642	308	15-OH-LTC_4
36	284(t)	642	324	LTC−S = O
38	280(t)	626	308	LTC_4
39	278(t)	n.d.†	n.d.	11-trans-LTC_4

* Triplet chromophore, λ_{max} reported.
† Not determined.
Reproduced from Westcott et al. *(1984) by permission of C. V. Mosby Company.*

2.8. CONCLUSION

Mass spectrometry has played a central role in the structure elucidation of the 5-lipoxygenase metabolites of arachidonic acid and promises to offer unparalleled sensitivity and specificity for the quantitative analysis of LTB_4 and 5-HETE. Fast atom bombardment mass spectrometry has been invaluable in studies of the structural changes of the sulphidopeptide leukotrienes produced either by metabolism or by spontaneous degradation. We have studied in detail the electron impact mass spectra of the mono- and dihydroxyeicosanoid derivatives and found them to be surprisingly complex, dominated by rearrangement processes. No doubt this is due to the conjugated diene and triene moieties adjacent to the derivatized hydroxyl groups, which act as the initial site of ionization as well as providing the molecular orbitals available for new bond formation in these rather flexible molecules. It is now quite clear, however, that what appears at first glance to be an unexpected vinyl bond cleavage is more likely a more straightforward allylic bond cleavage following a facile rearrangement with charge retention on an oxonium ion and a radical site stabilized over five or more carbon atoms.

ACKNOWLEDGMENT

This work was supported in part by a grant from the National Institutes of Health (HL 25785 and RR 01152).

REFERENCES

Blair, I. A., Barrow, S. E., Waddell, K. A., Lewis, P. J., and Dollery, C. T. (1983). Prostacyclin is not a circulating hormone in man, *Prostaglandins*, **23**, 579–589.

Borgeat, P., and Samuelsson, B. (1979). Transformation of arachidonic acid by rabbit polymorphonuclear leukocytes, *J. Biol. Chem.*, **254**, 2643–2646.

Borgeat, P., Hamberg, M., and Samuelsson, B. (1976). Transformation of arachidonic acid and homo-gamma-linolenic acid by rabbit polymorphonuclear leukocytes, *J. Biol. Chem.*, **251**, 7816–7820.

Burghuber, O. J., Strife, R. J., Zirrolli, J. A., Henson, P. M., Mathias, M. M., Reeves, J. T., Murphy, R. C., and Voelkel, N. F. (1985). Leukotriene inhibitors attenuate rat lung injury induced by hydrogen peroxide, *Am. Rev. Respir. Dis.*, **131**, 778–785.

Capella, P., and Zorzut, C. M. (1968). Determination of double bond position in monounsaturated fatty acid esters by mass spectrometry of their trimethylsiloxy derivatives, *Anal. Chem.*, **40**, 1458–1463.

Capella, P., Galli, C., and Fumagalli, R. (1968). Hydroxy fatty acids from cerebrosides of the central nervous system: GLC determination and mass spectrometric identification, *Lipids*, **3**, 431–438.

Draffan, G. H., Stillwell, R. N., and McCloskey, J. A. (1968). Electron impact-induced rearrangement of trimethylsilyl groups in long chain compounds, *Org. Mass Spectrom.*, **1**, 669–685.

Eneroth, P., Hellstrom, K., and Ryhage, R. (1964). Identification and quantification of neutral fecal steroids by gas-liquid chromatography and mass spectrometry: Studies of human excretion during two dietary regimens, *J. Lipid Res.*, **5**, 245–262.

Green, K., Hamberg, M., Samuelsson, B., Smigel, M., and Froelich, J. H. (1978). Measurement of prostaglandins, thromboxanes, prostacyclin and their metabolites by gas liquid chromatography-mass spectrometry, *Adv. Prostaglandin Thromboxane Res.*, **5**, 39–94.

Hamberg, M., and Samuelsson, B. (1974). Novel transformation of arachidonic acid in human platelets, *Proc. Nat. Acad. Sci. USA*, **71**, 3400–3404.

Harper, T. W. (1985). *Mass Spectrometry and Metabolism of Leukotriene* B_4, Doctoral Thesis, University of Colorado.

Harrison, A. G. (1983). *Chemical Ionization Mass Spectrometry*, CRC Press, Boca Raton.

Lewis, R. A., Drazen, J. M., Austen, K. F., Clark, D. A., and Corey, E. J. (1980). Identification of the C(6)-S-conjugate of leukotriene A with cysteine as a naturally occurring slow-reacting substance of anaphylaxis (SRS-A). Importance of the 11-cis geometry for biological activity, *Biochem. Biophys. Res. Comm.*, **96**, 271–277.

Mathews, W. R., Clay, K. L., and Murphy, R. C. (1981). In beam chemical ionization of peptides, *Proc. 29th Ann. Conf. Mass Spectrometry and Allied Topics, Minneapolis*, American Society for Mass Spectrometry, pp. 488–489.

Middleditch, B. S., and Desiderio, D. M. (1973). Mass spectra of prostaglandins. IV. Trimethylsilyl derivatives of prostaglandins of the F series, *Anal. Biochem.*, **55**, 509–520.

Morris, H. R., Taylor, G. W., Piper, P. J., and Tippins, J. R. (1980). Structure of slow-reacting substance of anaphylaxis from guinea pig lungs, *Nature*, **285**, 104–107.

Murphy, R. C., and Clay, K. L. (1982). Preparation of ^{18}O-derivatives of eicosanoids for GC/MS quantitative analysis, in *Methods in Enzymology* (Eds., W. E. M. Lands, and W. L. Smith), pp. 547–551, Academic Press, New York.

Murphy, R. C., Hammarstrom, S., and Samuelsson, B. (1979). Leukotriene C: A slow-reacting substance from murine mastocytoma cells, *Proc. Nat. Acad. Sci. USA*, **76**, 4275–4279.

Murphy, R. C., Mathews, W. R., Rokach, J., and Fenselau, C. (1982). Comparison of biological-derived and synthetic leukotriene C_4 by fast atom bombardment mass spectrometry, *Prostaglandins*, **23**, 201–206.

Samuelsson, B., Goldyne, M., Granstrom, E., Hamberg, M., Hammarstrom, S., and Malmsten, C. (1978). Prostaglandins and thromboxanes, *Ann. Rev. Biochem.*, **47**, 997–1029.

Sherman, W. R., Eilers, N. C., and Goodwin, S. L. (1970). Combined gas chromatography–mass spectrometry of the inositol trimethylsilyl ethers and acetate esters, *Org. Mass Spectrom.*, **3**, 829–840.

Strife, R., and Murphy, R. C. (1984a). Stable isotope labelled 5-lipoxygenase metabolites of arachidonic acid. Analysis by negative ion chemical ionization mass spectrometry, *Leukotrienes, Prostaglandins Med.*, **13**, 1–8.

Strife, R., and Murphy, R. C. (1984b). Preparation of pentafluorobenzyl esters of arachidonic acid lipoxygenase metabolites, *J. Chromatog.*, **305**, 3–11.

Westcott, J. Y., Clay, K. L., and Murphy, R. C. (1984). Decomposition of leukotriene C_4, *J. Allergy Clin. Immunol.*, **74**, 363–368.

Westcott, J. Y., Clay, K. L., and Murphy, R. C. (1985). *Biomed. Mass Spectrom*, **12**, 715–719.

Westcott, J. Y., Stenmark, K. R., and Murphy, R. C. (1986). Analysis of leukotriene B_4 in human lavage by HPLC and mass spectrometry, *Prostaglandins*, **31**, 227–237.

Zirrolli, J. A., and Murphy, R. C. (1976). Mass spectrometry of 3-methoxy fatty acid methyl esters, *Org. Mass Spectrom.*, **11**, 1114–1120.

Mass Spectrometry in Biomedical Research
Edited by S. J. Gaskell
© 1986 John Wiley & Sons Ltd.

Chapter 3

SAMPLE PREPARATION FOR FAB AND FD MASS SPECTROMETRY OF POLAR ANALYTES

Catherine E. Costello

3.1. INTRODUCTION

As each of the so-called 'soft ionization' methods for mass spectrometry has been introduced, there has been an initial period filled with enthusiastic reports of the method's versatility and high sensitivity based on the analysis of commercially-available samples. The second and longer period, during which the new methods have begun to be applied to the analysis of samples derived from biological or other complex matrices, has been characterized by a recognition that the ultimate sensitivity and indeed even the appearance of the mass spectrum is usually highly dependent on the purity of the sample and the presence (or absence) of salts, buffers, and other materials which may enhance or inhibit the ionization of the compound of interest. Field desorption (FD) and fast atom bombardment (FAB) or liquid secondary ion mass spectrometry (liquid SIMS) are particularly subject to these effects. For these techniques, therefore, sample preparation ultimately holds the key to the success or failure of the experiment. A variant of Murphy's law applies to FD and FAB: the smaller the amount of sample available, the greater the importance of a single sample, and the more steps involved in its isolation, the more likely it seems that some extraneous material will be present and will interfere with the ionization process. It is wise, therefore, for the careful investigator to anticipate this situation and take steps to optimize the instrument and matrix conditions for maximum sample ionization efficiency and to minimize the concentration and deleterious effects of contaminants. It is with the selection of sample preparation procedures that we shall be concerned here.

3.2. FINAL STEPS FOR SAMPLE ISOLATION

Most samples isolated from biological sources are subjected to a sequence of extractions and/or chromatographic procedures. The sample presented for mass spectral analysis may contain some of the salts, buffers, soluble supports or remainders of the mobile phases. Even trace impurities can assume significant roles, especially when the original solution containing the sample has to be highly concentrated before mass spectral analysis. (Consider, for example, that concentration to 10 µl of 10 ml of a 0.01% sodium chloride solution results in a 10% NaCl solution containing 1000 µg – 17.1 µmoles of salt. If only 2 nmoles [2 µg of a compound with M_r 1000] of the compound of interest are present, this represents a vast excess of salt!) These extraneous materials are often either present as charged species or are ionized more easily than the analyte. Their effects include both the suppression of the signal from the sample and the contribution of extra peaks to the spectrum. In FAB, interaction of contaminants with the matrix can produce extremely complex patterns that overlay or suppress sample signals entirely (Martin *et al.*, 1982). A tandem mass spectrometer may be employed to separate sample peaks from background and from impurities to simplify the spectra of mixtures before collisional activation, but if the sample ions themselves are suppressed, even elaborate mass analysis techniques are of no avail.

We have summarized the sample preparation steps for FAB in a flow diagram (Martin *et al.*, 1982). Przybylski (1983) has offered a similar scheme for FD sample preparation.

It is particularly important that isolation schemes for samples that are to be analysed using FD and FAB include steps designed to minimize the content of interfering species, and that suitable blanks be prepared and analysed before valuable samples are committed. Reversed phase HPLC with single solvents or solvent gradients (*e.g.* a gradient of acetonitrile/water on a C_{18} column for peptides) is an effective approach. When buffers are necessary, volatile buffers such as ammonium formate are obviously to be preferred. Simpler but similar separations can be achieved with solid-phase extraction cartridges, now available in a variety of chemistries, such as Sep-Paks® or Bond Elut®. In order to flush off possible contaminants, chromatographic columns may be washed with suitable solvents and plates for thin layer chromatography may be developed with the solvent system before application of the samples. Spots extracted from silica gel plates or collected from chromatographic columns may contain significant amounts of the silica gel, which is slightly soluble in solvents such as chloroform.

Cleanliness of sample containers is critical for success: glassware used for purified samples must be free of contaminants, including traces of detergents. (The appearance in FD or FAB spectra of an ion series spaced at 44 u [C_2H_4O]

intervals is usually an indication that detergent is present.) Acid-washing of glassware and the use of high quality solvents for final rinsing is strongly recommended.

The presence of salts, although troublesome in both FD and FAB, does not always affect both ionization methods to the same extent. Thus, in some cases, FD or FAB ionization may be suppressed totally, but analysis by the other method is still feasible for the same sample. It is therefore worth the effort, when both FD and FAB are available, to try by the other method a sample which has failed by one.

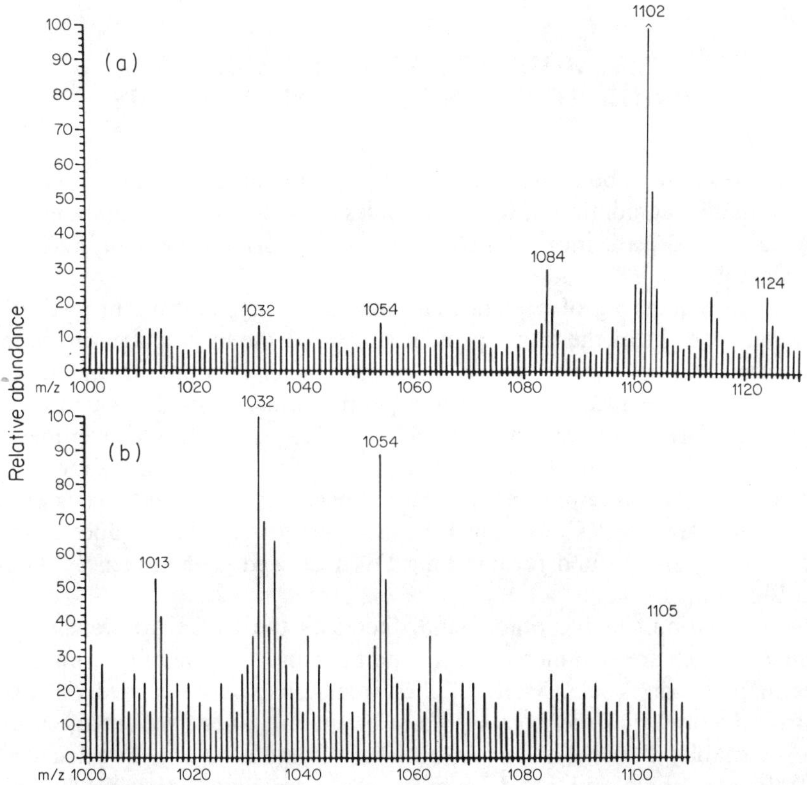

Figure 1: Molecular ion region of the (+)FAB mass spectrum of (mannose)$_5$-N-acetyl-glucosamine, M_r 1031. (a) Final purification step was HPLC using diaminobutane as a carrier. (b) Final purification step was thin-layer chromatography.

Compounds added during HPLC to facilitate the chromatography may remain with the sample and react before or during the ionization process. Figure 1a shows the molecular ion region of the (+)FAB mass spectrum obtained for a sample of (mannose)$_5$-N-acetyl glucosamine which had been purified by HPLC with diaminobutane added as a carrier, a routine procedure for compounds of this type. The spectrum shows an apparent $(M + H)^+$ ion at m/z 1102 with accompanying peaks that can be attributed to dehydration (m/z 1084) and sodium addition (m/z 1124), and only low abundance peaks for the expected $(M + H)^+$ at m/z 1032. The peak observed at m/z 1102 indicates the formation of a Schiff base between the carbohydrate and the carrier. When the same sample was purified by thin layer chromatography, the expected $(M + H)^+$ and its sodium addition peak were observed at m/z 1032 and 1054, respectively (Figure 1b).

3.3. MODIFICATION OF SAMPLES PRIOR TO MASS SPECTRAL ANALYSIS

More information about structural detail can sometimes be gained by chemical or enzymatic degradation of large molecules into their smaller components, or by selective derivatization of functional groups prior to FD or FAB mass spectral analysis.

For the sequencing of proteins, enzymatic cleavage is a useful tool. If the molecular weight of the intact protein is beyond the mass range of the mass spectrometer, this step is essential. Even for smaller peptides, degradation can be employed to reduce the weight to a peptide which is small enough to yield sequence information, information which is often incomplete for peptides that have molecular weights above 1500 daltons. The use of enzymatic rather than chemical cleavage is important so that the specificity of the cleavage is known. Enzymatic cleavage followed by FAB MS analysis of the product mixtures provides a rapid method for checking DNA-derived gene sequences (Gibson and Biemann, 1984.

Derivatization, on the other hand, increases the molecular weight by increments which are an indication of the number and location of reacting groups present in the molecule. Many of these reactions can be carried out in the matrix. Recording of the mass spectrum before and after derivatization of the sample enables the analyst to make direct comparisons and to calculate the shifts in mass of molecular and fragment ions. Thus, for example, the formation of methyl esters and N- and O-acetylated derivatives has been used to help characterize a series of model muramyl peptides and sleep-promoting factor components (Martin *et al.*, 1984).

Derivatization may also be used to introduce a preformed charge or site for facile charge localization into a neutral molecule (Busch *et al.*, 1982). This step can result in a substantial advantage for the compound of interest, since its response factor may be increased at the same time that matrix and impurity ions are suppressed.

3.4. PREPARATION OF THE ANALYTICAL SAMPLE

Glycerol was the first compound reported to be useful as a matrix for fast atom bombardment mass spectrometry (Barber *et al.*, 1981) and it remains the most popular. Compounds which are not soluble in glycerol may still be analyzed using this matrix if they are first dissolved in a solvent which is compatible with glycerol (Martin *et al.*, 1982). Alternative matrices or combinations of matrices can yield better results for certain classes of compounds that require non-polar matrices or matrices that are unreactive or which do not contain exchangeable hydrogen atoms. A recent review by Gower (1985) has catalogued the use of a variety of FAB matrices. Included in Gower's discussion are applications which have described the use of glycerol and its deuterated analogues, thioglycerol, aminoglycerol, thioglycolic acid, polyethylene glycol and related compounds, triethanolamine, diethanolamine, 1,2,4-butanetriol, thio-2,2-bis(ethanol), 2,4-di-*tert*-pentylphenol, crown ethers, citrate derivatives, phosphoric acid, *o*-nitrophenyl octyl ether, sulfolane and a 1:3 mixture of dithiothreitol/dithioerythritol. The reader is referred to this review for detailed references on uses of these materials. A reference collection of (+) and (−) FAB mass spectra of matrices has been assembled by R. C. Murphy (University of Colorado Medical Center, Denver), acting in his role as chairman of the American Society for Mass Spectrometry Biomedical Applications Committee.

In general, it has been our experience that better quality, more stable spectra are obtained for homogeneous solutions than for dispersions. We routinely employ vortexing and sonication to achieve homogeneity before application of the matrix/sample solution to the FAB probe tip. Other laboratories have reported satisfactory results for samples mixed with matrix on the FAB probe tip or precipitated onto the matrix surface (Zhang *et al.*, 1985), but the results of these approaches would seem to be less reproducible. Application of a solution (rather than a dispersion) to the FD probe tip also produces the best result.

The possibility of oxidation/reduction and other chemical reactions in some matrices must be considered. In particular, we and others have observed the oxidation of metal complexes and other easily oxidized species in *o*-nitrophenyl octyl ether and the reduction of peptide disulphide bridges by thiol matrices. Evidence of oxidation is provided by the appearance in the spectrum of peaks at $(MH - 2)^+$. The relative abundance of these peaks may not be an accurate indication of the extent of oxidation because of differences in ionization efficiencies of the oxidized and unoxidized species and differences in clustering behaviour. The (+)FAB MS of a mixture of glycerol and glyceraldehyde shows no excess of m/z 91 over pure glycerol but for the clusters $(93 + 92n)^+$ the relative abundance of the $(93 + 92n - 2)^+$ peak increases quickly as n increases (Costello, unpublished data).

Other reactions with the matrix which can modify the sample include esterification, which for acidified glycerol as matrix adds increments of $74n$, where $n = 1,2....$ up to the total number of carboxyl groups present. These reactions seem to occur as a function of time at room temperature even in the absence of the FAB beam.

More complex matrix–sample interactions occur upon irradiation. Field (1982) has described the formation of radiation products in neat glycerol and it seems likely that extensions of the reaction sequences include reaction of the glycerol radiation products with sample molecules. Metals with multiple accessible oxidation states may undergo oxidation, reduction and ligand exchange or loss. A notable radiation chemistry phenomenon is the replacement of aromatic halogen atoms by hydrogen atoms, as described by Sethi *et al.* (1984) and Dung *et al.* (1984). Another which we have investigated (Pang *et al.*, 1984) is the appearance of ions at $(MH + 12n)^+$ and related ions in the FAB mass spectra of primary and secondary amines. Lehmann *et al.* (1984) have also reported observation of the $(MH + 12)^+$ ion in the FAB mass spectra of peptides.

The matrix may be modified to increase the ion yield of the analyte, by variation of the pH or the addition of equimolar amounts of cations [(+) FAB] or anions [(−) FAB] to effect a pre-existing charge on the molecule. For example, when an active fraction of a chemo-attractant from the slime mould *P. violaceum* was first analyzed in our laboratory, the direct probe electron ionization mass spectrum was extremely complex and GC–MS also indicated that the sample was a mixture of many compounds. At that time, we obtained the FD mass spectrum shown in Figure 2a. Many of the peaks could be assigned to impurities such as silicones (*m/z* 281, 305, 327) and indeed it was not clear whether *any* of the observed peaks were related to the biological activity. The guidance provided by the identification of many of the initial impurities

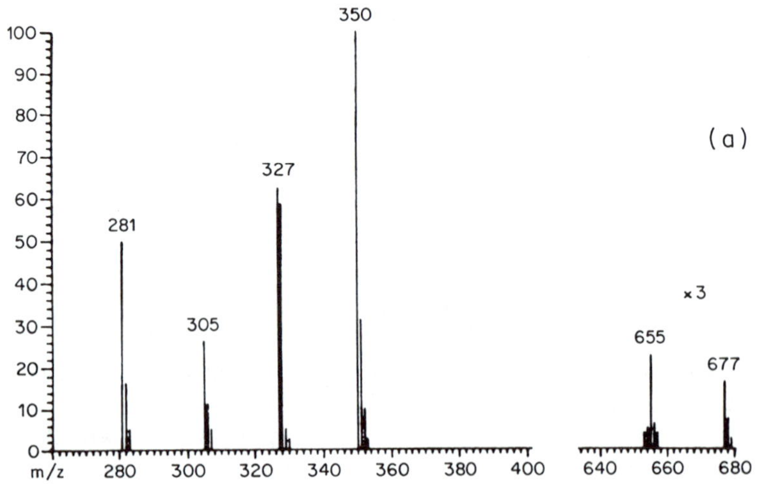

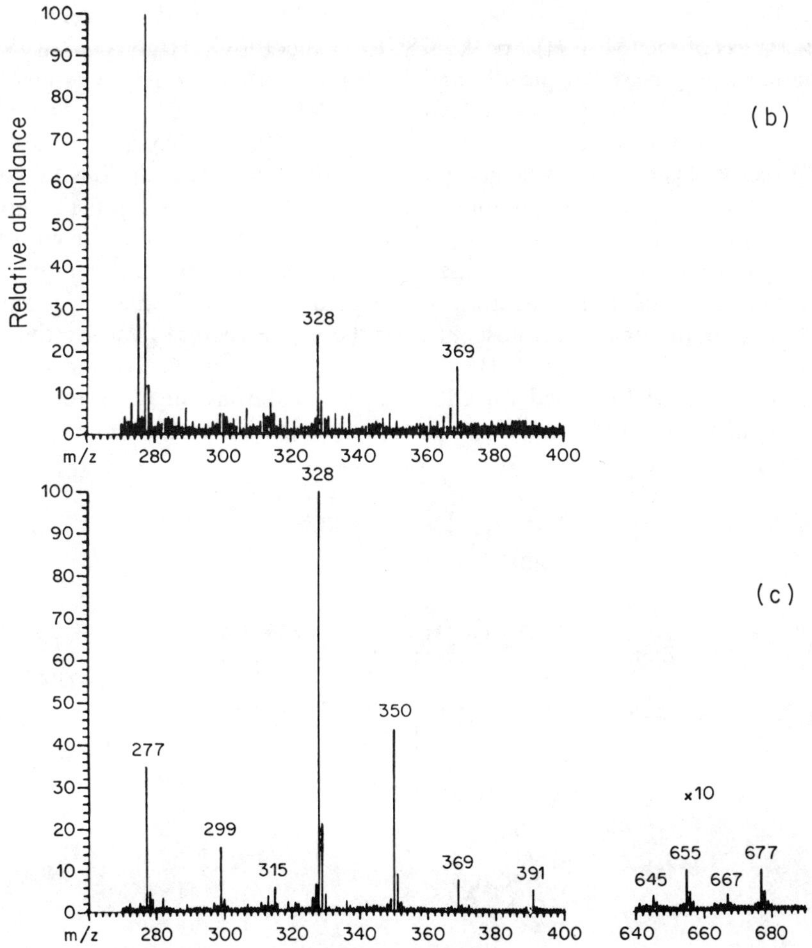

Figure 2: Mass spectra obtained for the chemo-attractant (Structure 1). (a) (+)FD mass spectrum of a sample containing silicones and other contaminants. (b) (+)FAB mass spectrum of a more highly purified sample dissolved in 1 : 1 glycerol/methanol. (c) (+)FAB mass spectrum of the solution used to obtain spectrum (b), after addition of small amounts of oxalic acid and sodium chloride.

helped in the further purification of the sample, however. FAB MS of 0.3 μl of a solution of 2 μg of the more highly purified fraction dissolved in 2 μl (1:1) methanol/glycerol gave the spectrum shown in Figure 2b, which suggested that the $(M + H)^+$ was m/z 328, one of the ions observed in the earlier FD mass spectrum of the crude fraction. Addition of small, approximately equimolar amounts of oxalic acid and sodium chloride to the solution used to obtain 2b produced a much clearer spectrum of the unknown, as shown in Figure 2c. Glycerol (G) peaks, although suppressed, were still of adequate abundance for peak-matching. The $(3G + Na)^+$ peak at m/z 299 was used to determine the exact mass of the $(M + H)^+$ peak, 328.1877 (calc. for $C_{15}H_{26}N_3O_5$ 328.1872). The remaining peaks in the FD and FAB mass spectra could be assigned as $(M + Na)^+$, m/z 350; $(2M + H)^+$, m/z 655; $(2M + Na)^+$, m/z 677. (The peaks in the FAB mass spectrum at m/z 277, 299, 315, 369, 391, 645, and 667 are due to glycerol clusters with attached proton, sodium or potassium ions.) Assignment of the structural details could then be made on the basis of the complete high resolution EIMS of the purified sample and on insight gained from NMR and amino acid analyses, so that the Structure **1** was determined (Shimomura *et al.*, 1982). Fragmentation assigned for the high resolution EI mass spectrum is indicated on the structure. Subsequent synthesis of Structure **1** verified that this structure was in fact correct. (It was possible to return most of the 17 μg of purified material furnished for MS analysis to the investigators for further biological testing and comparison with the synthetic peptide when the latter became available.)

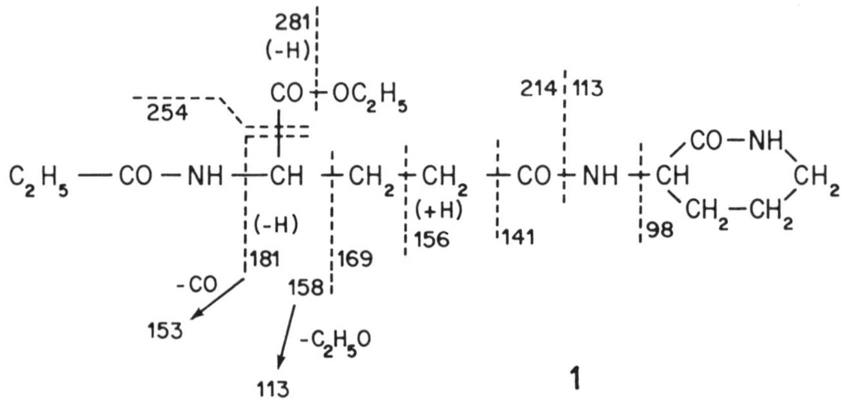

1

For certain compounds with low ionization energies, charge-transfer complexation can be employed as a simple, selective promotion reaction to enhance the abundance of $M^{+\cdot}$ (De Pauw, 1983).

Combined liquid chromatography–FAB MS offers the advantage of on-line analysis. This is an area in which extensive development work is being done at present. A small amount of matrix may be included in the HPLC mobile phase (Stroh *et al.*, 1985a) or matrix may be added after the chromatographic separation. Stroh *et al.* (1985b) have also demonstrated LC/MS FAB from a moving belt without use of a matrix. This approach provides spectra which are recorded with good sensitivity and are free of matrix ions.

3.5 CONCLUSION

Field desorption and fast atom bombardment mass spectrometry both offer the analyst an opportunity to obtain information about biomolecules whose negligible vapour pressures and/or thermal instability present formidable obstacles to the investigator. The Golden Fleece is not easily won, but the odds for success can be greatly improved by careful assessment of the likely situation and skillful selection of the course to be followed prior to the actual moment of encounter.

ACKNOWLEDGMENT

The MIT Mass Spectrometry Facility is supported by a grant from the National Institutes of Health, Division of Research Resources (Grant No. RR00317, to K. Biemann).

REFERENCES

Barber, M., Bordoli, R. S., Sedgwick, R. D., and Tyler, A. N. (1981). Fast atom bombardment of solids as an ion source in mass spectrometry, *Nature*, **293**, 270–275.

Busch, K. L., Unger, S. E., Vincze, A., Cooks, R. G., and Keough, T. (1982). Desorption ionization mass spectrometry: sample preparation for secondary ion mass spectrometry, laser desorption, and field desorption, *J. Am. Chem. Soc.*, **104**, 1507–1511.

De Pauw, E. (1983). Ion promotion in fast atom bombardment mass spectrometry by charge transfer complexation, *Anal. Chem.*, **55**, 2195–2196.

Dung, D. V., Marien, J., De Pauw, E., and Decuyper, J. (1984). Secondary ion mass spectrometry of some promazines and photochemistry of chloropromazine in glycerol, *Org. Mass Spectrom.*, **19**, 276–279.

Field, F. H. (1982). Fast atom bombardment study of glycerol: mass spectra and radiation chemistry, *J. Phys. Chem.*, **86**, 5115–5123.

Gibson, B. W., and Biemann, K. (1984). Strategy for the mass spectrometric verification and correction of the primary structures of proteins deduced from their DNA sequences, *Proc. Nat. Acad. Sci. USA*, **81**, 1956–1960.

Gower, J. L. (1985). Review paper: Matrix compounds for fast atom bombardment mass spectrometry, *Biomed. Mass Spectrom.*, **12**, 191–196.

Lehmann, W. D., Kessler, M., and Konig, W. A. (1984). Investigations on basic aspects of fast atom bombardment mass spectrometry: matrix effects, sensitivity and quantitation, *Biomed. Mass Spectrom.*, **11**, 217–222.

Martin, S. A., Costello, C. E., and Biemann, K. (1982). Optimization of experimental procedures for fast atom bombardment mass spectrometry, *Anal. Chem.*, **54**, 2362–2368.

Martin, S. A., Karnovsky, M. L., Krueger, J. M., Pappenheimer, J. R., and Biemann, K. (1984). Peptidoglycans as promoters of slow wave sleep. I. Structure of the sleep promoting factor isolated from human urine, *J. Biol. Chem.*, **259**, 12652–12658.

Pang, H., Costello, C. E., and Biemann, K. (1984). Fast atom bombardment induced adduct ions, *Proc. 32nd Ann. Conf. Mass Spectrometry and Allied Topics, San Antonio*; American Society for Mass Spectrometry, pp. 751–752.

Przybylski, M. (1983). Fast atom bombardment and field desorption mass spectrometry: comparative aspects of analytical development and bioanalytical application, *Fresenius' Z. Anal. Chem.*, **315**, 402–421.

Sethi, S. K., Nelson, C. C., and McCloskey, J. A. (1984). Dehalogenation reactions in fast atom bombardment mass spectrometry, *Anal. Chem.*, **56**, 1975–1977.

Shimomura, O., Suthers, H. L. B., and Bonner, J. T. (1982). Chemical identity of the acrasin of the cellular slime mold *Polysphondylium violaceum, Proc. Nat. Acad. Sci. USA*, **79**, 7376–7379.

Stroh, J. G., Cook, J. C., Milberg, R. M., Brayton, L., Kihara, T., Huang, Z., Rinehart, K. L., and Lewis, I. A. S. (1985a). On-line liquid chromatography/fast atom bombardment mass spectrometry, *Anal. Chem.*, **57**, 985–991.

Stroh, J. G., Cook, J. C., Rinehart, K. L., Kihara, T., Huang, Z., Lewis, I. A. S., and Keifer, P. A. (1985b). *Proc. 33rd Ann. Conf. Mass Spectrometry and Allied Topics, San Diego*; American Society for Mass Spectrometry, pp. 784–785.

Zhang, M-Y., Liang, X-Y., Chen, Y-Y., and Liang, X-G. (1985). Surface precipitation technique in fast atom bombardment mass spectrometry, *Anal. Chem.*, **56**, 2288–2290.

Mass Spectrometry in Biomedical Research
Edited by S. J. Gaskell
© 1986 John Wiley & Sons Ltd.

Chapter 4

DIRECT ANALYSIS OF BIOCHEMICAL REACTIONS IN AQUEOUS SOLUTIONS

Richard M. Caprioli

4.1. INTRODUCTION

The advent of ionization techniques which can produce ions from solutions containing polar and ionic compounds dissolved in aqueous or partially aqueous solutions has opened new areas of interest in using mass spectrometry in biochemistry. Although fast atom bombardment mass spectrometry (FAB MS) has perhaps been the most widely used technique to date in this regard (Caprioli, 1985), other methods have also been successfully employed including liquid secondary ion mass spectrometry, thermospray, and permeable membrane mass spectrometry. The unifying concept for all of these methods for the biochemist is the capability for molecular analysis of biological compounds with minimal sample preparation and little or no chemical derivatization. This is not simply for ease and convenience, but rather it provides molecularly specific analysis directly on reaction solutions, avoiding chemical manipulations which often alter the system to an extent which destroys the molecular dynamics of importance.

This chapter will focus on the use of mass spectrometry in the direct measurement of biochemical reactions and processes. Some of these applications measure chemical distributions in systems at equilibrium, while others involve reaction monitoring in a time-dependent manner to provide kinetic parameters of the system. Most of the reactions take place in aqueous or partially aqueous systems. Therefore, a key consideration is the extent to which the measurement of the ions which exist in the gas phase that is sampled by the mass spectrometer is truly a part of, or at least is representative of, the ionic distributions present in the condensed phase of the sample. From the work done thus far, it is clear that this may be achieved under certain conditions. However, a great deal of additional work is needed before we fully understand the parameters which control and affect these conditions. In the text which follows, several specific examples have been chosen to illustrate the types of samples and processes which have become amenable to analysis by mass spectrometry. The purpose of the chapter is to provide an overview and introduction to the subject, and not a comprehensive review. It will deal primarily with those applications which involve 'direct analysis', i.e., where reaction components are analyzed without derivatization or extensive preparation methods, and also with those which attempt to determine dynamic aspects of biochemical reactions.

4.2. PROTEOLYTIC ENZYME REACTIONS

The use of proteolytic enzymes for the sequence and structure analysis of polypeptides has been the focus of a number of investigators. Trypsin perhaps has been the most widely used enzyme for proteolysis that has been coupled with analysis utilizing mass spectrometry. FAB MS has played a major role in this regard in the identification of the various molecular species produced in the reaction. It is beyond the scope of this chapter to cite the work of many investigators who have utilized trypsin to specifically hydrolyze large polypeptides and proteins and have examined the molecular distributions of the products by FAB MS. These applications generally involve a single analysis rather than a monitoring of the time course of the reaction and clearly establish the method as a powerful tool in a sequencing strategy for proteins.

The use of mass spectrometry with exopeptidase digestions for the determination of sequence has also been explored by several investigators. With respect to following on-going enzyme digestions, two approaches have been employed for this purpose: analysis of aliquots of reaction mixtures taken at timed intervals from a batch reaction, and real-time analyses done *in situ* as the reaction proceeds inside the mass spectrometer. The latter application will be discussed later in this chapter. Tsugita and co-workers (1982) utilized both aminopeptidase M and carboxypeptidase A and B together with field desorption (FD) mass spectrometry for sequencing. In a typical analysis, they subjected peptides to exopeptidase digestion and removed 3–4 aliquots over several hours. Equal volumes of the aliquots were recombined and analyzed by FD and the $(M + H)^+$ ions for the various polypeptide molecules produced by the action of the enzyme were identified. Sample amounts of approximately 0.5 nmol were used. The sequence was then determined from the mass differences between the homologous series of molecular ions. Hong *et al.* (1983) have also employed FD mass spectrometry and carboxypeptidase digestion for sequencing. They used both carboxypeptidase Y and B in a sequential manner with analysis of the molecular polypeptide products as well as the free amino acids at the end of the reaction.

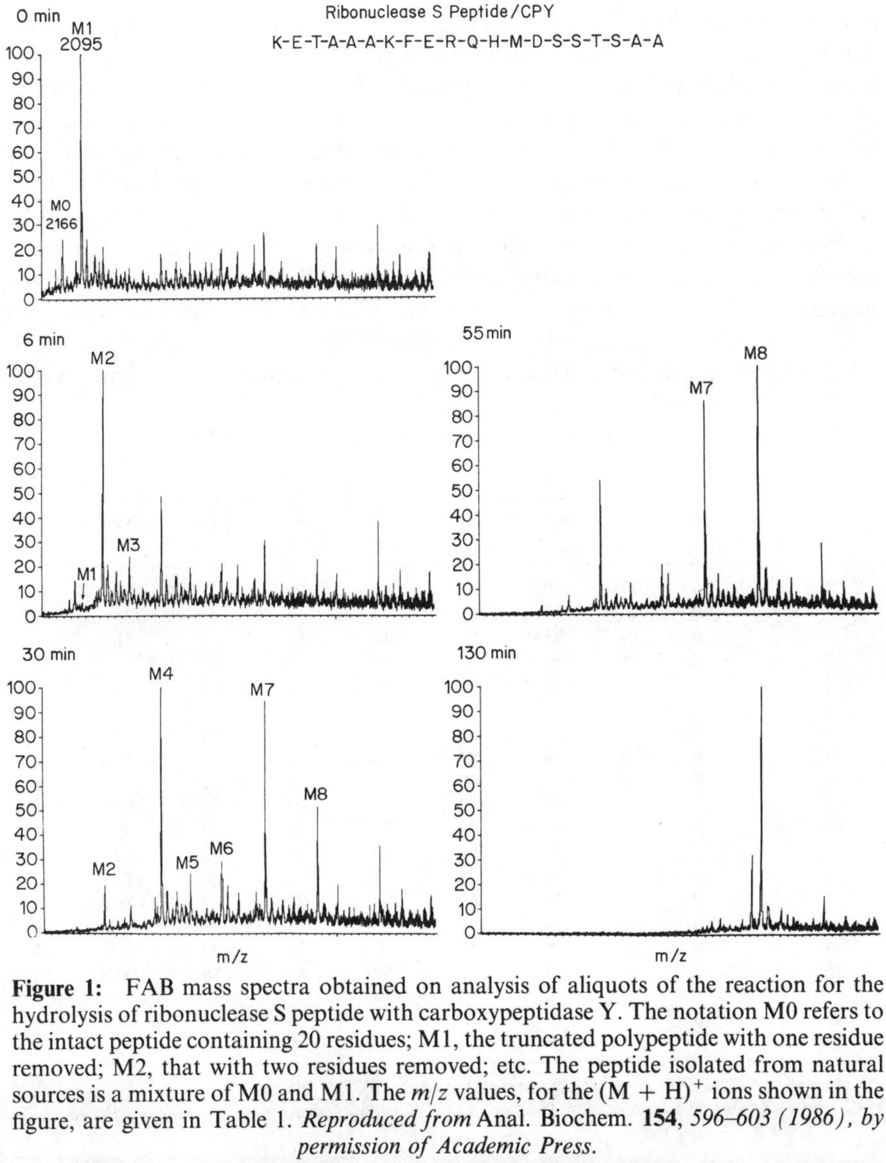

Figure 1: FAB mass spectra obtained on analysis of aliquots of the reaction for the hydrolysis of ribonuclease S peptide with carboxypeptidase Y. The notation M0 refers to the intact peptide containing 20 residues; M1, the truncated polypeptide with one residue removed; M2, that with two residues removed; etc. The peptide isolated from natural sources is a mixture of M0 and M1. The *m/z* values, for the $(M + H)^+$ ions shown in the figure, are given in Table 1. *Reproduced from* Anal. Biochem. **154**, *596–603 (1986), by permission of Academic Press.*

FAB MS has been widely used and is preferred by many investigators for producing ions from polar samples because of its reliability and ease of sample preparation. Bradley and Williams (1982) have used FAB MS for analysis of the shortened polypeptide after the Edman degradation. Caprioli and Fan (1986) have used carboxypeptidase Y and leucine aminopeptidase for the sequence determination of polypeptides. Figure 1 shows the FAB spectra of five of 20 aliquots removed with time on digestion of ribonuclease S peptide with carboxypeptidase Y. Ions corresponding to the homologous series of truncated polypeptides are clearly seen for removal of eight residues from the C-terminus. Although the sequence can be determined from the mass differences between these molecules, the time-course for production of these molecules can also be obtained from these data as shown in Table 1. Mass redundancies present two difficulties; Leu and Ile have exactly the same mass, and Gln and Lys have the same nominal mass, although strategies have been suggested to circumvent the latter. One of the major advantages of these approaches is the combining of the hydrolytic specificity of the carboxypeptidases and the molecular analytical specificity of mass spectrometry, providing a rapid, sensitive method for the sequencing of peptides from the C-terminus where no good classical technique now exists. Self and Parente (1983) also examined the use of exopeptidases with FAB MS for peptide sequencing and suggested that it may find particular value

TABLE 1. Hydrolysis of ribonuclease S peptide with carboxypeptidase Y.

	Relative Ion Intensities*								
Time (min) *m/z*	M0 2166	M1 2095	M2 2023	M3 1937	M4 1836	M5 1749	M6 1662	M7 1547	M8 1416
0	100	100	20	47	16	77	88	28	10
1	0	60	48	46	14	58	64	25	11
2	0	38	73	50	14	61	61	31	10
3	0	18	100	80	21	74	65	24	11
4	0	5	86	68	23	60	60	25	9
5	0	3	92	78	35	60	60	28	8
6	0	1	94	100	45	66	66	27	9
8	0	0	69	81	64	74	67	28	12
10	0	0	69	78	81	79	79	28	10
12	0	0	54	74	78	92	79	32	9
14	0	0	42	45	75	76	79	41	10
17	0	0	33	55	100	97	92	48	10
20	0	0	28	76	95	100	100	57	11
25	0	0	20	69	98	99	100	68	16
30	0	0	15	39	87	76	99	88	20
35	0	0	8	41	78	75	99	99	27
55	0	0	2	27	57	47	80	100	52
75	0	0	0	0	29	28	42	83	78
130	0	0	0	0	0	0	0	0	100
800	0	0	0	0	0	0	0	0	0

* Expressed as a percentage of the maximum for each ion. See Figure 1.
Reproduced from Anal. Biochem., **154**, *596–603 (1986), by permission of Academic Press.*

with N-blocked peptides and provide greater detail relating the molecular specificity of the enzymes with the proximity of particular amino acids at the site of cleavage on the substrate.

4.3. ENZYME REACTION RATES

Enzyme reaction rates are generally measured using synthetic substrates which contain chromophores for the ease and convenience of spectrophotometric assays. However, these substrates often have kinetic constants which can be quite different from those of natural substrates. Mass spectrometry offers a unique molecular specificity which has the potential to provide accurate kinetic constants for the action of enzymes with true substrates. Preliminary work by Smith and Caprioli (1984) has shown that FAB MS can be used for determining constants for the proteolytic enzyme trypsin with both the synthetic substrate TAME (tosylarginine methyl ester) and several natural peptide substrates. In these experiments, aliquots of a reaction mixture which contained trypsin, substrate, and buffer in 67% glycerol were removed at timed intervals and analyzed directly to determine the ratio of the abundance of the $(M + H)^+$ of

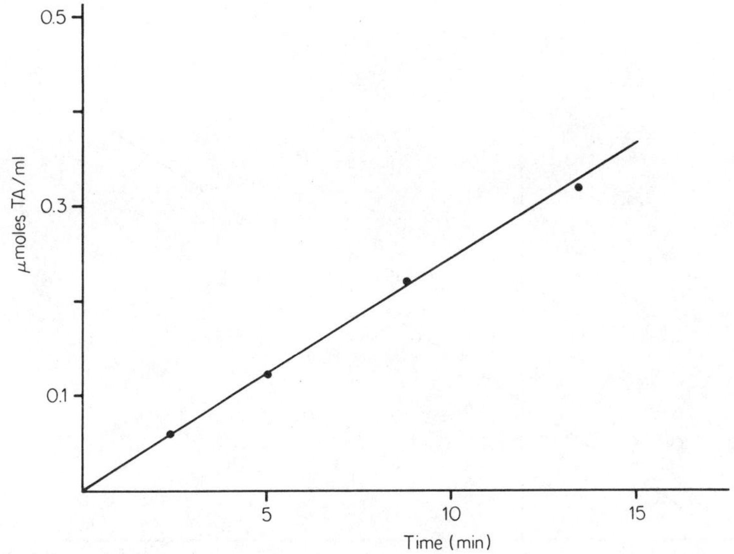

Figure 2: Tryptic hydrolysis of TAME to give TA (tosylarginine) monitored by FAB MS. Aliquots analyzed were taken from a batch reaction containing 1.0 mM TAME, 0.136 μg trypsin/ml in 0.05 M Tris buffer, pH 8.0, in 67% glycerol. *From* Biomed. Mass Spectrom., *11, 392–395 (1984).*

the substrate to that of the product. The concentration ratio was determined from a standard response curve for the two compounds and then concentrations calculated from the amount of substrate present at zero time. For example, for an enzyme concentration of 0.136 μg trypsin per ml in a solution containing 1.0 mM TAME, a reaction rate of 0.223 μmol product/ml/min was measured (Figure 2). In comparison, under the same reaction conditions, the spectrophotometric assay gave a rate of 0.23 μmol/ml/min, and with no glycerol present, a value of 0.29 μmol/ml/min. This work was extended (Caprioli and Smith, 1986) to measure the kinetic constants K_m and V_{max} for trypsin with several oligopeptides. As an example, the rate of hydrolysis V of the peptide Met-Arg-Phe-Ala was measured at several substrate concentrations $[S]$ and rates calculated. From a double reciprocal plot ($1/V$ versus $1/[S]$) of these data, one can obtain the value of 1.9 mM for the K_m and 0.30 μmol/ml/min for V_{max} of this reaction (Figure 3). In another approach to the measurement of rate constants for the reaction of trypsin with TAME, Murawski and co-workers (1985) have utilized a moving belt device for the introduction of samples into the FAB source. The sample solution was sprayed directly on the belt and dried before exposing it to the atom beam, eliminating the need for glycerol or other organic matrix. Using a 1.32 mM TAME solution in a volatile buffer with varying amounts of enzyme, the activity of trypsin was determined to be 89.2 μg of substrate hydrolyzed/min/mg enzyme, compared to a value obtained from a spectrophotometric assay of 104 μg/min/mg.

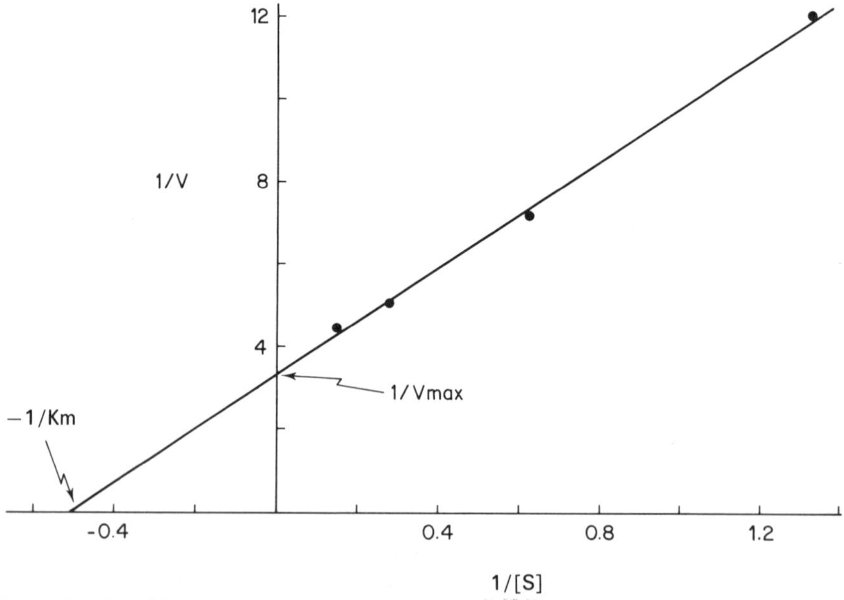

Figure 3: Double reciprocal plot of rates of hydrolysis of Met-Arg-Phe-Ala by trypsin. V: rate of hydrolysis (μmol/ml/min); [S] concentration of substrate.

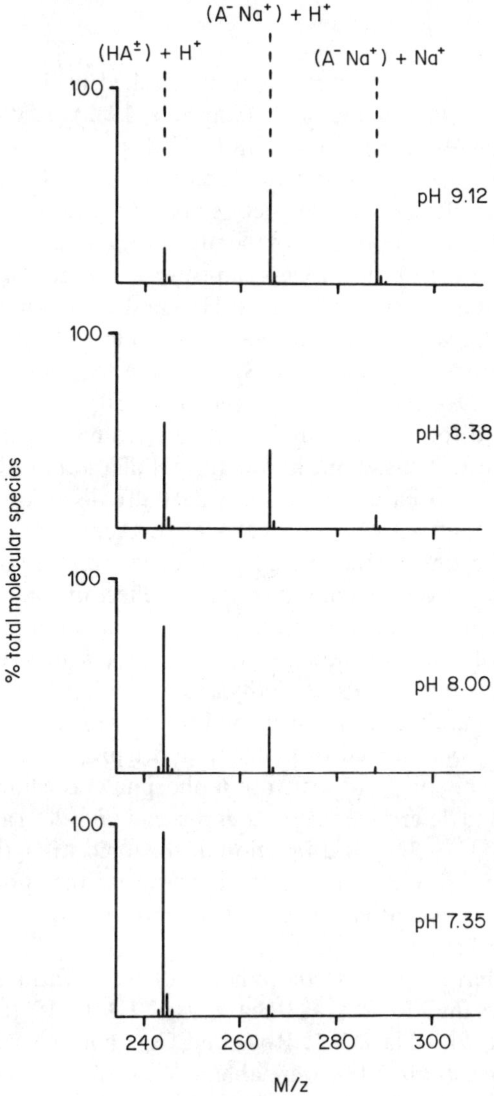

Figure 4: Effect of pH on the molecular ion distribution between the acid and conjugate base forms of the zwitterionic acid tris(hydroxymethyl)methylaminopropane-sulphonate (TAPS) measured by FAB MS. The amino group has a pK_a of 8.4. *Reprinted with permission from* Anal. Chem., *55, 2387. Copyright 1983 American Chemical Society.*

4.4. CHEMICAL EQUILIBRIA

One of the simpler chemical reactions to be studied by FAB MS is the acid dissociation reaction of organic acids (Caprioli, 1983). It was found that in aqueous glycerol solutions containing up to 70% glycerol, acid and conjugate base species which were quantitatively determined by FAB could be used directly in the Henderson–Hasselbalch equation relating the ratio of the concentrations of these components to the pH of the solution and the pK_a of the acid. The general nature of this process can be seen in Figure 4 which shows the relative ion intensities of the acid $(HA + H)^+$ and the conjugate base species, $(ANa + H)^+$ and $(ANa + Na)^+$, for the zwitterionic acid tris(hydroxymethyl)-methylaminopropanesulphonate (TAPS). As expected from consideration of solution chemistry, as the pH is lowered and falls below the pK_a of the compound, the acid form is favoured, while increasing the pH above the pK_a favours the base form. Thus, if one knows the pH of a solution and can measure the ratio of the acid to base species for a partially dissociated acid, then the apparent pK_a of that acid can be calculated. Actual pK_a values may be determined by correcting for the change in the dielectric constant of the medium due to the presence of glycerol and also for the effect of ionic strength. In this work, over twenty pK_a values were measured and compared to literature values for a wide range of acids, including simple aliphatic acids, zwitterionic acids, organophosphates, amino acids, dimethylarsinic acid, and others. The results showed that under carefully controlled conditions, the pK_a of these compounds could be accurately determined to better than 0.1 pK_a unit. For example, the determination of the pK_{a_2} of fructose-6-phosphate is shown in Figure 5. FAB MS was used to determine the ratio of the acid (*m*/*z* 283) and the conjugate base species (*m*/*z* 305 + 327) as a function of the ionic strength. Extrapolation of these curves to zero ionic strength gave a measure of the apparent constant in 50% glycerol/water, and after correcting to pure water, gave a pK_a of 6.05 (literature value: 6.11).

Other workers have looked at the binding of metal cations to a variety of ligands. Johnstone and Rose (1983) have used FAB MS to study complex formation between Li^+, Na^+, K^+, Rb^+, and Cs^+, and macrocyclic ligands in aqueous glycerol solutions. They were able to show that the abundances of the gas phase ions corresponding to metal–cation complexes closely reflected the calculated concentrations of these complexes in solution. They also examined the selectivity of the macrocyclic lactone antibiotic nonactin for metal cations. For example, with molar equivalents of nonactin and Na^+ and K^+ in a solvent of glycerol/ethanol (2 : 1 v/v), the ratio of ion abundances of the sodium complex to that of the potassium complex was 1 : 17, in good agreement with published values of 1 : 19 (at 30°C in methanol) and 1 : 23 (at 30°C in ethanol).

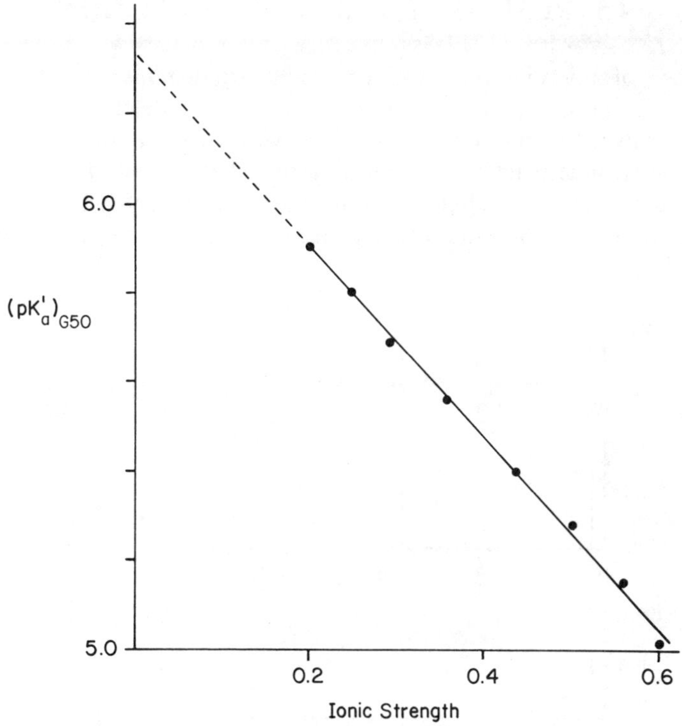

Figure 5: Effect of ionic strength on the apparent dissociation constant in 50% aqueous glycerol, $(pK_a')_{G50}$, for glucose 6-phosphate ($pK_{a_2} = 6.05$) measured by FAB MS. Ionic strength was increased by addition of 2M NaCl in 50% aqueous glycerol.

Enzyme equilibrium measurements have been made on enolase (Caprioli, 1985), an enzyme that catalyzes the reversible hydration of phosphoenolpyruvate (PEP) to form 2-phosphoglycerate (2PG). The equilibrium expression is simply the ratio of $[PEP^{-3}]$ to $[2PG^{-3}]$. However, since classical analytical methods measure the total of all ionized forms of PEP and 2PG, only an apparent equilibrium constant is measured by these methods. The actual K_{eq} is calculated from the apparent constant after evaluating the ionization constant for each substrate under the specific experimental conditions employed. FAB MS offers a unique advantage in that it is able to measure directly the ion ratios derived from the trianion species of each substrate in solution. The ions at *m/z* 235 and 257 represented $(PEP(Na_3)+H)^+$ and $(PEP(Na_3)+Na)^+$, respectively, and *m/z* 253 and 275 the corresponding ions for 2PG. Thus, the pH-independent constant was determined directly to be $K_{eq} = 5.0$, while that reported in the literature was 5.4 (Wold and Ballou, 1975).

4.5. REAL TIME CHEMICAL REACTIONS

A number of investigators have been able to demonstrate that chemical reactions can be maintained inside the mass spectrometer and can be continuously analyzed in real time using FAB MS. These *in situ* analyses include proteolytic enzyme reactions, chemical derivatization reactions, and detection of labile reaction intermediates. Smith and Caprioli (1983) have shown that under conditions of intermittent exposure to the atom beam, trypsin remains

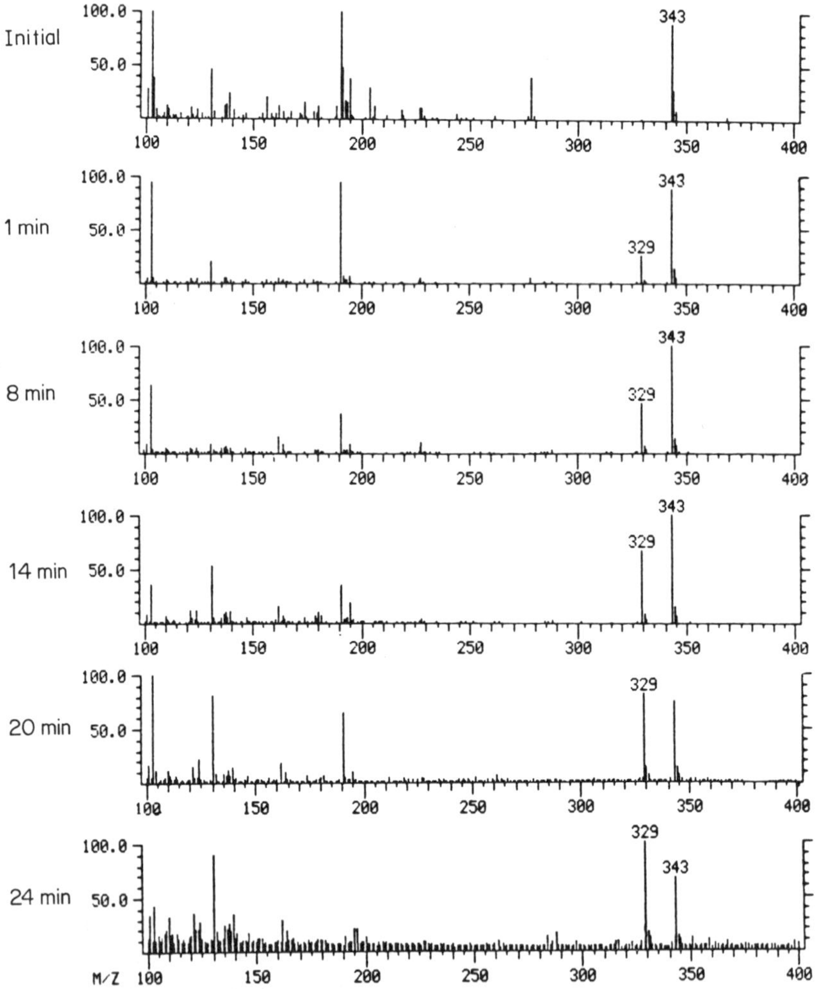

Figure 6: Hydrolysis of TAME by trypsin, determined in real-time *in situ* by FAB MS. The ion at m/z 343 is the $(M + H)^+$ ion of TAME and m/z 329 that of the $(M + H)^+$ ion of the hydrolyzed product, tosylarginine. *From* Biomed. Mass Spectrom., ***10**, 98–102 (1983)*.

fully active towards its assay substrate TAME on the sample probe. Figure 6 shows the result of the time-course digestion of this substrate over a period of 24 minutes. The reaction proceeds rapidly at first and then slows substantially as the relative concentration of water decreases from the reaction mixture. For all practical purposes, when the glycerol content exceeds 90%, the reaction stops, although the addition of water will initiate further hydrolysis. The rate of tryptic hydrolysis of the substrate TAME was also followed and reaction constants calculated for the *in situ* reaction (Smith and Caprioli, 1984). The results show that for the reaction taking place in 67% glycerol/water on the probe, rate constants were considerably lower after 2 minutes than that for the corresponding rate determined spectrophotometrically, i.e., 0.01 μmol product/min versus 0.48 μmol product/min, respectively. Nevertheless, the rate appeared relatively constant at this level for several minutes before stopping. It was also shown that carboxypeptidase Y can be employed in a 50% glycerol/water mixture and its activity can be maintained on the probe inside the mass spectrometer for 8-10 minutes. For example, Figure 7 shows the real time hydrolysis of the peptide substance P performed in this manner. Three spectra are shown in the figure from a total of 12 FAB spectra taken over a 10 minute period. As the reaction proceeded, the reaction rate decreased markedly due to the increased glycerol concentration. Further, evaporation of the matrix caused high noise levels as seen in the lowest spectrum of the figure. Nevertheless, sequence ions for several truncated molecular polypeptides can be readily identified. A number of other proteolytic enzymes can be used in the real-time mode, including dipeptidyl aminopeptidase I, proline specific endopeptidase, and chymotrypsin (Smith *et al.*, 1983). The relative ruggedness of these enzymes is surprising although not entirely unexpected. It remains a fortunate coincidence that many enzymes are routinely stabilized by the addition of substantial amounts of glycerol and that this same substance is one of the preferred liquid supports for FAB MS.

The reaction of boronic acids with trifunctional hydroxy compounds such as sugars has been studied on the probe tip *in situ* by Rose and co-workers (1983). Boronic acids react with compounds having the appropriate configuration to give boronate cage compounds that are negatively charged. For example, 4-tolueneboronic acid was used to characterize glycosidic natural products since the extent of the reaction and the behaviour of the resulting M^- ions are dependent on the orientation of the reacting hydroxyl groups. Complexes of *D*(−)ribose and *D*(−)xylose gave more intense M^- ions and thus were more stable than those produced from *D*(−)arabinose and *D*(+)xylose. This was rationalized on the basis that the first two compounds can react through stable chair conformations to give strain-free boronate cage compounds, whereas the latter compounds would be required to adopt unfavourable boat conformations.

Saito and Kato (1984) studied the glutathione conjugation reaction of the active metabolite of a potent carcinogen, (2-nitroso-6-methyl

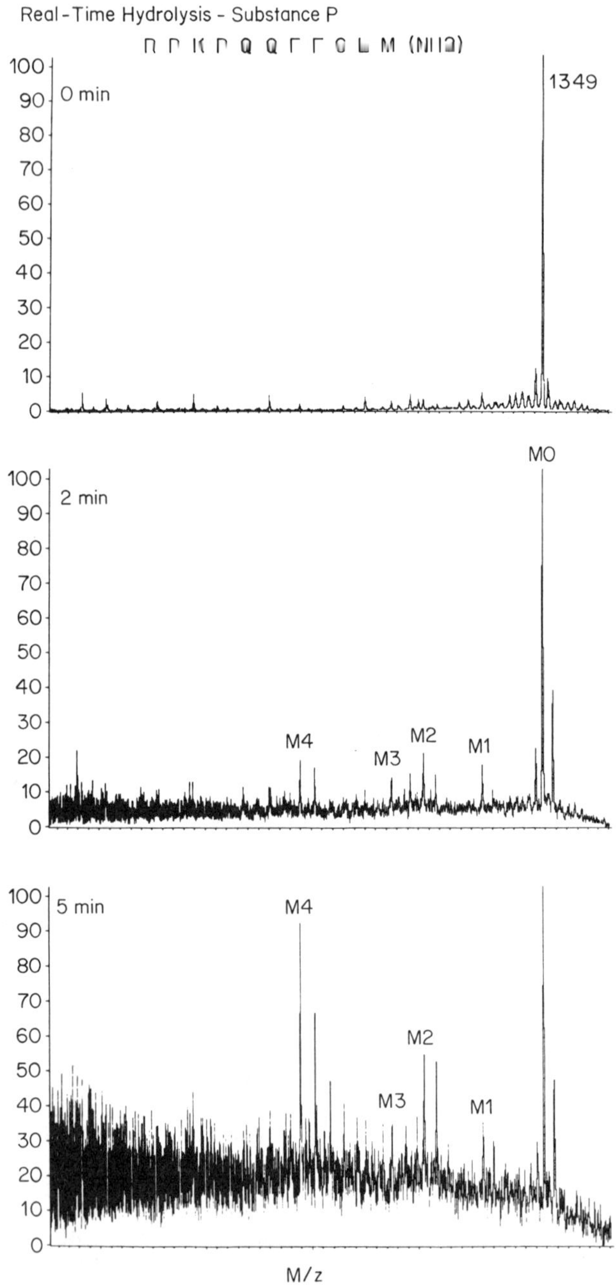

Figure 7: The real-time *in situ* hydrolysis of the C-terminal residues of the peptide substance P by carboxypeptidase Y. The notation M0, M1, M2, etc. is defined in the caption to Figure 1.

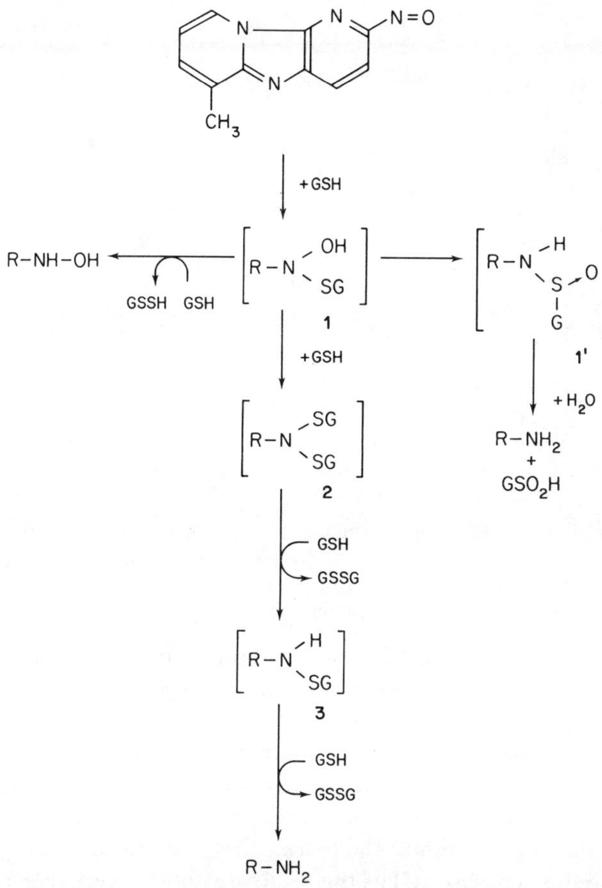

Figure 8: The proposed mechanism of glutathione conjugation with arylnitroso-compounds. *Reproduced from* Biochem. Biophys. Res. Commun., ***124**, 1–5 (1984), by permission of Academic Press.*

dipyrido[1,2-a : 3′,2′-]imidazole), in the glycerol matrix inside the mass spectrometer. The glutathione conjugates are short-lived and in the presence of reduced glutathione will give rise to the primary amine as shown in Figure 8, where compounds 1, 2 and 3 are postulated intermediates. Experimentally, ions were detected at m/z 520 (intermediate 1) and m/z 504 (intermediate 3), but not that corresponding to intermediate 2. Figure 9 shows the time course for the reaction by following the ratio of m/z 504/520 with time. These results demonstrated the existence of the glutathione-conjugate intermediate and also indicated the reaction sequence of the conjugates.

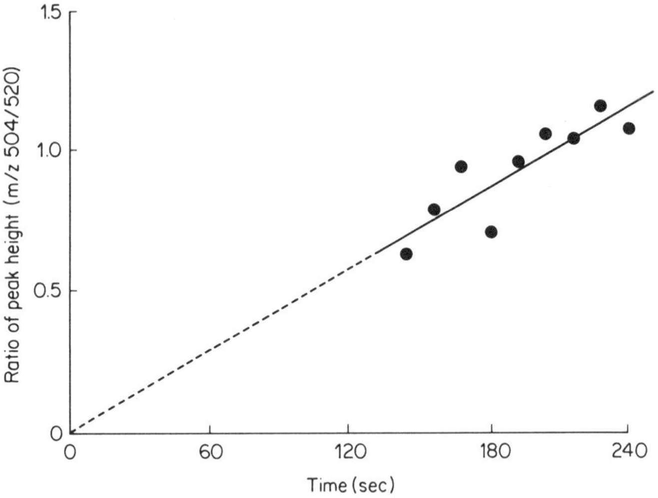

Figure 9: Real-time *in situ* measurements of the time course for the conjugation reaction containing glutathione and an arylnitroso-compound in 60% glycerol. *Reproduced from* Biochem. Biophys. Res. Commun., ***124****, 1–5 (1984), by permission of Academic Press.*

Thermospray has been used in combination with immobilized enzymes by Vestal and co-workers (Kim *et al.*, 1984; Pilosof *et al.*, 1984) for the analysis of peptides by directly passing the effluent of the reactions into the mass spectrometer. Carboxypeptidase Y was immobilized on glass beads and packed into an HPLC column. A peptide solution was injected in a small volume and the amino acids released from the C-terminus were directly transported by the continuously flowing buffer into the thermospray-equipped mass spectrometer. The various amino acids as well as the residual peptides were then detected and quantified as the $(M + H)^+$ or $(M - H)^-$ ions. Columns containing heavier or lighter immobilized enzyme concentrations may be used to control the extent of hydrolysis, an important aspect when large polypeptides are being hydrolyzed. Immobilized trypsin was used in the same manner. For example, Figure 10a shows the mass spectrum obtained from 2 nanomoles of dynorphin fragment (1–13), a tridecapeptide of 1604 dalton molecular weight. Doubly, triply and quadruply charged molecular ions are seen at m/z 803, 536 and 402, respectively. The singly charged molecular species was outside the mass range of the instrument used. After digestion by passing through a column containing 4.5×10^{-8} moles of immobilized trypsin, the tryptic peptides were analyzed with the thermospray ionization as shown in Figure 10b. Except for one, all the expected tryptic fragments could be identified as singly or doubly charged molecular species. Finally, these authors used three serial columns containing immobilized trypsin, a peptide separating column, and immobilized carboxypeptidase Y. The preliminary results towards developing this as a rapid,

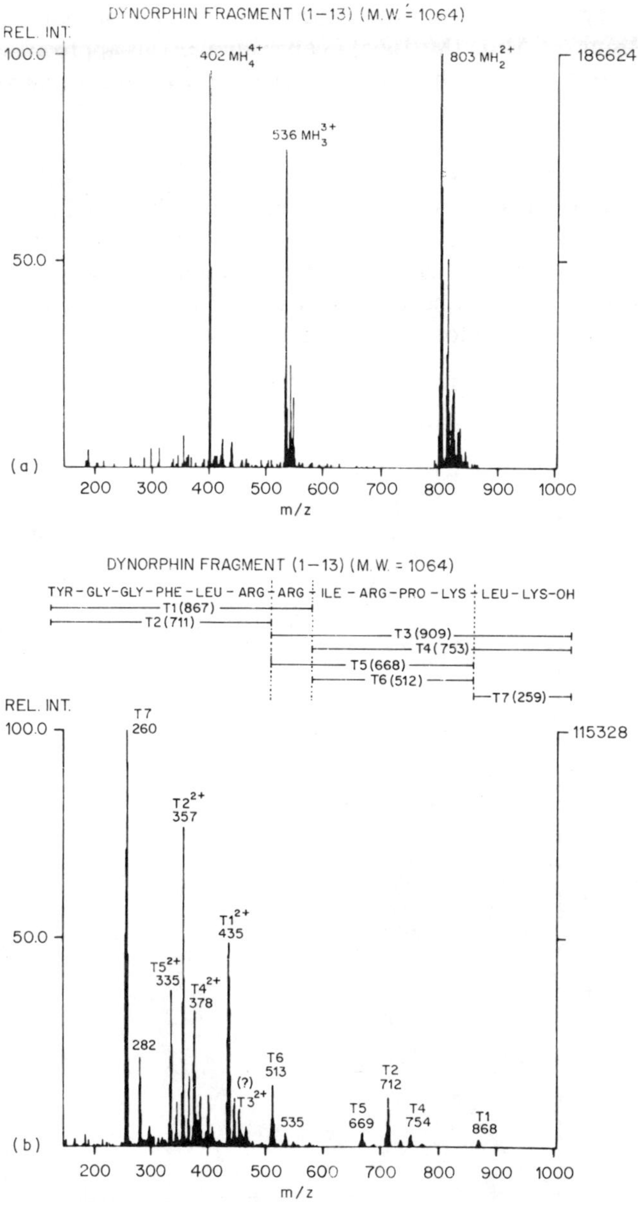

Figure 10: a) Positive ion thermospray mass spectrum of a 2 nmol sample of dynorphin fragment (1–13). b) Mass spectrum obtained after hydrolysis of the above peptide in a column containing 4.5×10^{-8} mol immobilized trypsin with a flow rate of 1.5 ml/min at room temperature. *Reprinted with permission from* J. Am. Chem. Soc., ***106**, 7304–7309.*
Copyright 1984 American Chemical Society.

automated method to obtain sequence information for polypeptides were presented using α – MSH (M_r 1665) as an example. This approach is promising and when coupled to high efficiency separation columns, may lead to an efficient sequencing method.

Permeable membrane mass spectrometry has also received more attention in the past few years because of its ability to provide real-time analyses for a variety of chemical reactions occurring in aqueous solutions. In one of the early experiments, Weaver and co-workers (1976) developed biochemical assays for immobilized enzymes such as urease and alcohol dehydrogenase. The method involves placing the enzyme next to a semi-permeable membrane where a reactant is converted into products, one of which must be volatile. This

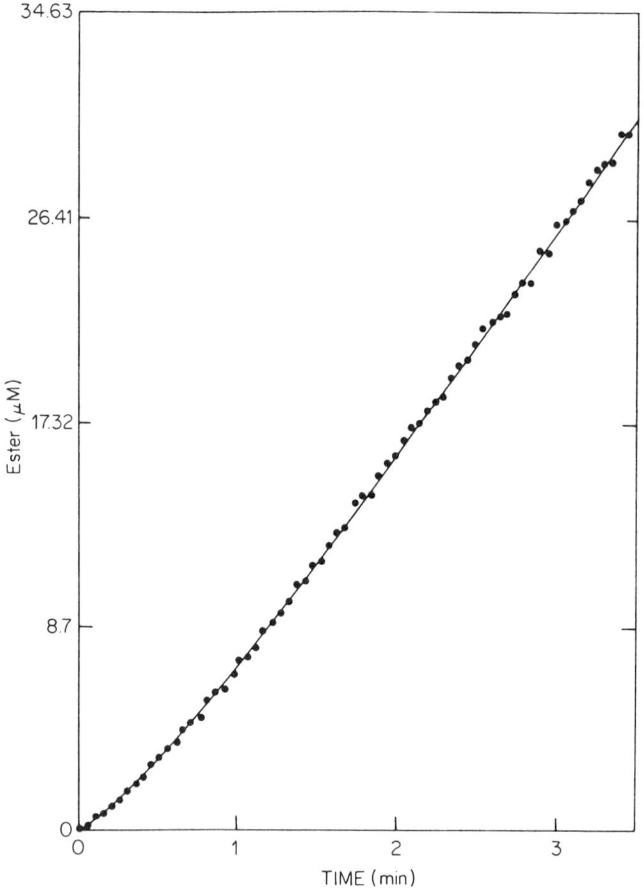

Figure 11: Real-time reaction monitoring of the α-chymotrypsin-catalyzed transesterification of p-nitrophenyl 2-furoate to ethyl 2-furoate in the presence of ethanol using permeable membrane mass spectrometry. *Reprinted with permission from* Anal. Chem., *53, 981–985. Copyright 1981 American Chemical Society.*

component then passes through the membrane and is analyzed by the mass spectrometer. For urease, which converts urea to carbon dioxide and ammonia, the carbon dioxide was monitored, while for alcohol dehydrogenase, levels of NADH were determined by following the production of ethanol from acetaldehyde. In later work, Weaver and Abrams (1979) showed that a variety of volatile components in aqueous buffered solutions could be measured by the permeable membrane method, including compounds such as acetic acid, ethylene glycol, dimethylsulphoxide, pyridine and pyruvic acid. Calvo *et al.* (1981) have used permeable membrane mass spectrometry for the measurement of the reaction kinetics of the enzyme α-chymotrypsin. In this case, the *in situ* measurement involved analysis of ethyl 2-furoate, the product of the α-chymotrypsin catalyzed transesterification of p-nitrophenyl 2-furoate. The mechanism of the reaction is multistep, involving the formation of an acyl-enzyme intermediate. The hydroxyl group of an active site serine residue on the enzyme is acylated by the substrate with concomitant release of p-nitrophenol. The second step involves deacylation to give the free enzyme and 2-furoate. In the presence of sufficient ethanol, there is a competing transesterification reaction with the formation of ethyl 2-furoate. This product is permeable through the membrane and is detected from either the parent ion, m/z 140, or the major fragment ion at m/z 112. Figure 11 shows the reaction curve for this enzymatic reaction and from these data, kinetic parameters were calculated. The same investigators used this technique to measure the enzymatic kinetic isotope effect of the α-chymotrypsin catalyzed transesterification reaction (Calvo *et al.*, 1983). Using [^{18}O] ethanol, the kinetic isotope effect for the formation of ethyl 2-furoate and ethyl 5-n-propyl-2-furoate from p-nitrophenyl esters reached constant values after approximately 2 minutes at 25°C, pH 8.5, of 1.009 ± 0.007 and 0.90 ± 0.010, respectively. The large inverse steady state kinetic isotope effect of the latter decreased when the pH was lowered and this pH dependence could be fitted with an apparent pK_a of 7.3. The mechanistic implications of these and other data were evaluated for this enzyme catalyzed reaction. Brodbelt and Cooks (1985) have used a semi-permeable membrane with a flow-through interface and a triple-quadrupole mass spectrometer for the molecular analysis of volatile and apolar substances. In addition, the reaction involving the methylation of cyclohexanone by methyl iodide was monitored over a 90 minute period.

4.6. CONCLUSIONS

In the past few years, there has been an increasing interest in the direct measurement of on-going reactions in aqueous solutions.This has important implications to the biochemist since most of the reactions of interest in his work occur in aqueous media, many under dynamic conditions which are easily perturbed by preparative procedures. It is essential that analytical procedures and instruments should be designed to deal directly with these aqueous solutions if the measurements are to be meaningful. With the appearance of soft ionization methods which are capable of producing gas phase ions from aqueous solutions, it is now possible to utilize the extraordinary molecular specificity of mass spectrometry for the measurements. Although we are in an early stage and much must still be learned and further innovations made, nevertheless, mass spectrometry will no doubt become a common and essential tool for the biochemist and molecular biologist.

REFERENCES

Bradley, C. V., and Williams, D. H. (1982). Peptide sequencing using the combination of Edman degradation, carboxypeptidase digestion and fast atom bombardment mass spectrometry, *Biochem. Biophys. Res. Commun.*, **104**, 1223–1230.

Brodbelt, J. S., and Cooks, R. G. (1985). An exceedingly simple mass spectrometer interface with application to reaction monitoring and environmental analysis, *Anal. Chem.*, **57**, 1153–1155.

Calvo, K. C., Weisenberger, C. R., Anderson, L. B., and Klapper, M. H. (1981). Permeable membrane mass spectrometer measurement of reaction kinetics, *Anal. Chem.*, **53**, 981–985.

Calvo, K. C., Weisenberger, C. R., Anderson, L. B., and Klapper, M. H. (1983). Permeable membrane/mass spectrometric measurement of an enzymatic kinetic isotope effect: α-chymotrypsin-catalyzed transesterification, *J. Am. Chem. Soc.*, **105**, 6935–6941.

Caprioli, R. M. (1983). Fast atom bombardment mass spectrometry for determination of dissociation constants of weak acids in solution, *Anal. Chem.*, **55**, 2387–2391.

Caprioli, R. M. (1985). Fast atom bombardment mass spectrometry: application to solution chemistry, in *Mass Spectrometry* (Specialist Periodical Report) vol. 8 (Ed. M. Rose), pp. 184–209, Royal Society of Chemistry, London.

Caprioli, R. M., and Fan, T. (1986). Peptide sequence analysis using exopeptidase with molecular analysis of the truncated polypeptides, *Anal. Biochem.*, **154**, 596–603.

Caprioli, R. M., and Smith, L. A. (1986). Determination of K_m and V_{max} for tryptic peptide hydrolysis using fast atom bombardment mass spectrometry, *Anal. Chem.*, **58**, 1080–1083.

Hong, Y. M., Takao, T., Aimoto, S., and Shimonishi, Y. (1983). Simultaneous analysis of C-terminal amino residues of peptide mixtures by combination of field desorption mass spectrometry and carboxypeptidase digestion, *Biomed. Mass Spectrom.*, **10**, 450–457.

Johnstone, R. A. W., and Rose, M. E. (1983). Macrocyclic ligands examined by fast atom bombardment mass spectrometry: direct observation of metal cation selectivity in complexation, *J. Chem. Soc. Chem. Commun.*, 1268–1270.

Kim, H. Y., Pilosof, D., Dyckes, D. F., and Vestal, M. L. (1984). On-line peptide sequencing by enzymatic hydrolysis, high performance liquid chromatography, and thermospray mass spectrometry, *J. Am. Chem. Soc.*, **106**, 7304–7309.

Murawski, S. L., Cook, J. C., and Rinehart, K. L. (1985). Studies of enzyme reactions using a moving belt and FABMS, *Proc. 33rd Ann. Conf. on Mass Spectrometry and Allied Topics, San Diego*; American Society for Mass Spectrometry; pp. 506–507.

Pilosof, D., Kim, H-Y., Vestal, M. L., and Dyckes, D. F. (1984). Direct monitoring of sequential enzymatic hydrolysis of peptides by thermospray mass spectrometry, *Biomed. Mass Spectrom.*, **11**, 403–407.

Rose, M. E., Longstaff, C., and Dean, P. D. G. (1983). Negative ion fast atom bombardment mass spectrometry: *in situ* reactions of boronic acids with triols and related compounds, sugars and nucleosides, *Biomed. Mass Spectrom.*, **10**, 512–527.

Saito, K., and Kato, R. (1984). Glutathione conjugation of arylnitroso compound: detection and monitoring labile intermediates *in situ* inside a fast atom bombardment mass spectrometer, *Biochem. Biophys. Res. Commun.*, **124**, 1–5.

Self, R., and Parente, A. (1983). The combined use of enzymatic hydrolysis and fast atom bombardment mass spectrometry for peptide sequencing, *Biomed. Mass Spectrom.*, **10**, 78–82.

Smith, L. A., and Caprioli, R. M. (1983). Following enzyme catalysis in real-time inside a fast atom bombardment mass spectrometer, *Biomed. Mass Spectrom.*, **10**, 98–102.

Smith, L. A., and Caprioli, R. M. (1984). Enzyme reaction rates determined by fast atom bombardment mass spectrometry, *Biomed. Mass Spectrom.*, **11**, 392–395.

Smith, L. A., Beckner, C. F., and Caprioli, R. M. (1983). Fast atom bombardment for real-time analysis of peptide fragments from protein digestions, *Int. J. Mass Spectrom. Ion Phys.*, **46**, 419–422.

Tsugita, A., Van Den Broek, R., and Przybylski, M. (1982). Exopeptidase digestion in combination with field desorption mass spectrometry for amino acid sequence determination, *FEBS Lett.*, **137**, 19–24.

Weaver, J. C., and Abrams, J. H. (1979). Use of variable pH interface to a mass spectrometer for the measurement of dissolved volatile compounds, *Rev. Sci. Instrum.*, **50**, 478–481.

Weaver, J. C., Mason, M. K., Jarrell, J. A., and Peterson, J. W. (1976). Biochemical assay by immobilized enzymes and a mass spectrometer, *Biochim. Biophys. Acta*, **438**, 296–303.

Wold, F., and Ballou, C. E. (1975). Studies of the enzyme enolase I. Equilibrium studies, *J. Biol. Chem.*, **227**, 301–312.

Mass Spectrometry in Biomedical Research
Edited by S. J. Gaskell
© 1986 John Wiley & Sons Ltd.

Chapter 5

NOVEL MS APPROACHES TO THE ANALYSIS OF FREE AND CONJUGATED BILE SALTS

J. O. Whitney

5.1. INTRODUCTION

Bile acids are synthesized in the liver from cholesterol, and are secreted into bile and then into the small intestine where they aid in the absorption of fat. The common bile acids have a 5-carbon side chain, *trans* orientation of rings A and B and 1–3 α-hydroxyl groups in the nucleus, and as shown in Figure 1 are customarily divided into primary (those synthesized in the liver) or secondary (those whose final structure is the result of bacterial action in the intestine). Glycine and taurine conjugation at the terminal C24 carboxylic acid residue in the side chain occurs readily so that bile acids in bile and serum are usually amidated. Hepatobiliary and intestinal disorders are often associated with deranged metabolism, resulting in the accumulation of a wide array of positional and geometric isomers as well as glucuronidated and sulphated conjugates. Since many of these forms are known hepatotoxins and since conjugation is thought to be protective against their action, accurate identification and quantification of bile acids in diseased states is imperative for diagnosis, therapeutic monitoring and clarification of pathophysiology.

Recent advances in 'wet' chemical analysis including simple but powerful extraction procedures on minicartridges (Whitney and Thaler, 1980; Setchell and Worthington, 1982) and efficient separation of both non-conjugated and conjugated forms by reversed phase high performance liquid chromatography (HPLC) (Onishi *et al*., 1982; Whitney and Thaler, 1984) have greatly improved purification and in many instances allowed quantification. Capillary gas chromatography–mass spectrometry (GC–MS) has been of particular value in

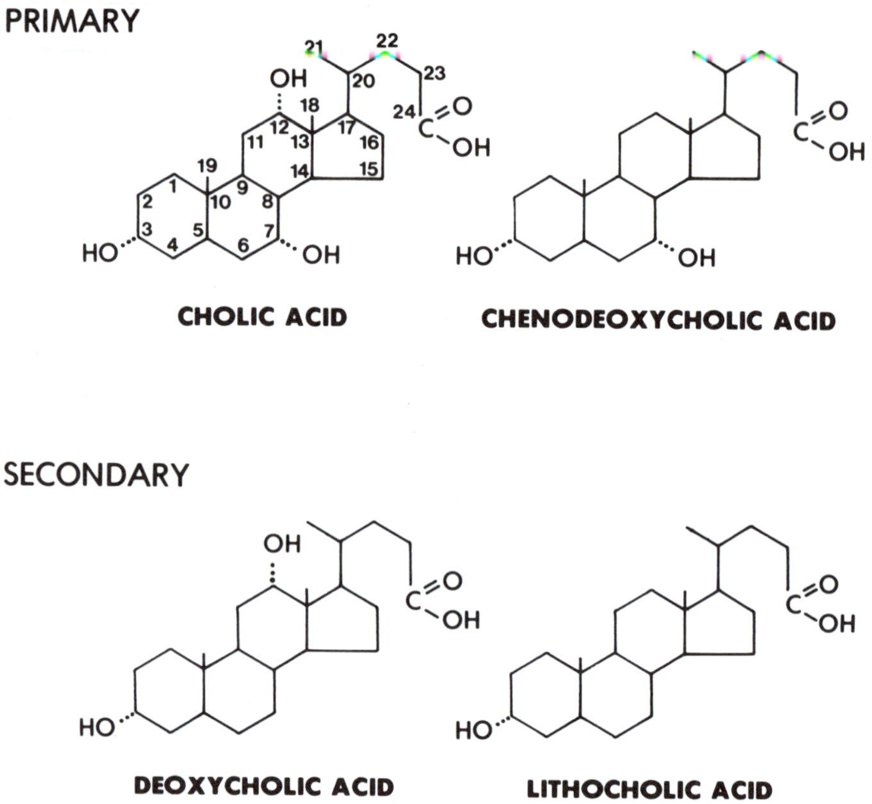

Figure 1: Structures of common primary and secondary bile acids. Secondary bile acids arise from bacterial dehydroxylation of the 7α-hydroxyl group of the primary bile acids.

elucidating unusual forms which appear in cholestatic diseases (Karlaganis and Paumgartner, 1978; Whitney and Burlingame, 1985) but information concerning the degree and type of conjugation is lost due to necessary hydrolysis and derivatization. Electron impact spectra of conjugated bile salts are of limited value. Glycine conjugates, even with great care, give molecular ions of very low intensity (Elliott, 1980) while taurine, glucuronide and sulphate conjugates undergo degradation. Chemical ionization techniques generally result in loss of the conjugated moiety (Elliott, 1980; Kenyon *et al.*, 1983). Field desorption (FD) spectra give cationized and doubly charged molecular ions for amidated conjugates, and have been used to characterize bile acid species occurring even in unchromatographed extracts of biological samples (Whitney *et al.*, 1981). FD, however, requires considerable expertise on the part of the operator, and an inordinate amount of mass spectrometer time must be expended for the determination of the large number of samples generated in biological and clinical investigations. In contrast, liquid secondary ion mass spectrometry

(LSIMS), using either a fast atom bombardment (FAB) (Whitney *et al.*, 1981) or a caesium ion source (Aberth *et al.*, 1982) results in molecular ion information on all intact non-conjugated and conjugated bile salts in a matter of minutes and can be used to identify bile acid species in HPLC fractions, unchromatographed extracts and in many cases in raw, unextracted biological specimens. The present report will outline the characteristics of LSIMS spectra of bile salts, discuss some of the problems encountered in their interpretation and utilization and demonstrate important applications.

5.2. LSIMS ANALYSES OF BILE SALTS

Positive FAB (a) and negative LSIMS (b) spectra of bile salts in 1 μl of raw unextracted rat bile are shown in Figure 2. When run in positive ion mode in slightly alkaline glycerol, protonated and cationized molecules of the original salts predominate and there is little fragmentation. Thus, the abundant ion at *m/z* 538 in Figure 2a is protonated sodium taurocholate, the major bile salt

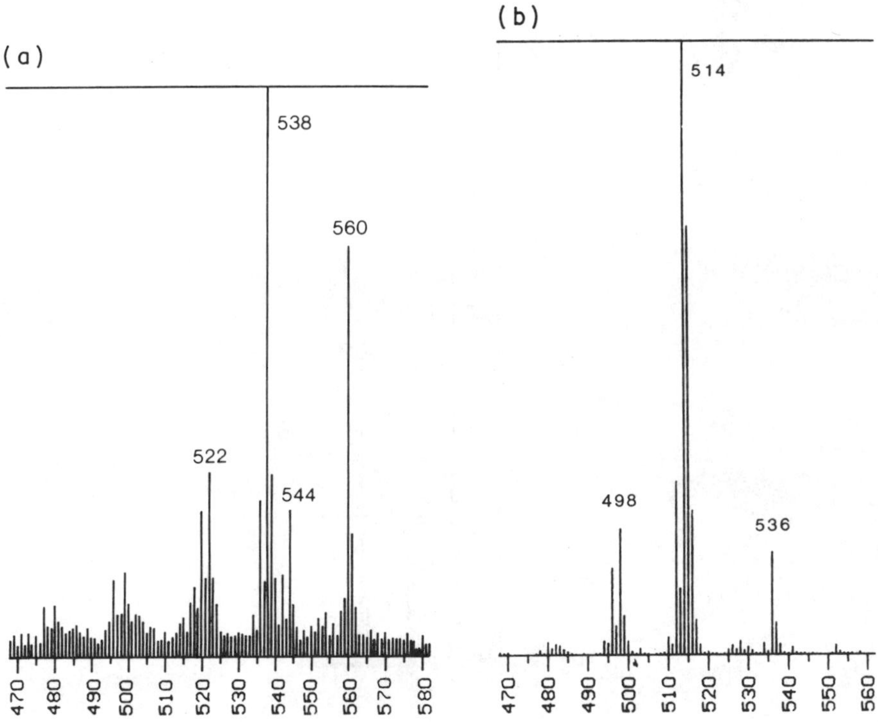

Figure 2: Positive ion FAB (a) and negative ion LSIMS (b) spectra of raw unextracted rat bile (1 μl).

found in the rat, and the ion at *m*/*z* 560 represents the (M + Na)$^+$ species. The ions at *m*/*z* 522 and 544 are protonated and sodiated sodium taurochenodeoxycholate, respectively. The negative ion spectrum identifies the same bile salts, taurocholate at *m*/*z* 514 and taurochenodeoxycholate at *m*/*z* 498. Cation loss from the salt is usual for free and amidated forms in negative ion analysis although weak ions containing the neutral salt, *m*/*z* 536 for sodium taurocholate – H), can be seen if enough material is used. Accompanying the [M – cation]$^-$ species is an intense ion 2 mass units lower, representing loss of H_2, a phenomenon which has also been noted in other LSIMS spectra. Since many bile salts contain ketonic and unsaturated functionalities, care must be taken not to confuse [M – cation – 2]$^-$ ions with [M – cation]$^-$ species of such compounds.

Addition of different cations (K^+, Li^+, etc.) to the matrix yields LSIMS spectra of bile salts which contain mixed cationized molecules. Dimer, trimer, tetramer formation with and without cation inclusion is common, and if a

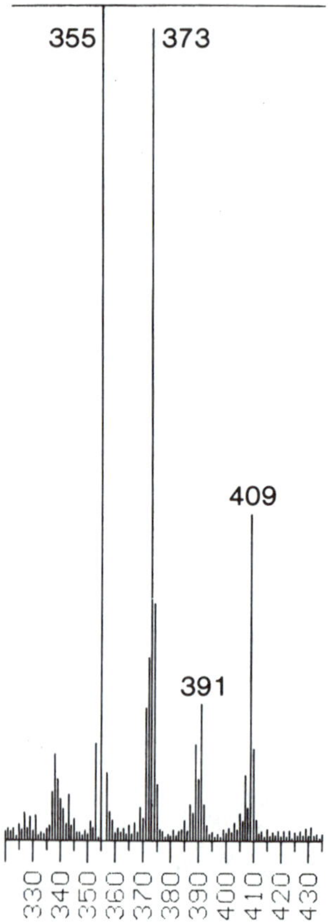

Figure 3: Positive ion FAB spectrum of β-muricholic acid run in acidified glycerol.

mixture of bile salts is present, mixed polymeric ion species result. Ability to cationize, the presence of polymeric ions and the appearance of an ion 2 mass units lower contribute evidence for the identification of ions retaining the intact molecule.

Judicious addition of acid to the matrix promotes water loss and confirms the number of hydroxyl groups present in an unknown. The FAB positive ion spectrum of β-muricholic (3α, 6β, 7β-trihydroxycholanoic) acid isolated from rat bile as the taurine conjugate, hydrolyzed and run in acidified glycerol is illustrated in Figure 3. The $(M + H)^+$ ion is at m/z 409 and ions of increasing

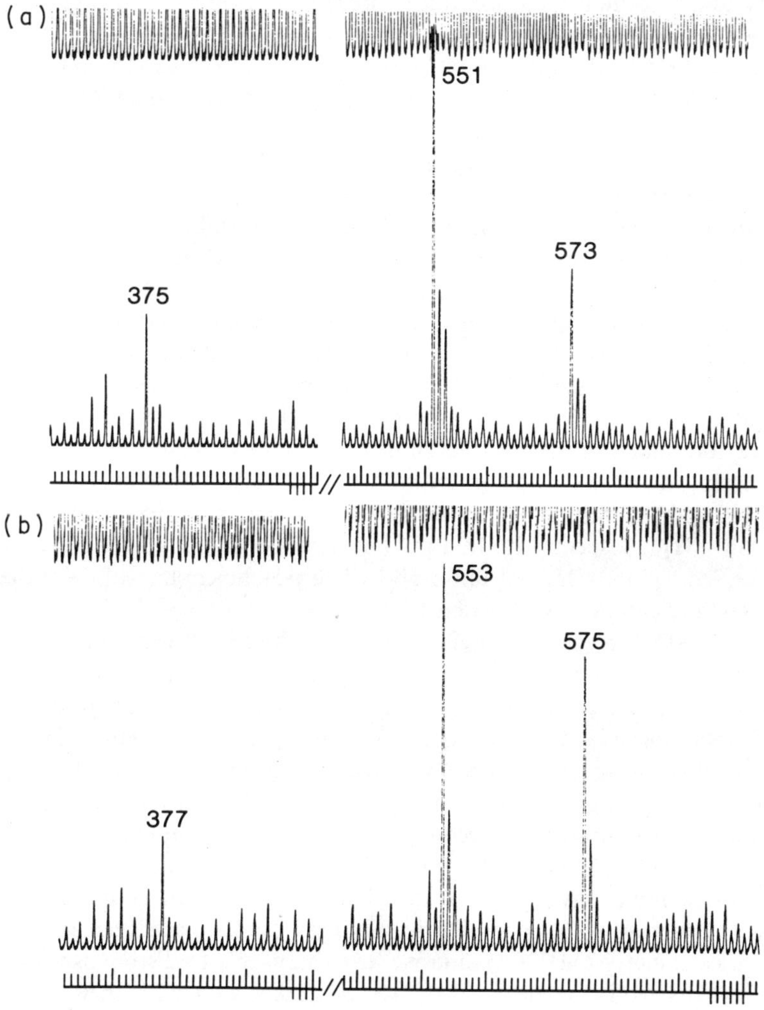

Figure 4: Negative ion LSIMS spectra of a keto, hydroxy-nor-bile acid glucuronide (a) and of a dihydroxy-nor-bile acid glucuronide (b).

intensity are found at m/z 391, 373 and 355 corresponding to the loss of 1, 2 and 3 molecules of water, respectively.

Bile salt glucuronides and sulphates and indeed other such conjugates have historically been identified on the basis of chromatographic mobility and susceptibility to appropriate enzymatic hydrolysis. LSIMS offers additional and conclusive proof of identity. Figure 4a is a partial negative ion LSIMS spectrum of a putative glucuronide of a keto, hydroxy-nor-bile acid isolated by thin layer chromatography from rabbit bile from an animal dosed with nor-chenodeoxycholic acid. In conformity with the predicted molecular weight, $(M - H)^-$ is at m/z 551 and a sodium inclusion ion is prominent at m/z 573. Fragmentation produces an additional confirming ion at m/z 375 $(M-176-H)^-$, corresponding to loss of the glucuronyl portion of the molecule. In contrast, the spectrum of a dihydroxy-nor-bile acid isolated from the same sample (Figure 4b), shows $(M - H)^-$ at m/z 553, the sodium inclusion ion at m/z 575 and $(M - 176 - H)^-$ at m/z 377, again confirming the identity on the basis of molecular weight consideration.

Negative ion spectra of sulphated, free and amidated, bile salts are presented in Figure 5. $(M - H)^-$ for 5-cholenoic acid-3-sulphate (Figure 5a) is at m/z 453 and a prominent ion at m/z 475 represents $(M + Na - H)^-$. There is very little fragmentation. When lithocholic acid-3-sulphate conjugated with glycine is analyzed (Figure 5b), $(M - H)^-$ appears, as expected, at m/z 512 and the sodium inclusion ion is at m/z 534, again with minimal fragmentation. Taurine conjugated sulphates, however, undergo major fragmentation. Figure 5c is a negative ion spectrum of tauro-5-cholenoic acid-3-sulphate. $(M - H)^-$ and particularly $(M + Na - H)^-$ are prominent, but very intense ions also appear at m/z 462 and 480, representing loss of sulphuric acid $(M - 98 - H)^-$ and SO_3 $(M - 80 - H)^-$, respectively. In a mixture, the ion at m/z 480 could easily be misidentified as $(M - H)^-$ of tauro-3β-hydroxy-5-cholenoic acid, with the ion at m/z 462 indicating loss of water.

Because LSIMS spectra give mainly molecular weight information, positional or geometric isomers cannot be distinguished and confusion can arise when fragmentations such as discussed above yield ions at masses corresponding to that of other molecular entities. Generally such problems can be obviated by use of simple chromatographic techniques available in most laboratories, supplemented by GC–MS or additional LSIMS confirmation of molecular identity. Alternatively, for those with access to appropriate instruments, MS/MS can solve such problems but at the price of increasing complexity. Liehr *et al.* (1985), for example, have demonstrated that positional isomers of the common bile acids can readily be identified in mixtures by mass-analyzed ion kinetic energy spectrometry (MIKES), and we have been able to distinguish unsaturated bile salts from the $(M - H - 2)^-$ ions associated with their saturated counterparts (Tomer *et al.*, 1986).

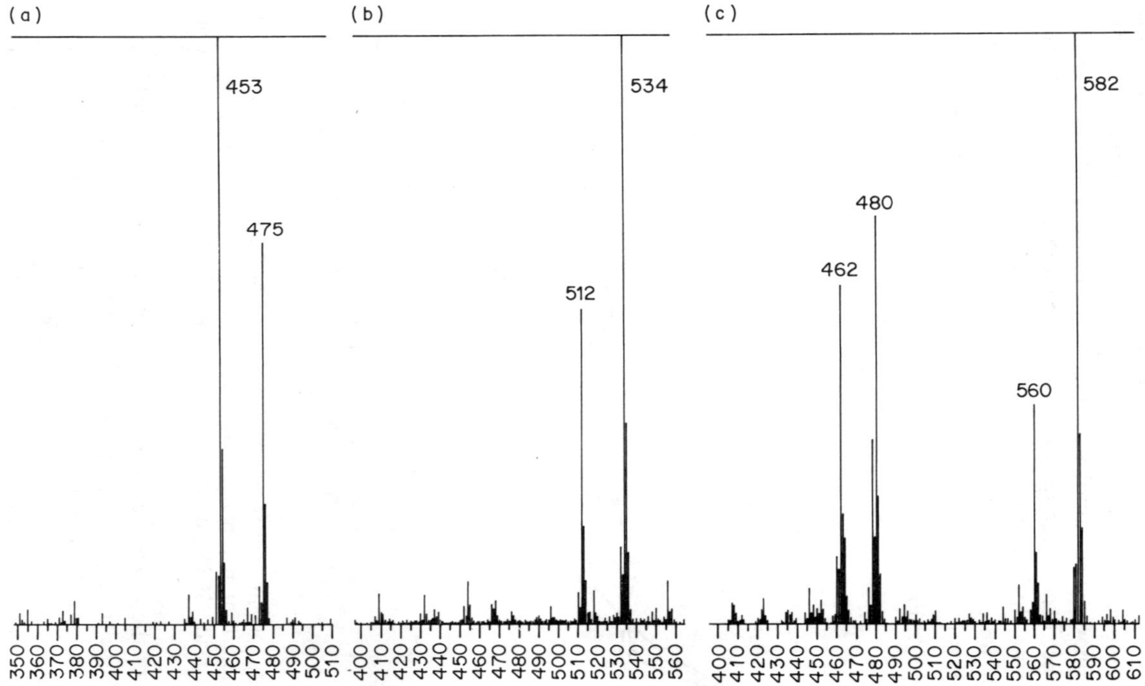

Figure 5: Negative ion LSIMS spectra of 5-cholenoic acid-3-sulphate (a), glycolithocholic acid-3-sulphate (b) and tauro-5-cholenoic acid-3-sulphate (c).

5.3. QUANTIFICATION OF BILE SALTS USING LSIMS

Identification of a compound in biological research invariably generates the question of quantification. We have attempted to quantify mixtures of bile salts by LSIMS of unextracted samples because of the obvious simplicity involved in not having to perform extraction and chromatography. Since most of our work deals with human clinical specimens, and since glycocholic acid is normally the major bile salt present in human bile, we chose [1-^{13}C]glyco[2,2,4,4,-$^{2}H_4$]cholic acid as an internal standard which was stirred directly into glycerol containing the sample on a copper probe tip.

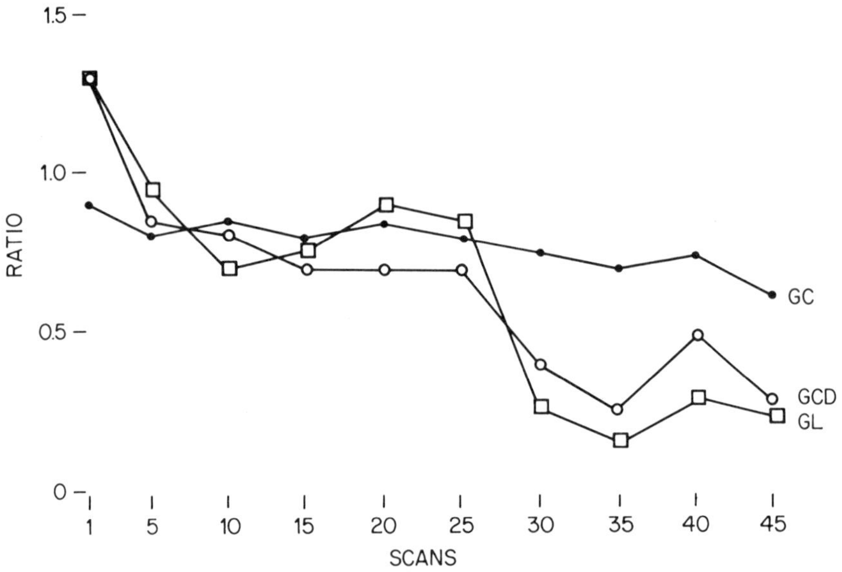

Figure 6: Ratio of intensities of [M-cation]$^-$ from 1 μg of major glyco-bile acid conjugates found in human bile and 1 μg of [1-^{13}C]glyco[2,2,4,4-$^{2}H_4$]cholic acid. GC, glycocholic acid; GCD, glycochenodeoxycholic acid; GL, glycolithocholic acid.

The ratios of the intensities of the signals from the (M − cation)$^-$ species of the major glycine conjugates found in human bile to the (M − cation)$^-$ from the internal standard, as a function of time of ion bombardment, is shown in Figure 6. While the ratio for glycocholate remained steady, significant variation and a decline in the ratios for both glycochenodeoxycholate and glycolithocholate occurred over 45 scans. For quantification, therefore, intensities obtained in the least variable portion of the curve (between scan 5 and 25) were averaged. Standard curves were generated for a variety of bile salt standards. Figure 7 shows that the standard curve for glycocholate itself was linear between 0.5 and

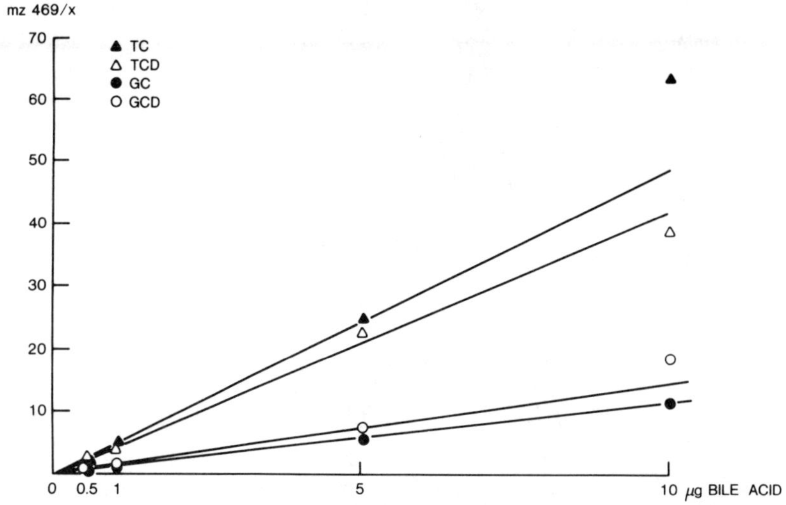

Figure 7: Standard curves for major amidated conjugates of bile salts measured against 0.5 µg of [1-^{13}C]glyco[2,2,4,4-$^2H_4$]cholic acid. GC, glycocholic acid; GCD, glycochenodeoxycholic acid; TC, taurocholic acid; TCD, taurochenodeoxycholic acid.

10 µg but that standard curves for other major bile acids deviated from linearity and tauroconjugates had extremely high response factors. Standard curves for monohydroxy bile salt sulphates (not shown) were not linear beyond 2 µg. The lower limit of detection obtained with glycocholate was approximately 37 ng. The great variation in response is probably related to differences in ionization in the matrix and to poorly understood competitive surface effects. Since the dynamic range was extremely limited, great care had to be taken to ensure that all components of a sample were within that range, an often impossible task. Reproducibility for bile salts other than glycoholate was poor. Repetitive determinations ($n = 6$) for rat bile (0.5 µl), for example, gave the following assay values: taurocholate, 3.8 ± 1.1 µmoles/l; taurochenodeoxycholate, 4.8 ± 1.2 µmol/l; taurolithocholate, 1.9 ± 0.4 µmol/l. In order to estimate the actual utility of LSIMS quantification, we measured total bile acids by both LSIMS and by enzymatic fluorimetry in bile obtained from a rat liver perfusion system at 10 min intervals for 1 h after the addition of lithocholic acid to the perfusion medium. Figure 8 shows that LSIMS described the general nature of the secretion of bile salts over this period, but values in most cases were much lower than those found by the enzymatic method. When LSIMS was compared to HPLC in ability to measure glycine to taurine (G/T) and dihydroxy to trihydroxy (D/T) ratios in human duodenal bile acids, LSIMS overestimated G/T by 23% but underestimated D/T by 49%. Discrepancies can be partially accounted for by the fact that the enzymic method measures all 3α-hydroxysteroids present in a sample and that HPLC and LSIMS are both selective processes

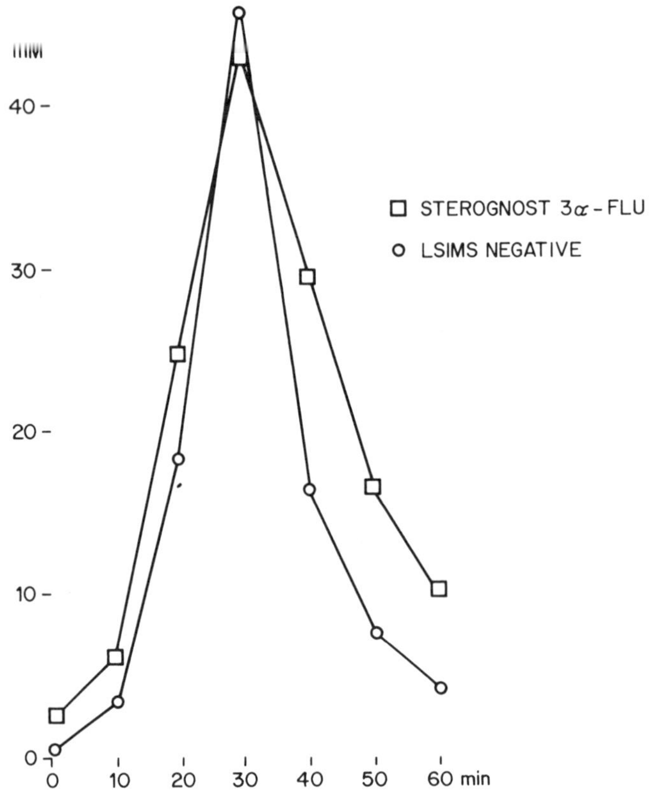

Figure 8: Total bile salt concentration in bile during perfusion of a bolus dose of lithocholic acid in a rat liver perfusion system as measured by enzymatic fluorimetry and negative LSIMS.

where the investigator chooses which peaks or which ions should be included in the measurement. Nevertheless, the data clearly indicate that quantification by LSIMS of a variety of individual bile salts using a single internal standard in an unextracted mixture is a hazardous undertaking which may give results in the correct range but which deviates seriously from attaining desirable accuracy.

Undoubtedly, the use of separate internal standards for each bile acid would greatly improve precision and accuracy. However, measurements would still be complicated by limited dynamic range, possible competitive effects, inability to distinguish isomers with different response factors as well as fragment ions appearing at the same nominal mass as ions retaining an intact molecule. MS/MS could solve some of these problems as could preliminary extraction, separation and purification, but with an increasing degree of complexity and expenditure of time. Since alternative methods of quantification are available (Street *et al.*, 1983), we believe that the greatest utility for LSIMS in the bile acid field is as a rapid screening device which can provide a respectable estimation of the range of free and conjugated bile acids present in a sample. Such information can be invaluable in the planning of both experimental and technical approaches to the solution of real problems.

5.4. FURTHER RESEARCH APPLICATIONS

In spite of limitations discussed above, LSIMS has been most useful in several research studies. Type and kinetics of conjugation of hepatotoxins such as lithocholic and 3β-hydroxy-5-cholenoic acid have been followed by LSIMS in unextracted bile (Whitney and Burlingame, 1982) and the methodology has been invaluable for identification of conjugates isolated by HPLC and TLC in product-precursor experiments (Whitney *et al.*, 1983). LSIMS spectra of unextracted hydrolysates were used to show conclusively that the enzyme cholylglycylhydrolase is inactive against synthetically conjugated tauronorcholic acid (Whitney and Vessey, 1984) and also provided evidence in unextracted samples that norcholic acid is excreted into bile unchanged (Vessey *et al.*, 1984).

Identification of bile acids in an untreated sample by LSIMS can have direct clinical utility. Figure 9a is a negative ion LSIMS spectrum of duodenal fluid obtained from a normal child by a relatively non-invasive Enterotest® device, generally used for sampling for intestinal parasites. A full complement of bile acids is present with ions at *m/z* 448 and 464 corresponding to glycodi- and tri-hydroxy conjugates and those at *m/z* 498 and 514 representing the corresponding tauroconjugates. The intense ion at *m/z* 469 is the internal standard. In contrast, the negative ion spectrum (Figure 9b) of duodenal fluid from an infant with surgically proven biliary atresia (complete absence of bile ducts) shows only the ion related to the internal standard. In a blind analysis of duodenal fluid obtained in this manner from 20 patients, we were able to distinguish three cases of proven biliary atresia, and found diminished quantities of bile salts in children with cholestasis due to inflammatory processes (Whitney *et al.*, 1985). LSIMS methodology made it possible to analyze the very small quantities of duodenal bile available (less than 2 μl in most cases). Such initial results should encourage the possibility of distinguishing between anatomical and metabolic cholestasis by examination of duodenal contents for biliary related metabolites.

In addition, we have been able to detect bile acids in unextracted gastric juice and are currently examining oesophageal fluid for their presence in an attempt to understand the pathophysiology of alkaline reflux. LSIMS spectra showing the presence of unconjugated bile salts in duodenal fluid, moreover, could provide additional evidence for a diagnosis of bacterial overgrowth or ileal dysfunction. The consideration that over 70 samples can be examined in a normal working day whereas extraction and conventional analysis of this many samples would take weeks also makes LSIMS an attractive tool for biological research.

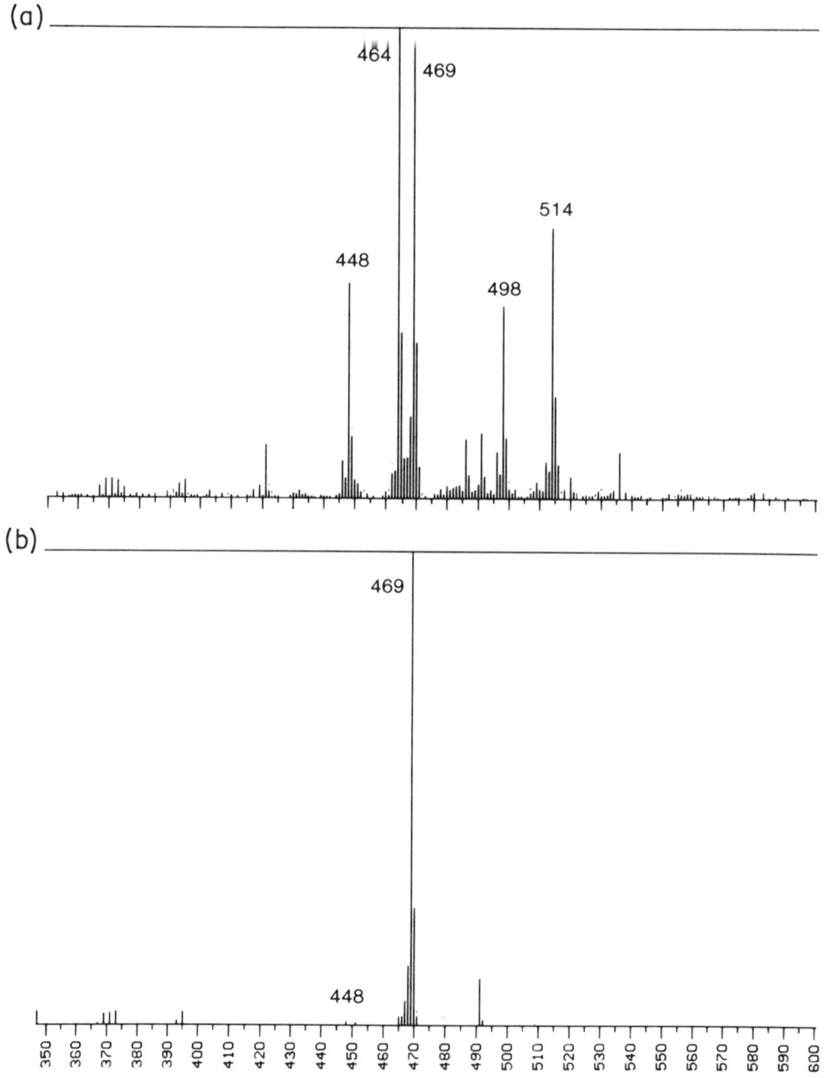

Figure 9: Negative ion LSIMS spectra of duodenal bile obtained by Enterotest® from a normal child (a) and a child with surgically proven biliary atresia (b).

ACKNOWLEDGMENT

The work discussed in this chapter was supported by NIH grants HD 13787 and RR 01614.

REFERENCES

Aberth, W., Straub, K., and Burlingame, A. L. (1982). Secondary ion mass spectrometry with cesium ion primary beam and liquid target matrix for analysis of bioorganic compounds, *Anal. Chem.*, **54**, 2029–2034.

Elliott, W. H. (1980). Mass spectra of bile acids, in *Biochemical Applications of Mass Spectrometry, 1st Suppl. Vol.* (Eds. G. R. Waller and O. C. Dermer), pp. 229–253, Wiley, New York.

Karlaganis, G., and Paumgartner, G. (1978). Analysis of bile acids in serum and bile by capillary gas-liquid chromatography, *J. Lipid Res.*, **19**, 771–774.

Kenyon, C. N., Goodley, P. C., Dixon, D. J., Whitney, J. O., and Faull, K. (1983). Direct liquid inlet liquid chromatography/mass spectrometry and negative chemical ionization mass spectrometry—specific and sensitive techniques for the analysis of molecules of biomedical importance, *Amer. Lab.*, Jan., 38–49.

Liehr. J. B., Kingston, E. E., and Beynon, J. H. (1985). Collision-induced dissociation-mass analysed ion kinetic energy (CID-MIKE) analysis of ions generated by fast atom bombardment of isomeric bile salts, *Biomed. Mass Spectrom.*, **12**, 95–99.

Onishi, S., Itch, S., and Ishida, Y. (1982). Assay of free and glycine- and taurine-conjugated bile acids in serum by high pressure liquid chromatography by using post-column reaction after group separation, *Biochem. J.*, **204**, 135–139.

Setchell, K. D. R., and Worthington, J. (1982). A rapid method for the quantitative extraction of bile acids and their conjugates from serum using commercially available reverse-phase octadecylsilane bonded silica cartridges, *Clin. Chim. Acta*, **125**, 135–144.

Street, J. M., Trafford, D. J. H., and Makin, H. L. J. (1983). The quantitative estimation of bile acids and their conjugates in human biological fluids. *J. Lipid Res.*, **24**, 491–511.

Tomer, K. B., Jensen, N. J., Gross, M. L., and Whitney, J. (1986). Fast atom bombardment combined with tandem mass spectrometry for determination of bile salts and their conjugates, *Biomed. Mass Spectrom.*, **13**, 265–272.

Vessey, D. A., Whitney, J. O., and Gollan, J. L. (1984). The role of conjugation reactions in enhancing biliary secretion of bile acids into bile, *Biochem J.*, **24**, 923–927.

Whitney, J. O., and Burlingame, A. L. (1982). Bile salt analysis by fast atom bombardment mass spectrometry, *Proc. Japan. Soc. Biomed. Mass Spectrom.*, **7**, 3–17.

Whitney, J. O., and Burlingame, A. L. (1985). Mass spectrometric analysis of bile acids in neonatal liver disease, in *Mass Spectrometry of Large Molecules* (Ed. S. Facchetti), pp. 185–207, Elsevier, Amsterdam.

Whitney, J. O., and Thaler, M. M. (1980). A simple liquid chromatographic method for quantitative extraction of hydrophobic compounds from aqueous solutions, *J. Liquid Chromatog.*, **3**, 545–556.

Whitney, J. O., and Thaler, M. M. (1984). Analysis of conjugated and unconjugated bile salts in bile by radial compression HPLC, in *Clinical Analysis of Exogenous and Endogenous Constituents by Liquid Chromatography* (Eds. L. Marton and P. Kabra), pp. 17–22, CRC Press, Boca Raton, FL.

Whitney, J. O., and Vessey, D. A. (1984). Inability of cholylglycine hydrolase to cleave the amide bond of tauronorcholic acid, *Steroids*, **44**, 77-83.

Whitney, J. O., Lewis, S., Straub, K., Thaler, M. M., and Burlingame, A. L. (1981). Analysis of conjugated bile salts in human duodenal bile using fast atom bombardment and field desorption mass spectrometry, *Proc. Japan. Soc. Med. Mass Spectrom.*, **6**, 33–44.

Whitney, J. O., Ling, V., Grunberger, D., Thaler, M. M., and Burlingame, A. L. (1983). Identification of hepatotoxic bile acid metabolites in bile by liquid secondary ion mass spectrometry, *Proc. Japan. Soc. Med. Mass Spectrom.*, **8**, 47–59.

Whitney, J. O., Rosenthal, P., Yang, Y. M., and Roitman, E. (1985). Measurement of intestinal bile salts by liquid secondary ion mass spectrometry (LSIMS): validation and clinical application, *Gastroenterology*, **88**, 1704.

Mass Spectrometry in Biomedical Research
Edited by S. J. Gaskell
© 1986 John Wiley & Sons Ltd.

Chapter 6

EXPERIMENTAL APPROACHES TO THE CHARACTERIZATION OF NUCLEIC ACID CONSTITUENTS BY MASS SPECTROMETRY

James A. McCloskey

6.1. INTRODUCTION

More than 80 different bases or nucleosides are presently known in RNA and DNA (Sprinzl *et al.*, 1985; Dirheimer, 1983; Warren, 1980). Approximately 60 have been discovered and characterized from transfer RNA (tRNA), one of the smallest nucleic acids, where they have been implicated in a remarkable variety of biological roles (Björk, 1984). In recent years, three factors have arisen which have made the characterization of new natural nucleosides more difficult: (i) the extension of work to sources of materials which are less plentiful (for example, much early work on tRNA was carried out on kilogram quantities of yeast); (ii) increased emphasis on radioisotopic postlabelling techniques for sequence studies, which require very little nucleic acid and thus leave insufficient material for chemically-based methods of characterization after new nucleosides have been discovered; (iii) the discovery of so-called 'hypermodified' nucleosides, which can often be isolated only in low microgram-level quantities, and which are structurally more elaborate and thus more difficult to characterize. As a result, mass spectrometry has evolved as the primary technique for structural studies on new nucleic acid constituents, due mainly to its relatively high sensitivity, and has played a prominent role in the characterization of most new components of RNA and DNA over the past 15 years.

In the present chapter an outline is given of the experimental approaches available for the structure elucidation of new natural constituents of nucleic acids by mass spectrometry. The material is presented from two independent but interrelated standpoints: considerations based on sample purity or matrix, and on the nature of structural information sought. Although it is generally true that the ease and certainty of structural characterization will increase as sample purity and amount increases, the efforts and risks associated with microscale isolation from cells or nucleic acids must be weighed against the nature of information sought, and the range of experimental mass spectrometric approaches available to the investigator. In extreme cases, the low levels of some forms of modifications in DNA and some RNAs may require prefractionation as an essential step, although increased sample handling may result in greater tendency for chemical degradation and adsorptive losses. In Section 6.2 the approaches described are governed by the desirability or necessity of working with mixtures, while Section 6.3 is oriented toward the characterization of isolated components of relatively high purity. In practice, a combination of the two approaches is often useful. For example, initial work with mixtures may be used to gain preliminary structural information and to define the complexity and composition of the mixture, as a prelude to isolation of the component of interest.

Discussion and literature citations are not intended to be comprehensive, and the reader is referred to earlier reviews of a general nature (McCloskey 1974; Hignite, 1980), and those dealing with nucleic acid photoproducts (Fenselau, 1976), modification by xenobiotics (Burlingame *et al.*, 1983), and fast atom bombardment (FAB) studies (McCloskey, 1985; Slowikowski and Schram, 1985).

6.2. EXPERIMENTAL APPROACHES BASED ON SAMPLE PURITY

Issues relating to sample purity and quantity are major considerations in experimental design, and often govern the experimental parameters chosen. The extent of modification of the four major bases in DNA (thymine, cytosine, adenine, guanine) ranges from total modification of one residue (e.g., 5-hydroxymethylcytosine replaces cytosine in some bacteriophages), down to levels of less than 1 in 10^5 bases. In messenger RNA and ribosomal RNA, certain modified constituents occur at a frequency of less than 1 in 10^3 for a particular base. Transfer RNA from a given source may contain 30–40 different tRNA species, each containing about 76 nucleotides, and each with about 5–10 different modified residues. For a nucleotide which occurs once per tRNA molecule, the nucleoside component will therefore constitute about 1% of the sample by weight. Isolation of a new nucleoside from unfractionated or isoaccepting tRNA may typically yield 2–50 μg, upon which characterization must be based unless considerable effort is expended to produce larger quantities. In some cases, only mixtures of tRNAs rather than a single (isoaccepting) tRNA will be available for analysis, in which case recognition of new nucleosides in enzymatic hydrolysates—and identification of known ones—may be a formidable problem. In such cases nucleoside concentrations may vary over 2 or 3 orders of magnitude, and assignment of the more than 60 known nucleosides when based solely on chromatographic mobility becomes increasingly difficult and subjective. Direct examination of mixtures of nucleosides in enzymatic digests may then serve to qualitatively confirm assignments based on chromatographic mobilities or other data, and to provide initial structural data on unassigned components.

6.2.1. CHROMATOGRAPHY–MASS SPECTROMETRY

Directly combined gas or liquid chromatography-mass spectrometry offers a powerful means for initial examination of enzymatic hydrolysates of DNA or RNA without the necessity of isolation of individual components. A summary is given in Table 1 of the two most useful approaches, namely, GC–MS of silyl derivatives, and thermospray LC–MS (Vestal, 1983).

Volatile derivatives of most of the nucleic acid bases and many of the simply-modified nucleosides and mononucleotides are amenable to gas chromatography. Many of the hypermodified nucleosides (for examples see Structures **7** and **8** in later discussion) are excluded from this approach, as are probably most of the non-methylated modified nucleotides, although they have not been systematically examined in this regard. Trimethylsilyl (TMS) derivatives (see following section) are generally preferred, and the resulting EI mass spectra will offer considerably more structural detail than the corresponding thermospray mass spectra derived from LC–MS. Enzymatic digests of DNA or RNA can be silylated and injected into the chromatograph without interference from hydrolytic nucleases. Enzymatic hydrolysis and derivatization procedures applicable to silylation of RNA hydrolysates are detailed in the literature (Hattox and McCloskey, 1974; Pang *et al.*, 1982c; Gehrke and Patel, 1977). For nucleosides, sensitivities in the range ≥ 0.05 ng per component for selected ion monitoring and 0.1–0.5 ng for full scans are generally attainable depending on derivatization yields for minor components and the degree of column adsorption encountered at low concentrations. Structure-retention characteristics for silylated nucleosides may be useful in predicting chromatographic properties, and occasionally for inferring certain structural features from chromatographic behaviour (Hattox and McCloskey, 1974; Quilliam *et al.*, 1980a).

For the purpose of examining nucleic acid modification patterns or for identifying new components in RNA, bases are less useful than nucleosides (see

TABLE 1. Chromatography–mass spectrometry for characterization of nucleic acid constituents.

	GC-ELECTRON IONIZATION MS	LC-THERMOSPRAY MS
Bases	TMS or *t*-butyldimethylsilyl (TBDMS) derivatives offer about 10^{-11} g sensitivity with packed or capillary columns. Excellent for methylated bases from DNA, by derivatization of HCO_2H hydrolysate. Greater structure information from TMS.	Simple spectra; abundant $(M + H)^+$. Sensitivity 0.1–1 ng with reversed phase chromatographic systems.
Nucleosides	TMS derivatives applicable to all major and most methylated and simply-modified nucleosides, but not to many hypermodified components. Therefore, suitable for certain targeted components. Generally 10–50 times more sensitive than LC–MS. Permethyl derivatives also useful but chromatographically less well studied.	Excellent for screening of enzymatic hydrolysates of tRNA. 0.1–1 ng/component for SIM and 10–50 ng for scanned spectra. Useful HPLC reversed phase systems available. Structure information represented by mass of base, sugar and molecule, with limited side chain fragmentation.
Nucleotides	Suitable for ng–µg level measurements as TMS derivatives for common nucleotides, but with tailing, reduced sensitivity when cytosine is present.	Same as above. $(M + H)^+$ very low abundance or absent. Spectrum similar to corresponding nucleoside but readily distinguished by elution position. Suitable for analysis of alkaline hydrolysates of RNA.

Section 6.3.1) due to difficulty in obtaining their release by acid hydrolysis, while bases are readily liberated from DNA by acid hydrolysis (HCO_2H, 175°C, 1 h). Cytosine methylation, the principal form of natural modification in many DNAs, is important in regard to gene regulation (Doerfler, 1983). A quantitative assay for 5-methylcytosine as the *t*-butyldimethylsilyl derivative has been developed which can in principle be extended to other structurally simple modified bases, at concentration levels below 1 in 10^4 DNA bases (Crain and McCloskey, 1983).

If TMS derivatives are used for examination of nucleic acid hydrolysates, they offer the advantage of considerable structural detail in their EI mass spectra, due to extensive systematic studies of models in the case of bases (White *et al.*, 1972), nucleosides (Pang *et al.*, 1982b), and nucleotides (Lawson *et al.*, 1971; Hignite, 1972).

Combined high performance liquid chromatography–mass spectrometry (LC–MS) in general constitutes a more useful approach than GC–MS when crude mixtures such as nucleic acid hydrolysates are involved. Although other LC introduction methods have been studied (Games and Lewis, 1980; Esmans *et al.*, 1983), the thermospray technique is most useful because of its applicability to polar compounds and to aqueous, buffered reversed phase HPLC systems in which the entire effluent can be passed into the mass spectrometer (Vestal, 1983). The necessity for derivatization is obviated, and the wide range of polarities encountered in modified constituents from RNA and DNA can be accommodated. When working with crude mixtures of nucleosides, reliable identification can be made based on three parameters acquired during a single experiment: mass of the molecule or base, chromatographic retention position, and UV absorbance ratios (Edmonds *et al.*, 1985). An added advantage is the availability of an excellent reversed phase HPLC system, for which the retention positions of a number of ribonucleosides have been catalogued (Buck *et al.*, 1983).

In a typical LC–MS experiment involving tRNA, an enzymatic digest of several μg (for isoaccepting tRNA) or 20–40 μg (unfractionated tRNA) can be examined by acquisition of repetitively scanned mass spectra, and assignments made based on the three parameters listed above (Edmonds *et al.*, 1985). Refinement of results obtained in this fashion can be obtained from selected ion monitoring experiments, generally requiring 10–20 times less material. LC–MS offers a powerful means of screening crude mixtures of tRNA for new natural nucleosides, by direct analysis of enzymatic digests. A striking example of this approach is represented by the characterization of N^4-acetyl-2′-*O*-methylcytidine (Structure **1**), carried out solely by LC–MS on approximately 50 ng of Structure **1** in a hydrolysate of unfractionated tRNAs from the thermophilic archaebacterium *Thermoproteus neutrophilus* (Edmonds *et al.*, 1986). The level at which the structure assignment was made is notable in being 2 orders of magnitude less than usually required by the approaches

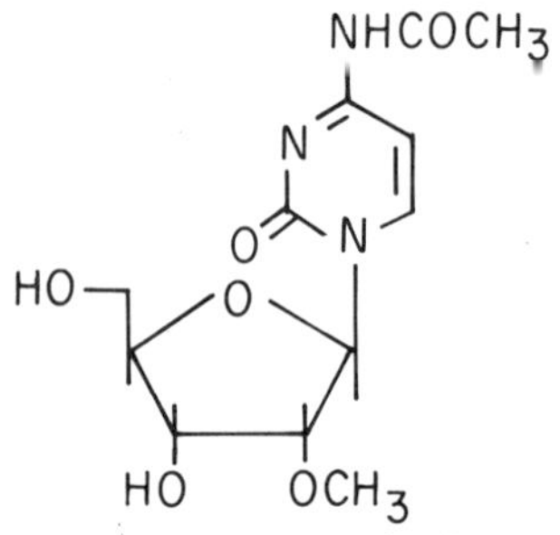

1

from *T. neutrophilus* tRNA

described in Section 6.3, which generally involve isolation of the component of interest.

Because the principal fragment ion from nucleosides is the protonated base, selected ion monitoring with LC–MS can be effectively used to search for a particular nucleoside when its elution position is unknown or it is chromatographically unresolved from other components by UV detection. Similarly, LC–MS is generally effective in following the isolation of a component through multichromatographic steps, prior to isolation and more detailed examination for structural characterization.

6.2.2. OTHER METHODS OF DIRECT ANALYSIS

Although combined chromatography–mass spectrometry has obvious advantages for examination of hydrolytic mixtures or crude isolates, two additional approaches may be useful under certain circumstances: direct probe analysis of volatile derivatives, and tandem mass spectrometry (MS/MS) following desorption ionization.

Direct probe analysis with electron ionization is advantageous in its speed and simplicity. If volatile derivatives (such as TMS) are used, molecular-region ions from the larger polyfunctional nucleosides will shift to the region over 600 u, thus avoiding most ions from background and from the major nucleosides, permitting recognition of characteristic $M^{+\cdot}/(M-15)^{+}$ mass differences. In the case of tRNA, success in detecting the less common nucleosides by this method will depend on cleanliness of the tRNA and hydrolytic enzymes in reducing background ions. Chemical ionization of derivatized RNA hydrolysates is a useful method (Wilson, 1974) because of the greatly improved signal to noise ratio in the molecular ion region if the components of interest each constitute about 1% or more of the total nucleoside content. Otherwise, LC–MS is generally more effective in accommodating wide concentration ranges. The direct analysis approach can also be valuable to confirm that the component of interest has not undergone chemical degradation during chromatography or sample handling during isolation. This issue was of concern in the structure determination of the hypermodified adenine derivative (Structure **2**) from bacteriophage Mu DNA, which was therefore examined as its 5′-phosphate from an unfractionated silylated digest from DNase I and snake venom phosphodiesterase hydrolysis (Swinton *et al.*, 1983).

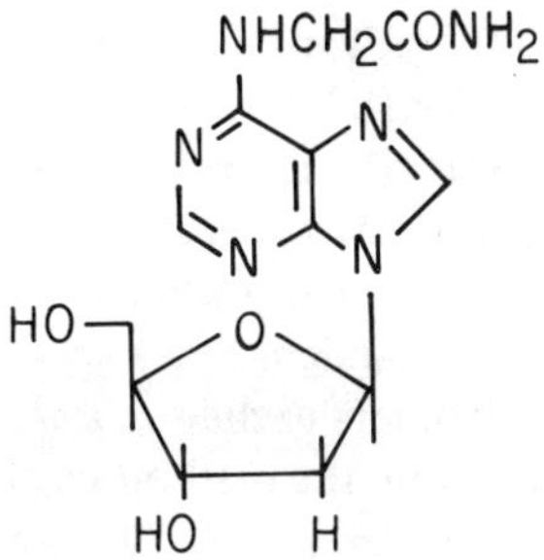

2

from bacteriophage Mu DNA

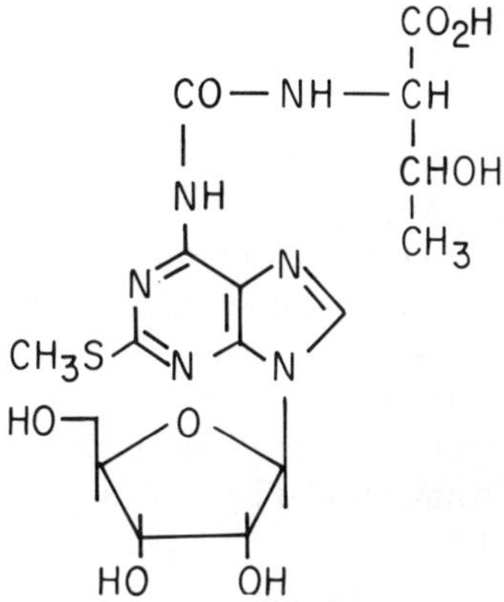

3

from *B. subtilis* $tRNA^{Lys}$

High resolution mass spectrometry can also be employed for identification of modified nucleosides in unfractionated RNA (Pang *et al.*, 1982c). The method is based on recognition of predetermined sets of exact masses which are characteristic of individual nucleosides, an extensive set of values of which has been published (Pang *et al.*, 1982c). The technique as applied to tRNA requires 1–4 μg of material prior to hydrolysis and derivatization. The method was designed primarily for identification of known nucleosides in crude mixtures; for example, presence of the adenosine derivative Structure **3** was established in *B. subtilis* $tRNA^{Lys}$ when conventional methods of detection failed to give conclusive results (Vold *et al.*, 1982). The method can in principle be applied to components of unknown structure and has the potential advantage of producing elemental composition data in the same experiment. In such cases, recognition of $M^{+\cdot}/(M\text{-}CH3)^+$ patterns (difference = 15.0235 u) is important, in particular when trace constituents are of interest. Ions from impurities can often be distinguished by absolute fractional mass values, which in the case of nucleosides fall in a relatively narrow range (Pang *et al.*, 1982c). As an alternative, desorption mass spectrometry may be used for direct analysis of nucleoside mixtures, particularly in combination with linked-scan or tandem measurements (Crow *et al.*, 1984 Tondeur *et al.*, 1984). The method should generally complement LC–MS, but to date insufficient data are available concerning the effects of hydrolytic enzymes and buffers on sensitivity to enable one to assess the direct applicability of desorption techniques to enzymatic hydrolysates. Spectral characteristics and some limited examples are given from reports on fast atom bombardment (FAB) (Eagles *et al.*, 1984) and FAB-tandem mass spectra of nucleosides (Crow *et al.*, 1984), and on identification of nucleoside antibiotics in fermentation broth (Tondeur *et al.*, 1984). Limited available data suggest that some hypermodified nucleosides give rise to more abundant parent ions from FAB mass spectrometry than from thermospray ionization, which may be due in part to the higher temperature associated with thermospray (>250° versus 25° for FAB). (Edmonds, Nelson, Phillipson and McCloskey, unpublished data). While negative ion thermospray of mononucleotides has not been investigated, FAB is predicted to be generally more useful than LC–MS in detecting parent ions of nucleotides in hydrolysates. However, for the detection of trace constituents of hydrolysates, the advantage of more abundant parent ions from FAB is offset in the case of LC–MS by the utility of chromatographic data. Another method of direct analysis applicable to mixtures from nucleic acids is use of mass-analysed ion kinetic energy spectrometry (MIKES) (Schoen *et al.*, 1979; Puzo and Wiebers, 1981). Particularly when used with collisionally-activated dissociation, either MIKES or tandem mass spectrometry can be used to distinguish methyl-substituted isomers (Schoen *et al.*, 1979; Unger *et al.*, 1981). However, in a published application, the method used for degradation of nucleic acids was pyrolysis (Schoen *et al.*, 1979), an unreliable technique because of the occurrence of thermal methyl transfer reactions which can form methylated bases not present in the undegraded biopolymer (Unger *et al.*, 1981; Kuntzel *et al.*, 1975; Smith, Crain and McCloskey, unpublished data, 1979).

A discussion of the potential merits of tandem mass spectrometry for the analysis of methylated deoxyribonucleosides was presented by Ashworth *et al.* (1985). Daughter ion spectra from chemical ionization of model deoxynucleosides at the 10 ng level clearly demonstrate that useful sensitivity for analysis of nucleosides in hydrolysates should be possible. However, it is noted that in the case of DNA, where bases can readily be liberated by acid hydrolysis, substantially greater sensitivity can be obtained by GC–MS (Crain and McCloskey, 1983). The future role of tandem mass spectrometry in this respect will probably be in the analysis of complex nucleosides or nucleotides, where desorption ionization will be particularly advantageous.

6.3. EXPERIMENTAL APPROACHES BASED ON THE NATURE OF STRUCTURAL MODIFICATION

6.3.1. GENERAL EXPERIMENTAL STRATEGY

For microscale characterization of natural nucleic acid constituents, nucleosides rather than bases or nucleotides generally represent the preferred level at which to work, for three main reasons.

(i) A variety of effective enzyme systems are available for the quantitative release of nucleosides from nucleic acids under mild conditions. Examples are venom phosphodiesterase or nuclease P_1 for RNA, and DNase I and nuclease P_1 for DNA, to produce mononucleotides, each followed by dephosphorylation by alkaline phosphatase, which can be carried out in the same tube in 10–20 μl volume (Pang *et al.*, 1982c; Edmonds *et al.*, 1985). Chemical degradation (release of bases from DNA by acid; formation of 2′ (3′) -mononucleotides from RNA by base) should be studiously avoided unless it is firmly established that unwanted chemical degradation will not result. For example, such degradation does not occur with most of the simply methylated bases.

(ii) In RNA only one form of modification is not found in the base: methylation at position O-2′ in ribose. Examination of the nucleoside rather than the corresponding base will therefore be required to detect sugar methylation.

(iii) Mass spectra of nucleotides usually give no more information than the corresponding nucleoside, and the undesirable side effect of diluting ion current levels with phosphate-containing ions is often encountered. As one exception to this general conclusion that nucleosides are preferred for structural work, questions of isomerism (2- versus 4-thiouracil, 1- versus N^6-methyladenine, etc.) can often more clearly be answered by the mass spectrum of the base (McCloskey, 1974).

In general, much more structural information will result from electron ionization than by desorption or chemical ionization methods. In many cases this will require preparation of a volatile derivative, although it is pointed out that several significant structural problems have been solved based in part on the underivatized material, a good example being Structure **4** and several of its

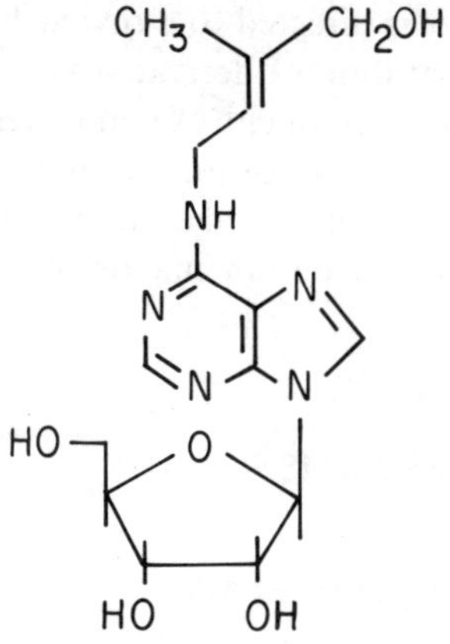

4

from wheat germ tRNA

derivatives (Burrows *et al.*, 1970). However, examination of the free nucleoside or base by EI mass spectrometry will not be possible for some of the more polar compounds (e.g., nucleoside Structure **5**), and work at lower molecular weight levels (<300) is often confounded by impurity peaks when small samples of isolated materials are involved. Lastly, in some (though not all) cases involving extensive modification, more structural detail will be provided by the TMS derivative than by the free nucleoside. The nucleoside of Structure **6** is an

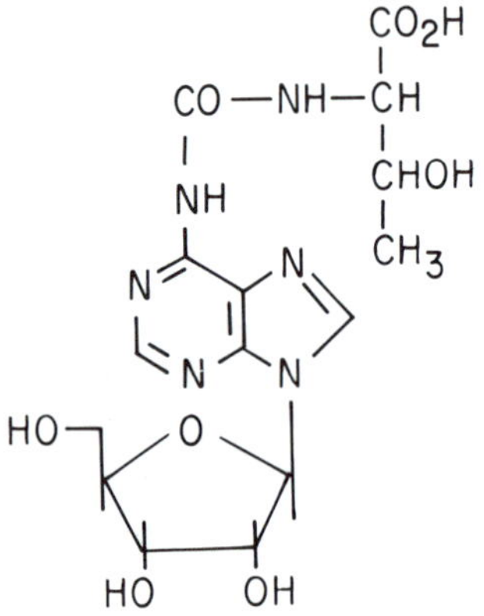

5

from tRNA

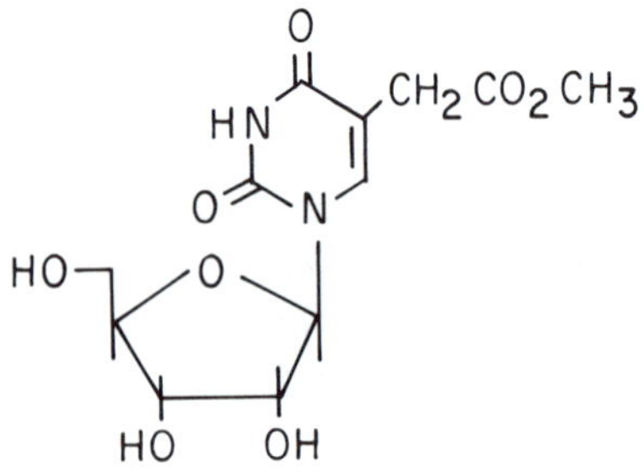

6

from *Saccharomyces cerevisiae* $tRNA^{Arg}$

interesting example in which thermal methyl transfer was found to occur during heating of the free nucleoside but was avoided by trimethylsilylation (Kuntzel *et al.*, 1975). A summary of characteristics of selected derivatives for nucleosides is given in Table 2. The reader is directed to a review by Schram and McCloskey (1979) for methods of preparation of derivatives.

In the case of an isolated component of completely unknown structure, the exact approach taken may differ according to individual circumstances. However, as a general procedure the scheme shown below represents a typical minimum-risk plan designed to obtain maximum initial information from a total of 10–15 μg of material.

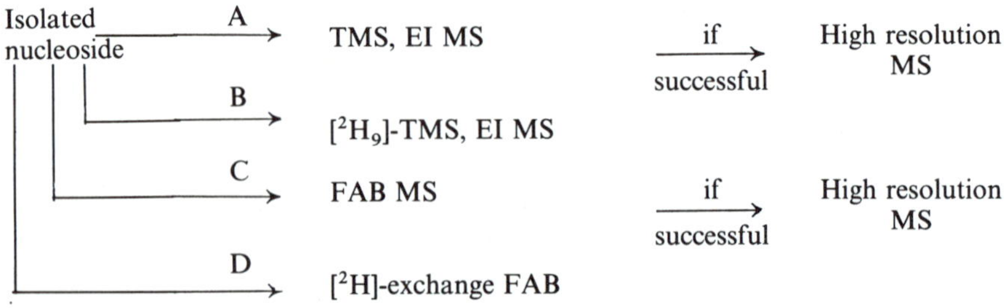

TABLE 2. Selected volatile derivatives of nucleosides for mass spectrometry.

Derivative	Comment	Reference
Trimethylsilyl (TMS)	Mass spectra exhibit great structural detail, although in limited cases (notably pyrimidine deoxyribonucleosides) non-informative TMS-related ions predominate. Extensive work published on mechanisms, and identities of ions. Numerous reference spectra available. Reaction yield and stability of derivative sensitive to traces of moisture. 7-Methylpurines characteristically oxidize at C-8 during silylation.	Pang *et al.* (1982b) von Minden *et al.* (1972)
t-Butyldimethylsilyl (TBDMS)	Complex mass spectra. Sensitive to structural variations in the sugar. Complex nucleosides not studied, but should exhibit intense high-mass peaks. O-TBDMS derivatives highly stable.	Quilliam *et al.* (1980b)
O-isopropylidene	Useful for test of presence of vicinal diol group in side chain. Abundant base-containing ions; models not extensively studied. Mixed isopropylidene-TMS derivative useful in some cases to increase volatility for GC–MS.	McCloskey (1974)
Acetyl	Abundant base and sugar ions but M-related ions of low abundance. Easily prepared. Models not extensively studied. Insufficient volatility for GC–MS.	McCloskey (1974)
Trifluoroacetyl	Abundant base-containing ions. Well-suited for deoxyribonucleosides.	Koenig *et al.* (1971)
O,N-permethyl	Useful alternative to TMS derivative. Extensive model studies. Derivative is highly stable; several methods of preparation, of which use of methylsulphinylcarbanion reagent is least convenient. 'Overmethylation' (two methyls) characteristically occurs at C-5 of 5,6-dihydropyrimidines.	von Minden and McCloskey (1973) Panzica *et al.* (1973) Teece *et al.* (1983)

In the four experiments shown, 2–4 μg may be consumed at each step. For experiment A a full, high resolution mass spectrum is preferred. Photographic recording is often employed in this laboratory because of its ability to record a very wide mass range simultaneously and thus consume minimal sample (Pang *et al.*, 1982c; Koenig *et al.*, 1971). For C, peak matching will usually suffice because of the relative simplicity of the FAB spectrum compared with EI. Permethylation or use of mixed derivatives may replace silylation but in either case the use of [^{2}H]-labelling (step B) serves several purposes: (i) to assist in interpretation of the EI spectrum (McCloskey *et al.*, 1968); (ii) to permit (by mass shifts) determination of the number of blocking groups introduced, thus giving a molecular weight value that should correlate with FAB data; (iii) to reduce the number of computer-generated elemental composition candidates, for example, by defining the number of Si atoms. A procedure for increasing the accuracy of the measured molecular mass value, and use of computer-implemented rules specific for nucleosides for reduction of the number of composition candidates is discussed by Crain *et al.* (1980). If a FAB mass spectrum is successfully acquired, the exact number of active hydrogens in the molecule can be measured (experiment D) by a deuterium exchange procedure (Sethi *et al.*, 1983). The establishment of an unambiguous value for the number of active hydrogens in a molecule of unknown structure (for example, see Isono *et al.*, 1984) provides a useful constraint in developing initial structural candidates, and in excluding some isomeric possibilities. For an account of the FAB mass spectrometry of nucleic acid constituents, the reader is referred to a recent review by Slowikowski and Schram (1985).

6.3.2. BASE MODIFICATION

The base is the principal site of natural modification in RNA and DNA. Determination of the mass of the base and its elemental composition (by measurement of exact mass) is therefore an important step in solution of the structure. This is accomplished through a series of ions which contain the intact base plus certain atoms from the sugar, in the mass spectra of both free (Shaw *et al.*, 1970) and blocked (McCloskey, 1974) nucleosides. In the case of simple modifications, such as methylation or substitution of O by S, the mass of the base alone may be sufficient to reduce the number of candidate structures to a set of isomeric possibilities which can then be tested. Placement of substituents on the purine or pyrimidine ring can often be made through mass shifts in the basic fragmentation patterns, as in establishment of the position of uracil thiation in uridine (Hecht *et al.*, 1969). Recognition of the commonly occurring N^6-methyladenine moiety can be made from the characteristic expulsion of methylene imine (von Minden *et al.*, 1974).

In the case of more extensive side chain substitution, the first step is recognition of the basic heterocyclic moiety from among the six presently known in RNA (uracil, cytosine, adenine, hypoxanthine, guanine, 7-deazaguanine) and four in DNA (thymine, cytosine, adenine, guanine). In tRNA this information can often be inferred from sequence location of the residue, if it is known. Otherwise, the fragmentation pattern and elemental composition data are the primary sources of such data. For example recognition of the deazapurine nucleus in queuosine, Structure **7**, the only non-purine or non-pyrimidine

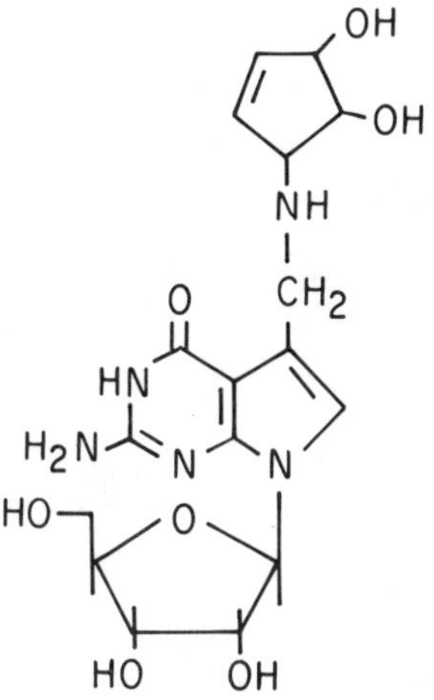

7

from $tRNA^{Asn}$, $tRNA^{Asp}$, $tRNA^{His}$, $tRNA^{Tyr}$

known in nucleic acid, first came from its high resolution mass spectrum by recognition that one of the five nitrogen atoms was in a side chain and not directly bound to the heterocyclic ring (Kasai *et al.*, 1975). If the heterocyclic moiety is known, the side chain composition can sometimes be inferred from the difference in composition between the intact molecule and the basic nucleoside skeleton. An example in which this approach was of key importance was establishment of presence of the carbamoylthreonine moiety in the complex adenosine derivative Structure **3**, which had been isolated from mammalian tRNA (Yamaizumi *et al.*, 1979).

Of the two remaining general approaches, the most frequently used is recognition of ions formed by side chain cleavage, several examples of which are shown by the cleavage points as noted in Structures **8** (Murao and Ishikura, 1978), **9** (Kuchino *et al.*, 1979), **10** (Ohashi *et al.*, 1974), and **11** (Kasai *et al.*, 1975). Side chain cleavage ions often occur in addition to the common base-series and sugar-series ions, and in many cases dominate the mass

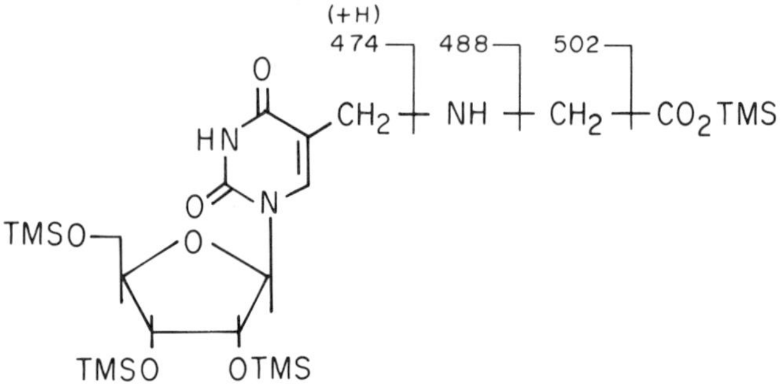

8

from tRNA$_1^{Gly}$ from *B. subtilis*
(TMS derivative)

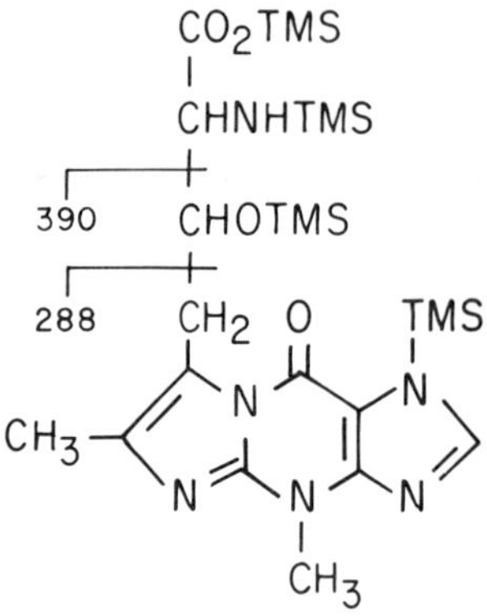

9

from Erlich ascites cell tRNA^{Phe}
(TMS derivative)

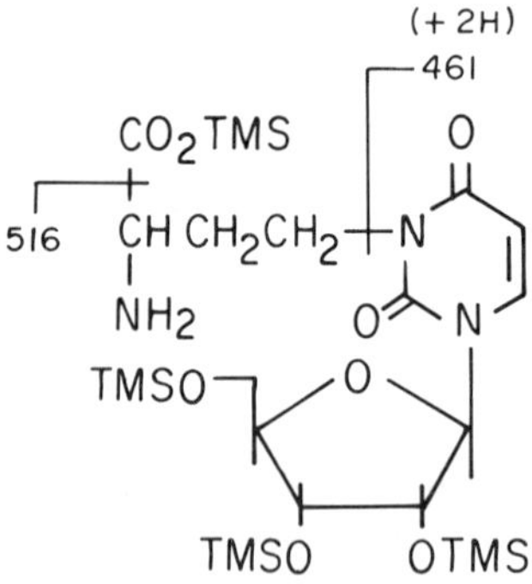

10

from *E. coli* tRNA^{Phe}
(TMS derivative)

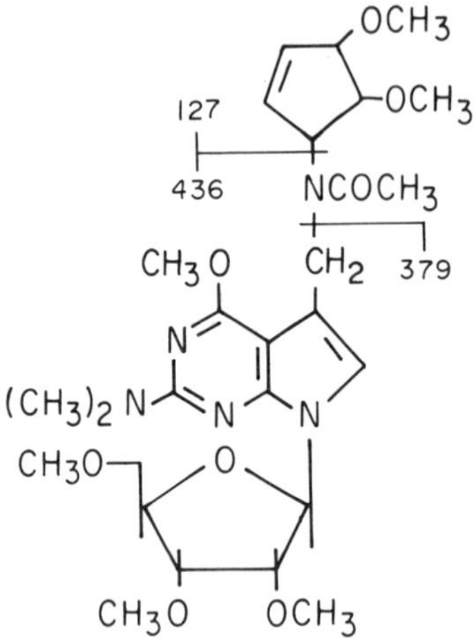

11

from *E. Coli* tRNA^{Tyr}
(methyl-acetyl derivative)

spectrum. Although the important base series of ions can be reduced to very low abundance in such circumstances, they can often be readily recognized by their exact mass interrelationships (Pang *et al.*, 1982b). In some instances side chain generated fragment ions are useful in confirming the identity of the basic nucleoside moiety to which the side chain is attached, in particular when substitution is made at heteroatoms in, or directly attached to, the heterocyclic ring. An example is given by *m/z* 461 in Structure **10**. Similarly, N^6-substituted adenosines give rise to adenosine and adenine ions by EI, or their protonated counterparts by FAB. Side chain fragmentation can also offer key information concerning the presence of unique or unexpected structural units. For example the cyclopentenediol moiety in Structure **7** had not previously been found in any natural nucleoside. Its presence was first indicated by ion *m/z* 127 in the high resolution mass spectrum of the derivative Structure **11**. Additionally, in the case of the family of tRNA nucleosides which bear the isopentenyl group at position N^6 of adenine (e.g., Structure **12**), characteristic loss of C_3H_7 from the side chain occurs (reviewed by McCloskey, 1974), by a ring closure mechanism involving N-1 of the purine ring (Hashizume *et al.*, 1976).

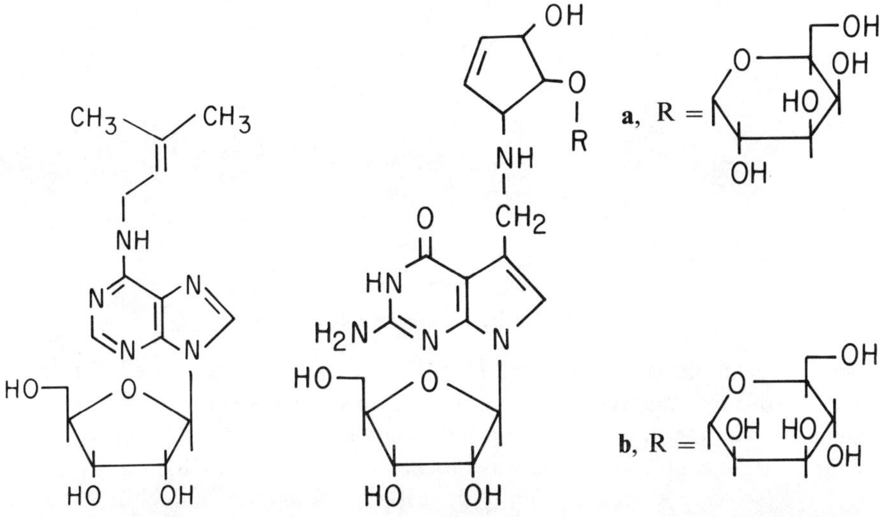

12 **13**

A less often employed but useful approach involves hydrolytic cleavage of the side chain, followed by analysis of the chemically produced products. The complex queuosine derivatives, Structures **13a** and **13b**, which occur widely in tRNA of higher eukaryotes, were hydrolyzed by 0.5M H_2SO_4 and the resulting hexoses, galactose (from **13a**) and mannose (from **13b**), were identified chromatographically (Kasai *et al.*, 1976b). The threonyl moiety in the family of nucleosides related to Structure **4** can similarly be detected by GC–MS of a silylated alkaline hydrolysate (Kasai *et al.*, 1976a).

No data have yet been produced to demonstrate the utility of tandem mass spectrometry with collisional activation for characterization of side chains of the types shown above. However, such studies involving FAB mass spectrometry of simpler models (Crow *et al.*, 1984) suggest that this method will be particularly valuable, and will reduce the dependence on sample purity for proper interpretation of the data.

6.3.3. SUGAR MODIFICATION

The only presently known sugar modification in nucleic acids is *O*-2′ methylation in RNA. Recognition of such methylation can be made indirectly, by the difference in mass between the molecular and base ion species, or by comparison of the common ions shown in Structure **14** as *a*, *b*, and *c*, which are usually produced by free and blocked nucleosides.

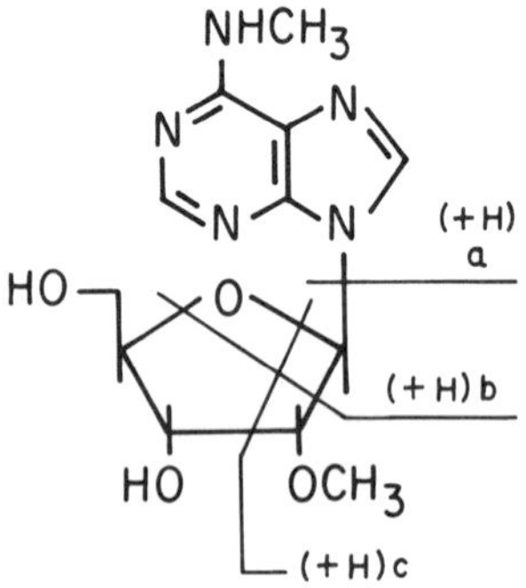

14

from nuclear RNA Of Novikoff hepatoma

In the example shown (Structure **14**), in which 2 μg was used for total characterization, methylation both in the base and sugar was established primarily from the three ions of the TMS derivative, and the fact that silylation of the amino group in the base was blocked (Crain *et al.*, 1978).

6.3.4. GLYCOSIDIC LINKAGE

Pseudouridine (Structure **15**), which occurs widely in RNA and is nearly ubiquitous in tRNA, is the basic member of a family of nucleosides which have

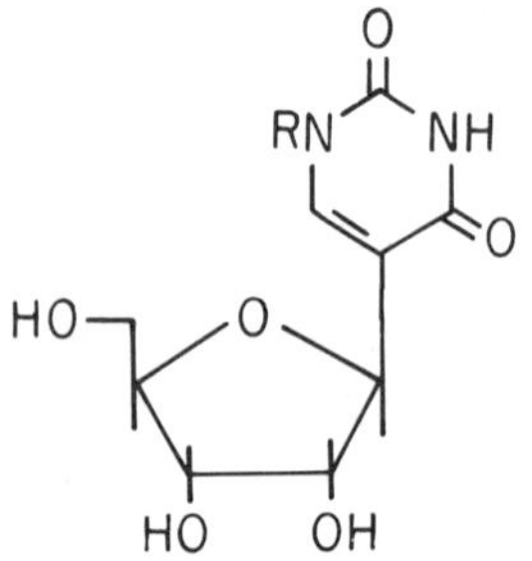

15, R = H

16, R = CH_3

from *H. morrhuae* tRNA

C–C rather than C–N glycosidic bonds. Recognition of this feature in mass spectra is gained by the very low abundances of all base or sugar ions resulting from glycosidic bond cleavage (McCloskey, 1974; Pang *et al.*, 1982b). Secondarily, if blocked derivatives are prepared, the availability of two active hydrogens in the base (versus one for uridine) is marked by incorporation of an additional blocking group. These factors were used in the characterization of 1-methylpseudouridine Structure **16**, which replaces thymine riboside in the tRNA of many archaebacteria (Pang *et al.*, 1982a). The mass spectrum of the TMS derivative of Structure **16** demonstrated the presence of a C–C glycosidic bond, while the incorporation of only one TMS function indicated methyl substitution on N or O rather than carbon. Analogous correlations regarding lower fragment ion abundances resulting from higher C–C glycosidic bond strength have also been made in mass spectra resulting from chemical ionization (Wilson and McCloskey, 1975), and FAB (Crow *et al.*, 1984).

6.4. SUMMARY

Mass spectrometry has played a major role in the structural characterization of naturally modified constituents of nucleic acids. Most effectively applied to nucleosides rather than bases or nucleotides, the utility of mass spectrometry derives from relatively high sensitivity compared with other chemically-based methods as well as the nature and extent of information provided, and the ability in many instances to carry out structural work directly from mixtures. Although significant successes to date have resulted from applications of high resolution electron ionization mass spectrometry—a technique unlikely to decline in importance—new experimental approaches involving LC–MS and desorption ionization–tandem mass spectrometry offer great potential for future applications. The continued expansion of new methodology for structural studies based on mass spectrometry is an important element in understanding structure–function relationships in nucleic acids, a biological topic of considerable interest (Björk, 1984).

ACKNOWLEDGMENTS

The author is grateful to the National Institute of General Medical Sciences for support through grants GM 21584 and GM 29812, and to P. F. Crain for a critical reading of this manuscript.

REFERENCES

Ashworth, D. J., Baird, W. M., Chang, C.-j., Ciupek, J. D., Busch, K. L., and Cooks, R. G. (1985). Chemical modification of nucleic acids. Methylation of calf thymus DNA investigated by mass spectrometry and liquid chromatography, *Biomed. Mass Spectrom.*, **12**, 309–318.

Björk, G. (1984). Modified nucleosides in RNA—their formation and function, in *Processing of RNA* (Ed. D. Apirion), pp. 291–330, CRC Press, Boca Raton.

Buck, M., Connick, M., and Ames, B. N. (1983). Complete analysis of tRNA-modified nucleosides by high-performance liquid chromatography: the 29 modified nucleosides of *Salmonella typhimurium* and *Escherichia coli* tRNA, *Anal. Biochem.*, **129**, 1–13.

Burlingame, A. L., Straub, K., and Baille, T. (1983). Mass spectrometric studies on the molecular basis of xenobiotic-induced toxicities, *Mass Spectrom. Rev.*, **2**, 331–387.

Burrows, W. J., Armstrong, D. J., Kaminek, M., Skoog, F., Bock, R. M., Hecht, S. M., Dammann, L. G., Leonard, N. J., and Occolowitz, J. (1970). Isolation and identification of four cytokinins from wheat germ transfer ribonucleic acid, *Biochemistry*, **9**, 1867–1872.

Crain, P. F., and McCloskey J. A. (1983). Analysis of modified bases in DNA by stable isotope dilution gas chromatography-mass spectrometry. 5-Methylcytosine, *Anal. Biochem.*, **132**, 124–131.

Crain, P. F., Choi Y. C., Busch, H., and McCloskey, J. A. (1978). Characterization of N^6, $O^{2'}$-dimethyladenosine from nuclear RNA of Novikoff hepatoma, *Nucleic Acids Res.*, **5**, 771–778.

Crain, P. F., Yamamoto, H., McCloskey, J. A., Yamaizumi, Z., Nishimura, S., Limburg, K., Raba, M., and Gross, H. J. (1980). Structure elucidation of modified nucleosides by high resolution mass spectrometry, *Adv. Mass Spectrom.*, **8**, 1135–1141.

Crow, F. W., Tomer, K. B., Gross, M. L., McCloskey, J. A., and Bergstrom D. E. (1984). Fast atom bombardment combined with tandem mass spectrometry for the determination of nucleosides, *Anal. Biochem.*, **139**, 243–262.

Dirheimer, G. (1983). Chemical nature, properties, location, and physiological and pathological variations of modified nucleosides in tRNAs, *Recent Results in Cancer Research*, **84**, 1–46.

Doerfler, W. (1983). DNA methylation and gene activity, *Ann. Rev. Biochem.*, **52**, 93–124.

Eagles, J., Javanaud, C., and Self, R. (1984). Fast atom bombardment mass spectrometry of nucleosides and nucleotides, *Biomed. Mass Spectrom.*, **11**, 41–46.

Edmonds, C. G., Vestal, M. L., and McCloskey, J. A. (1985). Thermospray liquid chromatography–mass spectrometry of nucleosides and of enzymatic hydrolysates of nucleic acids, *Nucleic Acids Res.*, **13**, 8197–8206.

Edmonds, C. G., Hashizume, T., McCloskey, J. A., Gupta, R., and Stetter, K. O. (1986). The role of LC/MS in the characterization of new nucleosides from transfer RNA, in *Proc. 34th Ann. Conf. on Mass Spectrometry and Allied Topics, Cincinnati,* American Society for Mass Spectrometry, paper FOA2.

Esmans, E. L., Luyten, Y., and Alderweireldt, F. C. (1983). Liquid chromatography–mass spectrometry of nucleosides using a commercially available direct liquid introduction probe, *Biomed. Mass Spectrom.*, **10**, 347–351.

Fenselau, C. (1976). Analysis by mass spectrometry of the photoproducts of nucleic acid bases, in *Photochemistry and Photobiology of Nucleic Acids* (Ed. S. Y. Wang), Vol. I, pp. 419–446, Academic Press, New York.

Games, D. E., and Lewis, E. (1980). Combined liquid chromatography mass spectrometry of glycosides, glucuronides, sugars and nucleosides, *Biomed. Mass Spectrom.*, **7**, 433–436.

Gehrke, C. W., and Patel, A. N., (1977). Gas–liquid chromatography of nucleosides. Effect of silylating reagents and solvents, *J. Chromatogr.*, **130**, 103–114.

Hashizume, T., McCloskey, J. A., and Liehr, J. G. (1976). Electron impact-induced reactions of N^6-(3-methyl-2-butenyl) adenosine and related cytokinins, *Biomed. Mass Spectrom.*, **3**, 177–183.

Hattox, S. E., and McCloskey, J. A. (1974). Structure-retention relationships in the gas chromatography of nucleosides, *Anal. Chem.*, **46**, 1378–1383.

Hecht, A. M., Gupta, A. S., and Leonard, N. J. (1969). Position of uridine thiation: the identification of minor nucleosides from transfer RNA by mass spectrometry, *Biochim. Biophys. Acta*, **182**, 444–448.

Hignite, C. (1972). Nucleic acids and derivatives, in *Biochemical Applications of Mass Spectrometry* (Ed. G. R. Waller), Chap. 16, Wiley-Interscience, New York.

Hignite, C. (1980). Nucleic acids and derivatives, in *Biochemical Applications of Mass Spectrometry* (Eds. G. R. Waller and O. C. Dermer), First Suppl. Vol., Chap. 16, Wiley-Interscience, New York.

Isono, K., Utamoto, M., Kusakabe, H., Miyata, N., Koyama, T., Ubukata, M., Sethi, S. K., and McCloskey, J. A. (1984). Ascamycin and dealanylascamycin, nucleoside antibiotics from *Streptomyces* Sp., *J. Antiobiotics*, **37**, 670–672.

Kasai, H., Ohashi, Z., Harada, F., Nishimura, S., Oppenheimer, N. J., Crain, P. F., Liehr, J. G., von Minden, D. L., and McCloskey, J. A. (1975). Structure of the modified nucleoside Q isolated from *Escherichia coli* transfer ribonucleic acid. 7-(4,5-cis-Dihydroxy-1-cyclopenten-3-ylaminomethyl)-7-deazaguanosine, *Biochemistry*, **14**, 4198–4208.

Kasai, H., Murao, K., Liehr, J. G., Crain, P. F., McCloskey, J. A., and Nishimura, S. (1976a). Structure determination of a modified nucleoside isolated from *Escherichia coli* transfer ribonucleic acid. *N*-[*N*-[9-β-*D*-Ribofuranosylpurin-6-yl)carbamoyl] threonyl]-2-amido-2-hydroxymethylpropane-1,3-diol, *Eur. J. Biochem.*, **69**, 435–444, and microfilm supplement.

Kasai, H., Nakanishi, K., Macfarlane, R. D., Torgerson, D. F., Ohashi, Z., McCloskey, J. A., Gross, H. J., and Nishimura, S. (1976b). The structure of Q* nucleoside isolated from rabbit liver transfer ribonucleic acid, *J. Am. Chem. Soc.*, **98**, 5044–5046.

Koenig, W. A., Smith, L. C., Crain, P. F., and McCloskey, J. A. (1971). Mass spectrometry of trifluoroacetyl derivatives of nucleosides and hydrolysates of deoxyribonucleic acid, *Biochemistry*, **10**, 3968–3979.

Kuchino, Y., Kasai, H., Yamaizumi, Z., Nishimura, S., and Borek, E. (1979). Undermodified Y base in a $tRNA^{Phe}$ isoacceptor observed in tumor cells, *Biochim. Biophys. Acta*, **565**, 215–218.

Kuntzel, B., Weissenbach, J., Wolff, R. E., Tumaitis-Kennedy, T. D., Lane, B. G., and Dirheimer, G. (1975). Presence of the methyl ester of 5-carboxymethyl uridine in the wobble position of the anticodon of $tRNA^{Arg}_{III}$ from brewer's yeast, *Biochimie*, **57**, 61–70.

Lawson, A. M., Stillwell, R. N., Tacker, M. M., Tsuboyama, K., and McCloskey, J. A. (1971). Mass spectrometry of nucleic acid components. Trimethylsilyl derivatives of nucleotides, *J. Am. Chem. Soc.*, **93**, 1014–1023.

McCloskey, J. A. (1974). Mass spectrometry, in *Basic Principles in Nucleic Acid Chemistry* (Ed. P. O. P. Ts'o), Vol. I, pp. 209–309, Academic Press, New York.

McCloskey, J. A. (1985). Mass spectrometry of nucleic acid constituents and related compounds, in *Mass Spectrometry in the Health and Life Sciences* (Eds. A. L. Burlingame and N. Castagnoli, Jr.), pp. 521–544, Elsevier, Amsterdam.

McCloskey, J. A., Stillwell, R. N., and Lawson, A. M. (1968). Use of deuterium labeled trimethylsilyl derivatives in mass spectrometry, *Anal. Chem.*, **40**, 233–236.

Murao, K., and Ishikura, H. (1978). A new uridine derivative located in the anticodon of $tRNA_{1}^{Gly}$ from *Bacillus subtilis*', *Nucleic Acids. Res., Spl. Publ.*, **5**, s333–s336.

Ohashi, Z., Maeda, M., McCloskey, J. A., and Nishimura, S. (1974). 3-Amino-3-carboxypropyl)uridine:a novel modified nucleoside isolated from *Escherichia coli* phenylalanine transfer RNA, *Biochemistry,* **13**, 2620–2625.

Pang, H., Ihara, M., Kuchino, Y., Nishimura, S., Gupta, R., Whoese, C. R., and McCloskey, J. A. (1982a). Structure of a modified nucleoside in archaebacterial tRNA which replaces ribosylthymine: 1-methylpseudouridine, *J. Biol. Chem.*, **257**, 3589–3592.

Pang, H., Schram, K. H., Smith, D. L., Gupta, S. P., Townsend, L. B., and McCloskey, J. A. (1982b). Mass spectrometry of nucleic acid constituents. Trimethylsilyl derivatives of nucleosides, *J. Org. Chem.*, **47**, 3923–3932, and microfilm supplement.

Pang, H., Smith, D. L., Crain, P. F., Yamaizumi, K., Nishimura, S., and McCloskey, J. A. (1982c). Identification of nucleosides in hydrolysates of transfer RNA by high resolution mass spectrometry, *Eur. J. Biochem.*, **127**, 459–471.

Panzica, R. P., Townsend, L. B., von Minden, D. L., Wilson, M. S., and McCloskey, J. A. (1973). Formation of 3,5,5-trimethyl derivatives of dihydrouracil nucleosides by reaction with methylsulfinyl carbanion and methyl iodide, *Biochim. Biophys. Acta*, **331**, 147–153.

Puzo, G., and Wiebers, J. L. (1981). Identification of modified nucleosides in intact transfer ribonucleic acid by pyrolysis–electron impact–collisional activation mass spectrometry, *Nucleic Acids Res.*, **9**, 4655–4667.

Quilliam, M. A., Ogilvie, K. K., Sadana, K. L., and Westmore, J. B. (1980a). Gas chromatographic retention data for silyl and acyl derivatives of nucleosides, *J. Chromatogr.*, **196**, 367–378.

Quilliam, M. A., Ogilvie, K. K., Sadana, K. L., and Westmore, J. B. (1980b). Mass spectra of sterically crowded trialkylsilyl derivatives of nucleosides, *Org. Mass Spectrom.*, **15**, 207–219.

Schoen, A. E., Cooks, R. G., and Wiebers, J. L. (1979). Modified bases characterized in intact DNA by mass-analyzed ion kinetic energy spectrometry, *Science*, **203**, 1249–1251.

Schram, K. H., and McCloskey, J. A. (1979). Nucleosides and nucleotides, in *GLC and HPLC Determination of Therapeutics Agents* (Ed. K. Tsuji), Part 3, Chap. 33, Marcel Dekker, New York.

Sethi, S. K., Smith, D. L., and McCloskey, J. A. (1983). Determination of active hydrogen content by fast atom bombardment mass spectrometry following hydrogen-deuterium exchange, *Biochem. Biophys. Res. Commun.*, **112**, 126–131.

Shaw, S. J., Desiderio, D. M., Tsuboyama, K., and McCloskey, J. A. (1970). Mass spectrometry of nucleic acid components. Analogs of adenosine, *J. Am. Chem. Soc.*, **92**, 2510–2522.

Slowikowski, D. L., and Schram, K. H. (1985). Fast atom bombardment mass spectrometry of nucleosides, nucleotides and oligonucleotides, *Nucleosides Nucleotides*, **4**, 309–345.

Sprinzl, M., Moll, J., Meissner F., and Hartmann, T. (1985). Compilation of tRNA sequences, *Nucleic Acids Res.*, **13**, r1–r49.

Swinton, D., Hattman, S., Crain, P. F., Cheng, C. S., Smith, D. L., and McCloskey, J. A. (1983). Purification and characterization of the unusual deoxynucleoside, α-*N*-(9-β-*D*-2′-deoxyribofuranosylpurin-6-yl)glycinamide, specified by the phage Mu modification function, *Proc. Nat. Acad. Sci., U.S.A.*, **80**, 7400–7404.

Teece, R. G., Slowikowski, D., and Schram, K. H., (1983). Synthesis and utility of deuterium labeled trimethylanilinium hydroxide in the mass spectral analysis of nucleosides, *Biomed. Mass Spectrom.*, **10**, 30–33.

Tondeur, Y., Shorter, M., Gustafson, M. E., and Pandey, R. C., (1984). Fast atom bombardment mass spectrometry and tandem mass spectrometry in antibiotics: identification of nucleoside antitumor antibiotic toyocamycin in fermentation broth, *Biomed. Mass Spectrom.*, **11**, 622–627.

Unger, S. E., Schoen, A. E., Cooks, R. G., Ashworth, D. J., Gomes, J. D., and Chang, C-j. (1981). Identification of modified nucleosides by secondary-ion mass spectrometry, *J. Org. Chem.*, **46**, 4765–4769.

Vestal, M. L. (1983). Ionization techniques for nonvolatile molecules, *Mass Spectrom. Reviews*, **3**, 447–480.

Vold, B. S., Keith, D. E. Jr., Buck, M., McCloskey, J. A., and Pang, H. (1982). Lysine tRNAs from *Bacillus subtilis* 168: structural analysis, *Nucleic Acids Res.*, **10**, 3125–3132.

von Minden, D. L., Stillwell, R. N., Koenig, W. A., Lyman, K. J., and McCloskey, J. A. (1972). Mass spectrometry of 7-methylpurine nucleosides. Studies of a unique oxygen incorporation reaction which occurs during trimethylsilylation, *Anal. Biochem.*, **50**, 110–121.

von Minden, D. L., and McCloskey, J. A. (1973). Mass spectrometry of nucleic acid components. N,O-Permethyl derivatives of nucleosides, *J. Am. Chem. Soc.*, **95**, 7480–7490.

von Minden, D. L., Liehr, J. G., Wilson M. H., and McCloskey, J. A. (1974). Mechanism of electron impact induced elimination of methylenimine from dimethylamino heteroaromatic compounds, *J. Org. Chem.*, **39**, 285–289.

Warren, R. A. J. (1980). Modified bases in bacteriophage DNA's, *Ann. Rev. Microbiol.*, **34**, 137–157.

White, E. (V), Krueger, P. M., and McCloskey, J. A. (1972). Mass spectra of trimethylsilyl derivatives of pyrimidine and purine bases, *J. Org. Chem.*, **37**, 430–438.

Wilson, M. S. (1974). *Ph.D. Thesis*, Baylor College of Medicine, Houston, TX, pp. 141–158. *Dissertation Abstracts* (1975) **36**, 196.

Wilson, M. S., and McCloskey, J. A. (1975). Chemical ionization mass spectrometry of nucleosides. Mechanisms of ion formation and estimations of proton affinity, *J. Am. Chem. Soc.*, **97**, 3436–3444.

Yamaizumi, Z., Nishimura, S., Limburg, K., Raba, M., Gross, H. J., Crain, P. F., and McCloskey, J. A. (1979). Structure elucidation by high resolution mass spectrometry of a highly modified nucleoside from mammalian transfer RNA. *N*-[(9-β-*D*-ribofuranosyl-2-methylthiopurin-6-yl)carbamoyl]threonine, *J. Am. Chem. Soc.*, **101**, 2224–2225.

Mass Spectrometry in Biomedical Research
Edited by S. J. Gaskell
© 1986 John Wiley & Sons Ltd.

Chapter 7

NEW METHODS FOR THE ANALYSIS OF ACYLCARNITINES AND ACYL-COENZYME A COMPOUNDS

David S. Millington

7.1. INTRODUCTION

Many essential steps in intermediary metabolism that involve the biochemical degradation of fats and amino acids occur in the mitochondria. For example, enzymes transfer long-chain fatty acyl groups in the cytosol from coenzyme A to *L*-carnitine, which is the vehicle for transporting them into the mitochondria (Figure 1). There, the acyl groups are transferred back to coenzyme A and cleaved by enzymes of the beta-oxidation pathway into sub-units of acetyl-coA, the building block of numerous essential compounds synthesized *in vivo* (Smith *et al.*, 1983; Pande and Parvin, 1980). Thus, *L*-carnitine (1) and coenzyme A (2) are vital mediators of metabolism.

$$(CH_3)_3\overset{+}{N}\text{–}CH_2\text{–}\underset{}{\overset{\overset{\displaystyle OH}{|}}{CH}}\text{–}CH_2\text{–}COO$$

(1)

In the group of inherited metabolic disorders known as the organic acidurias, enzymes or cofactors in the beta-oxidation pathway are defective or absent, leading to profound metabolic disturbances in affected individuals. The accumulation of an intermediate product results in the excretion of unusual metabolites, especially during a catabolic episode triggered by an infection, for example.

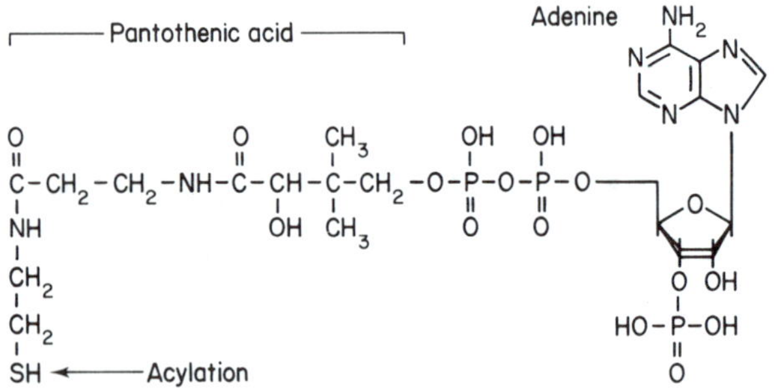

(2)

Many such metabolites are organic acids, easily detected by gas chromatography–mass spectrometry (GC–MS), hence the origin of the collective term, organic acidurias. Since the pioneering studies in this field began, almost 20 years ago (Tanaka *et al.*, 1966), GC–MS has detected and diagnosed most of the known metabolic diseases (Chalmers and Lawson, 1982).

For several reasons, a better understanding of the biochemistry of altered metabolism is desirable. Perhaps the most important of these is the development of more effective treatment strategies than the inadequate dietary restrictions currently used. In addition, certain disorders, especially those affecting the metabolism of fats, are not well understood and can be difficult to diagnose because the organic aciduria is either non-specific or is expressed only when the host is severely ill. GC–MS of organic acids is of limited value in these circumstances and more definitive or supplementary diagnostic procedures are therefore desirable.

A new insight into this problem has resulted from the recent observations that organic acidurias are generally associated with abnormally low carnitine levels in the serum and urine (Chalmers *et al.*, 1984). Significantly, most of the carnitine is acylated at the 3-hydroxy group (Structure 1). *L*-carnitine is usually assayed radio-enzymatically and the acylated fraction determined by difference after alkaline hydrolysis and re-assay (Brass and Hoppel, 1978). When *L*-carnitine was administered to a patient with propionic acidemia in an attempt to correct the deficiency, significant biochemical and clinical improvement was noted, in addition to the enhanced excretion of acylcarnitine (Roe and Bohan, 1982).

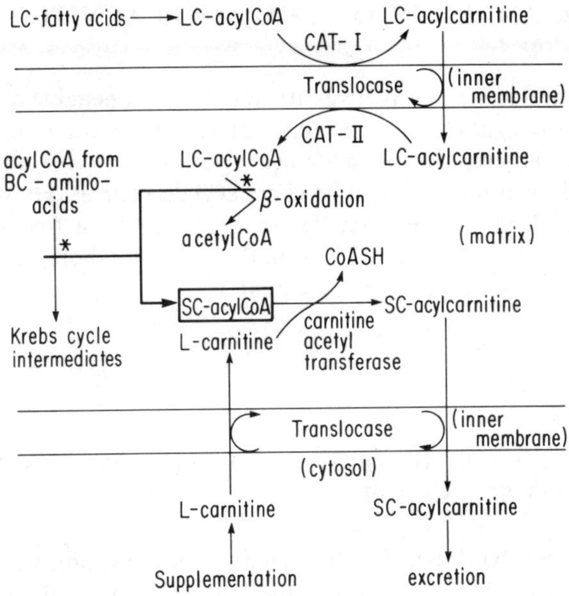

Figure 1: Role of carnitine in metabolic disorders characterized by the accumulation of acyl-coA in mitochondria. Abbreviations: LC, long-chain; BC, branched-chain, SC, short-chain; CAT, carnitine acyltransferase; coA, coenzyme A. Potential sites of metabolic disturbances are marked*.

An attractive hypothesis was postulated to account for these observations (Chalmers *et al.*, 1984; Roe *et al.*, 1985a), which is summarized in the lower part of Figure 1. The intramitochondrial sequestration of coenzyme A, in the form of short-chain acyl-coA (Corkey *et al.*, 1981), and carnitine deficiency are believed to be typical of and to account for the clinical manifestations in most of the organic acidurias. An adequate supply of exogenous L-carnitine promotes excretion of the toxic acyl-coA compounds as harmless acylcarnitines, liberating free coenzyme A for its vital functions and thus improving clinical and biochemical status.

Support for this concept and the impetus to study these diseases in greater detail required the capability to analyze the metabolic intermediates, the acyl-coA compounds, and their related end-products, the acylcarnitines. The lack of sensitive, specific methods for these compounds prompted the development and application of the new mass spectrometric methods described in this chapter.

7.2. ANALYSIS OF THE ACYLCARNITINES

L-carnitine (1) and its acyl derivatives are not directly amenable to conventional mass spectrometric, gas chromatographic or liquid chromatographic techniques of analysis. The only reports in the literature prior to 1983 describe pyrolysis EI mass spectra of carnitine and its short-chain acyl derivatives (Hvistendahl *et al.*, 1970) and a solid–SIMS spectrum (Unger *et al.*, 1981) of free carnitine. The following new methods were developed in the author's laboratory and those of certain collaborators (see acknowledgements).

7.2.1. FAST ATOM BOMBARDMENT WITH CONSTANT B/E LINKED-FIELD SCANNING

The fast atom bombardment (FAB) technique of ionization was immediately successful when applied to acylcarnitines (Millington *et al.*, 1984). Intense spectra were obtained from a glycerol matrix, dominated by $(M + H)^+$ ions and fragments of the carnitine moiety. Analysis of $(M + H)^+$ ions was achieved by the linked-field scan at constant magnetic (*B*) to electric (*E*) field ratio (*B*/*E* scan) without the addition of collision gas in the first field-free region. The characteristic daughter ions observed in the *B*/*E* scan spectra correspond to $(M + H - 59)^+$, due to loss of $N(CH_3)_3$, and a common ion at *m*/*z* 85 with the presumed structure ($^+CH_2$–CH=CH–COOH).

The first examples of the application of FAB with the *B*/*E* scan described the unequivocal identification of propionylcarnitine in propionic acidemia and in methylmalonic aciduria, and acetylcarnitine excretion in Reye's syndrome, an acquired or temporary disorder characterized by excessive lipolysis (Millington *et al.*, 1984; Roe *et al.*, 1983; Roe *et al.*, 1985a). In these cases, acylcarnitine concentrations in the urine were high (> 1 μmol/ml) and the urine was partially purified by elution through anion exchange columns (Dowex-1). An example from these early studies (Figure 2a) shows the detection of a peak corresponding to the propionylcarnitine $(M + H)^+$ ion in the FAB spectrum of urine from a child with propionic acidaemia, collected after a carnitine load. The accurate mass and the *B*/*E* scan (Figure 2b), which is identical to that from authentic material (Figure 2c), confirm the identity of propionylcarnitine.

Although it is known that normal children excrete predominantly acetyl-carnitine, even after a carnitine load (Roe *et al.*, 1984), the acylcarnitine signals are undetectable by this method in normal urine and in that of affected patients before carnitine is administered. The method's shortcomings include interference from alkali metal adducts with glycerol, urea, creatine and other matrix constituents. The acylcarnitine signals themselves are diluted by adduct formation; for example, the ion at *m/z* 256 in Figure 2a is the potassium adduct of propionylcarnitine. Nevertheless, accurate quantification of acetyl- and propionylcarnitine in these samples was achieved very effectively using isotope dilution (Millington *et al.*, 1984). The internal standards, [$^2H_3$]acetylcarnitine and [$^2H_5$]propionylcarnitine, were added to the urine before work-up and the $(M + H)^+$ ion intensity ratios determined by high resolution ($m/\Delta m$ 7000) selected-ion detection.

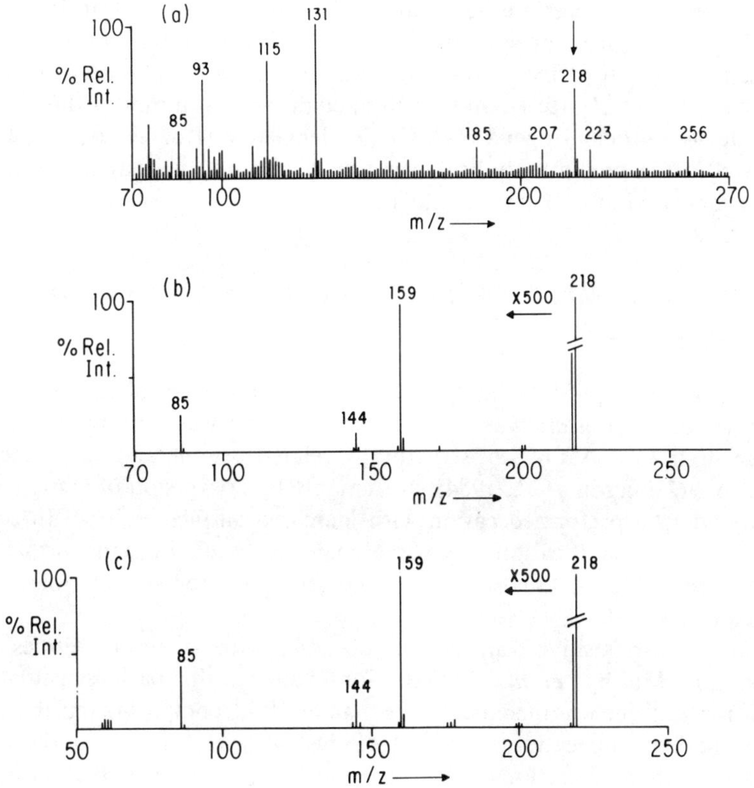

Figure 2: Identification of propionylcarnitine in urine from a patient with propionic acidaemia after a carnitine load. Full FAB spectrum of partially purified urine (a) shows ion at *m/z* 218 corresponding to $(M + H)^+$ of propionylcarnitine. The *B/E* scan of the *m/z* 218 ion in the urine (b) compares with that from the $(M + H)^+$ of authentic material (c).

7.2.2. ENHANCEMENT OF FAB SENSITIVITY FOLLOWING SAMPLE PURIFICATION

Significant improvement in the detectability of acylcarnitines by FAB MS was gained by applying modifications of the work-up procedure designed to remove alkali metals and urea. Urine was passed through a second ion-exchange column (Dowex-50) that retained the acylcarnitines while interfering compounds were washed out using dilute hydrochloric acid and water. The absorbed fraction, enriched in acylcarnitines, was eluted with 1M ammonium hydroxide in ethanol/water (1:4). After lyophilization, treatment with urease (1 U at 37°C) and re-lyophilization, the residue was dissolved in dilute HCl before FAB analysis.

This improved procedure has enabled the detection of acylcarnitines at or near the physiological normal range in urine (50–100 nmol/ml). It was employed in recent studies to detect isovalerylcarnitine, a new metabolite in isovaleric acidemia (Roe *et al.*, 1984), and medium-chain acylcarnitines in the medium-chain acyl-coA dehydrogenase (MCAD) deficiency (Roe *et al.*, 1985b). As observed in the previous studies, excretion of the unusual acylcarnitines was greatly increased after a carnitine load.

7.2.3. ENHANCEMENT OF FAB SENSITIVITY FOLLOWING ESTERIFICATION

In an effort to simplify the analytical procedure, the effect of converting acylcarnitines to alkyl esters was studied. Esterification was reported to enhance the sensitivity to FAB MS of structurally related betaine and its biochemical precursors (Jamieson *et al.*, 1984), presumably by conversion of a zwitterionic compound to a preformed cation with increased surface activity. In a urine extract, general esterification may also be expected to minimize the formation of alkali metal adducts, thus improving the selectivity for strongly basic compounds in the matrix.

Both of these positive benefits were observed with the acylcarnitines (Millington and Maltby, *et al.*, 1985a). For example, the basic sensitivity for propionylcarnitine was increased more than ten-fold, comparing the abundance of the methyl ester cation with that of the $(M + H)^+$ ion of underivatized material, enabling detection in raw urine containing as little as 50 nmol/ml total acylcarnitine after desiccation and methylation (3M HCl in dry methanol). This simple methylation procedure is now the basis for routine acylcarnitine profiling of urine samples in our laboratory.

The effect of methylation on the detection of acylcarnitines in unpurified urine can be judged from Figures 3a and 3b, which compare FAB MS of urine from a patient with suspected metabolic disease before and after derivatization.

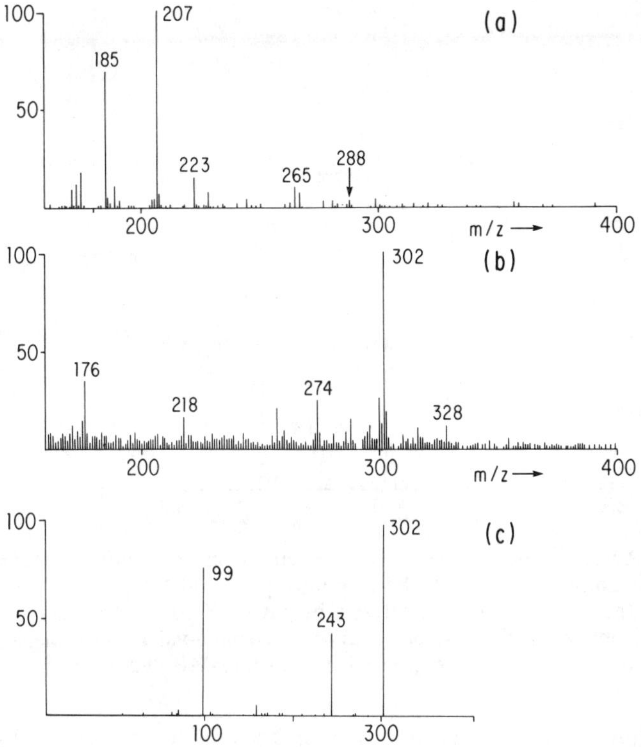

Figure 3: Advantage of methylation for acylcarnitine profiling by FAB MS. The $(M + H)^+$ peak for octanoylcarnitine (*m*/*z* 288) is barely discernible in the spectrum of raw urine from a MCAD patient (a), while the entire acylcarnitine profile is clearly revealed in the same sample after methylation (b). Tandem mass spectrometry (EB–QQ instrument, see text) shows the *m*/*z* 302 peak to be an acylcarnitine methyl ester (c).

Before methylation, it is difficult to assign any of the peaks to acylcarnitines. After methylation, the major signal at *m*/*z* 302 corresponds to the methyl ester of octanoylcarnitine, the diagnostic feature of MCAD deficiency (Roe *et al.*, 1985b). The daughter ion spectrum of *m*/*z* 302 obtained by low energy collisional activation (CA) in a tandem mass spectrometer of EB-QQ configuration contains the characteristic signals for an acylcarnitine corresponding to loss of 59 u and at *m*/*z* 99 (Figure 3c). Other signals in the conventional spectrum were similarly shown to correspond to free carnitine (*m*/*z* 176) and to other acylcarnitines: C_2(*m*/*z* 218), C_6 (*m*/*z* 274), $C_{8:1}$ (*m*/*z* 300), $C_{10:1}$ (*m*/*z* 328).

The structural information obtained from the *B*/*E* scan and the CA–MS/MS spectrum on the same instrument (VG 7070 E-Q) were equivalent, but the latter was more conveniently obtained under computer control.

Numerous previously unknown acylcarnitines, including some derived from dicarboxylic acids (Roe *et al.*, 1986), have been detected by this method in the

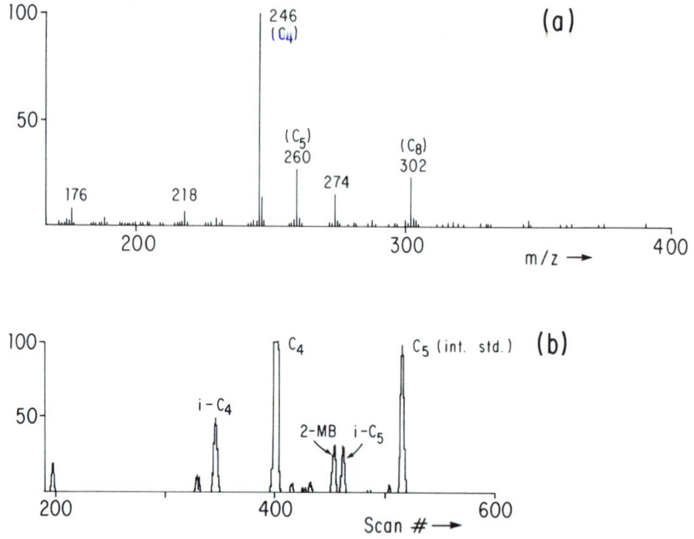

Figure 4: Acylcarnitine profile (methyl esters) from raw urine of a patient with ethylmalonic-adipic aciduria (a), showing mostly C_4 and C_5 acylcarnitines. After base hydrolysis, the liberated acids analyzed by GC–MS as TMS esters confirmed the presence of isomers (b). The *n*-valeric acid peak is from *n*-valerylcarnitine, a non-biological isomer added as the internal standard. (2-MB = 2-methylbutyric).

urine of children with as yet unknown or unexplained metabolic disturbances. For example, the profile in Figure 4a, from a patient with ethylmalonic-adipic aciduria, is dominated by short-chain acylcarnitines (mostly C_4), suggesting the likelihood of a defect in the short-chain acyl-coA dehydrogenase of beta-oxidation.

The major limitation of the FAB method is the inability to distinguish between isomeric acylcarnitines, such as the C_4 isomers *n*-butyryl- and *iso*-butyrylcarnitine. Three successful solutions to this problem are discussed in the following sections.

7.2.4. ALKALINE HYDROLYSIS AND ANALYSIS OF LIBERATED ACIDS BY GC–MS

A simple and effective procedure for increasing the information content of the acylcarnitine profile is to cleave the acyl groups by mild alkaline hydrolysis, extract the liberated acids into ether and analyze them as either methyl or trimethylsilyl ester derivatives by GC–MS (Roe *et al.*, 1985b). Isomers are resolved by use of a fused-silica capillary GC column (DB-1, J & W Scientific). The liberated acid GC–MS profile from the patient with ethylmalonic-adipic aciduria, for example (part of which is shown in Figure 4b), aided the interpretation of the FAB MS profile (Figure 4a) by revealing the acyl groups as propionyl, *iso*-butyryl, *n*-butyryl (mostly), *iso*-valeryl plus 2-methylbutyryl, hexanoyl and octanoyl.

It is important to prepare an acylcarnitine-enriched fraction from the urine (as described in Section 7.2.2) before the hydrolysis and to use pure solvents and reagents. The method is apparently trustworthy except for acetylcarnitine at levels approaching 100 nmol/ml, for which consistent, low blanks have been difficult to achieve. Quantitative data for propionylcarnitine and isovalerylcarnitine, obtained by addition of analogues labelled in the acyl groups as internal standards, were comparable with values determined by high-resolution FAB MS on the same samples.

Qualitatively, the GC–MS and FAB MS profiles have agreed without discrepancy, but it remains to be seen whether the GC–MS method is quantitatively valid at low concentration. If so, this would probably become the method of choice for routine analysis of acylcarnitines in laboratories equipped with GC–MS.

7.2.5. THERMOSPRAY LC–MS ANALYSIS

The recently developed thermospray interface for liquid chromatography–mass spectrometry (LC–MS) has been successfully applied to the analysis of short-chain acylcarnitines (Yergey *et al.*, 1984; Liberato *et al.*, 1985). The thermospray mass spectra, which are quite different from those obtained by FAB MS, characteristically show prominent ions corresponding to $(M + H)^+$, $(M + H - 14)^+$ and $(M + H - 42)^+$ as shown for *n*-butyrylcarnitine in Figure 5b. Ions at *m*/*z* 102 and 130 are derived from the carnitine moiety. Most of the fragments probably arise by thermal degradation, their relative and absolute abundances being critically dependent upon the conditions at the interface. However, the overall sensitivity is well within the required range for biological sample analysis. Excellent agreement with high resolution FAB MS has been achieved for quantitative analysis in selected samples using isotope dilution (Liberato *et al.*, 1985).

The ability to resolve chromatographically isomeric acylcarnitines is a significant advantage of this method. For example, *n*-butyryl- and *iso*-butyrylcarnitine, whose thermospray mass spectra (Figure 5b) are indistinguishable, were resolved by reverse-phase LC on a C_{18}-column using 0.1 M ammonium acetate in water/methanol (4 : 1) as the eluant (Figure 5a). Thermospray LC–MS was also employed to confirm the excretion of valproylcarnitine, a previously unknown metabolite, by children receiving valproic acid (2-propylpentanoic acid) for seizure control (Millington *et al.*, 1985a). The ability to resolve the target compound from octanoylcarnitine, a biological isomer, was crucial.

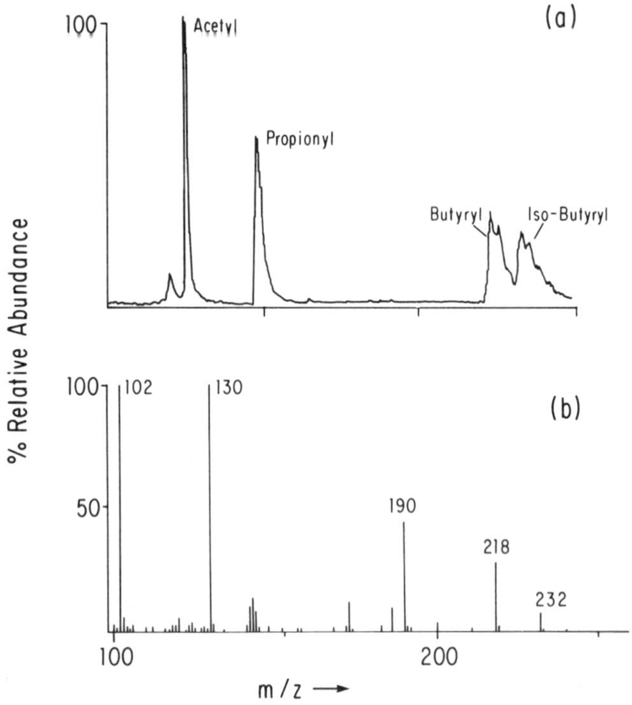

Figure 5: Analysis of short-chain acylcarnitines by LC–MS. The isomeric C_4-acylcarnitines are resolved (a). The thermospray ionization mass spectra of both isomers (b) are indistinguishable, showing $(M + H)^+$ at *m/z* 232 with characteristic fragments at $(M + H - 14)^+$, $(M + H - 42)^+$, *m/z* 102 and *m/z* 130.

Although *iso*-valerylcarnitine can be resolved from the non-biological isomer, *n*-valerylcarnitine, the LC column was not able to resolve it from 2-methylbutyrylcarnitine.

These examples represent the first reported applications of a quantitative chromatographic analytical method for the intact acylcarnitines.

7.2.6. TANDEM MASS SPECTROMETRY USING HIGH ENERGY COLLISIONAL ACTIVATION

The application of either the previously described *B/E* scan or low energy CA as a supplement to FAB MS for acylcarnitine analysis is limited by the failure to generate ions characteristic of the acyl group. Using a tandem double-focusing mass spectrometer of *BE-EB* configuration (VG Analytical ZAB-4F) and 8keV CA energy at the intermediate focal point, more informative daughter ion spectra have been observed (Gaskell *et al.*, 1986). Specifically, additional fragments corresponding to aliphatic bond cleavages in the acyl group are produced, enabling the clear differentiation of isomers such as *n*-butyryl- and *iso*-butyrylcarnitine, as shown in Figure 6a and 6b. The $(M + H)^+$ ion corresponding to C_4-acylcarnitine in the FAB MS profile of ethylmalonic-adipic aciduria was shown to be predominantly that of *n*-butyrylcarnitine (Figure 6c).

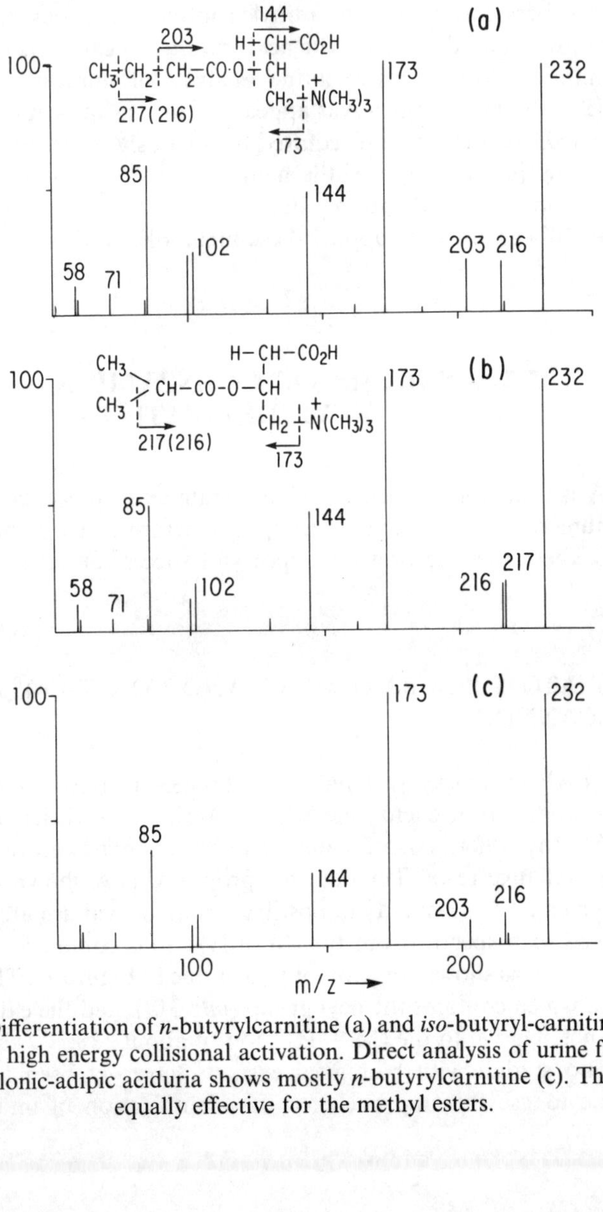

Figure 6: Differentiation of *n*-butyrylcarnitine (a) and *iso*-butyryl-carnitine (b) by FAB MS/MS with high energy collisional activation. Direct analysis of urine from a patient with ethylmalonic-adipic aciduria shows mostly *n*-butyrylcarnitine (c). The technique is equally effective for the methyl esters.

The two biological C_5-isomers, *iso*-valeryl- and 2-methylbutyrylcarnitine, are also easily differentiated by this method. Since LC–MS is not able to differentiate the two, tandem mass spectrometry with high energy CA is the only method applicable to the intact molecules. Other successful applications include the differentiation of valproylcarnitine and octanoylcarnitine, and of 3-methylglutarylcarnitine and sebacylcarnitine. The latter example is interesting because 3-methylglutarylcarnitine has been detected in 3-hydroxy-3-methylglutaryl-coA dehydrogenase deficiency (Roe *et al.*, 1986), while sebacylcarnitine and the homologous suberylcarnitine and adipylcarnitine were observed in some patients with pronounced dicarboxylic aciduria due to carnitine deficiency.

The technique of direct acylcarnitine analysis of unpurified urine using FAB MS/MS with high energy CA is appealing for its simplicity, but access to the sophisticated instrumentation required is obviously too limited for routine applications. The judicious use of the method for structural analysis of new acylcarnitines and for the validation of less specific methods has already proved valuable and will continue to support these metabolic studies.

7.3. ANALYSIS OF COENZYME A AND ITS ACYL DERIVATIVES

Coenzyme A is a complex molecule of moderate size (molecular weight 766) containing numerous sensitive, polar groups (Structure 2). Until the application of FAB MS, there were no previous reports of successful mass spectrometric analysis.

7.3.1. FAST ATOM BOMBARDMENT AND CONSTANT *B*/*E* RATIO LINKED SCANNING

Informative FAB mass spectra have been obtained in both the positive and negative ion modes from coenzyme A and a series of acyl derivatives (Millington and Maltby, 1984). Alkali metal ions were removed before the analyses using cation exchange resin. The data for propionyl-coA shown in Figure 7a and 7b are typical. The sensitivity in positive ion mode was apparently greater, and *B*/*E* linked scan spectra from the $(M + H)^+$ ions contained the essential structural features, as shown for propionylcarnitine in Figure 7c. The dominant fragment ion, which contains the acyl group (*m*/*z* 317), had the extraordinarily high abundance relative to the $(M + H)^+$ ion of about 5%.

Although no studies with biological extracts have yet been attempted, it seems feasible to use this approach for the identification of intact acyl-coA

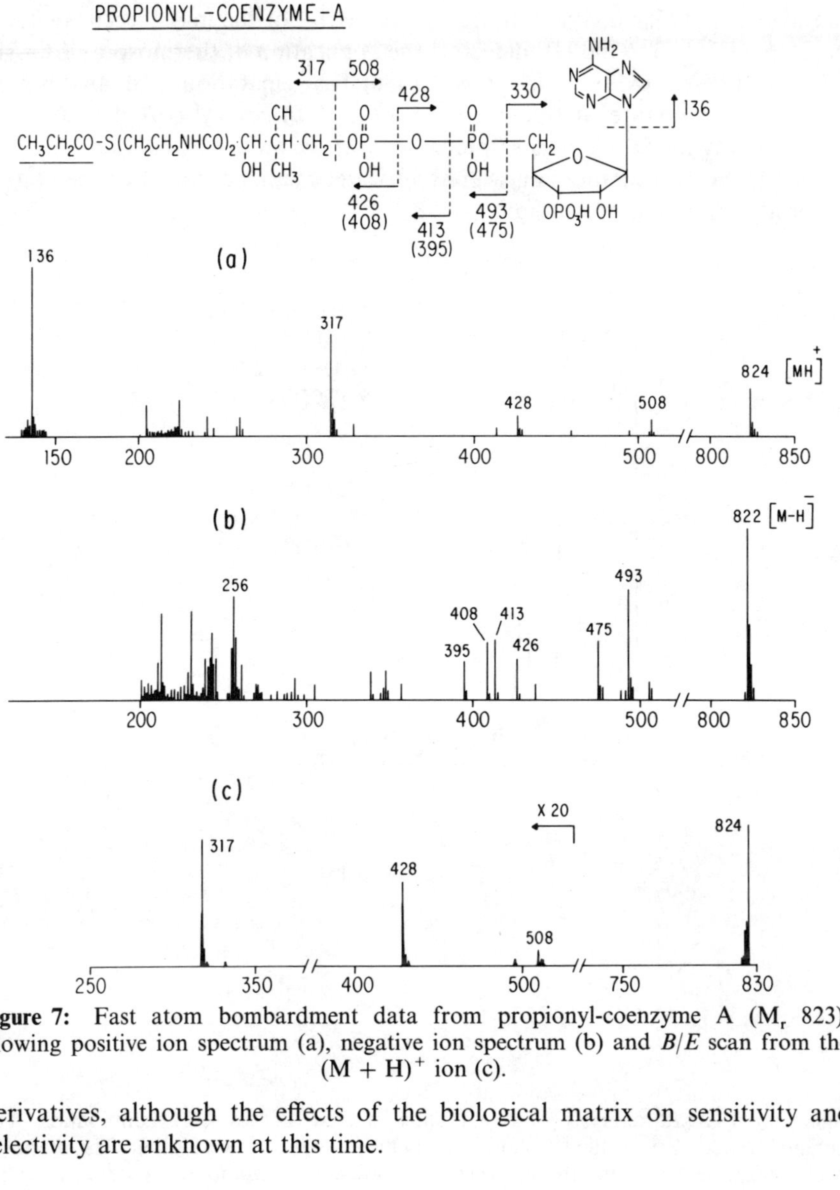

Figure 7: Fast atom bombardment data from propionyl-coenzyme A (M_r 823), showing positive ion spectrum (a), negative ion spectrum (b) and *B*/*E* scan from the $(M + H)^+$ ion (c).

derivatives, although the effects of the biological matrix on sensitivity and selectivity are unknown at this time.

7.3.2. THERMOSPRAY LC–MS

Unlike the acylcarnitines, acyl-coA compounds are amenable to LC with conventional ultraviolet (UV) detectors. Their analysis in rat liver mitochondrial preparations and other tissues has been reported (Corkey *et al.*, 1981;

DeBuysere and Olson, 1983). Since the selectivity of the mass spectrometer is more appealing than the UV detector, the application of thermospray LC–MS was investigated (Millington *et al.*, 1986).The separation and analysis by LC–MS of a mixture of free coenzyme A and three acyl-coA derivatives is shown in Figure 8a. A reverse-phase separation on a C_{18}-column was performed using 1 M ammonium acetate in water/methanol (4 : 1) as the eluting buffer at a flow rate of 1 ml/min.

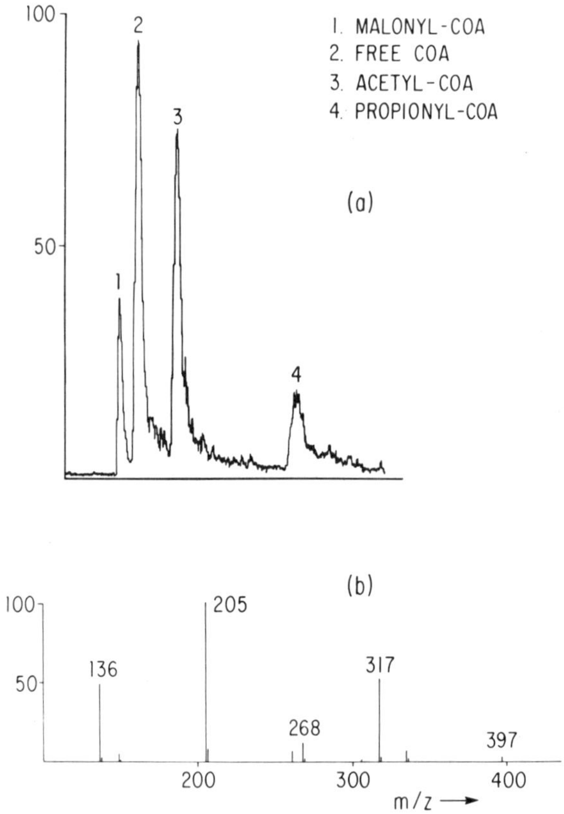

Figure 8: LC–MS analysis of a standard mixture of acyl-coA compounds. The summed ion-current profile (a) shows good chromatographic resolution. The thermospray ionization spectrum (b) of peak 4, propionyl-coenzyme A, contains essential structural information.

These compounds exhibit consistent behaviour in thermospray ionization. No $(M + H)^+$ ion is observed, but signals characteristic of the acyl group (*m*/*z* 205, 317, 397) and the adenosine sub-unit (*m*/*z* 268, 136) are especially prominent, as shown for propionyl-coA in Figure 8b. The *m*/*z* 205 ion has the probable structure 3, while the derivation of the other ions can be appreciated by reference to Figure 7.

$$CH_3\text{-}CH_2\text{-}CO\text{-}S\text{-}CH_2\text{-}CH_2\text{-}NH\text{-}CO\text{-}CH_2\text{-}CH_2\text{-}NH_3{+}$$

(3)

The spectrum of malonyl-coA is very similar to that of acetyl-coA, suggesting decarboxylation in the interface, but the chromatographic separation is adequate for differentiation.

Good quality full scan mass spectra were obtained from about 20 nmol injected on-column. Using the selected ion detection mode, sufficient sensitivity for projected biological applications is expected.

7.4. IMPLICATIONS FOR FUTURE RESEARCH

The main goals of the current research leading to the development of these new analytical methods were to investigate the biochemical role of *L*-carnitine in mitochondrial metabolic disease and to assess the value of *L*-carnitine therapy and other novel treatment strategies. In consideration of the fact that little progress has been made in the treatment of most inherited metabolic diseases since their discovery, the impact of the research so far has been quite dramatic. Acylcarnitine profiling has provided key evidence for the secondary but important role of mitochondrial detoxification for *L*-carnitine, and has confirmed the logic of carnitine therapy in serveral of the organic acidurias. The diagnostic value of acylcarnitine profiling *in vivo* has been demonstrated, especially in disorders of fat metabolism.

Suspected biochemical lesions can be incisively probed *in vivo* using appropriately labelled precursors to the metabolic block and detection of labelled acylcarnitines in the urine. An excellent example of this (Roe, Millington and Maltby, to be published) is the discovery of a new metabolic disease which results in the mitochondrial accumulation of *iso*-butyryl-coA and excretion of *iso*-butyrylcarnitine when adequate carnitine is available. When $[^2H_8]$*L*-valine was administered, $[^2H_6]$*iso*-butyrylcarnitine was detected in the urine, suggesting that the *iso*-butyryl-coA dehydrogenase was affected.

One of the requirements in the basic biochemical investigation of metabolic diseases is the need for specific enzyme assays. The ability to analyze acyl-coA compounds, the biochemical intermediates, will enable the development of new, selective enzyme assays based on mass spectrometric analysis of the substrate or product. In addition, this capability will facilitate studies of enzyme-substrate specificity and the secondary effects of acyl-coA compounds on other metabolic processes. A productive and valuable role for mass spectrometry in these future studies seems assured.

ACKNOWLEDGMENTS

The author wishes to thank Drs. A. L. Yergey and D. J. Liberato (N.I.H., Bethesda, MD) for their expertise in the development and performance of LC–MS analyses of the acylcarnitines, Dr. S. J. Gaskell (N.I.E.H.S., Research Triangle Park, NC) for the high-energy CA spectra from the ZAB-4F and VG-MassLab (Altrincham, UK) for the loan of thermospray LC–MS instrumentation. Gratitude is also expressed to Dr. C. R. Roe, Duke University Medical Center, for insight, leadership and support and to D. A. Maltby for excellent technical assistance. This research was supported in part by grants from the National Reye's Syndrome Foundation, Bryan, OH, the Reye's Syndrome Research Fund-SFA, Duke University Medical Center and the RR-30 General Clinical Research Centers Program, Division of Research Resources, N.I.H., Bethesda, MD.

REFERENCES

Brass, E. P., and Hoppel, C. L. (1978). Carnitine metabolism in the fasting rat, *J. Biol. Chem.*, **235**, 2688–2693.

Chalmers, R. A., and Lawson, A. M. (1982). *Organic Acids in Man*, Chapman and Hall, London.

Chalmers, R. A., Roe, C. R., Stacey, T. E., and Hoppel, C. L. (1984). Urinary excretion of L-carnitine and acylcarnitines by patients with disorders of organic acid metabolism: Evidence for secondary insufficiency of L-carnitine, *Ped. Res.*, **18**, 1325–1328.

Corkey, B. E., Brandt, M., Williams, R. J., and Williamson, J. R. (1981). Assay of short-chain acyl-coenzyme A intermediates in tissue extracts by high pressure liquid chromatography, *Anal. Biochem.*, **118**, 30–41.

DeBuysere, M. S., and Olson, M. S. (1983). The analysis of acyl-coenzyme A derivatives by reverse-phase high-performance liquid chromatography, *Anal. Biochem.*, **133**, 373–379.

Gaskell, S. J., Guenat, C., Millington, D. S., Maltby, D. A., and Roe, C. R. (1986). Differentiation of isomeric acylcarnitines using tandem mass spectrometry, *Anal. Chem.*, in press.

Hvistendahl, G., Undheim, K., and Bremer, J. (1970). Mass spectrometry of onium compounds—II. Carnitine and O-acyl derivatives, *Org. Mass Spectrom.*, 3, 1433–1438.

Jamieson, G. C., Myers, A. C., Fitch, W. L., and Rhodes, D. (1984). Determination of betaines and precursors by stable isotope dilution and fast atom bombardment mass spectrometry, *Proc. 32nd Ann. Conf. on Mass Spectrometry and Allied Topics, San Antonio*; American Society for Mass Spectrometry, pp. 830–831.

Liberato, D. J., Millington, D. S., and Yergy, A. L. (1985). Analysis of acylcarnitines in human metabolic disease by thermospray liquid chromatography–mass spectrometry, in *Mass Spectrometry in the Health and Life Sciences*, Analaytical Symposia Series, Vol. 24 (Eds., A. L. Burlingame and N. Castagnoli), pp. 333–348, Elsevier, Amsterdam.

Millington, D. S., and Maltby, D. A. (1984). Characterization of coenzyme A and its acyl derivatives by FAB–MS/MS, *Proc. 32nd Ann. Conf. on Mass Spectrometry and Allied Topics, San Antonio*; American Society for Mass Spectrometry, pp. 398–399.

Millington, D. S., Roe, C. R., and Maltby, D. A. (1984). Application of high resolution fast atom bombardment and constant *B*/*E* ratio linked scanning to the identification of acylcarnitines in metabolic disease, *Biomed, Mass Spectrom.*, **11**, 236–241.

Millington, D. S., Maltby, D. A., and Roe, C. R. (1985a). Diagnostic and therapeutic implications of acylcarnitine profiling in organic acidurias, *Proc. Japan Soc. Med. Mass Spectrom.*, **10**, 125–130.

Millington, D. S., Bohan, T. P., Roe, C. R., Yergey, A. L., and Liberato, D. J. (1985b). Valproylcarnitine: a novel drug metabolite identified by fast atom bombardment and thermospray liquid chromatography–mass spectrometry, *Clin. Chim. Acta*, **145**, 69–76.

Millington, D. S., Maltby, D. A., and Farrow, P. (1986). The analysis of intermediary metabolities using thermospray LC–MS, in *Advances in Mass Spectrometry, 1985*, (Ed. J. F. J. Todd), pp. 591–592, Wiley and Sons, Chichester, U.K.

Pande, S. V., and Parvin, R. (1980). Carnitine-acylcarnitine translocase-mediated transport of fatty acids into mitochondria: its involvement in the control of fatty acid oxidation in liver, in *Carnitine Biosynthesis, Metabolism and Functions* (Eds. R. A. Frenkel and D. J. Mcgarry), pp. 143–157, Academic Press, New York.

Roe, C. R., and Bohan, T. P. (1982). L-Carnitine therapy in propionic acidemia, *Lancet*, **1**, 1411.

Roe, C. R., Hoppe;, C. L., Stacey, T. E., Chalmers, R. A., Tracey, B. M., and Millington, D. S. (1983). Metabolic response to carnitine in methylmalonic aciduria: an effective stragegy for elimination of propionyl groups, *Arch. Dis. Child.*, **58**, 916–920.

Roe, C. R., Millington, D. S., Maltby, D. A., Bohan, T. P., and Hoppel, C. L. (1984).L-Carnitine therapy in isovaleric acidemia, *J. Clin. Invest.*, **74**, 2290–2295.

Roe, C. R., Millington, D. S., Maltby, D. A., and Bohan, T. P. (1985a). Relative carnitine insufficiency in Reye's syndrome and related metabolic disorders, *J. Nat. Reye's Synd. Found.*, **5**, 201–215.

Roe, C. R., Millington, D. S., Maltby, D. A., Bohan, T. P., Kahler, S. G., and Chalmers, R. A. (1985b). Diagnostic and therapeutic implications of medium-chain acylcarnitines in the medium-chain acyl-coA dehydrogenase deficiency, *Ped. Res.*, **19**, 459–466.

Roe, C. R., Millington, D. S., and Maltby, D. A. (1986). Identification of 3-methylglutarylcarnitine, a new diagnostic metabolite of 3-hydroxy-3-methylglutaryl-coA dehydrogenase deficiency, *J. Clin. Invest.*, **77**, 1391–1394.

Smith, E. L., Hill, R. L., Lehman, I. R., Lefkowitz, R. J., Handler, P., and White, A. (1983). *Principles of Biochemistry: General Aspects*, 7th Edition, pp. 513–518, McGraw-Hill, New York.

Tanaka, K., Budd, M. A., Efron, M. L., and Isselbacher, K. J. (1966). Isovaleric acidemia:a new defect of leucine metabolism, *Proc. Nat. Acad. Sci. U.S.A.*, **56**, 236–242.

Unger, S. E., Day, R. J., and Cooks, R. G. (1981). Positive and negative secondary ion mass spectra and mass analyzed ion kinetic energy mass spectra of some amides, amines and related compounds: mechanisms in molecular SIMS, *Int. J. Mass Spectrom. Ion Phys.*, **39**, 231–255.

Yergey, A. L., Liberato, D. J., and Millington, D. S. (1984). Thermospray liquid chromatography/mass spectrometry for the analysis of L-carnitine and its short-chain acyl derivatives, *Anal. Piochem.*, **139**, 278–283.

Mass Spectrometry in Biomedical Research
Edited by S. J. Gaskell
© 1986 John Wiley & Sons Ltd.

Chapter 8

'METABOLIC MAPPING' OF DRUGS: RAPID SCREENING FOR POLAR METABOLITES USING FAB MS/MS

Kenneth M. Straub

8.1. INTRODUCTION

Elucidation of the pathways of biotransformation of xenobiotic substances is central to much of biochemical and pharmacological research. The biotransformation of a parent structure to a metabolite which possesses different biological properties is often responsible for determining both the magnitude and duration of activity of the parent compound. Characterization of the structures of these metabolites is complicated by the fact that they are usually present in trace amounts (parts per million or parts per billion) within a complex sample matrix such as tissue, urine, or plasma. The inherent difficulty of this task is somewhat reduced by the fact that the structure of the parent drug is usually known, and that the metabolites will retain at least some of the original substructural features of the parent compound.

Metabolites are traditionally divided into two broad classes: phase 1 or 'non-synthetic', which are formed by metabolic oxidation, reduction, or hydrolysis reactions of the parent structure, and phase 2 or 'synthetic' metabolites, which include conjugation reactions of the parent drug or its phase 1 metabolites with various polar moieties such as carbohydrates (usually glucuronic acid), peptides, or sulphate (Mayer *et al.*, 1980). Structural characterization of phase 1 metabolites is usually achieved by a combination of mass spectrometry and other techniques (primarily NMR), either on pure components that have been isolated chromatographically, or by GC–MS. Derivatization methodologies are usually employed to assist in the analysis of polar or thermally labile metabolites by GC–MS. Derivatization of phase 2 metabolites is complicated either by the types of functionality present, or by molecular weight considerations, so that their identification is most often achieved by chromatographic (HPLC) isolation, followed by enzymatic hydrolysis of the conjugating group and subsequent mass spectral characterization of the released primary drug metabolite. More recently, particle induced desorption ionization methods such as FAB MS or liquid SIMS have become a standard tool for obtaining molecular weight and limited fragmentation information on polar, thermally labile drug metabolites and conjugates after isolation from the sample matrix (Barber *et al.*, 1981; Aberth *et al.*, 1982; Shackleton and Straub, 1982; Jardine *et al.*, 1984). Spectra obtained using these ionization techniques typically exhibit an abundant protonated $[(M + H)^+]$ or cationized $[(M + A)^+]$ molecule, with little or no fragmentation. Thus molecular weight information, but relatively little structural data, are provided by this technique. In addition, the liquid matrix that is used to enhance ion formation contributes a relatively high background signal across the entire mass range.

Identification of all of the metabolites derived from a given drug that are present in a complex sample matrix is clearly a difficult task, and typically requires a lengthy series of studies using a variety of analytical techniques. In particular, radiolabelled (^3H or ^{14}C) analogues are frequently used to aid in the isolation and identification of drug-related components.

An alternative approach to these conventional chromatography-based methods of complex mixture analysis is to use tandem mass spectrometry techniques, where the mass spectrometer serves as both a separation device for distinguishing between components of a mixture, and as a tool for structure elucidation (McLafferty, 1980; Yost and Fetterolf, 1983; McLafferty, 1983). An example of the application of MS/MS methodologies to the analysis of primary drug metabolites was first reported by Yost and co-workers (Perchalski *et al.*, 1982). The strategy outlined in this study made use of sample introduction via a heated solids probe, where the heating rate was used to effect partial fractionation of the various components in a crude sample extract into the ion source under the conditions of chemical ionization (CI). The extension of this particular technique to the analysis of polar or thermally labile metabolites and conjugates is problematic, since these types of structures would not survive the sample introduction process.

This article will outline some of the recent results that have been obtained in our laboratory using MS/MS in conjunction with FAB/ liquid SIMS ionization as a technique for characterizing polar drug metabolites present in complex sample matrices. One of the long-term goals of these studies is to devise a general strategy for rapidly assessing the overall biotransformation of candidate drugs ('metabolic mapping'); the examples presented here will serve to demonstrate the utility of MS/MS for this purpose.

8.2. ANALYTICAL STRATEGY

The overall strategy that we have experimented with is based on obtaining a sample extract that contains the metabolites of interest, and then carrying out an MS/MS analysis which incorporates specific scan sequences that can be used to identify particular substructural features which are characteristic of either the drug or of a conjugating group. An essential requirement of the method is that it should incorporate rapid and efficient sample preparation techniques, since one of the objectives of this methodology is an ability to screen rapidly biological samples for the presence of particular substructural features.

While the therapeutic dose ranges for different types of drugs vary enormously, doses in the range of 0.1–10 mg/kg are common for a large number of conventional drug types. For those cases where elimination of drug-related material occurs principally via the kidney, the resulting urine sample will contain major metabolites in the range of 10^{-6} to 10^{-9} (ppm to ppb). Similar considerations apply to bile samples for those drugs which are extracted into bile by the liver. Typically, then, one can expect to encounter urine or bile samples which contain drugs and their metabolites in the μg/ml range. Plasma samples may contain the drug or metabolites at substantially lower concentrations (ng/ml).

While direct analysis of complex mixtures at these concentrations is sometimes possible (*vide infra*), in most cases a pre-concentration or extraction step facilitates subsequent mass spectrometric analysis. The most generally useful technique for pre-concentrating urine or plasma samples involves the use of small volume, pre-packed solid phase extraction cartridges. For lipophilic drugs and their metabolities, an octadecyl (C_{18}(silica-based packaging has been found to be of general utility, while anion or cation exchange silica-based packings can be employed for drugs which contain a polar moiety. Sample 'cleanup' with this methodlogy is thus limited to a simple extraction step, the main function of which is to assist in the removal of proteinaceous material and to facilitate subsequent concentration of the extract. These extraction procedures work well on limited volumes (less than 1 ml of urine or plasma) of sample, and the resulting sample extract is in a form that is suitable for direct analysis by FAB MS/MS.

8.3. METHODOLOGY

The instrument that was used for the studies described in this article consisted of a Finnigan/MAT Triple Stage Quadrupole (TSQ-15) mass spectrometer, equipped with a 4500 series ion source. An Ion Tech 11NF saddle field (FAB) source was used for the primary ionizing beam, using xenon (Xe^0) at an energy of 8 keV. The atom gun is mounted on a 4.5 in. Conflat-type flange, at right angles to the ion source block. A copper stage is used for introduction of the sample, using either glycerol or thioglycerol as the liquid matrix. Collision spectra are generated within the second quadrupole analyser (Q2, operated in RF-only mode), using argon as a target gas at pressures of between 0.85 and 1.5 mTorr (0.11–0.20 Pa). The collision energy used for FAB MS/MS analysis is typically between ± 15 and ± 30 eV; the collision energy and target gas pressure used for a particular series of experiments vary, depending on the kind of information desired as well as the type of compound that is under study. In general, both positive and negative ion mass spectra are recorded by switching the ion source and analyzer potentials on alternate scans. The results obtained from such a scan sequence are complementary, and provide valuable information about the type and polarity of functional groups present in individual components of the sample.

An important feature of the triple quadrupole device is the ability to switch rapidly between the various types of scans, and to construct lists of particular sequences of scans using the instrument data system. This is important for the analysis of samples where the ion current changes rapidly with time. For example, many types of drug conjugates may give an ion current which drops in intensity by a factor of ten or more over the course of 0.1–1 min. A similar situation is encountered when volatile liquid matrices such as thioglycerol are employed. The ability of the quadrupole device to scan rapidly, and to carry out a lengthy series of daughter, precursor, and neutral loss scans of alternating polarity is a key factor in obtaining structural data on individual components of complex mixtures. In principle, however, the types of experiments described here could be carried out on other types of MS/MS instruments. A limitation of the QQQ type of instrument is the rather restricted range of collision energies which are available ([E_{lab}] *ca.* 20–100 eV on commercially available systems, corresponding to a centre-of-mass energy [E_{cm}] range of 1.8–9.0 eV for m/z 400 using argon as the target gas). Collision energies of 25–30 eV (E_{lab}) have been found to give structurally significant fragmentations for a wide variety of compounds, however.

A typical experiment, then, consists of giving an oral dose of the drug under study to a test species (usually a rat or dog), collecting urine over a 0–8 h time period, and then obtaining an extract of the urine that contains the metabolites of interest. For relatively lipophilic drugs, the urine is treated by applying an aliquot to a C_{18} solid phase extraction cartridge, and eluting with water to remove salts and macromolecules. The cartridge is then washed with successive aliquots of water containing some percentage of an organic solvent such as methanol or acetonitrile; these aliquots are then either analysed directly or are further concentrated and then analyzed.

These general considerations are best illustrated by referring to a number of specific examples.

8.4. ALKYLAMINE DRUGS

Drugs which contain an alkyl-substituted amine functionality typically undergo metabolism by a combination of reactions. Oxidation can occur at the nitrogen to yield an *N*-oxide, or at a carbon alpha to the nitrogen to yield a carbinol-amine. The carbinolamine can then undergo hydrolysis to yield a desalkyl metabolite. These two pathways are often the major primary metabolic reactions responsible for biotransformation of drugs which contain an amine functionality, so that a technique for assessing their relative contributions will help to define the overall metabolism. *N*-oxide metabolites are relatively polar substances, and readily undergo thermal degradation when analyzed by conventional ionization techniques. FAB MS, in contrast, gives an intense signal corresponding to the protonated molecule $(M + H)^+$, with little apparent fragmentation. Collisional activation can then be used to produce structurally significant fragments.

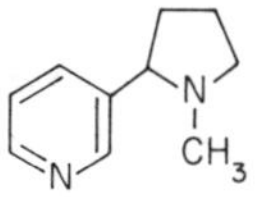

1

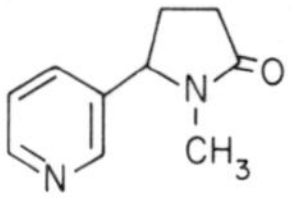

2

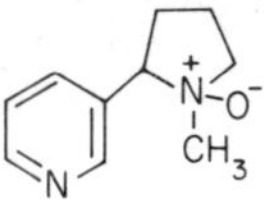

3

An example of this is shown in Figures 1–3. Nicotine (Structure **1**), undergoes oxidative metabolism to yield two major metabolites: the lactam cotinine (Structure **2**) and the 1′*N*-oxide (Structure **3**) (McKennis *et al.*, 1957; Booth and Boyland, 1970). FAB MS analysis of nicotine yields an intense signal for $(M + H)^+$ at m/z 163, with essentially no fragmentation. CAD analysis of this ion yields two prominent signals at m/z 80 and 84, corresponding to the pyridine and *N*-methylpyrrolidine moieties; additional fragments are also observed at m/z 106, 117, 130, and 134 (Figure 1a, 1b). These six ions, then, form the basis for a series of precursor ion scans to search for metabolites that are related structurally to the parent compound, nicotine.

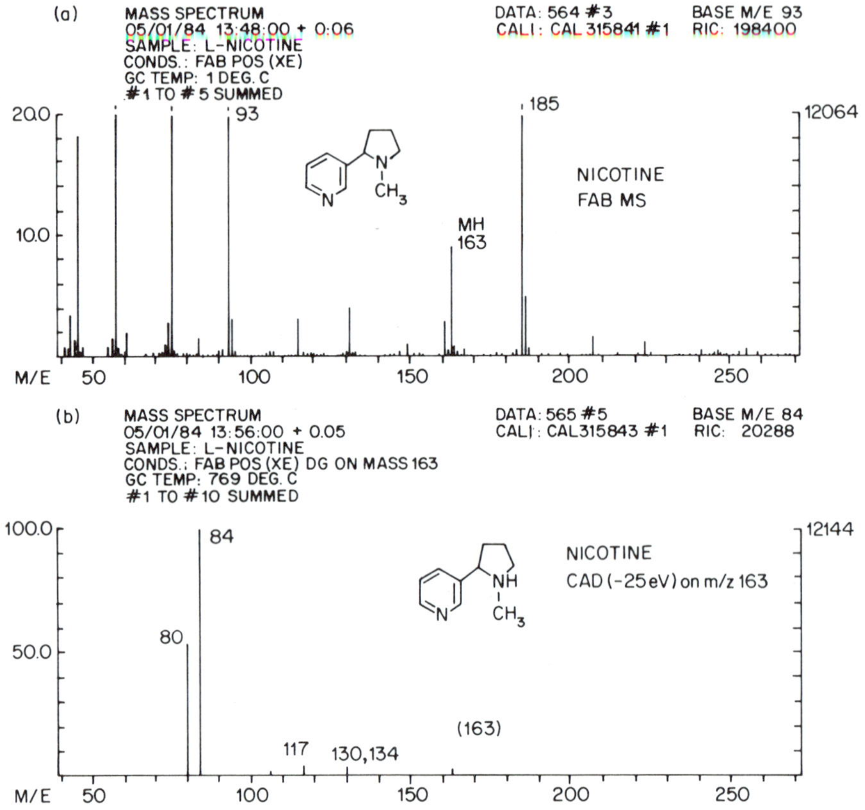

Figure 1: a) Positive ion FAB mass spectrum of (-)nicotine (glycerol matrix); b) CAD mass spectrum of *m/z* 163: -25eV, 0.85 mTorr argon.

When a methanolic SepPak™ extract of a 0–4 h urine sample obtained from a male Sprague Dawley rat that received an oral dose (2 mg[base]/kg) of nicotine is analyzed by FAB MS using a glycerol matrix, a relatively uninformative spectrum is obtained, with a continuum of signals over the mass range *m/z* 40–500 (Figure 2). The ion at *m/z* 115 is derived from a sodium-glycerol species $(M + Na)^+$. Despite this rather nondescript result, a search for precursor ions of the nicotine-specific fragments identified above yields *m/z* 177 and 179 as likely nicotine-related products. A daughter ion scan of *m/z* 179, obtained on the same crude SepPak™ extract that produced the mass spectrum shown in Figure 2, yields a set of relatively intense fragment ions at *m/z* 117, 130, and 134

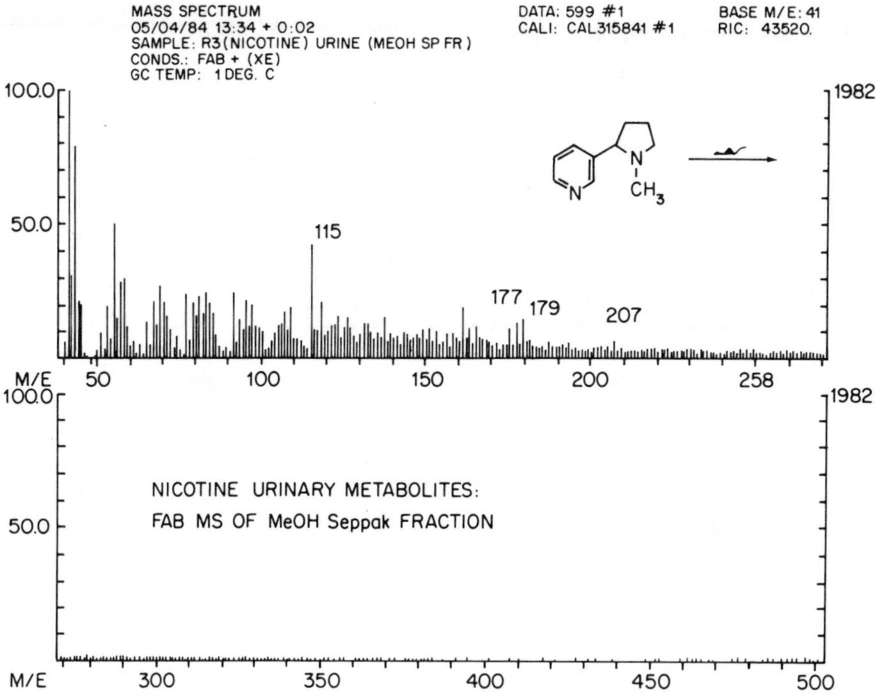

Figure 2: Positive ion FAB mass spectrum of methanolic SepPak™ extract of rat urine obtained after dosing with (-)nicotine (2 mg[base]/kg p.o.); glycerol matrix.

(Figure 3b); as shown in Figure 3a, this set of fragment ions matches the major fragments obtained from CAD analysis of $(M + H)^+$ of authentic nicotine-1′-*N*-oxide. Daughter ion analysis of m/z 177 yields a spectrum that is identical to that obtained with authentic cotinine. This example, then, shows that it is possible to carry out a 'target compound analysis' via FAB MS/MS for an *N*-oxide metabolite, directly on a relatively crude urine extract; what is required for positive identification is a reference CAD spectrum obtained on an authentic sample of the *N*-oxide.

The possibility of devising a specific set of scan sequences that would serve to confirm unambiguously the presence of an *N*-oxide metabolite in biological extracts was investigated by examining the FAB/CAD mass spectra of a number of representative *N*-oxides. For simple aliphatic or alicyclic *N*-oxides, the CAD spectra are dominated by formation of an imminium ion (loss of 18 u).

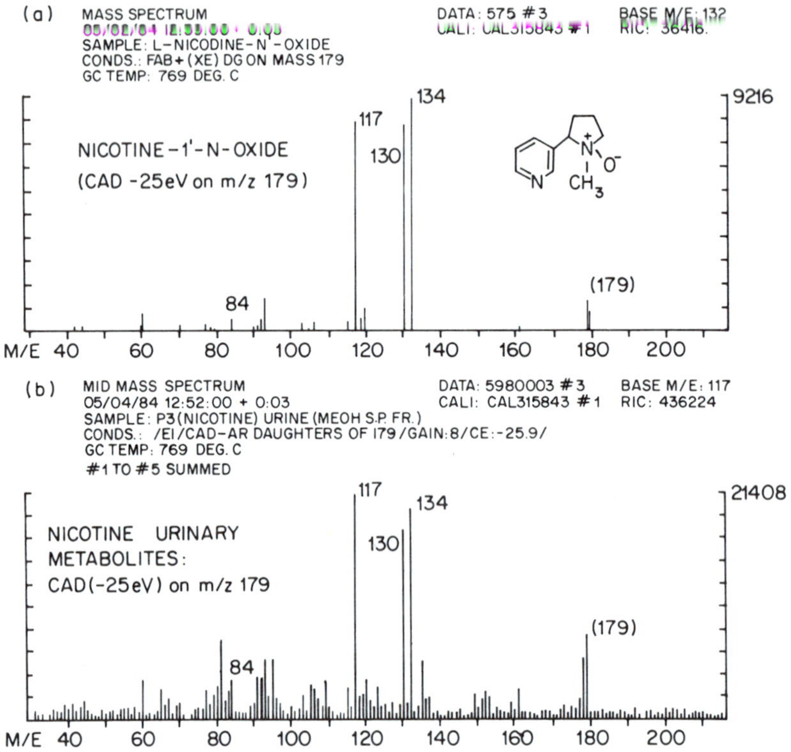

Figure 3: a) CAD mass spectrum of $(M + H)^+$ (m/z 179) of nicotine-1′-*N*-oxide: glycerol matrix, −25eV, 0.85 mTorr argon; b) CAD mass spectrum of m/z 179 of urine extract from Figure 2: −25eV, 0.85 mTorr argon.

The fragmentation of structurally more complex *N*-oxides, however, is frequently dominated by processes which involve other substructural features of the drug. In particular, it is not always possible to distinguish *N*-oxides from alcohols in instances where authentic reference spectra are not available. When such standards are available, identification of an *N*-oxide metabolite in a urine or plasma extract can often be made simply by examining the CAD spectrum obtained on a mass that is 16 u higher than $(M + H)^+$ of the parent drug.

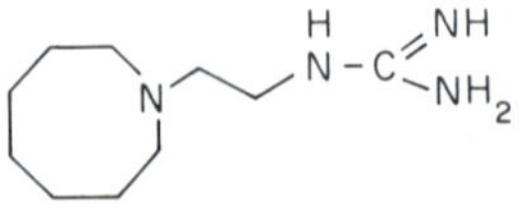

4

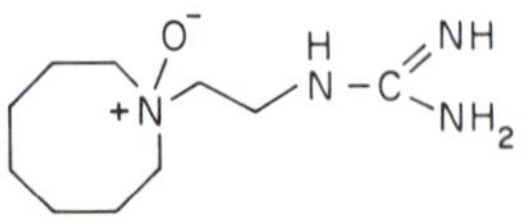

5

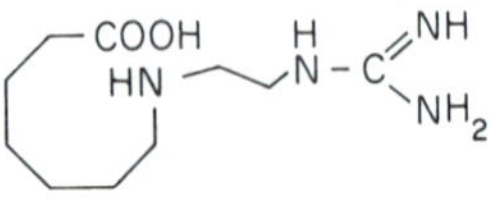

6

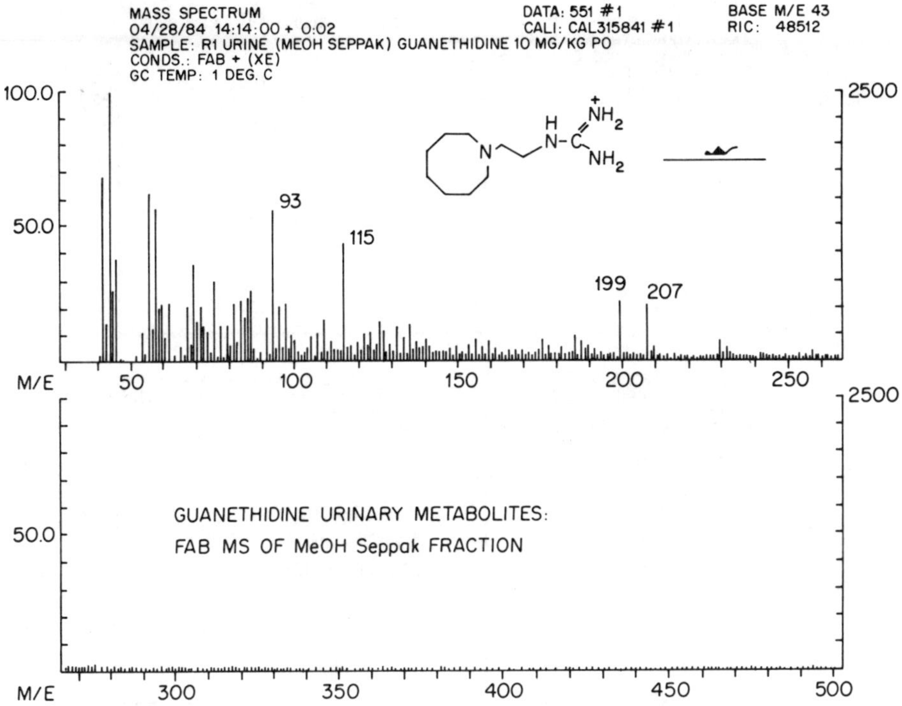

Figure 4: Positive ion FAB mass spectrum of methanolic SepPak™ extract of rat urine obtained after dosing with guanethidine (10 mg[base]/kg p.o.); glycerol matrix.

This approach is illustrated by the example shown in Figures 4–6. The drug guanethidine ([2-(octahydro-1-azocinyl)ethyl]guanidine; Structure **4**) is widely used as an antihypertensive, and has been shown to undergo rapid transformation *in vivo* to both the *N*-oxide (Structure **5**) and the alpha oxidation product, 2-(6-carboxyhexylamino)ethyl guanidine (Structure **6**) (McMartin *et al.*, 1970). A FAB mass spectrum of the methanolic SepPak™ fraction obtained from rat urine (0–4 h; 10 mg[base]/kg dose) is shown in Figure 4. Ions derived from the glycerol matrix occur at *m/z* 93, 115, 185, and 207. CAD analysis of the ion at *m/z* 199 confirms that unchanged guanethidine is present in the sample (Figure 5a, 5b). Despite the absence of any obvious signal at the mass expected for an

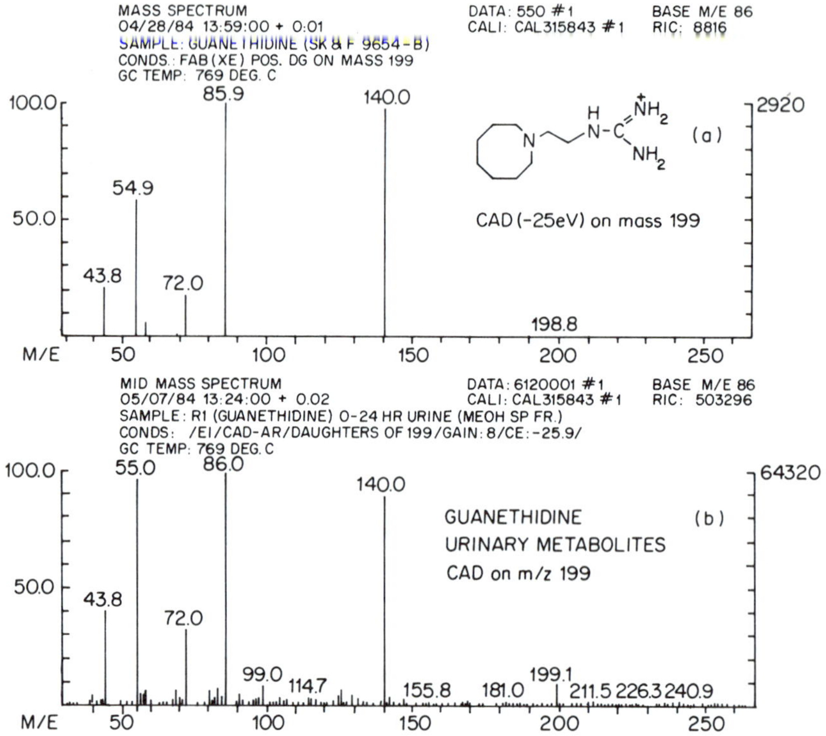

Figure 5: a) CAD mass spectrum of $(M + H)^+$ (m/z 199) of guanethidine: glycerol matrix, −25eV, 0.85 mTorr argon; b) CAD mass spectrum of m/z 199 of urine extract from Figure 4: glycerol matrix, −25eV, 0.85 mTorr argon.

N-oxide, CAD analysis of m/z 215 yields two prominent fragments at m/z 86 and 156 (Figure 6b). The fragment at m/z 86 is common to both the metabolite and the parent drug, and corresponds to cleavage of the exocyclic N–C bond, with charge retention by the guanidinium moiety. The fragment at m/z 140 observed in the CAD spectrum of guanethidine (Figure 5a) corresponds to cleavage of the N–C guanidinium bond; in the CAD mass spectrum of m/z 215 this fragment is shifted by 16 u to m/z 156, and confirms that oxidation has occurred on the azocinyl ring. A comparison of this CAD spectrum with that obtained on authentic guanethidine *N*-oxide confirms that this metabolite is present in the urine extract (Figure 6a, 6b).

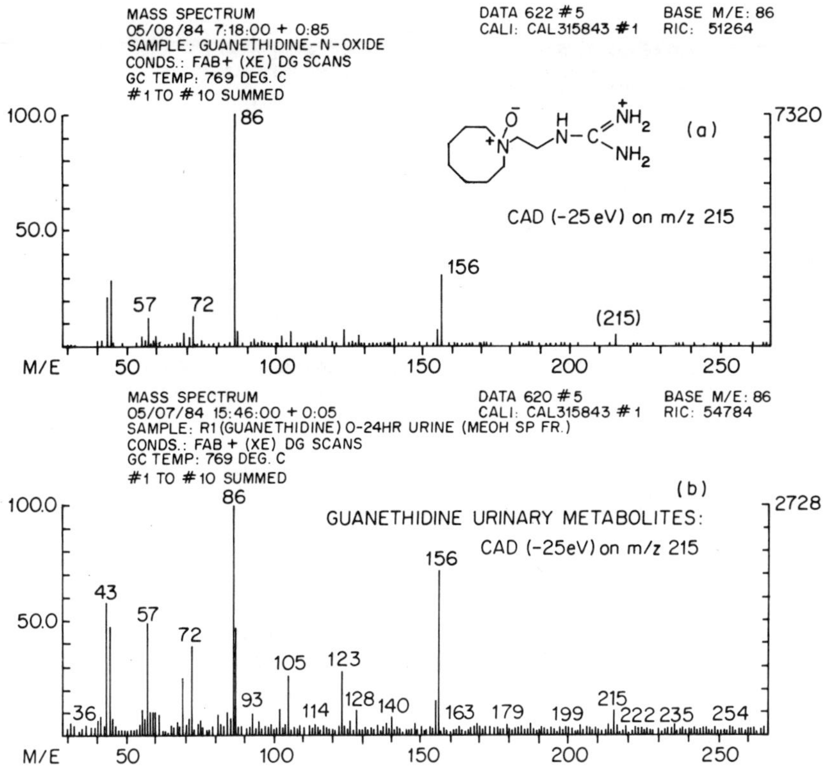

Figure 6: a) CAD mass spectrum of $(M + H)^+$ (m/z 215) of guanethidine-N-oxide: glycerol matrix, -25eV, 0.85 mTorr argon; b) CAD mass spectrum of m/z 215 of urine extract from Figure 4: glycerol matrix, -25eV, 0.85 mTorr argon.

8.5. MS/MS ANALYSIS OF CONJUGATES

FAB or liquid SIMS is particularly suited for the analysis of phase 2 metabolites such as sulphate esters, glucuronides, and peptide conjugates. Glucuronides, for example, yield intense signals in positive ion mode corresponding to $(M + H)^+$ or $(M + A_x - H_{x-1})^+$, where 'A' is typically sodium or potassium. CAD analysis of the glucuronide at collision energies of 15–30 eV (E_{lab}) yields a variety of fragment ions, depending on the types of substructural features that are present in the drug moiety. In virtually all cases that have been examined to date, however, a prominent signal corresponding to loss of the glucuronic acid moiety (176 u) is observed. Since the charge is retained on the drug fragment, a neutral loss scan for 176 u can be used to screen a sample extract for the presence of glucuronide conjugates. This fragmentation reaction is observed for

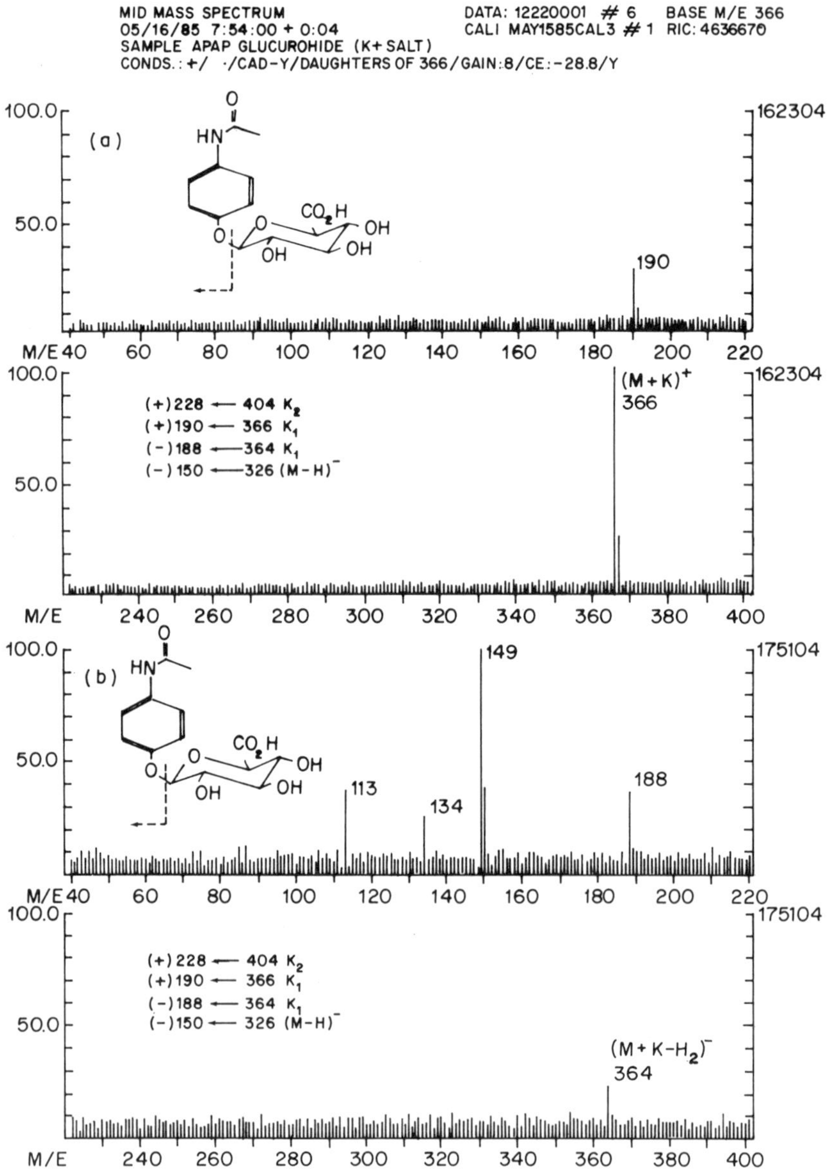

Figure 7: CAD mass spectra of acetaminophen-*O*-glucuronide (potassium salt): a) *m*/*z* 366, $(M + K)^+$, positive ion mode, −29eV, 0.85 mTorr argon; b) *m*/*z* 364, $(M + K - H_2)^-$, negative ion mode, +29eV, 0.85 mTorr argon [glycerol matrix].

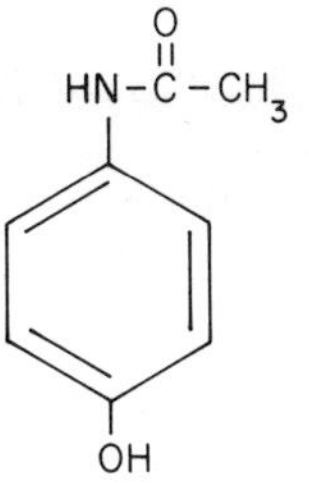

7

many types of glucuronides in both positive and negative ion modes. An example of such a result is shown in Figure 7, for the *O*-glucuronide of the drug acetaminophen (APAP, Structure **7**). Daughter ion scans on either $(M + K)^+$ or $(M + K - H_2)^-$ yield a number of fragment ions, including *m/z* 190(+) and 188(−) which correspond to loss of the glucuronic acid moiety. The negative ion neutral loss scan for 176 u is shown in Figure 8; *m/z* 150 and 188 correspond to the precursor ions at *m/z* 326 $(M - H)^-$ and 364 $(M + K - H_2)^-$.

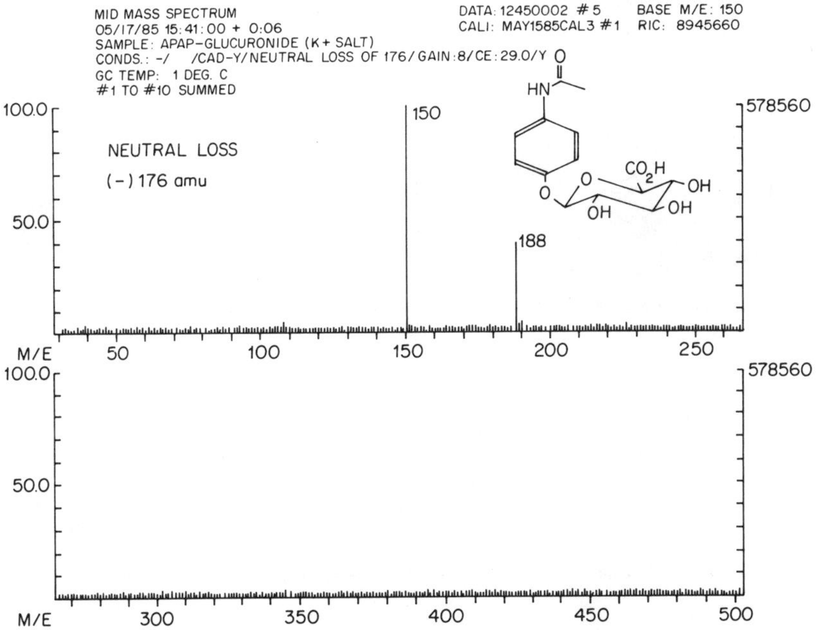

Figure 8: Negative ion neutral loss scan (176 u) of acetaminophen-*O*-glucuronide (potassium salt): glycerol matrix, +29eV, 0.85 mTorr argon.

Sulphate esters undergo CAD by cleavage of either an S–O or C–O bond, depending on the stability of the resulting fragment ion. Aryl *O*-sulphates, for example, undergo cleavage of the S–O bond with retention of charge by the phenolic ion; a neutral loss scan of 80 u can therefore be used to identify this type of sulphate ester. In contrast to this, alicyclic sulphate esters undergo CAD by cleavage of a C–O bond to generate a prominent fragment ion at *m/z* 97 $(HSO_4)^-$ in negative ion mode. Since the charge is retained by the sulphate fragment, a neutral loss scan is not appropriate for identification of this type of conjugate. Instead, a precursor ion scan for *m/z* 97 can be used.

Certain types of peptide conjugates can be identified by a combination of neutral loss and precursor ion scans. Aryl glutathione adducts, for example, undergo cleavage of the cysteinyl C–S bond, with retention of charge by the aryl system. A neutral loss scan corresponding to the mass of the tripeptide (275 u) can therefore be used to search for this type of conjugate. The fragmentation behaviour observed in FAB MS/MS for the 3-glutathionyl conjugate of acetaminophen is shown schematically in Figure 9.

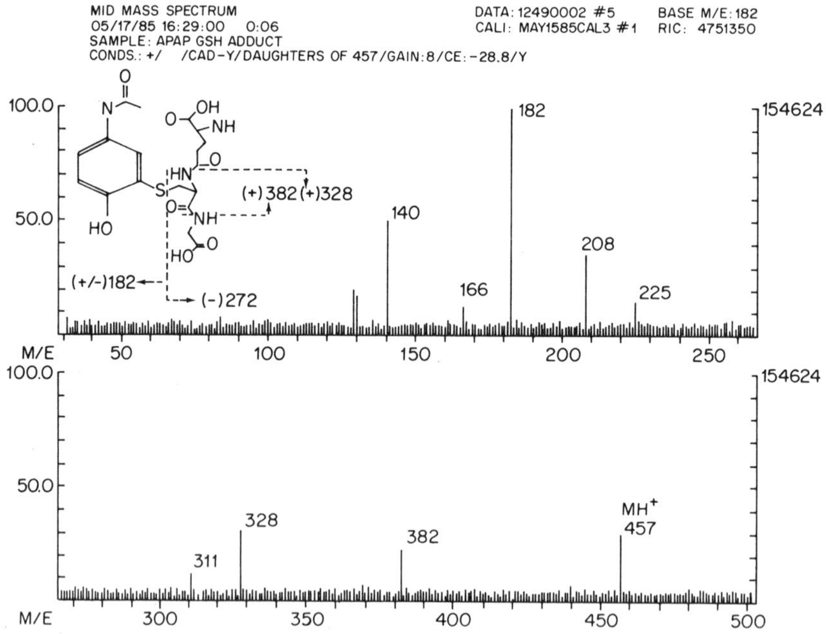

Figure 9: CAD mass spectrum of $(M + H)^+$ (*m/z* 457) of acetaminophen-glutathione conjugate: thioglycerol matrix, -29eV, 0.85 mTorr argon.

Analysis of a biological sample for these types of conjugates, then, involves carrying out a sequence of alternating positive and negative ion mode scans for the appropriate neutral loss (176 u for glucuronides, 80 u for aryl sulphates) or precursor ion (*m*/*z* 97 for alicyclic sulphate esters) scans. In certain favourable situations, this type of experiment can be carried out directly on a biological sample. Figure 10a shows the results of FAB MS/MS analysis of a 0.5 µl rat urine sample obtained from an animal that received a large (750 mg/kg) dose of acetaminophen. A positive ion neutral loss scan for 80 u yielded an intense signal for *m*/*z* 152, indicating the presence of an aryl sulphate with a precursor mass of [152 + 80] = 232. CAD analysis of this precursor ion gave the spectrum shown in Figure 10a, and the identity of the component responsible for this result was confirmed by comparison with a reference CAD spectrum of acetaminophen 4-sulphate (Figure 10b).

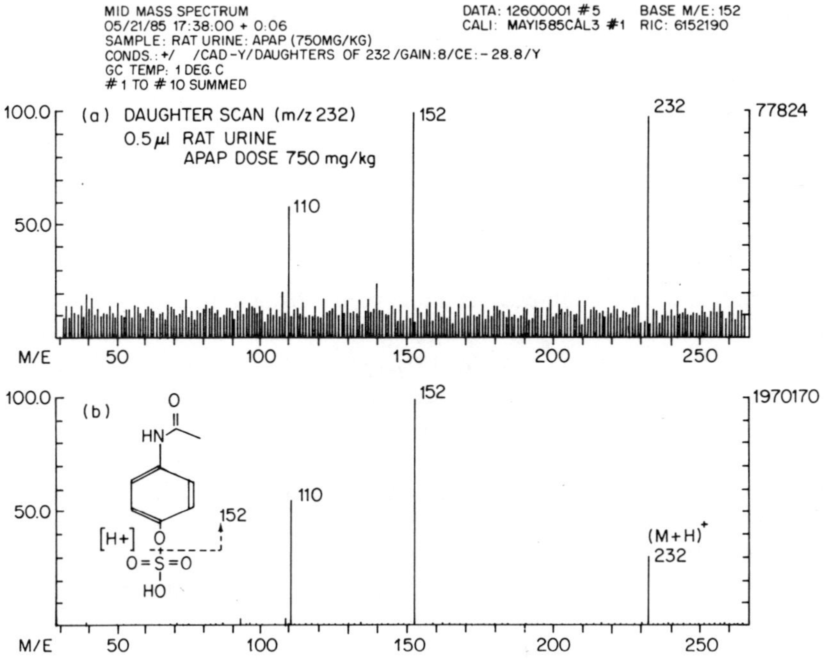

Figure 10: a) Positive ion CAD mass spectrum of *m*/*z* 232 of rat urine sample, obtained after dosing with acetaminophen (750 mg/kg p.o.): glycerol matrix, −29eV, 0.85 mTorr argon; b) CAD mass spectrum of $(M + H)^+$ (*m*/*z* 232) of acetaminophen-4-sulphate: glycerol matrix, −29eV, 0.85 mTorr argon.

8.6. METABOLIC MAPPING

An example of the use of this methodology to assess the overall metabolism of a drug is outlined in Figures 11 and 12. SK & F 86466 (6-chloro-3-methyl-2,3,4,5-tetrahydro-1H,3-benzazepine, Structure **8**), a selective alpha-adrenergic receptor antagonist, is currently under investigation as a

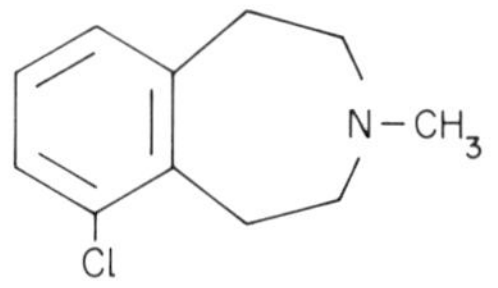

8

cardiovascular agent (De Marinis *et al.*, 1983). Urine obtained after oral administration of this agent to a male Sprague-Dawley rat (2mg[base]/kg) was extracted with the solid phase cartridge (C_{18}) methodology outlined above, and the resulting methanol fraction was analyzed directly by FAB MS/MS. The spectrum obtained on this fraction is shown in Figure 11; the ions at m/z 182, 196, and 212 were not observed in control urine. Neutral loss and precursor ion scans that are diagnostic for the usual conjugating groups (sulphate, glucuronic acid) did not reveal the presence of any components at masses greater than m/z 212, and a comparison of the collision spectrum of m/z 196 with that of SK &F 86466 demonstrated the presence of intact parent drug in the urine extract. The ion at m/z 182 was observed to have a number of CAD fragments in common with the parent structure, and the 14 u mass difference suggested that this component was the desmethyl analogue. This was confirmed by comparison with the synthetic *N*-desmethyl compound, as well as by GC–MS.

The ion at m/z 212 was found to undergo a neutral loss of 18 u, suggesting the presence of either a hydroxylated species or an *N*-oxide. The fragmentation pattern obtained by CAD on m/z 212 was found to be distinct from that of a number of phenols of similar structure, suggesting that this ion was not derived from a hydroxylated metabolite. This result strongly indicated the presence of an *N*-oxide, and this was confirmed by comparison with the CAD spectrum obtained on synthetic material (Figure 12a, 12b). Interestingly, this *N*-oxide was subsequently shown to be extremely thermally labile; attempts at obtaining conventional EI, CI or desorption CI spectra of this material were unsuccessful (Straub and Levandoski, 1984).

This analysis suggested that these three components (parent compound, *N*-desmethyl and *N*-oxide metabolites) were the only major drug-related

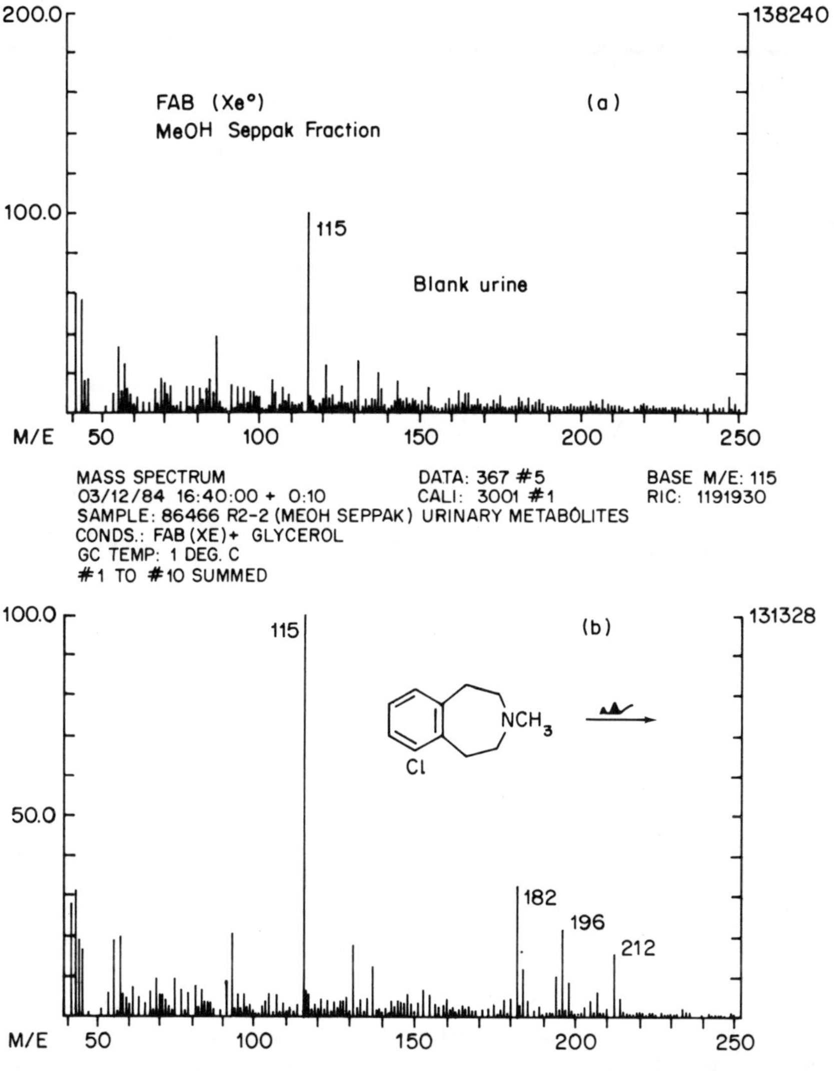

Figure 11: Positive ion FAB mass spectra of: a) methanolic SepPak™ extract of control rat urine; b) methanolic SepPak™ extract of rat urine obtained after dosing with SK & F 86466 (2 mg[base]/kg p.o.); glycerol matrix.

products present in the urine extract. These results were subsequently confirmed by studies utilizing ^{14}C-labelled SK & F 86466 along with conventional HPLC and GC–MS methodologies; in the rat these three components account for the bulk of drug-related material present in urine.

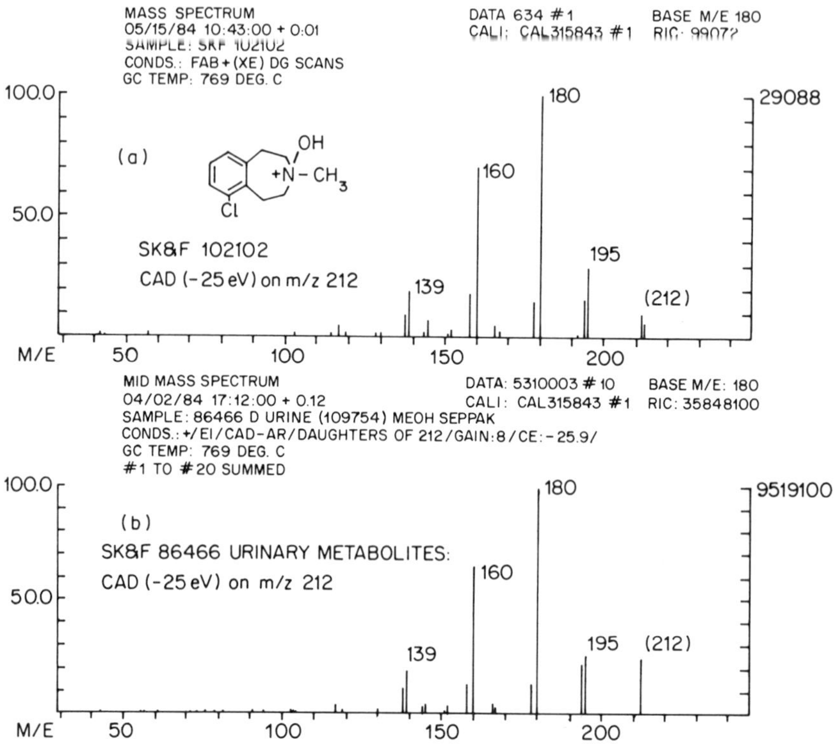

Figure 12: a) CAD mass spectrum of $(M + H)^+$ (m/z 212) of *N*-oxide of SK & F 86466: glycerol matrix, −29eV, 0.85 mTorr argon; b) CAD mass spectrum of m/z 212 of urine extract from Figure 11(b): glycerol matrix, −29eV, 0.85 mTorr argon.

8.7. CONCLUSIONS

MS/MS techniques, in conjunction with particle-induced desorption ionization methods such as FAB or liquid SIMS, are a powerful analytical tool for the analysis of complex biological samples. The application of these methodologies to problems in xenobiotic metabolism is particularly appropriate, since in most instances the structure of the parent compound is known. A knowledge of the fragmentation behaviour of the parent compound can therefore be exploited to design particular types of scan sequences that are diagnostic for specific substructural features of the drug or its metabolites.

The examples discussed in this article are for the most part semiquantitative. Quantitative data can be obtained using desorption ionization techniques (Desiderio *et al.*, 1983; Clay *et al.*, 1984; Millington *et al.*, 1984; Straub and

Levandoski, 1985), but the choice of internal standards is problematic. In general, stable isotope analogues are the best type of internal standard for use in this type of analysis. Since the MS/MS analysis as outlined in the above examples is most often applied in situations where the structures of the metabolites are not known, suitable internal standards will not in general be available. In certain favourable situations, however, FAB MS/MS as a quantitative method offers some intriguing possibilities. In the 'metabolic mapping' experiment described above, for example, the addition of isotopically labelled analogues of the two metabolites and the parent drug as internal standards and the use of the selected reaction monitoring mode of operation would allow simultaneous quantification of all three species, despite the appreciable differences in polarity and thermal lability.

ACKNOWLEDGMENTS

It is a pleasure to acknowledge the contributions of my colleagues in the Department of Drug Metabolism at SK & F Laboratories, especially R. K. Lynn, C. Garvie, P. Levandoski, and P. Rudewicz.

REFERENCES

Aberth, W., Straub, K. M., and Burlingame, A. L. (1982). Secondary ion mass spectrometry with cesium ion primary beam and liquid target matrix for analysis of bioorganic compounds, *Anal. Chem.*, **54**, 2029–2034.

Barber, M., Bordoli, R. S., Sedgwick, R. D., and Tyler, A. N. (1981). Fast atom bombardment of solids as an ion source in mass spectrometry, *Nature*, **293**, 270–275.

Booth, J., and Boyland, E. (1970). The metabolism of nicotine into two optically active stereoisomers of nicotine-1′-oxide by animal tissues *in vitro* and by cigarette smokers, *Biochem. Pharmacol.*, **19**, 733–742.

Clay, K. L., Stene, D. O., and Murphy, R. C. (1984). Quantitative analysis of platelet activating factor (AGEPC) by fast atom bombardment mass spectrometry, *Biomed. Mass Spectrom.*, **11**, 47–49.

De Marinis, R. M., Hieble, J. P., and Matthews, W. D. (1983). 6-Chloro-2,3,4,5-tetrahydro-3-methyl-1H-3-benzazepine: a potent and selective antagonist of α2-adrenoceptors, *J. Med. Chem.*, **26**, 1213–1214.

Desiderio, D. M., Katakuse, I., and Kai, M. (1983). Measurement of leukine enkephalin in caudate nucleus tissue with fast atom bombardment–collision activated dissociation-linked field scanning mass spectrometry, *Biomed. Mass Spectrom.*, **10**, 426–429.

Jardine, I., Scanlan, G. F., Mattox, V. R., and Kumar, R. (1984). Analysis of steroid and vitamin D glucuronides and sulfates by fast atom bombardment mass spectrometry, *Biomed. Mass Spectrom.*, **11**, 4–9.

Mayer, S. E., Melmon, K. L., and Gilman, A. G. (1980). Introduction: the dynamics of drug absorption, distribution, and elimination, in *The Pharmacological Basis of Therapeutics*, (Eds. A. G. Gilman, L. S. Goodman, and A. Gilman), pp. 12–20, Macmillan, New York.

McKennis, H. Jr., Turnbull, L. B. and Bowman, E. R. (1957). γ-(3-Pyridyl-γ-methyl-aminobutyric acid as a urinary metabolite of nicotine, *J. Am. Chem. Soc.*, **79**, 6342–6343.

McLafferty, F. W. (1980). Tandem mass spectrometry (MS/MS): a promising new analytical technique for specific component determination in complex mixtures, *Acc. Chem. Res.*, **13**, 33–38.

McLafferty, F. W. (Ed.) (1983). *Tandem Mass Spectrometry*, Wiley, New York.

McMartin, C., Rondel, R. K., Vinter, J., Allan, B. R., Humberstone, P. M., Leishman, W. D., Dandler, G., and Thirkettle, M. B. (1970). The fate of guanethidine in two hypertensive patients, *Clin. Pharmacol. Therap.*, **11**, 423–431.

Millington, D. S., Roe, C. R., and Maltby, D. A. (1984). Application of high resolution fast atom bombardment and constant B/E ratio linked scanning to the identification and analysis of acylcarnitines in metabolic disease, *Biomed. Mass Spectrom.*, **11** 236–241.

Perchalski, R. J., Yost, R. A., and Wilder, B. J. (1982). Structural elucidation of drug metabolites by triple quadrupole mass spectrometry, *Anal. Chem.* **54**, 1466–1471.

Shackleton, C. H. L., and Straub, K. M. (1982). Direct analysis of steroid conjugates: the use of secondary ion mass spectrometry, *Steroids*, **40**, 35–52.

Straub, K. M., and Levandoski, P. (1984). FAB–MS/MS: direct analysis of N-oxide metabolites in biological samples, *Proc. 32nd Ann. Conf. on Mass Spectrometry and Allied Topics, San Antonio*; American Society for Mass Spectrometry, pp. 785–786.

Straub, K. M., and Levandoski, P. L., (1985). Quantitative analysis of an N-oxide metabolite by fast atom bombardment tandem mass spectrometry, *Biomed. Mass Spectrom.*, **12**, 338–343.

Yost, R. A., and Fetterolf, D. D. (1983). Tandem mass spectrometry (MS/MS) instrumentation, *Mass Spectrom. Rev.*, **2**, 1–45.

PART II

ANALYSES AT HIGH MASS

Mass Spectrometry in Biomedical Research
Edited by S. J. Gaskell
© 1986 John Wiley & Sons Ltd.

Chapter 9

OVERVIEW TO PART II

Simon J. Gaskell

9.1 INTRODUCTION

Until recent years, mass spectrometry was a technique which was best suited to the analysis of compounds of, in biochemical terms, low molecular weight. Applications to analyses of compounds with molecular masses in excess of 1000 daltons were rare (though reference should be made to the elegant work of Karlsson and co-workers (summarized by Karlsson, 1977) which included analyses of glycolipids with $M_r > 2500$). Prior to the advent of the desorption ionization techniques (Chapter 1), the mass range of the mass spectrometer was rarely the limiting factor. The development of field desorption (FD) and, more recently and significantly, of fast atom bombardment (FAB), however, provided the capability for generation of ions from molecules of several thousand molecular mass. Major developments have followed in extending the mass range of double focusing sector instruments and in exploring the advantages and limitations of alternative mass analyzers, notably time-of-flight (TOF) and Fourier transform–ion cyclotron resonance (FT–ICR) instruments. From the discussion which follows, and from the contributions to this Part, it will be apparent that, following developments in mass analyzers, the factors limiting the useful mass range are now the efficiency of ion production and the sensitivity of ion detection.

For the purposes of this Part, 'high mass' techniques are those of demonstrable utility above 1000 daltons and which also have been, or promise to be, of value above 5000 daltons. Thus, quadrupole mass filters are not included because their utility in biomedical applications, at present and in the near future, does not extend beyond 3000–4000 daltons.

A discussion of tandem MS is also included in this Overview chapter. The importance of tandem MS goes well beyond high mass analyses but the potential of the technique in structure elucidation of large molecules indicates that the inclusion of the discussion here is appropriate.

9.2 MASS ANALYZERS

Table 1 indicates salient features of three types of mass analyzer in current use for high mass analyses. Such a compilation is obviously open to criticism for omissions and oversimplifications but should serve as a general guide to present capabilities. True comparisons of the different analyzers, with respect to the relevant parameters, are difficult to make without qualification. Nevertheless, a number of comments are worthwhile.

(i) Mass range.

The mass range accessible with double-focusing sector instruments has been dramatically improved in recent years (Chapter 13). It may be noted that much higher masses than the 15,000 quoted in Table 1 are accessible at reduced accelerating voltage (and hence reduced sensitivity) but the limitation then becomes the efficiency of ionization and ion detection. Impressive high mass results (at low resolution, see below) have been reported for TOF instruments (Table 1) and higher values should be possible (again, if ionization and detection permit). The FT–ICR instrument should be well suited to high mass analyses; its capabilities and potential are discussed in Chapter 17.

(ii) Resolution.

In comparisons of maximum achievable resolution, the double focusing sector and FT–ICR instruments appear to be clearly superior but, in practice, the advantage is less substantial. The generally weak ion currents obtained with high mass molecules usually dictate that sector instruments be operated at relatively high transmission (i.e. low resolution). The impressive resolution figures (albeit exaggerated by the general use of the 50% valley definition) achieved with FT–ICR are possible only with very low pressures in the analyzer cell. Recent designs which separate the ionization region from the analyzer cell represent developments crucial to the productive use of FT–ICR in biomedical applications. Two approaches have been taken, namely, a tandem quadrupole–ICR instrument (Hunt *et al.*, 1985a,b) and a dual ICR cell design (Ghaderi and Littlejohn, 1985). Practical limitations on resolution are also observed for the TOF instrument in that the use of longer flight tubes (and therefore flight times) to improve resolution also increases the proportion of ions undergoing

TABLE 1. Mass spectrometry instrumentation for high mass analysis[a].

Instrument	Mass range[b]	Maximum resolution[c]	Principal ionization techniques	Coupled chromatography/MS	MS/MS	See chapters	Other references
Double-focusing (B,E)	15,000[d]	100,000[e]	FAB/SIMS EI, CI FD	GC–MS LC–MS	Yes	10–14 16	Dell *et al.*, 1980 Matsuda, 1983 Dell and Taylor, 1984 Facchetti, 1985
TOF	> 50,000	3000[f]	SIMS PD	—	—	15,16	Torgerson *et al.*, 1974 Benninghoven, 1983 Sundqvist *et al.*, 1984
FT–ICR	> 15,000	> 500,000[f]	SIMS LD EI	GC–MS	Yes	17	Wilkins and Gross, 1981 Johlman *et al.*, 1983 Gross and Rempel, 1984

[a] Abbreviations. B, magnetic sector; E, electric sector; TOF, time-of-flight; FT–ICR, Fourier transform ion cyclotron resonance; EI, electron impact; CI, chemical ionization; FD, field desorption; FAB, fast atom bombardment; SIMS, secondary ion mass spectrometry; PD, plasma desorption; LD, laser desorption; GC–MS, gas chromatography–mass spectrometry; LC–MS, liquid chromatography–mass spectrometry.

[b] Presently demonstrated capabilities; approximate values only.

[c] Approximate values only; see the important qualifying comments in the text.

[d] At full accelerating voltage (i.e. maximum sensitivity).

[e] Conventionally quoted with the 10% valley definition.

[f] Conventionally quoted with the 50% valley definition.

metastable decomposition, with consequent broadening of peaks in the spectrum (Chapter 15, Section 15.1). Alternative approaches to improved resolution in TOF MS are discussed in Chapter 16, Section 16.3.

It may be noted that, for molecules of a few thousand daltons, it is unrealistic to expect to achieve resolution sufficient to obtain an accuracy of mass measurement which would permit the determination of elemental composition. Unit resolution defines the isotope pattern for the molecular ion region but centroid determination of an unresolved isotope cluster is sufficient (with the assumption of analyte purity and the absence of hydrogen losses from, for example, $(M + H)^+$ species) for calculation of the 'chemical' or 'average isotopic' molecular mass. These considerations are discussed further in Chapter 13 (Section 13.3).

(iii) IONIZATION TECHNIQUES AND SAMPLE INTRODUCTION.

Sector instruments provide the broadest range of available ionization techniques and means of sample introduction (Table 1). Both TOF and FT–ICR analyzers are particularly suited to pulsatile ionization methods such as laser desorption and ^{252}Cf-plasma desorption. Indeed the major contributions of TOF MS to the analysis of high mass biomolecules have arisen from the combination with ^{252}Cf-plasma desorption (Chapter 15).

Finally, in comparison of instrument configurations, note should be made of the basic differences in the means of recording spectra and of detecting ions. Both the TOF and FT–ICR instruments permit the detection of all ions produced, with consequent theoretical advantages of sensitivity. The detection system in FT–ICR results from image currents generated in the receiver plates of the analyzer cell; it is therefore fundamentally different from the detection system generally used in sector and TOF instruments, which is based on an electron multiplier.

The evaluation of instrument types for the analysis of high mass biomolecules is best made on the basis of results achieved so far. The contributions to Part II attest to the dramatic progress that has been made in recent years.

9.3. TANDEM MASS SPECTROMETRY

The fundamental objectives of the techniques which are now known as 'tandem mass spectrometry' or 'MS/MS' are (i) to abstract data otherwise obscured by interferences, and (ii) to generate new structural information following the inducement of fragmentation. The essential feature is the ability to define a single fragmentation, or a series of fragmentations, in terms of both parent and daughter ion mass/charge ratios. It may be noted that such possibilities existed (albeit with limitations) before terms such as 'tandem MS' and 'MS/MS' became current. Thus, for example, an accelerating voltage scan (following decoupling of the accelerating and electric sector voltages) produces a spectrum of all parent ions which decompose in the region between the ion source and the electric sector of a conventional (EB) double focusing instrument, yielding a specific common daughter ion (Barber and Elliot, 1964). The facility for studying fragmentations in this region was substantially improved by the introduction of linked scan techniques. The most common types are summarized in Table 2; useful surveys of linked scanning have been given by Boyd and Beynon (1977) and by Jennings (1978, 1983). The review by Jennings (1983) also addresses the problem of 'artifact' peaks which may be observed (generally at low intensity) in linked scan spectra and do not correspond to the collection of daughter ions formed in the field-free region which follows the ion source.

TABLE 2. Linked scans of double-focusing mass spectrometers yielding information on fragmentations occurring in the first field-free region[a].

SCAN	CONSTANT	INFORMATION	REFERENCES
V	B,E	Parents of chosen daughter	Barber and Elliot, 1964; Futrell *et al.*, 1965; Jennings, 1965
E,V	B,E^2/V	Daughters of chosen parent	Weston *et al.*, 1976
B.E	V,B/E	Daughters of chosen parent	{ Boyd and Beynon, 1977; Millington and Smith, 1977
B,E	V,B^2/E	Parents of chosen daughter	Haddon, 1979 }
B,E	V,B^2(1-E)/E^2	Constant neutral loss	Haddon, 1980

[a] Not a list of all possibilities: those listed have found application in biomedical analyses. V, accelerating voltage; E, electric sector voltage; B, magnetic field.

The common limitation to linked scanning methods for studying fragmentations in the first field-free region is the poor effective resolution (and hence imprecise definition) of the parent ion. Reversal of the configuration of a double focusing instrument (to give BE geometry) and study of fragmentations in the field-free region between the sectors reverses the problem with respect to mass resolution. Thus, resolution of the parent ion beam (achieved by the magnetic sector) is good but mass resolution of the daughter ions, during an electric sector scan, is now poor. (This mode of operation, in common with some linked scans, does provide information concerning energy release during fragmentation but such data have received little attention in biomedical applications.) Linked scanning, for studying fragmentations in the first field-free region, may also be performed on a reverse-geometry instrument but the limitations of parent ion resolution do of course apply, as with the EB configuration.

TABLE 3. Configurations of sector and quadrupole instruments for the study of parent/daughter ion relationships.

		Principal mode of operation				
Configuration[a]	Fragmentation regions accessible to study	Fragmentation region	Parent ion resolution[b]	Daughter ion resolution[b]	Collision energy	References
EB	↓ EB	EB	200	2000	keV	(see Table 1)
BE	↓ ↓ B–E	↓ B–E	2000	200	keV	Beynon *et al.*, 1973; Maurer *et al.*, 1971 Kondrat and Cooks, 1978
QQQ	↓ QQQ	↓ QQQ	Unit	Unit	eV	Yost and Enke, 1978, 1979 (and references there cited)
EBQQ	↓ ↓ EBQQ	↓ EBQQ	< 20,000	Unit	eV	Bateman *et al.*, 1982 Gaskell *et al.*, 1985
BEQQ	↓ ↓ ↓↓ B–E–QQ	↓ BEQQ	< 20,000	Unit	eV	Glish *et al.*, 1982
		↓ BE–QQ	< 20,000	Unit	keV	Ciupek *et al.*, 1984 Louris *et al.*, 1985
EBE	↓ ↓ EB–E	↓ EB–E	< 100,000	200	keV	Gross and Russell, 1983
BEB	↓ ↓ ↓ B–E–B	↓ BE–B	< 100,000	2000	keV	
BEEB	↓ ↓ ↓ B–E–EB	↓ BE–EB	< 100,000	< 10,000	eV–keV	McLafferty *et al.*, 1980 Hass *et al.*, 1984

[a] E, electric sector; B, magnetic sector; Q, quadrupole.
[b] Approximate resolution figures only.
[c] Laboratory frame-of-reference.

Thus the simultaneous achievement of at least unit resolution of both parent and daughter ions requires a multi-analyzer instrument (but note the comments below concerning Fourier transform–ion cyclotron resonance (FT–ICR) and quadrupole ion trap instruments). Table 3 summarizes established instrument configurations for the study of parent/daughter ion relationships using sector and quadrupole instruments. Others have been reported and many others are possible but those listed have already seen significant biomedical application.

Though the reactions studied may be spontaneous (that is, the decomposition of metastable ions), inducement of fragmentation, generally by collision with a neutral gas, is more frequently employed. 'Collisional activation (CA)', 'collisionally-activated decomposition (CAD)' and 'collision-induced dissociation (CID)' are equivalent terms in current usage and each is used in contributions to this book. The collision energy in the laboratory frame of reference (E_{lab}) may be in the eV to KeV range (Table 3) and is determined by the kinetic energy of the incident parent ion beam. The collision energy in the centre-of-mass frame of reference, E_{cm}, representing the maximum energy available for uptake from collision, is given by:

$$E_{cm} = E_{lab} \cdot \frac{m_c}{m_p + m_c}$$

where m_c is the mass of the collision gas atoms or molecules, and m_p is the mass of the parent ion. Thus, as an example, if a parent ion of m/z 1000 is accelerated through 8 kV and subjected to CAD, using helium as collision gas, the E_{cm} is approximately 15 eV. Detailed discussions of CAD processes are presented by Cooks (1978), Dawson and Douglas (1983), and Todd and McLafferty (1983).

The limitations of CAD for the production of diagnostically useful fragments from large biomolecules are discussed by Sheil and Derrick in Chapter 14. In particular, the proportion of the incident parent ion beam converted to detectable daughter ions following high energy collision in a sector instrument is frequently small (usually less than 5%) and this combines with other factors (modest yield of parent ions, distribution of daughter ion current between many different ion species) to dictate that, in general, the daughter ion analyzer must be operated at close to maximum transmission, regardless of its resolution capabilities.

Much higher daughter ion yields (frequently > 50%) may usually be achieved using an R_f-only quadrupole as the collision cell (Yost and Enke, 1978); the higher yield is attributable to the collimation effect of the quadrupole and the usual experimental conditions which favour multiple collisions. Collision energies, however, are low (< 200 eV (E_{lab}) for both hybrid sector/quadrupole and triple quadrupole instruments). Some compensation may be made by increasing the mass of the collision gas, with a proportional increase in E_{cm}.

Recent evidence suggests that, even for small molecules, this limitation may be important in biochemical analyses. Thus, for example, differentiation of isomeric acylcarnitine methyl esters (discussed by Millington in Chapter 7) may be achieved by tandem MS on a BEEB instrument by CAD, with E_{lab} in the keV range (Gaskell *et al.*, 1986). When E_{lab} is reduced to 200 eV, the diagnostically-important fragmentations are not observed.

Alternative means of inducing decomposition of selected parent ions include surface-induced decomposition (Cooks *et al.*, 1985), providing access to much higher E_{cm} values, and laser-induced photodissociation (Mukhtar *et al.*, 1981), with the potential for selectivity of induced decomposition. The procedures require further study before their utility in biochemical analyses can be assessed.

Important new developments may also be expected in the study of consecutive defined fragmentations using FT–ICR and quadrupole ion trap instruments, where there is temporal, rather than spatial, separation of the ionization and fragmentation processes (Cody *et al.*, 1982; McIver and Bowers, 1983; Kelley *et al.*, 1985).

The history, development and applications of tandem MS have been reviewed in the volume edited by McLafferty (1983). Biomedical applications of tandem MS fall into two categories, of structure elucidation and mixture analysis. Thus, CAD may be used to generate diagnostically useful daughters from selected parents, when the fragments are not apparent in the 'conventional' spectrum (which may result from a 'soft' ionization technique such as FAB). Furthermore, the unequivocal relationships established between parents and daughters facilitate the interpretation of fragmentations. In Chapter 14, Sheil and Derrick discuss CAD of peptides following ion production by FD or FAB. Reinhold (Chapter 11, Section 11.2.2.4) presents data obtained by tandem MS in structural characterization of oligosaccharides. While the value of the technique is apparent, both the Reinhold and Sheil/Derrick contributions indicate clearly that much remains to be understood concerning the mechanisms and limitations of CAD of large molecules.

The power of tandem MS in characterization of polar drug metabolites has been illustrated in Chapter 8 by Straub. Again, little evidence for structure may be apparent in the 'conventional' FAB spectra but here the importance of the tandem technique relates also to the facility for mixture analysis. The selectivity of detection by tandem MS is valuable in trace analysis and this aspect is specifically considered in contributions to Part III (Chapters 19, 25 and 26).

REFERENCES

Barber, M., and Elliot, R. M. (1964). In *Proc. 12th Ann. Conf. on Mass Spectrometry and Allied Topics, Montreal*; American Society for Mass Spectrometry.

Bateman, R. H., Green, B. N., and Smith, D. C. (1982). A new instrument for the analytical application of MS–MS, in *Proc. 30th Ann. Conf. on Mass Spectrometry and Allied Topics, Honolulu*; American Society for Mass Spectrometry, pp. 516–517.

Benninghoven, A., Ed. (1983). *Ion Formation from Organic Solids, Springer Series in Chemical Physics*, vol. 25, Springer, Berlin.

Beynon, J. H., Cooks, R. G., Amy, J. W., Baitinger, W. E., and Ridley, T. Y. (1973). Design and performance of a mass-analyzed ion kinetic energy (MIKE) spectrometer, *Anal. Chem.* **45**, 1023A–1031A.

Boyd, R. K., and Beynon, J. H. (1977). Scanning of sector mass spectrometers to observe the fragmentations of metastable ions, *Org. Mass Spectrom.*, **12**, 163–165.

Ciupek, J. D., O'Lear, J. R., Cooks, R. G., Dobberstein, P., and Schoen, A. E. (1984). The BEQQ tandem mass spectrometer, in *Proc. 32nd Ann. Conf. on Mass Spectrometry and Allied Topics, San Antonio*; American Society for Mass Spectrometry, pp. 378–379.

Cody, R. B., Burnier, R. C., and Freiser, B. S. (1982). Collision-induced dissociation with Fourier transform mass spectrometry, *Anal. Chem.*, **54**, 96–101.

Cooks, R. G., Ed. (1978). *Collision Spectrometry*, Plenum, New York.

Cooks, R. G., DeKrey, M. J., and Mabud, Md. A. (1985). Polyatomic ion/surface collision phenomena, in *Proc. 33rd Ann. Conf. on Mass Spectrometry and Allied Topics, San Diego*; American Society for Mass Spectrometry, pp. 161–162.

Dawson, P. H., and Douglas, D. J. (1983). Collisionally activated dissociation of low kinetic energy ions, in *Tandem Mass Spectrometry* (Ed. F. W. McLafferty), pp. 125–148, Wiley, New York.

Dell, A., and Taylor, G. W. (1984). High-field magnet mass spectrometry of biological molecules, *Mass Spectrom. Rev.*, **3**, 357–394.

Dell, A., McDowell, R. A., Morris, H. R., and Taylor, G. W. (1980). High mass chemical ionisation and field desorption studies of biopolymers, in *Soft Ionisation Biological Mass Spectrometry* (Ed. H. R. Morris), pp. 69–84, Heyden, London.

Facchetti, S., Ed. (1985). *Mass Spectrometry of Large Molecules*, Elsevier, Amsterdam.

Futrell, J. H., Lancaster, K., Ryan, R., and Sieck, L. W. (1965). Study of metastable transitions in mass spectra using a double-focusing instrument. *J. Chem. Phys.*, **43**, 1832–1833.

Gaskell, S. J., Porter, C. J., and Green, B. N. (1985). Trace analysis by gas chromatography–mass spectrometry (GC/MS) using a tandem double focusing/quadrupole instrument, *Biomed. Mass Spectrom.*, **12**, 139–141.

Gaskell, S. J., Guenat, C., Harvan, D. J., Millington D. S., and Maltby D. A. (1986). Comparative studies of tandem mass spectrometers of BE–EB and EB–QQ design in biochemical applications, in *Advances in Mass Spectrometry, 1985* (Ed. J. F. J. Todd), pp. 1447–1448, Wiley, Chichester.

Ghaderi, S., and Littlejohn, D. P. (1985). Fourier transform mass spectrometry using multiple cell configuration, in *Proc. 33rd Ann. Conf. on Mass Spectrometry and Allied Topics, San Diego*; American Society for Mass Spectrometry, pp. 727–728.

Glish, G. L., McLuckey, S. A., Ridley, T. Y., and Cooks, R. G. (1982). A new 'hybrid' sector/quadrupole mass spectrometer for mass spectrometry/mass spectrometry, *Int. J. Mass Spectrom. Ion Phys.*, **41**, 157–177.

Gross, M. L., and Rempel, D. L. (1984). Fourier transform mass spectrometry, *Science*, **226**, 261–268.

Gross, M. L., and Russell, D. H. (1983). Triple sector instruments for tandem mass spectrometry, in *Tandem Mass Spectrometry* (Ed. F. W. McLafferty), pp. 255–270, Wiley, New York.

Haddon, W. F. (1979). Computerized mass spectrometer linked scan system for recording metastable ions, *Anal. Chem.*, **51**, 983–988.

Haddon, W. F. (1980). The constant neutral linked magnetic field-electric sector scan, *Org. Mass Spectrom.*, **15**, 539–543.

Hass, J. R., Green, B. N., Bateman, R. H., and Bott, P. A. (1984). The design and performance of a tandem double-focusing mass spectrometer, in *Proc. 32nd Ann. Conf. on Mass Spectrometry and Allied Topics, San Antonio*; American Society for Mass Spectrometry, pp. 380–381.

Hunt, D. F., Shabanowitz, J., McIver, R. T., Hunter, R. L., and Syka, J. E. P. (1985a). Ionization and mass analysis of non-volatile compounds by particle bombardment tandem-quadrupole Fourier transform mass spectrometry, *Anal. Chem.*, **57**, 765–768.

Hunt, D. F., Shabanowitz, J., Yates, J. R., McIver, R. T., Hunter, R. L., Syka, J. E. P., and Amy, J. (1985b). Tandem quadrupole-Fourier transform mass spectrometry of oligopeptides, *Anal. Chem.*, **57**, 2728–2733.

Jennings, K. R. (1965). Metastable transitions in the mass spectrum of benzene, *J. Chem. Phys.*, **43**, 4176–4177.

Jennings, K. R. (1978). Observation of metastable transitions in a high performance mass spectrometer, in *High Performance Mass Spectrometry : Chemical Applications* (Ed. M. L. Gross), pp. 3–17, American Chemical Society Symposium Series, Washington DC.

Jennings, K. R. (1983). Linked scans, in *Proc. 31st Ann. Conf. on Mass Spectrometry and Allied Topics, Boston*; American Society for Mass Spectrometry, pp. 1–13.

Johlman, C. L., White, R. L., and Wilkins, C. L. (1983). Applications of Fourier transform mass spectrometry, *Mass Spectrom. Rev.*, **2**, 389–416.

Karlsson, K.-A. (1977). Mass-spectrometry sequence studies of lipid-linked oligosaccharides, blood-group fucolipids, gangliosides, and related cell-surface receptors, *Prog. Chem. Fats Lipids*, **16**, 207–230.

Kelley, P. E., Stafford, G. C., Syka, J. E. P., Reynolds, W. E., Louris, J. N., Amy, J. W., and Todd, J. F. J. (1985). New advances in the operation of the ion trap mass spectrometer, in *Proc. 33rd Ann. Conf. on Mass Spectrometry and Allied Topics, San Diego*; American Society for Mass Spectrometry, pp. 707–708.

Kondrat, R. W., and Cooks, R. G. (1978). Direct analysis of mixtures by mass spectrometry, *Anal. Chem.*, **50**, 81A–92A.

Louris, J. N., Wright, L. G., Cooks, R. G., and Schoen, A. E. (1985). New scan modes accessed with a hybrid mass spectrometer, *Anal. Chem.*, **57**, 2918–2924.

Matsuda, H. (1983). High-resolution high-sensitivity mass spectrometers, *Mass Spectrom. Rev.*, **2**, 299–326.

Maurer, K. H., Brunée, C., Kappus, G., Habfast, K., Schroder, U., and Schulze, P. (1971), in *Proc. 19th Ann. Conf. on Mass Spectrometry and Allied Topics, Atlanta*; American Society for Mass Spectrometry.

McIver, R. T., and Bowers, W. D. (1983). Fourier-transform ion cyclotron resonance instrument for tandem mass spectrometry, in *Tandem Mass Spectrometry* (Ed. F. W. McLafferty), pp. 287–302, Wiley, New York.

McLafferty, F. W., Ed. (1983). *Tandem Mass Spectrometry*, Wiley, New York.

McLafferty, F. W., Todd, P. J., McGilveray, D. C., and Baldwin, M. A. (1980). High-resolution tandem mass spectrometer (MS/MS) of increased sensitivity and mass range. *J. Am. Chem. Soc.*, **102**, 3360–3363.

Millington, D. S., and Smith J. A. (1977). Fragmentation patterns by fast linked electric and magnetic field scanning, *Org. Mass Spectrom.*, **12**, 264–265.

Mukhtar, E. S., Griffiths, J. W., Harris, F. M., and Beynon, J. H. (1981). Photodissociation of mass selected ions investigated using a modified ZAB-2F spectrometer. *Int. J. Mass Spectrom. Ion Phys.*, **37**, 159–166.

Sundqvist, B., Kamensky, I., Hakansson, P., Kjelberg, J., Salehpour, M., Widdiyasekera, S., Fohlman, J., Peterson, P. A., and Roepstorff, P. (1984). Californium-252 plasma desorption time of flight mass spectroscopy of proteins, *Biomed. Mass Spectrom.*, **11**, 242–257.

Todd, P. J., and McLafferty, F. W. (1983). Collisionally activated dissociation of high kinetic energy ions, in *Tandem Mass Spectrometry* (Ed. F. W. McLafferty), pp. 149–174, Wiley, New York.

Torgerson, D. F., Skowronski, R. P., and Macfarlane, R. D. (1974). New approach to the mass spectrometry of non-volatile compounds, *Biochem. Biophys. Res. Commun.*, **60**, 616–621.

Weston, A. F., Jennings, K. R., Evans, S., and Elliott, R. M. (1976). The observation of metastable transitions in a double-focusing mass spectrometer using a linked scan of the accelerating and electric sector voltages, *Int. J. Mass Spectrom. Ion Phys.*, **20**, 317–327.

Wilkins, C. L., and Gross, M. L. (1981). Fourier transform mass spectrometry, *Anal. Chem.*, **53**, 1661A–1676A.

Yost, R. A., and Enke, C. G. (1978). Selected ion fragmentation using a tandem quadrupole mass spectrometer, *J. Am. Chem. Soc.*, **100**, 2274–2275.

Yost, R. A., and Enke, C. G. (1979). Triple quadrupole mass spectrometry for direct mixture analysis and structural elucidation, *Anal. Chem.*, **51**, 1251A–1264A.

Mass Spectrometry in Biomedical Research
Edited by S. J. Gaskell
© 1986 John Wiley & Sons Ltd.

Chapter 10

HIGH MASS FAB STUDIES OF PEPTIDES, OLIGOSACCHARIDES AND GLYCOCONJUGATES

Anne Dell and Maria Panico

10.1. INTRODUCTION

This chapter describes FAB MS strategies developed in our laboratory for peptide and carbohydrate structural analysis. No attempt has been made to review comprehensively either our own work or that of others in these areas. Instead we have selected a limited number of studies and these are discussed in some depth in order to illustrate the types of peptide/carbohydrate problems which can best be solved by FAB MS and the most appropriate strategies for tackling such problems.

The work described here was carried out on the first of the first generation of high field ZAB instruments (VG Analytical) which has a mass range at full accelerating voltage (V_{acc} = 8 kV) of just over 3300 u (Dell and Taylor, 1984). For sensitivity reasons it is convenient to divide the experimentally viable mass range of the instrument as follows: (A) up to approx. 3800 (V_{acc} = 8 kV or 7 kV), (B) 3800–6600 (V_{acc} = 6–4 kV) and (C) over 6600 (V_{acc} below 4 kV). For the purposes of this chapter 'high mass' is somewhat arbitrarily taken to mean over 1000 u and therefore studies in all three mass ranges will be discussed.

As a rule of thumb, most peptides, saccharides and glycoconjugates having molecular weights within range (A) will give FAB spectra without too much difficulty, provided they are pure and are present at the microgram level. In this mass range we normally assign masses by counting the spectra which are recorded on oscillographic paper. We rarely use computer acquisition except for very high mass studies. Samples having molecular weights in range (B) are more difficult to run and the quality of data is very dependent on the structural type. To preserve sensitivity, samples are normally run in this mass range at low resolution and the unresolved clusters are mass assigned using a mass marker calibrated with caesium iodide. Two types of experiments will be considered for biopolymers of range (C) molecular weights, firstly the acquisition of molecular ions by scanning at low accelerating voltages, and secondly, the acquisition of fragment information by scanning at full accelerating voltage.

10.2. PEPTIDE AND PROTEIN STUDIES

10.2.1. GENERAL

Peptides were among the first compounds to be analyzed by the exciting new technique of FAB MS. Barber and his colleagues (Barber *et al.*, 1981) showed that both molecular weight and sequence information could be derived from the positive and negative spectra of underivatized peptides. These initial observations were rapidly exploited by mass spectrometrists involved in peptide analysis and it was soon clear that FAB MS would play a vital role in protein chemistry in the eighties. In the following sections we describe the strategies which have been developed and exploited at Imperial College for solving important problems.

We began our programme of FAB work in the peptide area by analyzing a variety of synthetic and protein-derived peptides (Morris *et al.*, 1981). This allowed us to delineate the major fragmentation pathways operating in FAB, to determine the limits of sensitivity for molecular weight assignment (10 picomolar level in favourable cases) and for sequencing (greater than one nanomolar level), and to establish that mixtures could be examined but that all components might not be observed. We discovered that the best sensitivity was achieved using xenon as the bombarding gas and loading the sample in 0.1 M HCl to ensure full protonation.

Importantly, we also learnt that the method could not be relied upon routinely to give sequence ions and that, even if present, sequence ions might be difficult to assign or might not define the full sequence. To assist sequencing we introduced derivatization procedures and devised conjoint FAB MS/protein chemical strategies which would yield the desired data. The following studies have been selected to highlight the different strategies we now employ for peptide and protein characterization.

10.2.2. CARDIOACTIVE PEPTIDE

A major thrust of our work has been the development of micro-sequencing strategies applicable to novel biologically active peptides which are difficult to purify in large quantities. A cardioactive substance (CPB) isolated from *Aplysia brasiliana* provided the first important challenge for our methods. Only a few precious micrograms were available, it was not completely pure, its structure was unknown (except for an amino acid analysis) and its sequence determination had eluded the classical protein chemists. The approach we used for the analysis of this cardioactive peptide can be tailored to suit any structural problem of this kind.

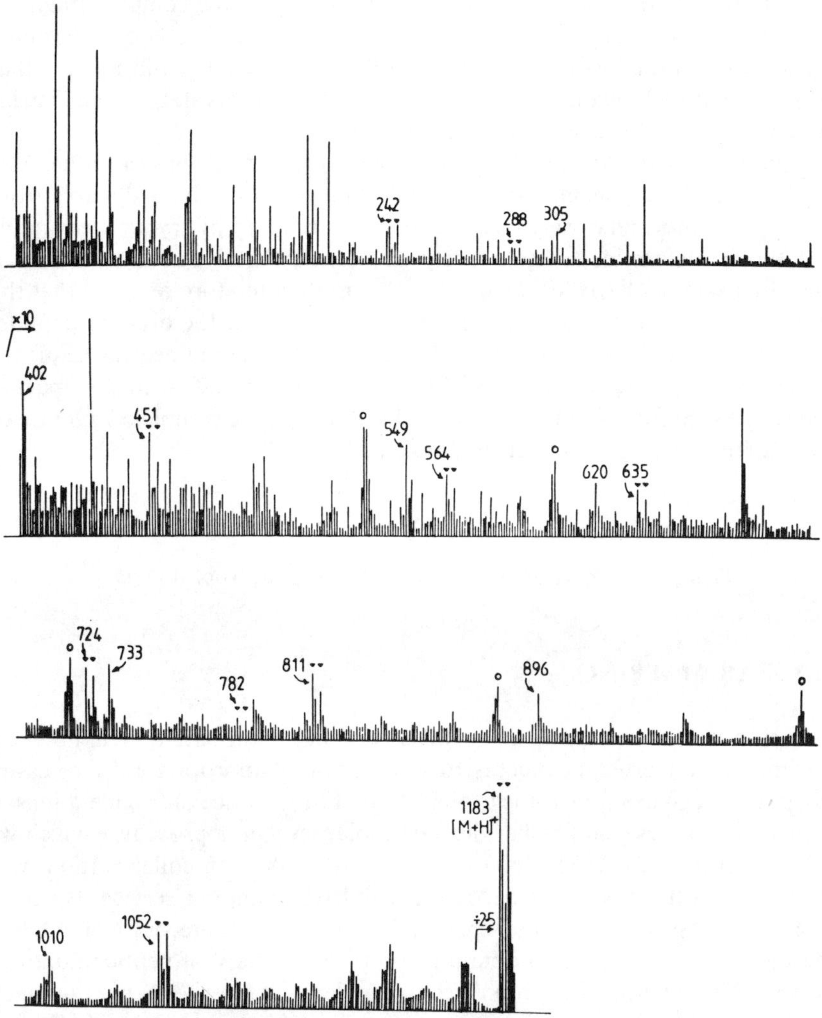

Figure 1: Positive FAB mass spectrum of a cardioactive substance CPB after conversion to the acetyl : deuteroacetyl (1 : 1) derivative. Doublets are indicated with solid arrow heads. Singlets which are accompanied by a signal 15 u lower (labelled with circles) are probably C-terminal sequence ions.

The FAB spectrum of CPB, which was obtained from 1 μg of impure material, was extremely complex, and no firm assignment, other than molecular weight of 1140, could be made. Accordingly we prepared the 1 : 1 acetyl : deuteroacetyl derivative. The purpose of the isotopic labelling was to assist the assignment of sequence ions containing the N-terminus which should all appear as 1 : 1 doublets, 3 u apart. The spectrum of this derivative is shown in Figure 1. The molecular ion is present as an intense doublet at m/z 1183, 1186 and a number of weak doublets can be 'extracted' from the complex fragmentation pattern. We were able tentatively to assign other signals as C-terminal sequence ions on the basis of the absence of an isotopic label and the fact that they were separated by amino acid mass differences. At this stage the molecular weight and a partial sequence had been established.

The amino acid analysis of CPB had indicated that a single basic residue, Arg, was present and we reasoned that the full sequence might be assignable from our data if the molecular weights of the peptides released by tryptic cleavage at Arg could be determined. Accordingly we carried out tryptic digestion on a further 1 μg sample. FAB MS analysis of the product mixture revealed that the tryptic digestion had removed a -Met-NH_2 residue from the original peptide. This experiment revealed that the cardioactive peptide contained the sequence Arg-Met-NH_2 at the C-terminus. Combining this result with the partial sequence gleaned from Figure 1 we were able to assign the complete structure of CPB (Morris *et al.*, 1982b). This is shown in Figure 2.

Met-Asn-Tyr-Leu-Ala-Phe-Pro-Arg-Met-NH_2

Figure 2: Sequence of the cardioactive peptide from *Aplysia*.

10.2.3. FAB MAPPING

It is a stimulating characteristic of FAB MS that its apparent weakness can sometimes be converted into successful applications if appropriate strategies are devised. One such example is the exploitation of the absence of fragment ions in mixtures of peptides examined at the nanomolar level, in a procedure which we have christened FAB MAPPING (Morris *et al.*, 1983). In collaboration with M-Scan Ltd, who now offer a commercial FAB mapping service, we have devised procedures for rapidly screening the products of recombinant DNA technology. The concept is to obtain a spectrum or 'map' of peptide mixtures produced by enzymic digestion. Very complex mixtures can be examined without fear of the spectra becoming uninterpretable because the absence of fragment ions results in a 'clean' spectrum of molecular ions. The majority of components in the sample digest will give molecular ion signals and the remainder can normally be observed by varying the matrix conditions (pH, hydrophobicity etc.). Since the probability of two peptides from any digest of a protein having the same molecular weight is very low, the FAB map produced is an accurate diagnostic for the presence (and thus correctness) or absence of the components of the protein sequence. Point mutations will show up as mass shifts and the numerical value of this shift usually reveals the identity of the replacement amino acid. Each ion assigned in the spectrum can be confirmed by chemically or enzymically manipulating the digest in a predictable manner (e.g. with carboxypeptidases, aminopeptidases, acetylation, the Edman degradation etc.), followed by additional FAB maps. All signals in the first map should shift in a predicted manner to give the signals in the other maps.

A definite advantage of the FAB mapping procedure is that there is an equal probability of observing C-terminal as well as N-terminal regions of the protein. Alternative methods of checking protein sequences, e.g. the gas phase sequencer, which rely on sequential cleavage from the N-terminus, rarely give firm data past the first 50 residues. We have now applied FAB mapping with complete success to a variety of genetically engineered proteins including insulins and interferons.

10.2.4. HUMAN CALCITONIN GENE RELATED PEPTIDE (hCGRP)

Our work on human Calcitonin Gene Related Peptide (hCGRP) provides an excellent example of FAB–MS work at the interface of protein chemistry and DNA technology. It is in this particular area, further examples of which are given later, that some of the most exciting future applications of FAB MS are certain to take place.

Prior to our studies on hCGRP it had been shown that the rat calcitonin gene coded for a putative peptide which was called CGRP (Amara *et al.*, 1982). There was, however, no protein chemical evidence for the existence of such a peptide in rat tissue. Using antibodies raised to rat CGRP, MacIntyre and colleagues succeeded in isolating cross-reacting material from human medullary thyroid carcinoma tissue. Approximately 10 nanomoles was obtained after extensive HPLC purification and fortunately proved sufficient for a complete structure elucidation. To sequence this peptide we used a combination of FAB analysis of the intact material, FAB mapping and FAB analysis of isolated tryptic peptides (Morris *et al.*, 1984).

The molecular weight of the intact peptide was shown to be 3785 (Figure 3) which was 217 u lower than that expected for the rat CGRP molecule as deduced from the gene sequence (Amara *et al.*, 1982). A tryptic FAB map of hCGRP (Figure 4) gàve four $(M + H)^+$ ions uniquely assignable to tryptic peptides, only two of which (m/z 546 and 729) corresponded in mass to tryptic peptides in the rat sequence. The other signals in Figure 4 were assigned as

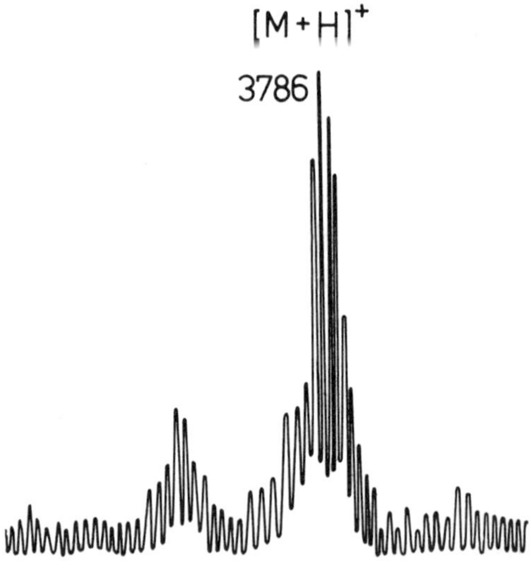

Figure 3: Molecular ion region of the positive FAB spectrum of hCGRP.

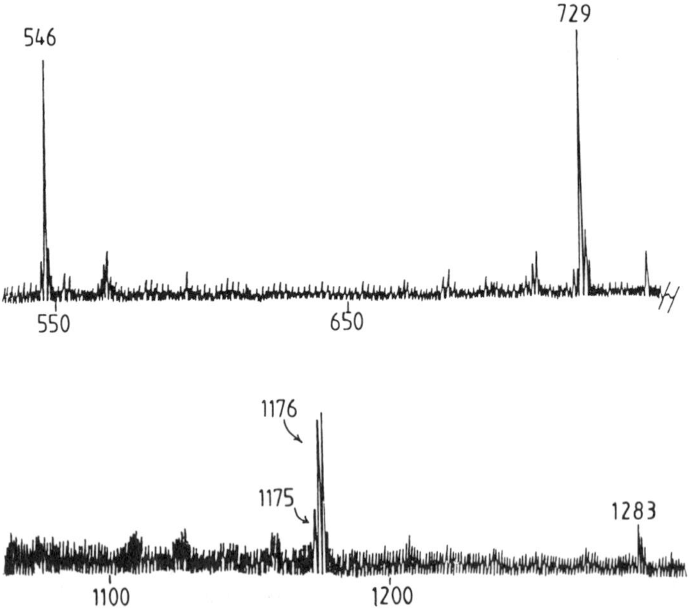

Figure 4: Tryptic FAB map of hCGRP. Four tryptic peptides are present at *m/z* 546, 729, 1175 and 1176.

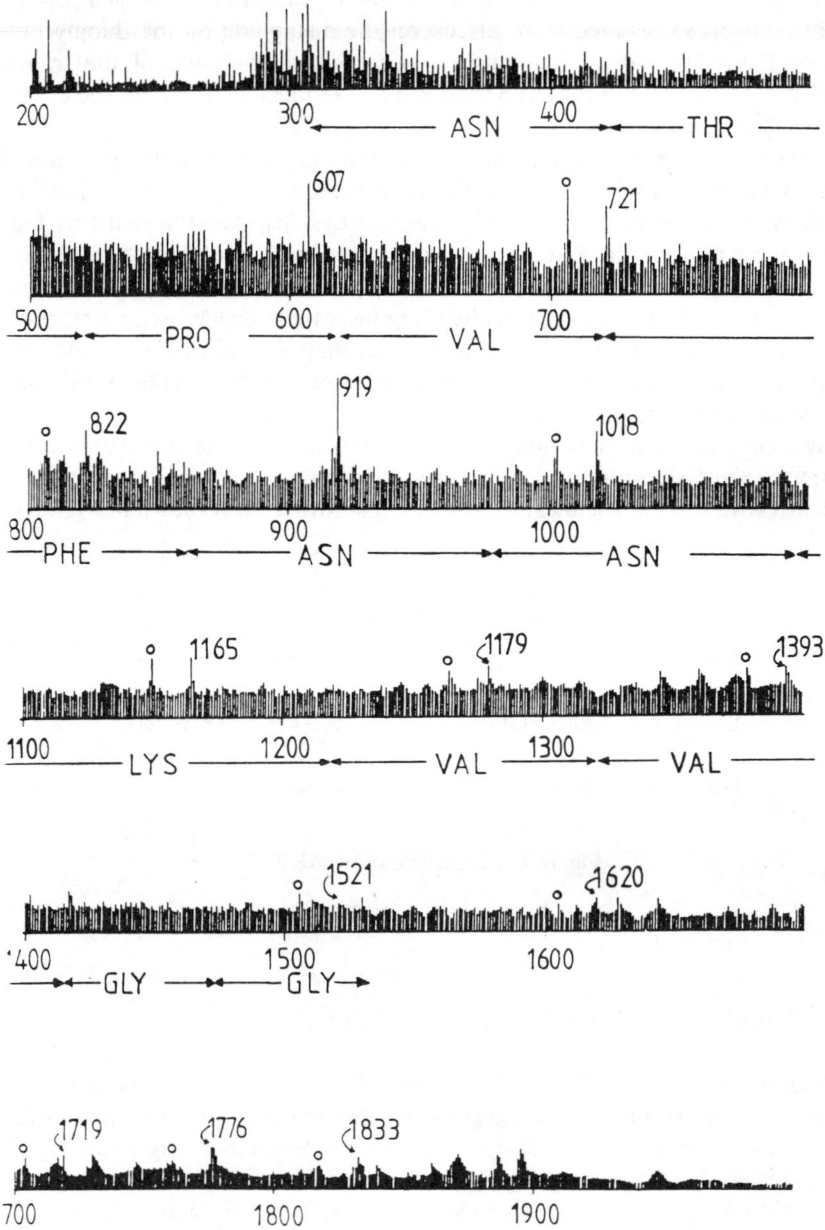

Figure 5: FAB spectrum below m/z 2000 of hCGRP showing C-terminal sequence ions. The circles denote signals 15 u below the labelled signals.

follows: *m/z* 1177 was deduced to be the dithiol of an intra S–S bridged peptide at 1175; *m/z* 1283 resulted from attack on the disulphide by the thioglycerol used as the FAB matrix. Reduction and carboxymethylation of the tryptic digest shifted the 1175 in the FAB map to 1291, confirming the presence of two cysteines in the molecule.

Further structural evidence came from the low mass region of the spectrum of intact hCGRP (Figure 5). A series of C-terminal ions were present from which we derived the sequence -Gly-Gly-Val-Val-Lys-Asn-Asn-Phe-Val-Pro-Thr-Asn-X, where X is the C-terminus of the molecule and has a molecular weight of 605. The structure was completed by analysis of the tryptic peptides after HPLC purification. As individual peptides, these yielded sequence information which was not forthcoming in the FAB map. In addition we located the dipeptide Ala-Phe-NH_2 which had not been seen in the tryptic FAB map because of suppression effects.

The complete structure of human CGRP including the C-terminal amide, the intramolecular S–S bridge and the sequence was thus established by combining FAB MS with protein chemical techniques. The structure is shown in Figure 6.

S - S

Ala-Cys-Asp-Thr-Ala-Thr-Cys-Val-Thr-His-Arg-Leu-Ala-Gly-

Leu-Leu-Ser-Arg-Ser-Gly-Gly-Val-Val-Lys-Asn-Asn-Phe-Val-

Pro-Thr-Asn-Val-Gly-Ser-Lys-Ala-Phe-amide

Figure 6: Structure of hCGRP.

10.2.5. GLUCOSE TRANSPORTER PROTEIN

An even more direct interfacing of protein and DNA technologies is illustrated by our recent work on the glucose transport protein of human erythrocytes in which we have been collaborating closely with molecular biologist colleagues (Mueckler *et al.*, 1985). Two parallel investigations were carried out:

(a) DNA sequencing of a cDNA clone, derived from a human hepatoma cell line, and containing the gene for the liver glucose transporter.

(b) Protein sequencing studies of the human erythrocyte glucose transporter. These studies combined the powerful techniques of FAB mapping and gas phase Edman degradation.

The FAB mass spectrum of the CNBr/tryptic digest of the erythrocyte protein yielded a number of strong signals corresponding to $(M + H)^+$ ions (Figure 7). A computer search for the peptide masses within the predicted amino acid sequence of the hepatoma protein (see Figure 8) revealed that most of the ions had masses expected for CNBr/tryptic fragments of this protein. These results suggested that the two proteins were very similar. Signal assignments were confirmed by comparing the CNBr/tryptic map with maps produced after one and two cycles of Edman degradation. The changes in mass allowed the identification of the N-terminal and subsequent amino acids in many cases and enabled the position of the peptides within the predicted sequence to be firmly established.

Tryptic digestion of the native transporter reconstituted into lipid vesicles (Cairns *et al.*, 1984) produced two peptides which were purified by HPLC. These peptides gave molecular ions at *m*/*z* 1142 and 1444 corresponding to residues 459–468 and 233–245 respectively in the hepatoma sequence (Figure 8). Each peptide was sequenced fully using Imperial College's gas phase sequencer and were shown to have the precise sequences that are underlined in Figure 8.

In summary, the FAB maps and the gas phase sequencing data confirmed the identity of the hepatoma clone and demonstrated that the liver and erythrocyte transporters are highly homologous, if not identical. A schematic representation

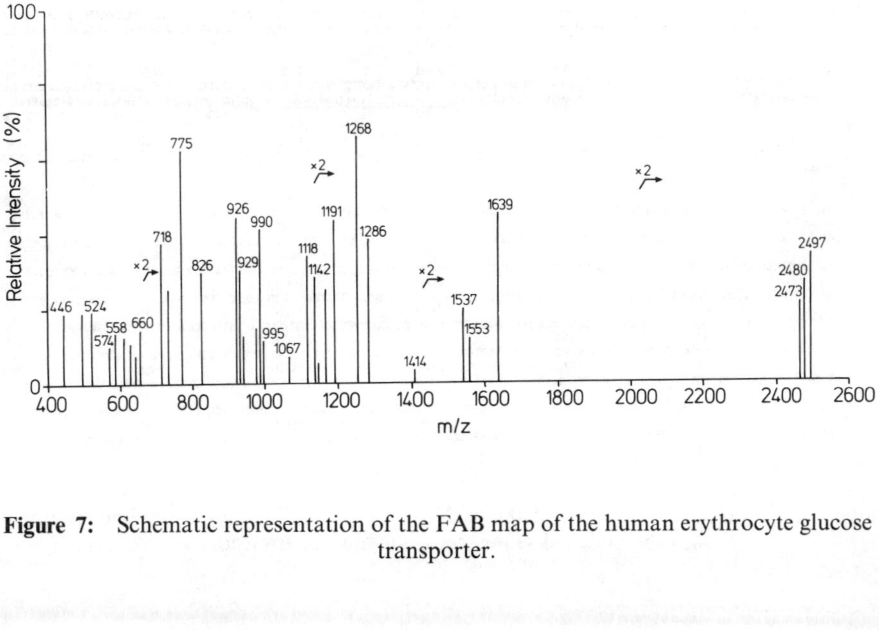

Figure 7: Schematic representation of the FAB map of the human erythrocyte glucose transporter.

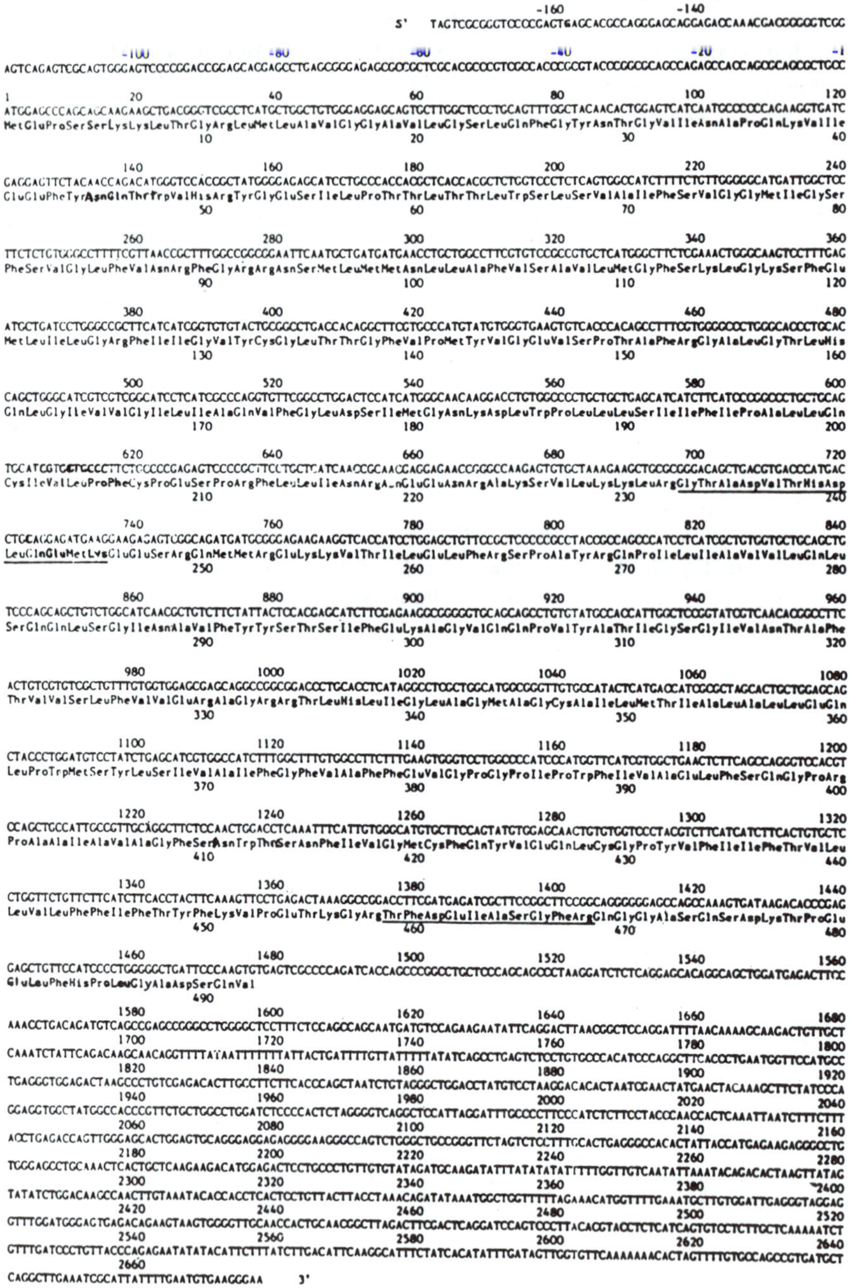

Figure 8: Nucleotide sequence of the cDNA of a human hepatoma cell line together with the deduced sequence of the glucose transporter.

for the membrane orientation of the glucose transporter is given in Figure 9 together with the regions of the protein which were mapped by FAB. Note that the FAB-mapping shows that the protein has no cleaved signal sequence at the amino-terminus and there is also no proteolytic processing at the carboxy-terminus. This protein is not as extensively mapped as most others we have studied, partly because only one type of digest was examined and partly a result of heavy contamination with lipid and carbohydrate. Despite this, sufficient information was obtained for firm conclusions on the protein's identity to be drawn.

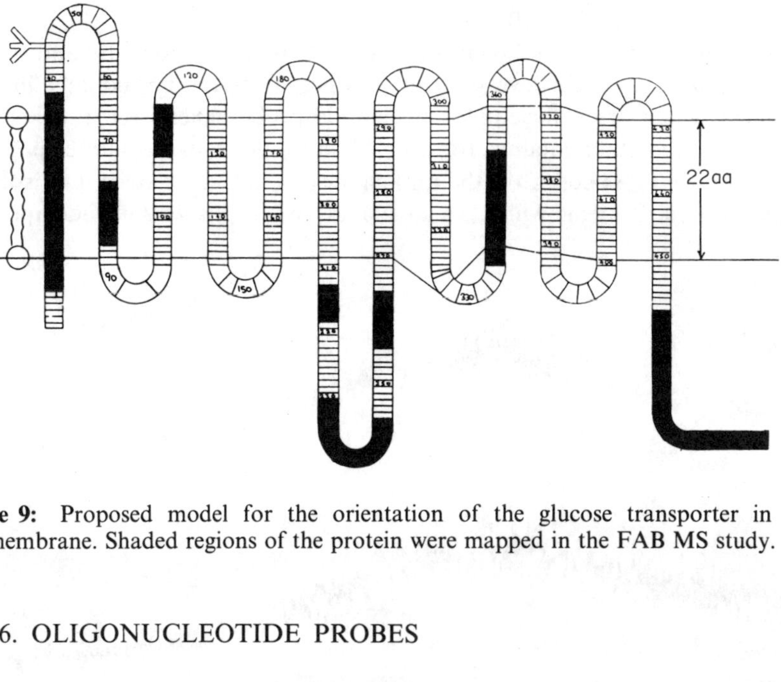

Figure 9: Proposed model for the orientation of the glucose transporter in the membrane. Shaded regions of the protein were mapped in the FAB MS study.

10.2.6. OLIGONUCLEOTIDE PROBES

During the second half of the 1970s there was an increasing tendency for molecular biologists to regard protein sequencing as a moribund subject. After all, was it not simpler and faster to sequence the gene rather than its protein product? This attitude is slowly changing, because the molecular biologists have recognised that an important mode of access to many interesting genes is via synthetic oligonucleotide probes. Synthesis of such probes requires a knowledge of at least part of the protein sequence.

We have developed a strategy for microsequencing proteins to yield sufficient information for probe generation. The protein is digested with a suitable proteolytic enzyme and the digest is purified by HPLC. Trp-containing peptides are located by comparing the UV absorbance of the fractions at 280 nm and 254 nm (a procedure which is sensitive down to about 500 picomole) and

Met-containing peptides are located by screening all fractions by amino acid analysis (sensitive to 50 picomole). Trp and Met are favoured for probe generation because, unlike other amino acids, they do not have degenerate codes. Potentially interesting fractions are then screened by FAB MS to determine the molecular weights and the state of purity of the peptides. Samples which appear to contain a single component are then sequenced by a combined FAB MS/gas phase sequencing strategy. FAB MS provides the molecular weight and also some of the N- and C-terminal residues after analysis of the products of Edman cleavages and carboxypeptidase digestions. Gas phase sequencing gives partial or complete sequences of the pure peptides which can be corroborated by the FAB data.

Our recent studies have shown that the method is very powerful and that at the 1 nanomole protein level, it can produce a series of randomly located sequences which are sufficient for nucleotide probe generation. Further, the molecular weight data obtained from a FAB map and from the purified peptides are a valuable cross-check on the authenticity of the clone once it is isolated and sequenced. Proteins which we have analyzed in this way include apolipo-

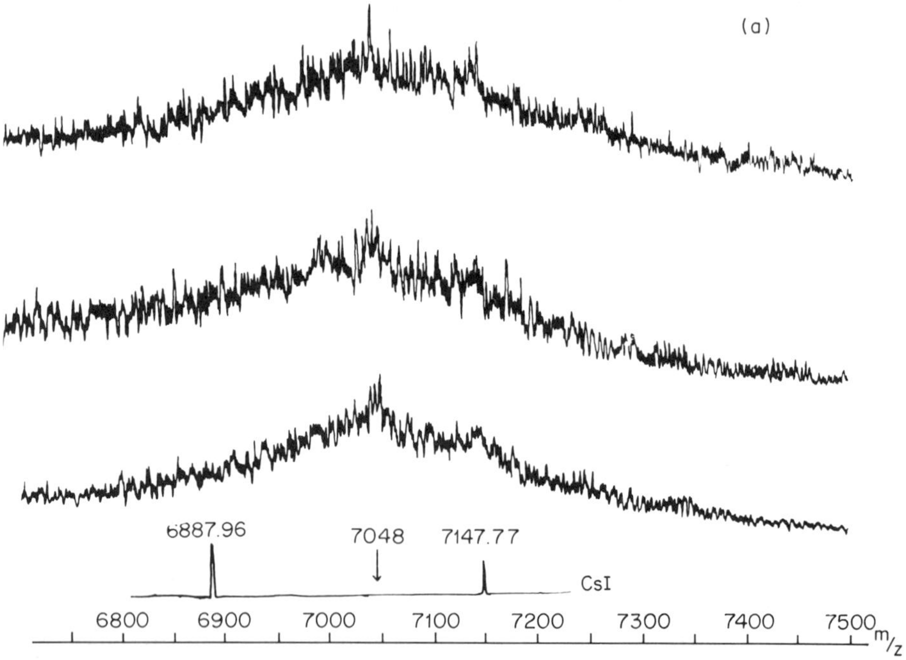

Figure 10: a) Molecular ion region of three separate full scale scans of dendrotoxin. The mass scale was obtained by calibrating against the spectrum of CsI, whose partial spectrum is reproduced. The scan rate was 300 s for coverage of the mass range 9000–100.

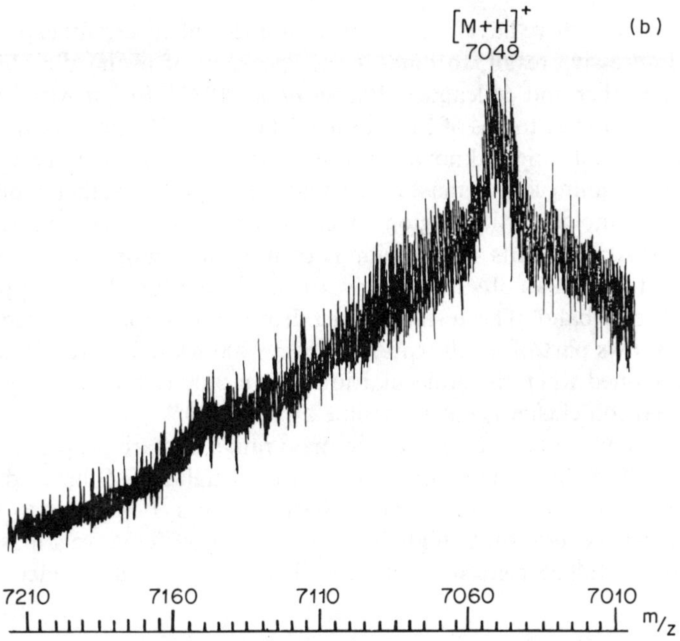

Figure 10: b) The molecular ion region of dendrotoxin obtained by scanning across 200 u in 100 s.

protein B (Morris *et al.*, 1986a; Knott *et al.*, 1985) and lactase (Morris *et al.*, 1986b).

10.2.7. HIGH MASS PROTEIN STUDIES

All the examples described so far have involved the scanning of mass range A (see Introduction). Oligopeptides and proteins larger than M_r of about 4000 have been brought within this mass range by enzymatic modification. There are occasions, however, when it is desirable to analyze large polypeptides intact, e.g. if key portions of the sequence are missing from a FAB map. It is of interest, therefore, to address the potential of the high field ZAB for analyzing peptides and proteins in mass ranges B and C (see Introduction).

We, and others, have had considerable success in the analysis of polypeptides in mass range B (e.g. insulin (Dell and Morris, 1982; Barber *et al.*, 1982), ACTH (Morris *et al.*, 1982a), monellin (Dell and Morris, 1982), CNBr digest of myoglobin (Morris, 1982)), and we conclude that most peptides of this size will give results provided they are pure and at least 10–20 nanomol is available. We recommend, nevertheless, that FAB mapping should be the first experiment carried out because it requires less sample (1–5 nanomol) and gives guaranteed data.

In mass range C there have been a limited number of successful experiments. The most impressive result, to date, is the spectrum of proinsulin (M_r 9390) obtained by Barber and colleagues (Barber *et al.*, 1983). In our work we have examined a number of toxins of M_r between 7000 and 8000 and have developed a protocol for analyzing unknowns in this mass range (Bateman *et al.*, 1985). This involves scanning a wide mass range relatively rapidly and then 'homing in' on the putative molecular ion region which appears as a broad band spanning several hundred mass units. This region is then scanned more slowly and data are accumulated until an obvious 'hump', the molecular ion cluster, appears on top of the broad band. The results on dendrotoxin exemplify this technique. Figure 10a shows part of a full scan from *m*/*z* 9000 while Figure 10b shows a spectrum obtained when the molecular ion region only is scanned very slowly. The molecular ion cluster is clearly visible at *m*/*z* 7049.

The proteins which have given data in mass range C share properties which, we believe, will facilitate desorption. They are all tightly S–S bridged, giving compact molecules, and all have several basic residues. It remains to be seen whether the new generation of high field instruments, which have a mass range of over 10,000 at full accelerating voltage, will be able to handle a wider range of samples.

10.3. OLIGOSACCHARIDES AND GLYCOCONJUGATES

10.3.1. GENERAL

In principle, FAB MS can provide two types of information which will assist the structure determination of oligosaccharides and glycoconjugates. These are (i) the mass of the molecular ion and (ii) the masses of the fragment ions. The former defines the sugar and peptide or lipid composition and confirms the presence or absence of modifying groups such as acetyl, pyruvyl, phosphate, sulphate etc. The latter may permit the sequence, including branch points, to be assigned.

It is usually desirable to define molecular weights by analyzing underivatized samples because chemically labile functional groups will then be observed. However, if no such groups are anticipated, it may be preferable to prepare derivatives prior to analysis. Partial or complete protection of the hydroxyl groups in carbohydrates greatly enhances sensitivity, an observation which we first made when studying the methylglucose polysaccharide (MGP) of *Mycobacteria smegmatis* (Figure 11) (Forsberg *et al.*, 1982). This substance yielded intense molecular ion signals at *m*/*z* 3413 (negative) and 3437 (positive) with incredible ease. Comparable data were very difficult to obtain from glucose polymers of this size. We attributed the excellent FAB properties of MGP to the partial *O*-methylation and we concluded that this resulted in diminished *H*-bonding between the sample and the matrix and thus facilitated desorption. Whatever the reason, it is now well established that peracetyl and permethyl derivatives run considerably more easily on FAB MS than their underivatized counterparts, and samples having molecular weights above 3800 (mass range B, Introduction) are best examined as their permethyl derivatives despite the increase in mass upon permethylation.

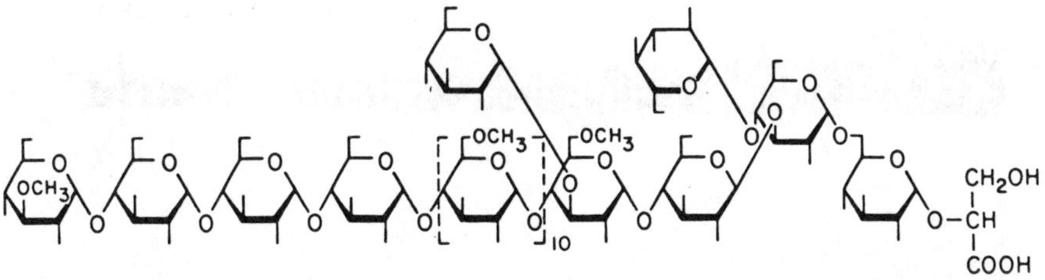

Figure 11: Structure of the methylglucose polysaccharide of *Mycobacteria smegmatis*.

We routinely derivatize in the following situations:

(i) SMALL QUANTITIES AND IMPURE SAMPLES. If sample quantities are limited to 1 μg or less and/or the sample is contaminated with salt, the peracetyl derivative is prepared using trifluoracetic anhydride and acetic acid (Dell *et al.*, 1983a). This derivative can be run directly or, if salt contamination is anticipated, it can be purified by extraction into chloroform and washing with water. Further purification is not required, making this an excellent derivative for studies in the sub-microgram area. Oligosaccharides which have been reduced with sodium borohydride are rarely salt-free even after column chromatography and are best examined as their acetyl derivatives. In our work on the structure determination of granulocyte lactosaminoglycans, the compositions of the oligosaccharides which were released by endo-β-galactosidase digestion and then reduced by sodium borohydride, were easily assigned from the abundant molecular ions present in the spectra of their acetylated derivatives (Fukuda *et al.*, 1984c). The quality of the data obtained is indicated by Figure 12 which shows the positive spectrum of the 'pentasaccharide fraction' from a Bio-Gel P-2 column. $(M + H)^+$ ions from fully acetylated molecules are present at *m*/*z* 1527 (Fuc.Hex.Hex*N*Ac$_2$.HexOH), 1815 (Fuc.Hex$_2$.Hex*N*Ac$_2$.HexOH) and 2045 (Fuc$_2$.Hex$_2$.Hex*N*Ac$_2$.HexOH). The last is well above background despite the fact that the difucosylated component is very minor. This sample gave only background ions when it was examined without acetylation.

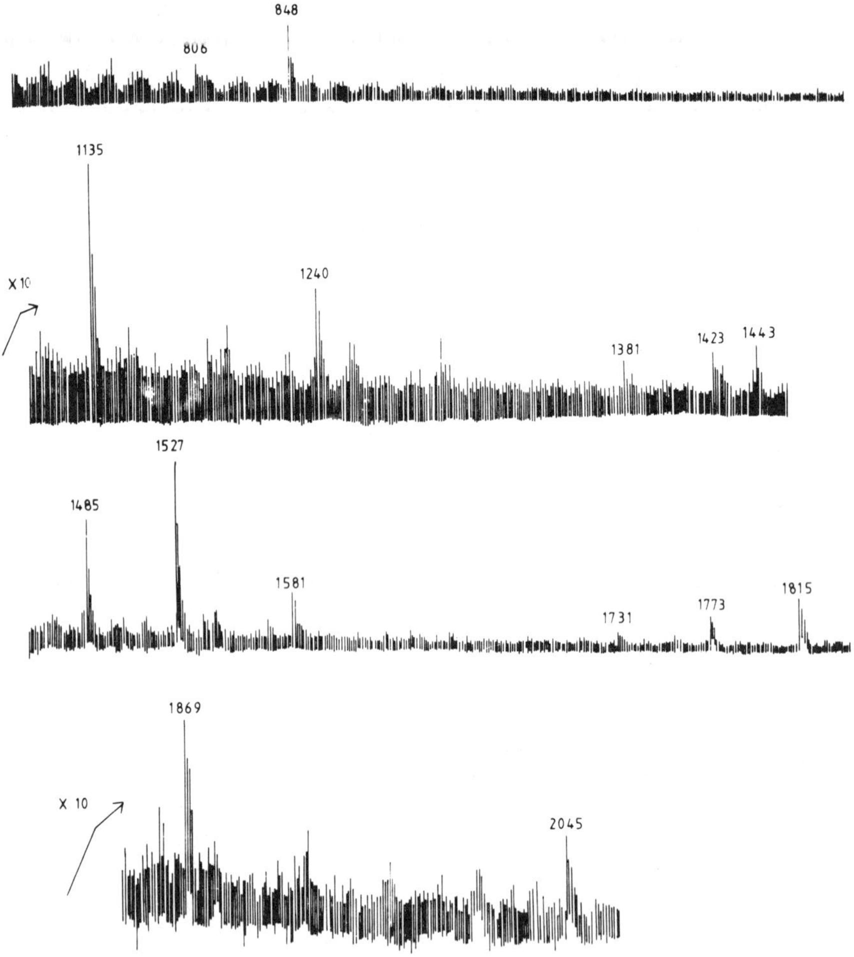

Figure 12: Positive FAB spectrum of the peracetyl derivative of a mixture of oligosaccharides from human granulocyte lactosaminoglycan.

(ii) SEQUENCING. With very few exceptions, derivatives must be prepared if unambiguous sequence data are to be obtained from oligosaccharides. There are two main reasons for this requirement. Firstly, fragmentation with H-transfer can occur on either side of the glycosidic bond (Figure 13). Unless the sample is chemically modified in some way it is not possible to establish whether the resulting fragment ions are derived from the non-reducing or reducing end of the oligosaccharide. Secondly, 'double cleavages', first observed in the MGP study (Dell and Ballou, 1983) are common. For example fragment ions frequently result from a combination of β-cleavage and A_1-type cleavage

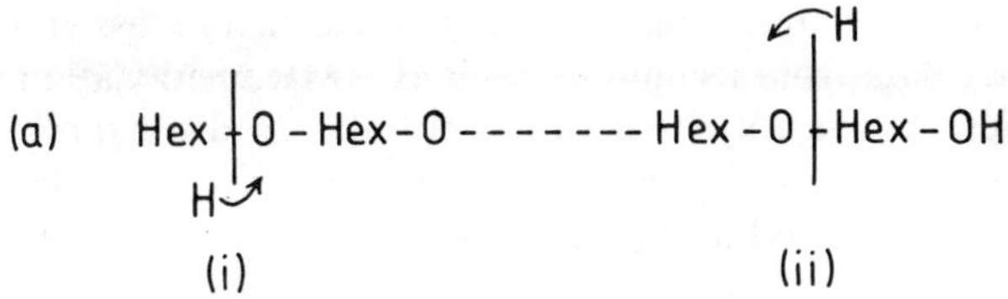

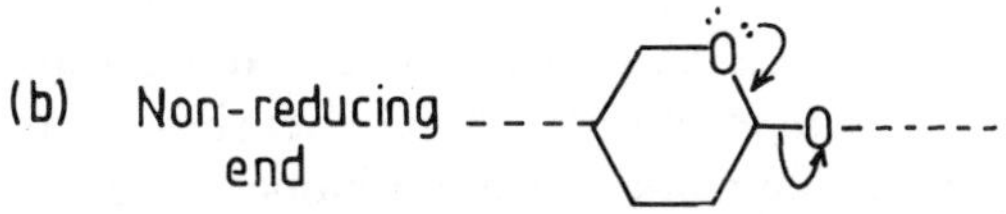

Figure 13: Schematic representation of three of the major fragmentation modes observed in FAB MS of oligosaccharides. a) glycosidic cleavages without charge being involved; cleavage (i) is referred to as β-cleavage. b) glycosidic cleavage (A_1-type) resulting in a positive charge on the non-reducing fragment.

(Figure 13) further along the chain. This gives apparent non-reducing end sequence ions which are not, in fact, derived from the non-reducing terminus. Both these ambiguities are resolved if derivatized samples are examined. Derivatization has the added advantage of directing fragmentation along a limited number of well defined fragmentation pathways. For example, in the positive mode, only one major fragmentation pathway operates when acetylated or permethylated samples are subjected to FAB MS. This is glycosidic cleavage to give A_1-type ions (Figure 13) from which the sequence and branch points can be readily assigned. If β-cleavage does occur, the non-reducing terminus will have one fewer methyl or acetyl groups than a genuine non-reducing terminus and can therefore be readily recognized.

As an aid to sequence assignment we frequently employ derivative-forming reactions such as acetolysis or methanolysis which effect partial cleavage of the glycosidic bonds. Two applications of methanolysis are covered in later examples (Sections 10.3.3 and 10.3.4). An important application of acetolysis is the preliminary screening of novel glycoproteins for sugar type (Naik *et al.*, 1985). In this procedure the intact glycoprotein is subjected to an acetolysis time course and the peracetylated oligosaccharides released from the glycoprotein are extracted into chloroform and analyzed by FAB MS without further purification.

The molecular weights of the acetolysis fragments are diagnostic of the type of glycan present, e.g. high mannose, complex, lactosaminoglycan, etc.

The studies described in the following sections are a guide to the strategies used in our laboratory for carbohydrate structural analysis by FAB MS. Each exploits particular strengths of FAB MS technique and together they illustrate the scope of problems which can be assisted by FAB MS. It should be remembered however, that in all cases complete structural assignment required complementary data from techniques such as NMR, methylation analysis and enzyme studies. At present, FAB MS is not able to distinguish isomers (e.g. Man, Gal, and Glc), or to assign linkage positions except in a few exceptional cases, e.g. 3-linked fucosyl in the Le^X antigen (Fukuda *et al.*, 1984c).

10.3.2. OLIGOGALACTURONIDES

Recent studies have suggested that plant cell wall derived oligogalacturonides of the general structures shown in Figure 14 have elicitor activity, i.e. they stimulate the production of the plant's own antibiotics. These substances are

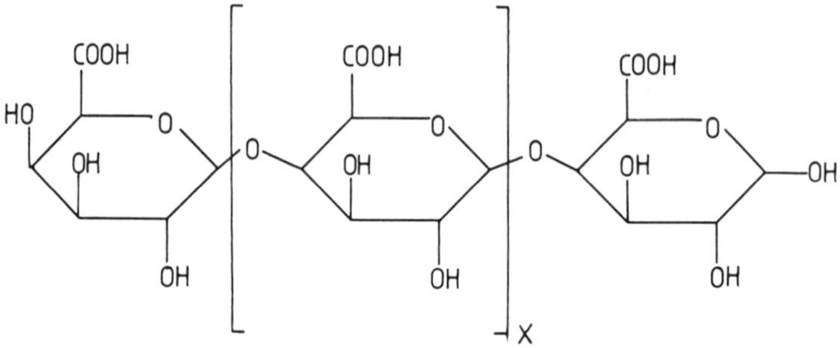

Figure 14: Oligogalacturonide structures; samples produced by lyase digestion differ from these structures by the presence of a dehydrated residue at the non-reducing end.

unbranched homopolymers of well defined anomeric stereochemistry and linkage. Molecular weight data are all that is required for complete structural assignment and they are therefore ideal problems for the FAB MS technique. Without FAB MS it is very difficult to define precisely the size of a large oligogalacturonide; with FAB MS, determination of the size is, in principle, a trivial problem. A correct strategy for studying these molecules is nevertheless important if success is to be guaranteed at the low microgram level. Firstly, pH control of the matrix is essential (Figure 15) (Davis *et al.*, 1986). Secondly, the larger the oligomer, the more difficult it is to dissolve in a suitable solvent for loading into the matrix. One solution to this problem is to convert the samples to their pentafluorobenzyloxime derivatives which have improved solubility properties and also desorb well from the matrix in negative FAB MS. We used this approach in the first characterisation of an active oligogalacturonide elicitor from soybean cell walls (Nothnagel *et al.*, 1983).

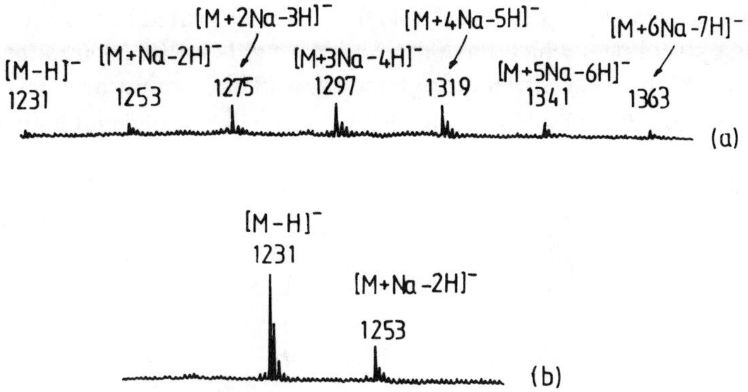

Figure 15: a) Negative FAB spectrum of a heptagalacturonide obtained after lyase digestion. The sample was dissolved in 5% acetic acid prior to loading into the glycerol matrix. b) Spectrum obtained after adding 0.1 M HCl to the matrix.

10.3.3. GLYCOLIPIDS

Glycolipids are perhaps the most amenable class of carbohydrate-containing biopolymers for FAB MS analysis. Their lipid component acts as a natural 'derivative' providing a handle for the interpretation of fragmentation patterns and giving the molecule excellent desorbing properties which result in abundant $(M + H)^{\pm}$ ions. Some glycolipids fragment extensively, allowing the structure of the lipid and the sequence of the glycan to be assigned. For example, the pyruvylated glycolipids of *Mycobacterium smegmatis* fragment in the negative mode at the positions shown in Figure 16. The fragment data defined the sugar sequence and the location of the fatty acid groups (Kamisango *et al.*, 1985).

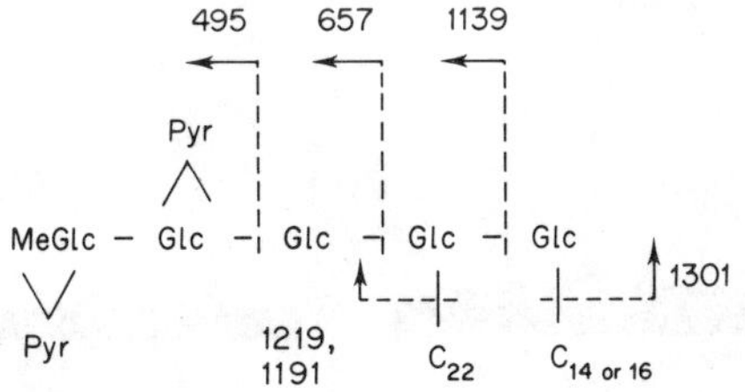

Figure 16: Structure and major negative FAB cleavages of the pyruvylated glycolipids of *Mycobacterium smegmatis*.

Other glycolipids afford more reliable sequence information at the low microgram level when they are analyzed as their permethyl derivatives. Glycosphingolipids fall into this category. Impressive data have been obtained by

Egge and co-workers (Egge *et al.*, 1984) from permethylated lactosaminoglycan containing glycolipids which have molecular weights in the 6000 range. These substances exhibit very specific fragmentation at the amino acid residues to produce A_1-type non-reducing sequence ions. Smaller glycosphingolipids, e.g.

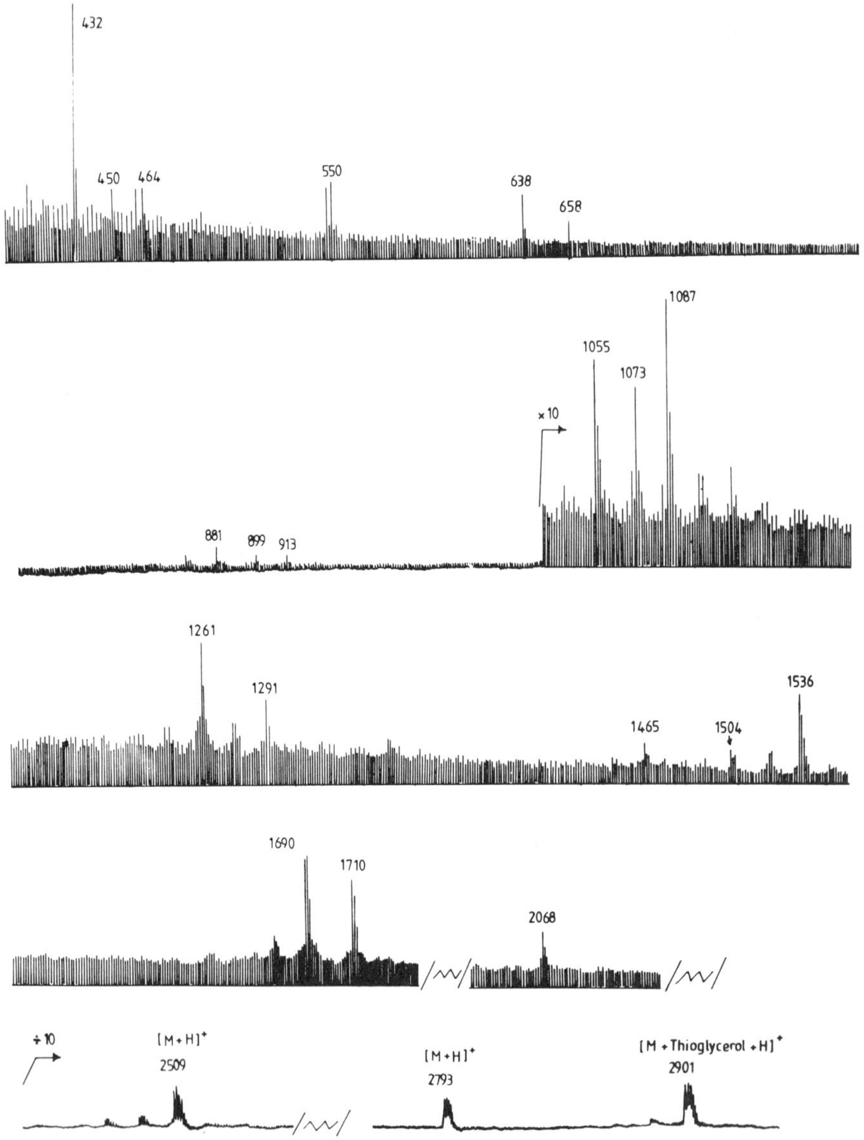

Figure 17: Positive FAB spectrum of a permethylated glycolipid fraction from human granulocytes.

those we have characterized from human granulocytes (Fukuda *et al.*, 1985c), show a similar preference for cleavage at amino sugars but these cleavages become less specific as the size of the molecule decreases. The FAB spectrum of one of the glycosphingolipid fractions which we isolated from human granulocytes is show in Figure 17. It contains all the characteristic features of a permethylated glycosphingolipid spectrum, i.e. very abundant $(M + H)^+$, A_1-type fragment ions, and several characteristic ions derived from the lipid. Assignments are given in Figure 18. It is of interest that only the unsaturated fatty acid containing-glycolipid forms a thioglycerol adduct.

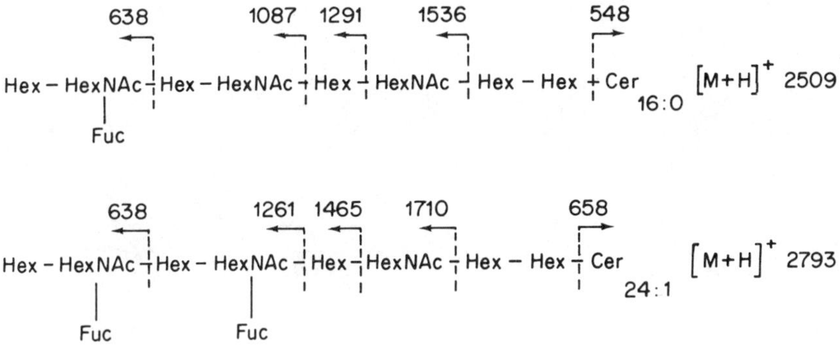

Figure 18: Structures and cleavages of glycolipids giving the spectrum shown in Figure 17.

If necessary, the sequencing of glycolipids can be assisted by carrying out a time course methanolysis on the previously permethylated sample. In our recent studies on the glycolipids of a human embryonal carcinoma cell line (Fukuda *et al.*, 1986) we have indentified the presence of a novel disialosyl glycolipid which has the following structure at the non-reducing end: Neu*N*Ac, Neu*N*Ac-(Hex-Hex*N*Ac). This moiety gives a characteristic A_1-type fragment ion at *m/z* 1186 in the spectrum of the permethyl derivative. Unfortunately the fragmentation pattern did not reveal whether the Neu*N*Ac residues were directly linked as Neu*N*Ac-Neu*N*Ac- or whether they were separately attached to the lactosaminyl moiety. To distinguish these possibilities we used FAB MS to monitor a time course methanolysis. The results showed unambiguously that Neu*N*Ac residues were attached separately since their removal produced a Hex-Hex*N*Ac$^+$ fragment ion which contained two free hydroxyl groups (*m/z* 436). Removal of Neu*N*Ac-Neu*N*Ac- would have only generated a single free hydroxyl group.

10.3.4. CYCLIC POLYSACCHARIDES

The existence of novel cyclic polysaccharides in certain types of bacteria has been revealed by FAB MS. Rhizobia and Agrobacteria secrete cyclic glucans of the type shown in Figure 19 and the common antigen of the enterobacteria (ECA) has also been confirmed as a mixture of cyclic molecules containing predominantly four or five repeats of the trisaccharide (ABC) shown in Figure 20 (Dell *et al.*, 1983b, 1984).

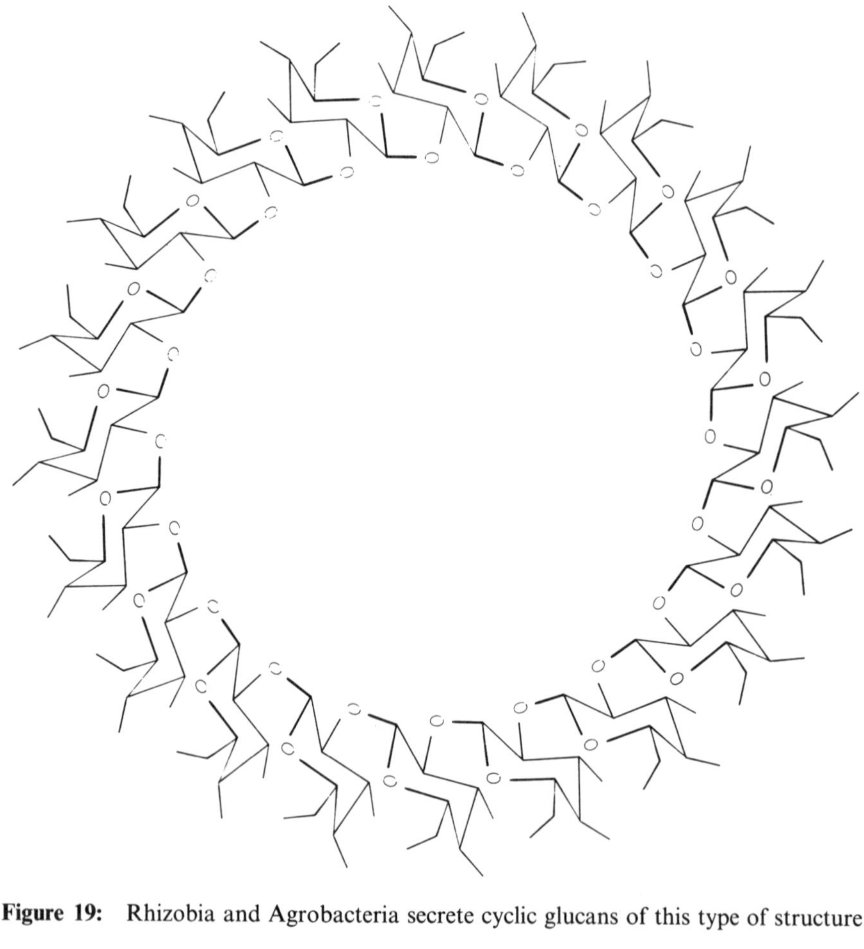

Figure 19: Rhizobia and Agrobacteria secrete cyclic glucans of this type of structure.

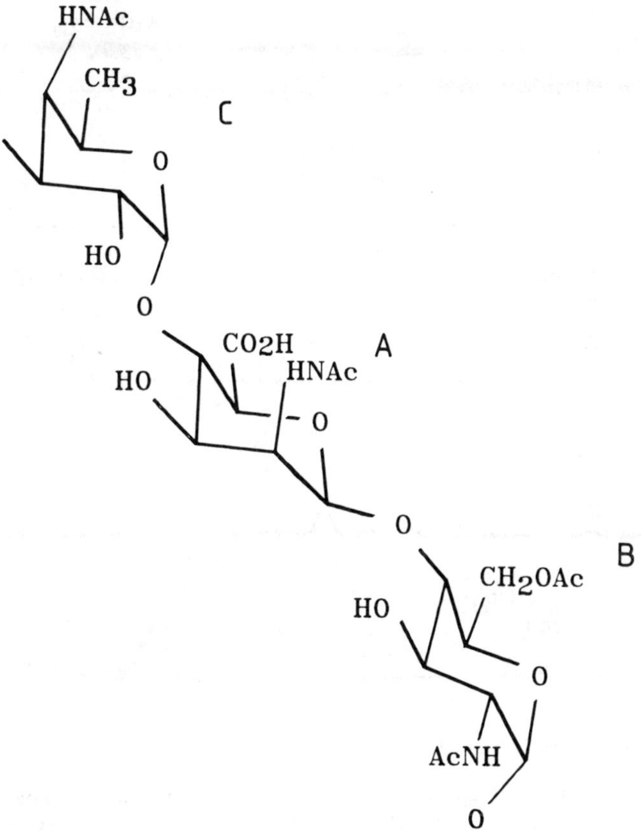

Figure 20: The repeating unit of the common antigen of enterobacteria (ECA) is made up of the sugars shown in this figure. They are called A, B and C for convenience. Note that in the intact antigen some of the B residues are not O-acetylated.

Strategies which we devised to prove these cyclic structures are generally applicable to any novel oligosaccharide. They are as follows: (i) FAB MS of the native substance (ii) FAB MS of permethyl and/or peracetyl derivatives, and (iii) FAB MS of hydrolysis, methanolysis and acetolysis reaction products. All these procedures can be carried out below the 10 μg level.

Data from FAB MS of the underivatized polysaccharides provided the initial evidence for their cyclic nature (all molecular ions were 18 u less than those expected for linear molecules) but did not rule out dehydrated or lactonized linear species. Definitive evidence for intact circles required the analysis both of derivatives and of the products of partial hydrolysis and partial methanolysis.

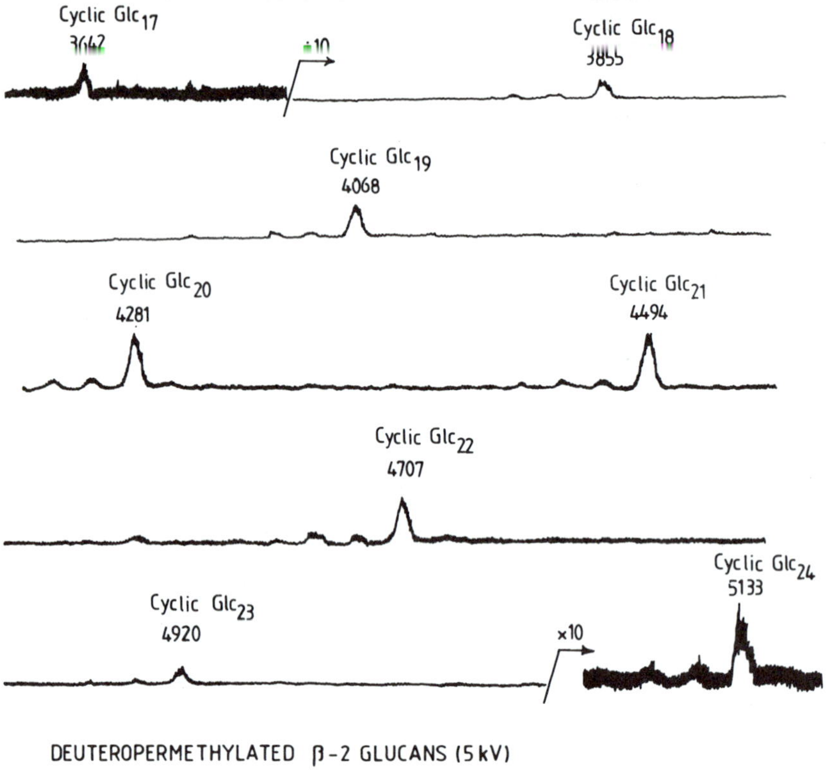

Figure 21: Positive FAB spectrum of a deuteropermethylated sample of glucans isolated from Rhizobia.

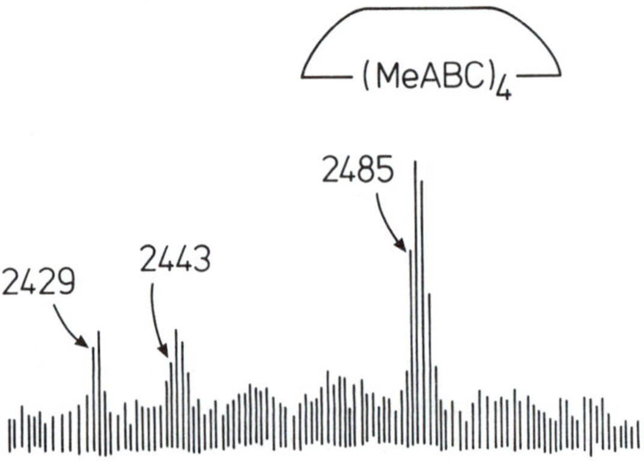

Figure 22: Molecular ion region of the smaller component of the common antigen of enterobacteria (ECA). The sample had first been de-*O*-acetylated and then treated with methanolic HCl at 20°C for 20 min. The $(M + H)^+$ ion corresponds to a fully esterified intact cycle of four repeating units.

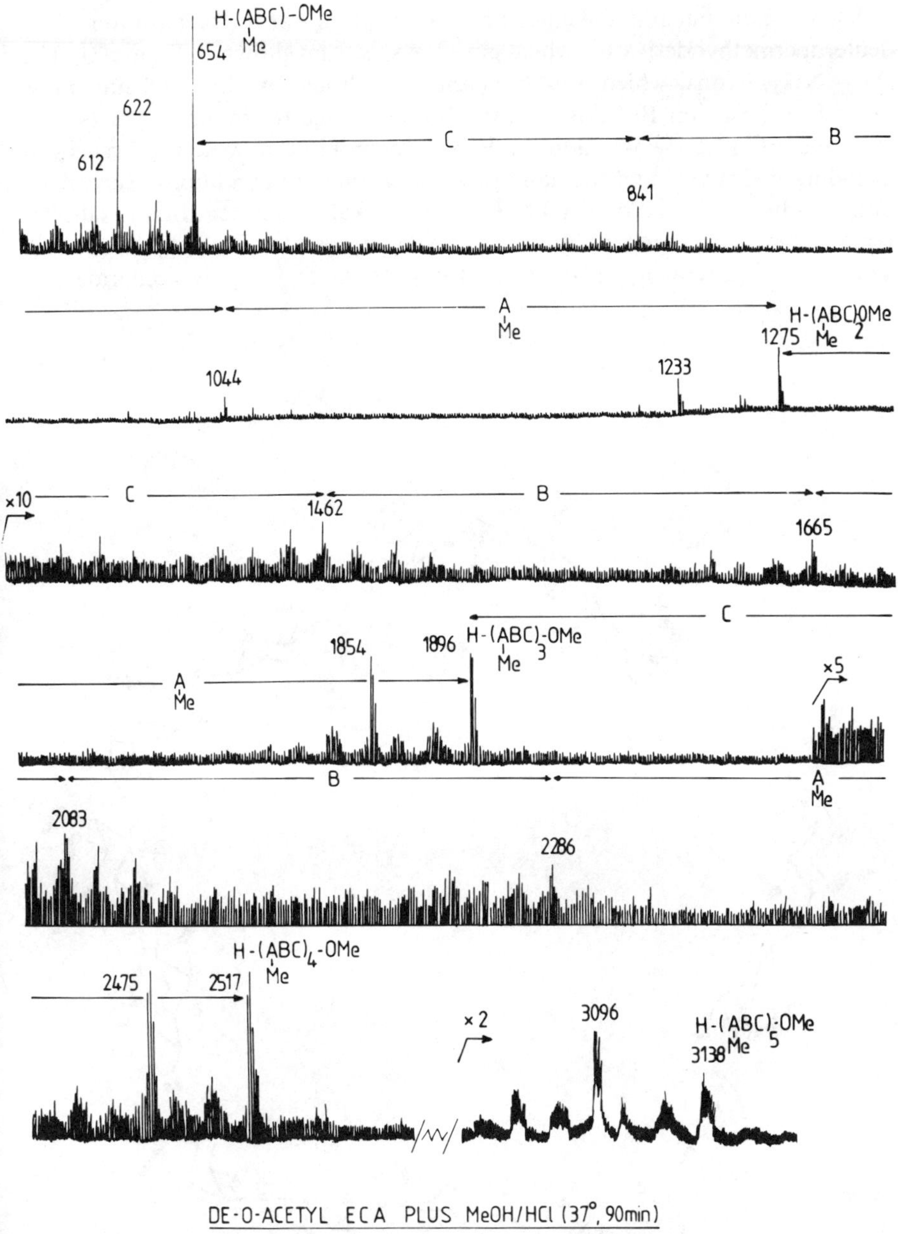

Figure 23: Complete spectrum of the sample described in Figure 22 after further reaction (37°C for 90 min). $(M + H)^+$ ions are present for linear methyl glycosides as indicated.

The key experiment in the glucan characterization was the preparation of the deuteropermethyl derivative which gave the spectrum shown in Figure 21. The $(M + NH_4)^+$ ions, which are abundant even though we are well into mass range B (see Section 10.1), are exactly those expected for cyclic polymers.

In the ECA study we used FAB MS to monitor a variety of reactions including hydrolyses and methanolyses under different conditions. Such reactions can be monitored by FAB MS with remarkable ease because the salt-free reaction media are compatible with the FAB matrix and no work-up is required. They are of general applicability in FAB strategies for polysaccharide and

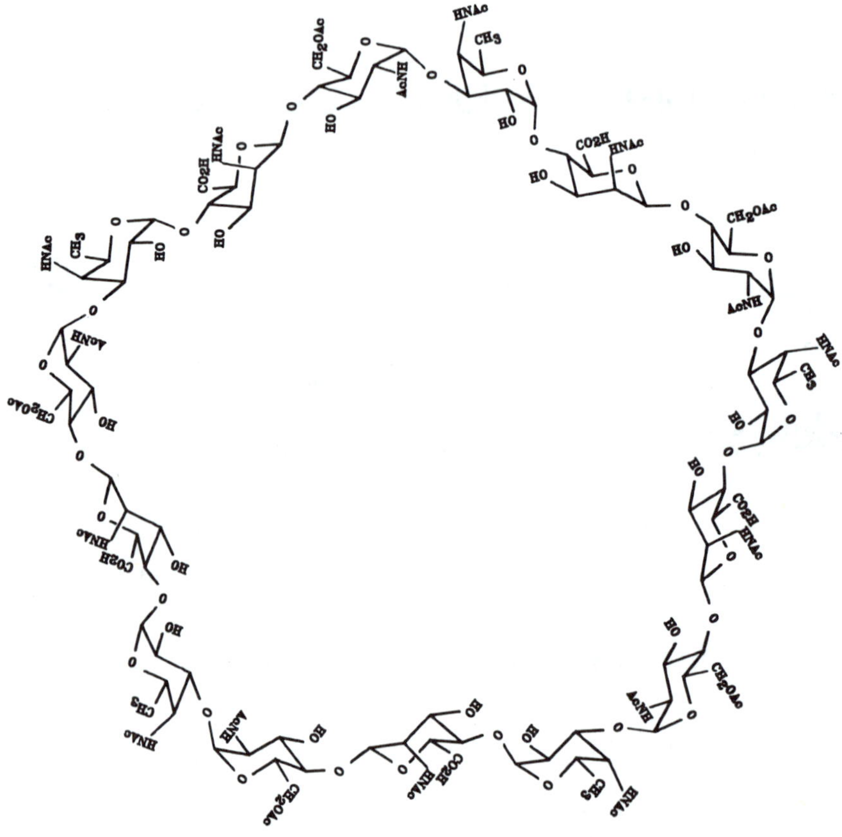

Figure 24: Structure of one of the components of the common antigen of enterobacteria (ECA).

glycoconjugate analysis. The type of data which can be obtained is illustrated in Figures 22 and 23. Mild methanolic conditions effected complete esterification of the uronic acid residues (Figure 22) while more severe conditions led to ring opening and cleavage between each repeating unit in ECA. Methanolysis of the glycosidic linkages within each repeating unit was less favoured (see Figure 23). The structure of one of the components of ECA is shown in Figure 24.

10.3.5. MAPPING OF NON-REDUCING STRUCTURES IN GLYCOCONJUGATES

FAB MS has played an important role in our structural studies on the lactosaminoglycans in cell surface glycoproteins of human erythrocytes (Fukuda *et al.*, 1984a,b), granulocytes (Spooncer *et al.*, 1984), embryonal carcinoma cells (Fukuda *et al.*, 1985b) and myelocytic leukemia cells (Fukuda *et al.*, 1985a). During the course of this work we made the following two important discoveries:

(i) Permethylated glycopeptides which are so large that their predicted molecular ions are far in excess of the high field ZAB's capabilities for analysis can, nevertheless, be examined as intact molecules because they produce abundant fragment ions within the working mass range of the instrument. We have successfully examined samples of M_r 10,000–20,000 by recording spectra in mass range A (see Introduction). We anticipate that even larger molecules will be amenable to similar analyses.

(ii) Large permethylated lactosaminoglycan-containing glycopeptides fragment very predictably and reproducibly to yield a diagnostic 'map' of fragment ions in which is contained information on all the non-reducing structures present in the glycopeptides. This 'mapping' procedure is potentially useful for the rapid indentification of antigenic moieties in glycoproteins and we further suggest that it constitutes an appropriate strategy for ascertaining the fidelity of carbohydrate synthesis in genetically engineered glycoproteins.

Figure 25 shows a typical 'map' of a permethylated glycopeptide. The sample was isolated from human embryonal carcinoma cells and has a permethylated M_r of about 14,000. The following non-reducing structures are present; Neu*N*Ac (376), HexHex*N*Ac (464), Neu*N*AcHexHex*N*Ac (825), Hex_2Hex*N*Ac_2 (913), Neu*N*Ac_2HexHex*N*Ac (1186), Neu*N*AcHex_2Hex*N*Ac_2 (1274), Hex_3Hex*N*Ac_3 (1362), Neu*N*$Ac_2$$Hex_2$Hex*N*$Ac_2$ (1635), Neu*N*AcHex_3Hex*N*Ac_3 (1723), Hex_4Hex*N*Ac_4 (1811), Neu*N*$Ac_2$$Hex_3$Hex*N*$Ac_3$ (2084), Neu*N*AcHex_4Hex*N*Ac_4 (2172), Neu*N*$Ac_2$$Hex_4$Hex*N*$Ac_4$ (2533), Neu*N*AcHex_5Hex*N*Ac_5 (2621), Neu*N*$Ac_2$$Hex_5$Hex*N*$Ac_5$ (2982), and Neu*N*AcHex_6Hex*N*Ac_6 (3070). The map shows very clearly that there is no fucosylation of this glycopeptide and that disialosyl-chains are abundant.

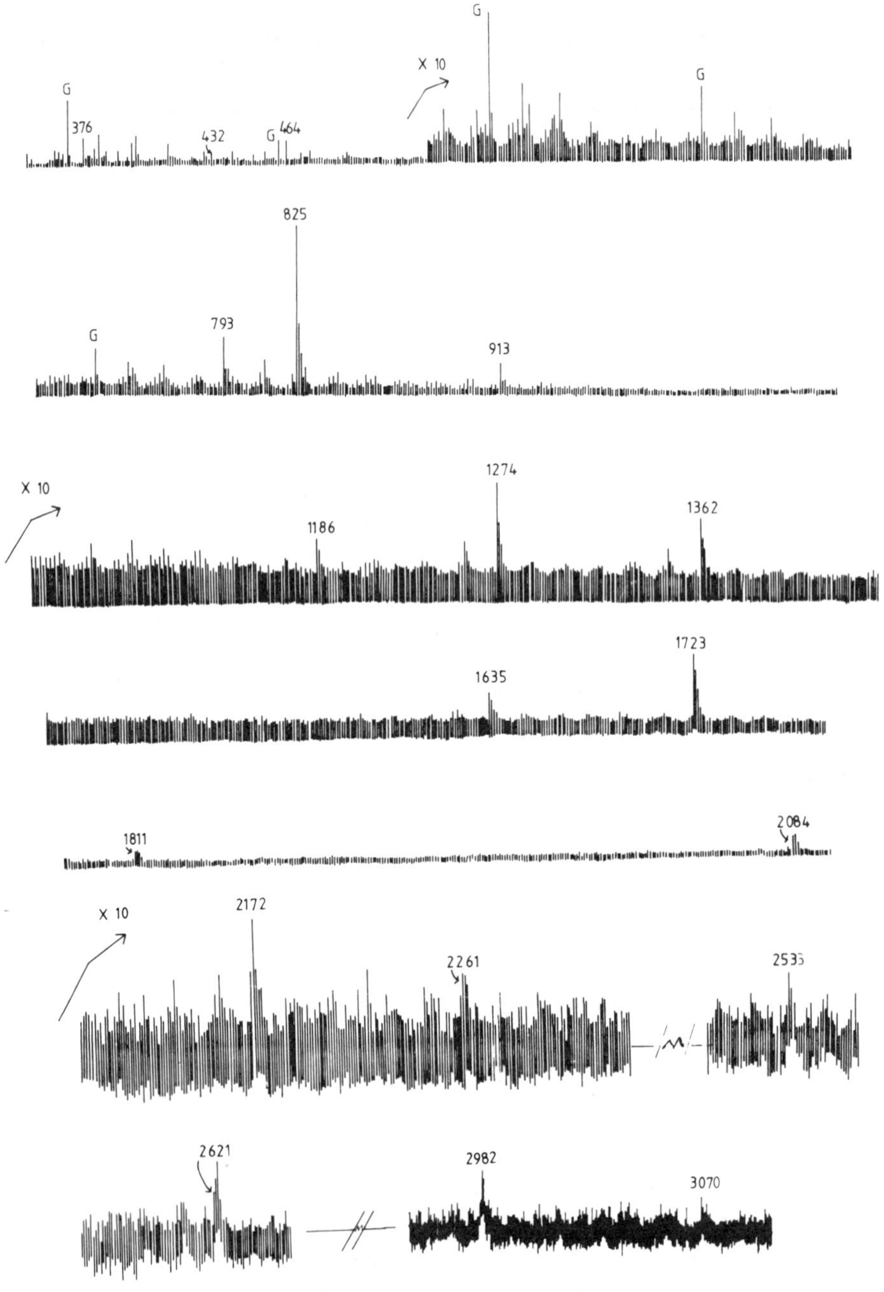

Figure 25: FAB spectrum of a permethylated glycopeptide (M_r about 14,000) isolated from human embryonal carcinoma cells. A_1-type fragment ions are present up to *m/z* 3070.

10.4. CONCLUSION

Future prospects for biopolymer characterization by FAB MS are very promising indeed. The studies described in this chapter should assist protein and carbohydrate chemists in their choice of appropriate protocols for structural analyses. It is very important that FAB MS is properly integrated into strategies encompassing the well established techniques of protein, carbohydrate and DNA technologies. We look forward to many more exciting applications of high field FAB MS.

ACKNOWLEDGMENTS

The protein chemical projects in this review were carried out under the direction of Professor H. R. Morris. We thank the MRC and SERC for financial support.

REFERENCES

Amara, S. G., Jonas, V., Rosenfeld, M. G., Ong, E. S., and Evans, R. M. (1982). Alternative RNA processing in calcitonin gene expression generates mRNAS encoding different polypeptide products, *Nature*, **298**, 240–244.

Barber, M., Bordoli, R. S., and Sedgwick, R. D. (1981). Fast atom bombardment mass spectrometry, in *Soft Ionisation Biological Mass Spectrometry* (Ed. H. R. Morris), pp. 137–152, Heyden, London.

Barber, M., Bordoli, R. S., Elliott, G. J., Sedgwick, R. D., Tyler, A. N., and Green, B. N. (1982). Fast atom bombardment mass spectrometry of bovine insulin and other large peptides, *J. Chem. Soc. Chem. Commun.*, **1982**, 936–938.

Barber, M., Bordoli, R. S., Elliott, G. J., Horoch, N. J., and Green, B. N. (1983). Fast atom bombardment mass spectrometry of human proinsulin, *Biochem. Biophys. Res. Commun.*, **110**, 753–757.

Bateman, A., Dell, A., and Morris, H. R. (1985). Fast atom bombardment mass spectrometry of large peptides: A rapid method for molecular weight analysis, *J. Appl. Biochem.*, **7**, 126–132.

Cairns, M. T., Elliot, D. A., Scudder, P. R., and Baldwin, S. A. (1984). Proteolytic and chemical dissection of the human erythrocyte glucose transporter. *Biochem. J.*, **221**, 179.

Davis, K. R., Darvill, A. G., Albersheim, P., and Dell, A. (1986). Oligogalacturonides released from sodium polypectate by endopolygalacturonic acid lyase are elicitors of phytoalexins in soybean, *Plant Physiol.*, **80**, 568–577.

Dell, A., and Ballou, C. E. (1983). Fast atom bombardment negative ion mass spectrometry of the mycobacterial methylglucose polysaccharide and lipopolysaccharides, *Carbohydr. Res.*, **120**, 95–111.

Dell, A., and Morris, H. R. (1982). Fast atom bombardment, high field magnet mass spectrometry of 6000 Dalton polypeptides, *Biochem. Biophys. Res. Commun.*, **106**, 1456–1462

Dell, A., and Taylor, G. W. (1984). High field magnet mass spectrometry of biological molecules, *Mass Spectrom. Rev.*, **3**, 357–394.

Dell, A., Oates, J. E., and Ballou, C. E. (1983a). FAB–MS of high mass oligosaccharides and glycoconjugates, in *Proc. 7th Int. Symp. on Glycoconjugates*, pp. 127–138, Lund, Sweden.

Dell, A., York, W. S., McNeil, M., Darvill, A. G., and Albersheim, P. (1983b). Structure determination of cyclic β-2 linked glucans secreted by Rhizobia and Agrobacteria, *Carbohydr. Res.*, **117**, 185–200.

Dell, A., Oates, J. E., Lugowski, C., Romanowska, E., Kenne, L., and Lindberg, B. (1984). The Enterobacterial common antigen, a cyclic polysaccharide, *Carbohydr. Res.*, **133**, 95–104.

Egge, H., Dabrowski, J., and Hanfland, P. (1984). Structural analysis of blood-group ABH, I, i, Lewis and related glycosphingolipids. Application of FAB mass spectrometry and high resolution proton NMR, *Pure Appl. Chem.*, **56**, 807–819.

Forsberg, L. S., Dell, A., Walton, D. J., and Ballou, C. E. (1982). Revised structure for the 6-O-methylglucose polysaccharide of Mycobacteria smegmatis, *J. Biol. Chem.*, **257**, 3555–3563.

Fukuda, M., Dell, A., and Fukuda, M. M. (1984a). Structure of fetal lactosaminoglycan, the carbohydrate moiety of band 3 isolated from human umbilical cord erythrocytes, *J. Biol. Chem.*, **259**, 4782–4791.

Fukuda, M., Dell, A., Oates, J. E., and Fukuda, M. N. (1984b). Structure of branched lactosaminoglycan, the carbohydrate moiety of band 3 isolated from adult human erythrocytes, *J. Biol. Chem.*, **259**, 8260–8273.

Fukuda, M., Spooncer, E., Oates, J. E., Dell, A., and Klock, J. C. (1984c). Structure of sialylated fucosyl lactosaminoglycan isolated from human granulocytes, *J. Biol. Chem.*, **259**, 10925–10935.

Fukuda, M., Bothner, B., Ramsamooj, P., Dell, A., Tiller, P. R., Varki, A., and Kock, J. C. (1985a). Structures of sialylated fucosyl polylactosaminoglycan isolated from chronic myelogenous leukaemia cells, *J. Biol. Chem.*, **260**, 12957–12967.

Fukuda, M. N., Dell, A., Oates, J. E., and Fukuda, M. (1985b). Embryonal lactosaminoglycan, *J. Biol. Chem.*, **260**, 6623–6631.

Fukuda, M. N., Dell, A., Oates, J. E., Wu, P., Klock, J. C., and Fukuda, M. (1985c). Structures of glycosphingolipids from human granulocytes, *J. Biol. Chem.*, **260**, 1067–1082.

Fukuda, M. N., Bothner, B., Lloyd, K. O., Tiller, P. R., and Dell, A. (1986). Structures of glycosphingolipids isolated from human embryonal carcinoma cells, *J. Biol. Chem.*, **261**, 5145–5153.

Kamisango, K., Saadat, S., Dell, A., and Ballou, C. E. (1985). Pyruvylated glycolipids from Mycobacterium smegmatis: nature and location of the lipid components, *J. Biol. Chem.* **260**, 4117–4121.

Knott, T. J., Rall, S. C., Jacobson, S. F., Scott, J., *et al.*, (1985). Human apolipoprotein B: structure of carboxyl terminal domains, sites of gene expression, and chromosomal localization, *Science*, **230**, 37–43.

Morris, H. R. (1982). Biomolecular mass spectrometry, *Int. J. Mass Spectrom. Ion Phys.*, **45**, 331–342.

Morris, H. R., Panico, M., Barber, M., Bordoli, R. S., Sedgwick, A. D., and Tyler, A. N. (1981). Fast atom bombardment: a new mass spectrometric method for peptide sequence analysis, *Biochem. Biophys. Res. Commun.*, **101**, 623–631.

Morris, H. R., Dell, A., Etienne, A. T., Judkins, M. B., McDowell, R. A., Panico, M., and Taylor, G. W. (1982a). New methods for the characterisation of biopolymers, *Pure Appl. Chem.*, **54**, 267–279.

Morris, H. R., Panico, M., Karplus, A., Lloyd, P. E., and Riniker, B. (1982b). Elucidation by FAB/MS of the structure of a new cardioactive peptide from Aplysia, *Nature*, **300**, 643–645.

Morris, H. R., Panico, M., and Taylor, G. W. (1983). FAB-MAPPING of recombinant-DNA protein products, *Biochem. Biophys. Res. Commun.*, **117**, 299–305.

Morris, H. R., Panico, M., Etienne, A. T., Tippins, J. R., Girgis, S. I., and MacIntyre, I. (1984). Isolation and characterisation of human calcitonin-gene-related peptide, *Nature*, **308**, 746–748.

Morris, H. R., Panico, M., Blench, I., and Etienne, A. T. (1986a). Manuscript in preparation.

Morris, H. R., Panico, M., Blench, I., Etienne, A. T., Ho, M. W., and Povey, S. (1986b). Manuscript in preparation.

Mueckler, M., Caruso, C., Baldwin, S. A., Panico, M., Blench, I., Morris, H. R., Lienhard, G. E., Allard, W. J., and Lodish, H. F. (1985). Sequence and structure of a human glucose transporter, *Science*, **229**, 941–945.

Naik, S., Oates, J. E., Dell, A., Taylor, G. W., Dey, P. M., and Pridham, J. B. (1985). A novel mass spectrometric procedure for the rapid determination of the types of carbohydrate chains present in glycoproteins, *Biochem. Biophys. Res. Commun.* **132**, 1–7.

Nothnagel, E. A., McNeil, M., Albersheim, P., and Dell, A. (1983). Galacturonic acid oligosaccharides from plant cell walls elicit phytoalexins, *Plant Physiol.*, **71**, 916–926.

Spooncer, E., Fukuda, M., Klock, J. C., Oates, J. E., and Dell, A. (1984). Isolation and characterisation of poly-fucosylated lactosaminoglycan from human granulocytes, *J. Biol. Chem.*, **259**, 4792–4801.

Mass Spectrometry in Biomedical Research
Edited by S. J. Gaskell
© 1986 John Wiley & Sons Ltd.

Chapter 11

STRUCTURAL ELUCIDATION OF COMPLEX CARBOHYDRATES

Vernon N. Reinhold

11.1. INTRODUCTION

Studies of the molecular reactions occurring in living systems offer an enormous opportunity for the discovery of basic scientific principles and for the application of these principles to the well-being of mankind. Over the past decades we have benefited greatly from studying cell-free and simple cellular systems, and have thereby gained much insight into metabolic processes. Clarification of these events has been accomplished by a detailed structural analysis at the molecular level.

Contemporary cell biology is now focusing increasing attention beyond the confines of the cell, on to factors that relate to the interplay of cells with their environment. Immune surveillance, angiogenesis, cellular differentiation and adhesion are events associated with multicellular systems. Cell–cell interactions and the integrated development and differentiation so characteristic of living tissues are manifested within an extracellular matrix which depends on membranes, tissue surfaces and the ordered transport of signals by and through these barriers. Close examination of this complex cellular milieu reveals that it is resplendent with a vast array of glycosylated molecules. This highly specific molecular surface represents the biological 'lock' against random events and unspecified body response. The need to know the detailed structural features of these molecules is compelling. An understanding of compound class or component composition has little meaning when one is trying to ferret out the biochemistry of cellular events or to devise synthetic strategies to augment, complement or retard these processes.

The relatively recent understanding that glycosylated molecules are important intercedents in the proper functioning of multicellular organisms is consistent with what we know of physiological demands for high biological specificity. There is, however, a serious lack of analytical capability to determine these structures. Carbohydrate analytical chemistry is still a 'cottage industry' when compared to the advances and great sophistication available for other biopolymers. The reasons for this can be related to the fact that oligosaccharide structures involve a multiplicity of linkage type, chain branching, structural isomerisms, and great diversity of conjugate structures. These are features which provide great molecular specificity to the organism but complicate structural analysis.

The recent technological developments for high molecular weight analysis by mass spectrometry (MS) would appear to be coming to fruition at a most propitious time. The ionization of very large polar molecules, together with ion transmission, resolution and detection at great sensitivity are making MS instrumentation a necessary component in carbohydrate research. Thus, two important factors have emerged within the past few years, namely the tremendous need to know carbohydrate structural detail and the knowledge that state-of-the-art mass spectrometry instrumentation will add new dimensions to biopolymer research.

Finally, in contrast to the determination of structure for nucleic acid and amino acid biopolymers, carbohydrate oligomers have, as yet, no effective chemical method with the required sensitivity and specificity to adequately approach the topics described above. It would appear that mass spectrometry and its adjunct techniques will have a decisive role in clarification of the structure–function relationship for many carbohydrate materials.

In this chapter we will briefly describe recent developments pertinent to the determination of an oligosaccharide structure and point out those contributions which show significant promise. Because of space limitations there has been no attempt to be comprehensive and the selection of topics reflects only the bias of the author.

11.2. STRUCTURAL CHARACTERIZATION

A complete polysaccharide structural analysis requires the determination of molecular weight and sequence, the linkage points (positional identification of the hydroxyl group(s) involved in glycosidic linkages), branching and anomeric configuration.

11.2.1. LINKAGE AND BRANCHING DETERMINATION

Problems of linkage determination have classically been approached by methylation analysis; sample permethylation is followed by the structural identification of hydrolyzed, reduced and acetylated (Lindberg, 1972) or methanolyzed and acetylated (Fournet *et al.*, 1981) products. The essential features of methylation analysis using the alditol acetate derivative (Lindberg, 1972) entail acid hydrolysis of the permethylated sample, hydride reduction and acetylation. Component sugars are determined by chromatographic comparison with standards, by mass spectrometry, or by a combination of both. These chemical procedures, however, lack the sensitivity, and in some cases the specificity, to approach many current polysaccharide problems. Particular problems are : (i) the inability to differentiate 4-linked aldopyranosyl from 5-linked aldofuranosyl moieties; (ii) the selective destruction of specific moieties during acid-catalyzed polymer degradation; (iii) the extensive sample processing necessary to generate alditol acetates is inconsistent with the demands for sensitivity. Methanolysis, because of the limited sample preparation, offers improvements in sensitivity but this advantage appears to be offset by the generation of new isomeric components (increasing chiral complexity at C-1) and the resistance to methanolic transglycosidation at amino sugar linkages.

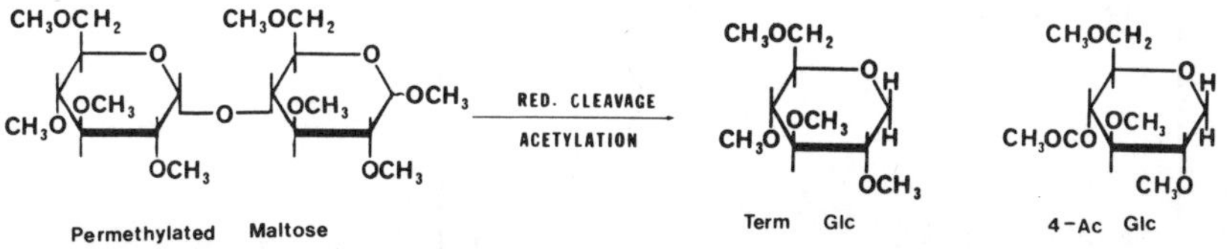

Scheme 1

An attractive alternative approach to linkage analysis is the direct reduction of the glycosidic bonds of permethylated samples using triethylsilane (Rolf and Gray, 1982). Following acetylation these products possess excellent gas chromatographic and mass spectral characteristics. Scheme 1 shows an example of this approach. GC–MS allows direct structural assignment of sugar components and type of linkage. Application of this procedure to a tetrasaccharide (stachyose) with three different hexoses (galactose, glucose, fructose) linked by two different linkages (1→6, 1→2) is demonstrated in Figure 1 (Van Langenhove and Reinhold, unpublished data). GC–MS shows five major peaks. The first three of these peaks (1–3) were found to be fully methylated deoxyhexitols (terminal residues) and the latter two peaks (4 and 5) monoacetylated, methylated moieties (internal sequence residues). Peaks four and five were identified as the 6-*O*-acetyl derivatives of 1-deoxyglucitol and galactitol, respectively. The

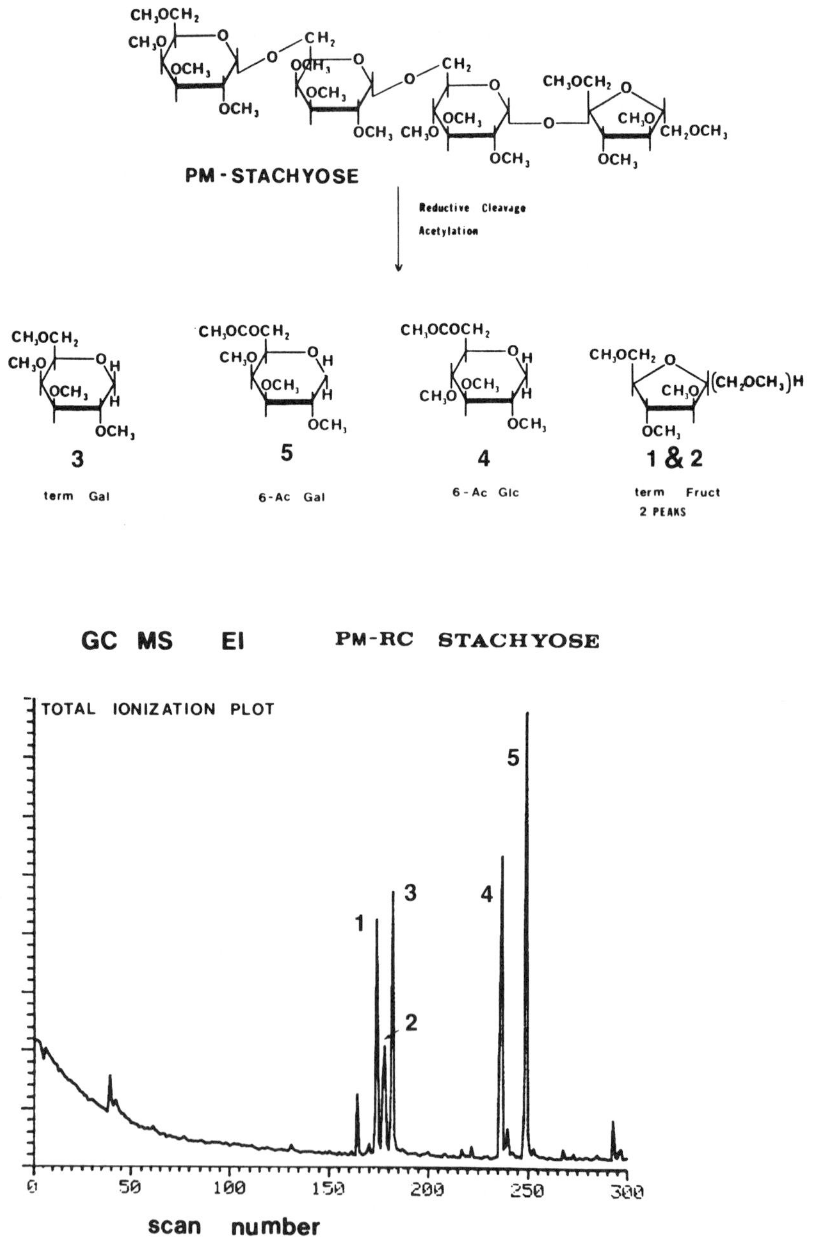

Figure 1: Total ion plot of permethylated, reductively cleaved and acetylated stachyose. Data generated by GC–MS on DB-225 capillary columns using electron ionization (Van Langenhove and Reinhold, unpublished results).

importance of GC retention time is exemplified by the fact that the first two peaks provided identical mass spectra (as did peaks 4 and 5). These two peaks have been shown in separate experiments to arise from a single component, permethylated fructose. The production of two components from fructosyl residues has been reported (Rolf and Gray, 1984). We have observed a similar effect for neuraminic acid-containing oligosaccharides (Van Langenhove and Reinhold, unpublished data). The GC and MS characteristics of a series of non-acetylated, mono- and di-acetylated, methylated 1-deoxy hexitols have been published (Van Langenhove and Reinhold, 1985). The GC analysis used a 30 metre DB-225 capillary column which provided adequate chromatographic resolution of all components. As expected and demonstrated above in Figure 1, stereoisomeric structures cannot be differentiated by their electron ionization (EI) mass spectra alone and structural assignment requires knowledge of the GC retention time. However, positional isomers (acetylated hydroxyl position, i.e., linkage point) are easily resolved chromatographically and readily determined by their EI mass spectra. This EI-specificity of four monoacetyl-1,5-anhydro trimethyl glucitols is demonstrated in Figure 2 for the 2-*O*-, 3-*O*-, 4-*O*- and 6-*O*-acetyl analogues.

A standard approach we use for the determination of oligosaccharide linkage and branching composition is to make a survey GC–MS run using chemical ionization. This provides molecular weight information for the determination of terminal, linear and branching residues (non-acetylated, mono-, di- and tri-*O*-acetyl derivatives) while the GC retention indices provide information about the specific carbohydrate moiety. These preliminary data are combined with information gained from a second GC–MS run using EI and relying on fragmentation to determine the linkage position and the degree of branching.

Boron trifluoride etherate has been reported to be a less effective catalyst than trimethylsilyl trifluoromethanesulphonate for reduction of certain linkages, and these features could be used to advantage in structural characterization (Rolf *et al.*, 1985). Such specificity may, however, reflect compound solubility (especially for large polysaccharides) rather than kinetics of linkage reduction; these more subtle features require further investigation. Limited studies of linkage specificity using high-performance liquid chromatography with fluorescently labelled disaccharides (which allowed focus on only one linkage type) have been reported with boron trifluoride as a catalyst (Reinhold *et al.*, 1982).

11.2.2. SEQUENCE DETERMINATION

Sequence information obtained by MS is dependent on a periodic polymer fragmentation that is characteristic of monomer array. For oligosaccharides, the major fragments from these processes can be explained by cleavage on either side of the glycosidic oxygen followed by a proton transfer. These cleavages

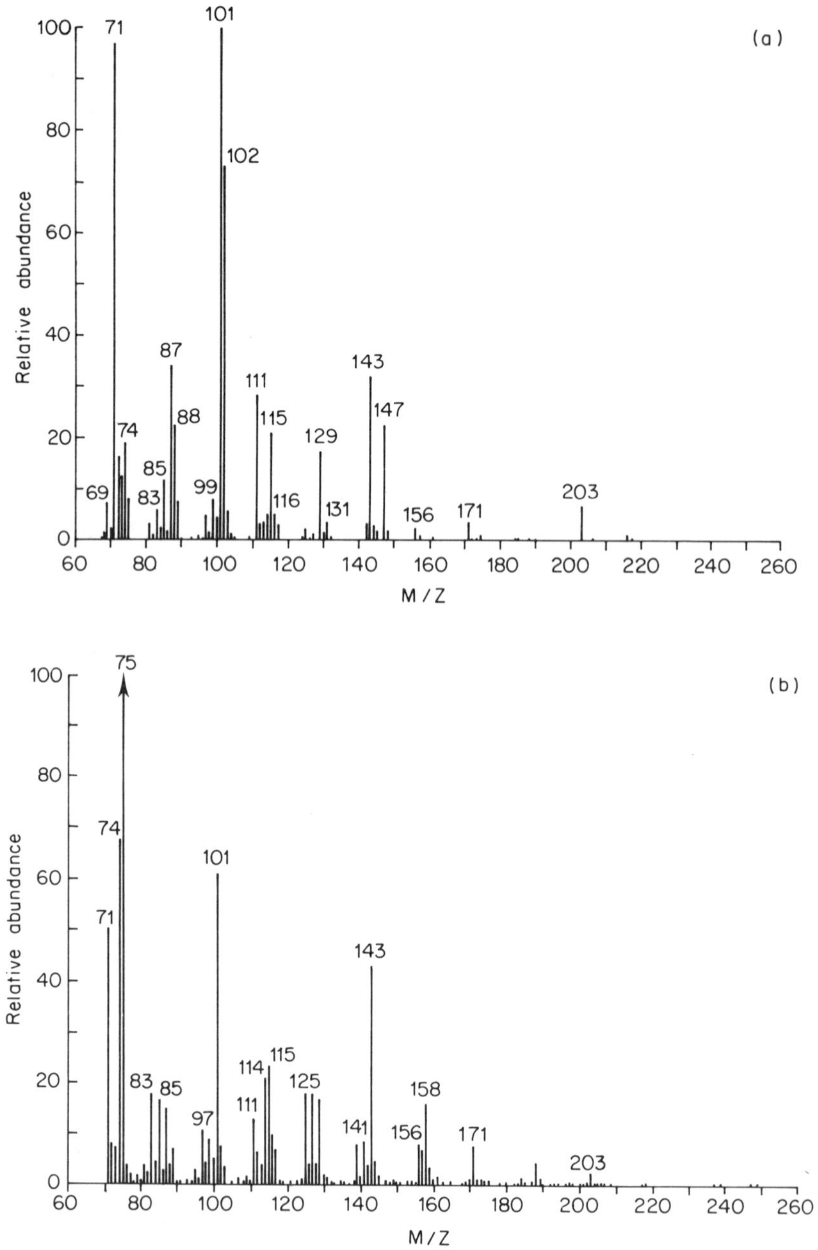

Figure 2: Electron ionization MS of reductive cleavage products of oligosaccharides (Van Langenhove and Reinhold, 1985). *Reproduced by permission of Elsevier Scientific Publishers.*

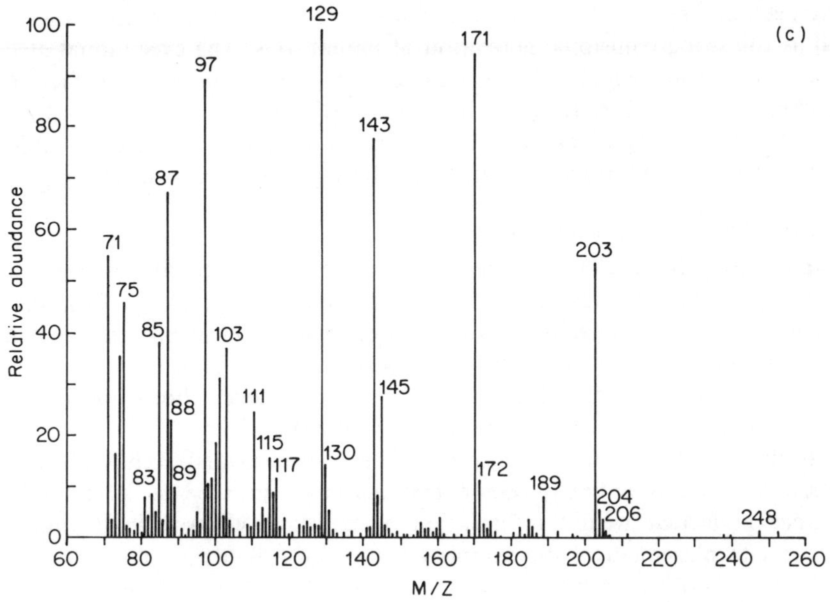

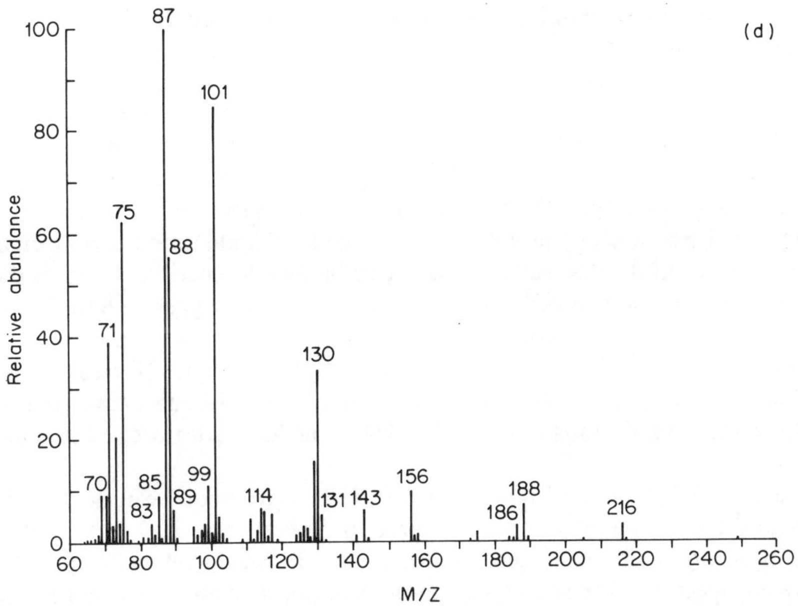

a) 1,5-anhydro-2-*O*-acetyl-3,4,6-tri-*O*-methyl-*D*-glucitol, from 2-*O*-linked saccharides.
b) 1,5-anhydro-3-*O*-acetyl-2,4,6-tri-*O*-methyl-*D*-glucitol, from 3-*O*-linked saccharides.
c) 1,5-anhydro-4-*O*-acetyl-2,3,6-tri-*O*-methyl-*D*-glucitol, from 4-*O*-linked saccharides.
d) 1,5-anhydro-6-*O*-acetyl-2,3,4-tri-*O*-methyl-*D*-glucitol, from 6-*O*-linked saccharides.

have been observed under chemical and fast atom bombardment ionization, as well as following collisional activation of parent ions. The exceptional glycosidic lability can probably be related to bond energetics and the fact that intervening ring rupture requires the cleavage of two bonds for depolymerization. Hydroxyl group blocking by derivatization would be expected to stabilize the ring and thereby provide spectra that show greater glycosidic cleavage.

11.2.2.1. Direct chemical ionization

Chemical ionization (CI) mass spectrometry provides a technique to maximize molecular and high mass information by replacing the highly exothermic process of electron ionization with a low energy, chemical process involving ion–molecule reactions. The topic has been reviewed by one of the major contributors to this area (Munson, 1977) and an excellent book covering the topic has recently been published (Harrison, 1983).

The development of 'direct' or 'in-beam' mass spectral analysis has followed a somewhat sinuous pathway, being first discussed in 1958 as a method to enhance molecular weight information, when using electron ionization, by appropriate placement of the sample (Reed, 1958). Baldwin and McLafferty (1973) have provided a basis for much of the subsequent work leading to the development of direct chemical ionization (DCI). There has been some confusion in the literature about the name DCI (direct or desorptive chemical ionization) and the actual placement of sample. In the discussions that follow we will use the term 'direct' to mean placement of sample on an electrically resistant wire (emitter) which can be inserted directly into an ion plasma. The materials are 'distilled' from this emitter by a programmed heating current. Samples analyzed by this direct method frequently provide mass spectra with enhanced high mass fragments. Simple resistance wires can replace commercial emitters for a fraction of the cost and these can be coated with a protective surface to augment the desorptive process (see below). Reports have shown the DCI technique to be successful on a variety of sample types: nucleic acids (Virelizier *et al.*, 1983), glycolipids (Carr and Reinhold, 1984; Reinhold and Carr, 1983), peptides (Carr and Reinhold, 1982), phospholipids (Dessort *et al.*, 1983), and carbohydrates (Reinhold and Carr, 1982; Harada *et al.*, 1983a; Takeda *et al.*, 1983; de Jong *et al.*, 1979). The topic has been reviewed (Cotter, 1980).

The essential features and limitations of the DCI technique can be directly related to the heat initiated desorptive process. Spectra acquired early during sample desorption (minimal heat) show molecular weight-related ions only. Subsequent spectra, taken at a higher emitter heating currents, show an increase in low mass fragments with decreasing abundance of molecular weight-related ions. This increasing pyrolytic component of the DCI method is advantageous in that it provides additional structural detail within a single sample analysis. However, as one proceeds to the analysis of higher molecular weight and more polar samples, the thermal energy needed to initiate desorption becomes excessive and the spectra exhibit a proponderance of pyrolytic fragments.

The precise mechanism of ion generation by DCI remains unclear. However, in the light of current trends to demonstrate concordance between different ionization techniques, we would like to integrate an attractive model recently discussed (Vestal, 1983) with observations by others (Cotter, 1980; Stillwell *et al.*, 1983) and ourselves. Much evidence has been presented to support a process involving evaporation of polar neutrals and preformed ions from heated surfaces (Stoll and Rollgen, 1980, 1981; Ohashi *et al.*, 1981; Cotter and Yergey, 1981; van der Peyl *et al.*, 1981; Cotter, 1980). These processes appear to fall into two categories: those that are assisted by particle bombardment and those effected by superheating a condensed phase. Both processes cause disruption of the surface and ejection of droplets, clusters, or packets. In the DCI context, this surface disruption could be initiated by heating, inducing bubble formation at the sample–support interface (superheating). Bubbles so formed would migrate to the surface and eventually spray droplets into the ion plasma of the CI source to become collisionally declustered and ionized. The energy resident with the sample following droplet disintegration and the thermodynamics of ion–molecule interactions are reflected in the resultant spectrum. These considerations presuppose two important factors: little sample interaction with its environment and a finite molecular depth for bubble formation. However, when the thermal energy required to disrupt the interactions between molecules is greater than that of intramolecular bonding, or when desorption continues to monolayers, other explanations must be embraced. As discussed above, it has been our observation that with samples of higher molecular weight and polarity, fragmentation increases with a corresponding decrease in molecular weight-related ions and this is inconsistent with a cluster-only desorption process. This suggests a pyrolytic contribution which may augment the resultant spectra. We (Reinhold *et al.*, 1982) and others (Baldwin and McLafferty,1973; Hunt *et al.*, 1977; Hansen and Munson, 1978; Thenot *et al.*, 1979; Constantin *et al.*, 1980; Sisseret *et al.*, 1983; Takeda *et al.*, 1985) have taken measures to diminish DCI-related pyrolysis and this has extended the structural information available for more polar and higher molecular weight samples. It is with these analytes that the ion plasma proximity ('direct') is critical and the greater molecular stability observed in DCI may simply be a more rapid sample ionization and extraction previous to the effects of pyrolytic decomposition, as previously suggested (Cotter, 1980).

For carbohydrates, we have found DCI most successful on permethylated materials when using ammonia as the reagent gas. Excellent sequence information can be obtained for analytes up to 3000 daltons and the technique appears to be three to five times more sensitive than fast atom bombardment mass spectrometry. The principal features and advantages of ammonia DCI are illustrated in Figure 3 for a permethylated heptasaccharide. The molecular

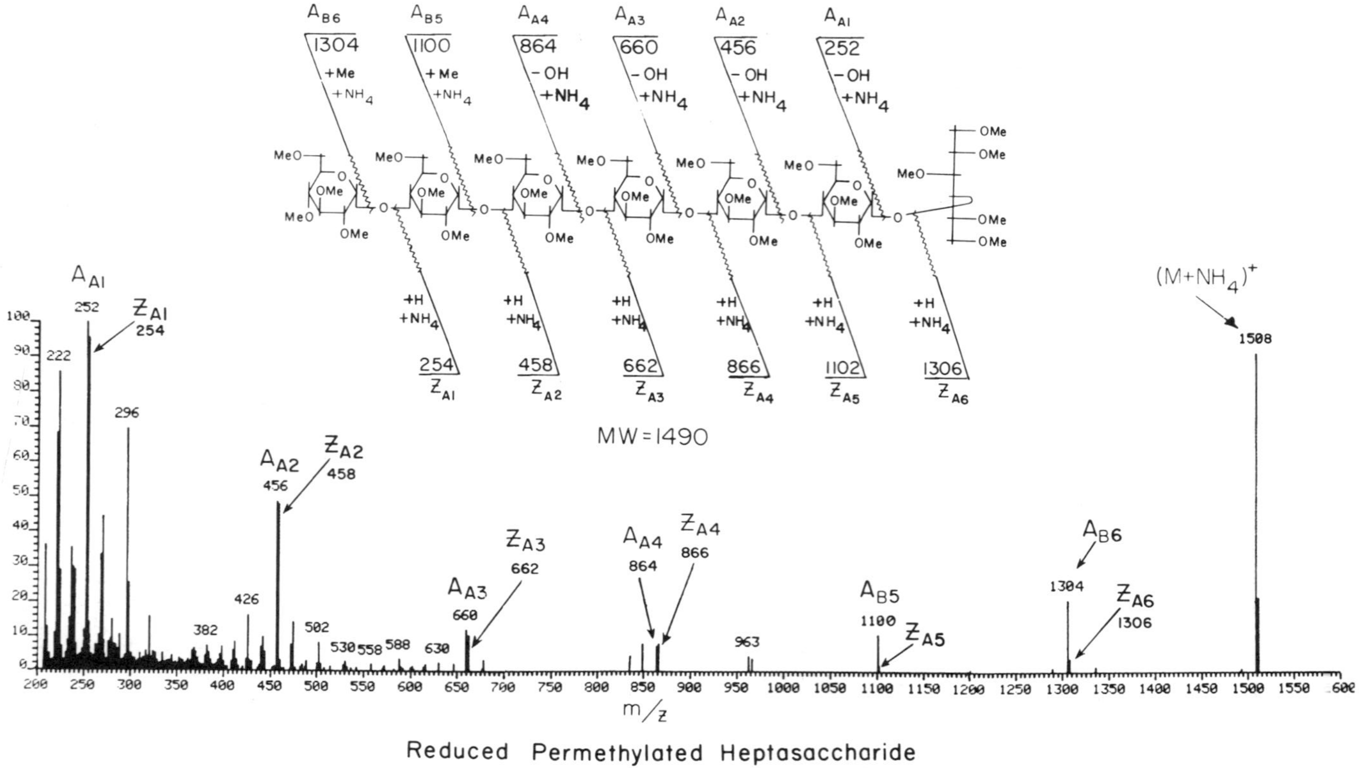

Figure 3: Direct chemical ionization MS of a reduced and permethylated glucoseptulose a(1–4)glycan using ammonia as a reagent gas. *Reprinted with permission from Reinhold and Carr (1982). Copyright © 1982 American Chemical Society.*

weight can be determined from the ammonium adduct ion at *m/z* 1508, and sequence ions are observed for each saccharide moiety (Reinhold *et al.*, 1982). Some of these fragments are isomeric but can be differentiated by reduction of the glycose end of the polysaccharide with NaB^2H_4. These studies, in combination with the use of isotopically labelled reagent gases ($^{15}N^1H_3$, $^{14}N^2H_3$), have provided a detailed understanding of oligosaccharide fragmentation. The $^{15}N^1H_3$ experiments have consistently shown that all ion fragments are incremented by one mass unit. The clustering of NH_4^+ with neutral species provides a situation in which a large portion of desorbed sample and its fragments are observed. This fact has been illustrated earlier (Reinhold and Carr, 1983) and is in contrast with many other MS techniques where neutral species are undetected. A summary of neutral oligosaccharide fragmentation is presented in Scheme 2. From this Scheme and the fragment ion abundance in Figure 3 it can be seen that all major fragments in ammonia DCI analysis are related to glycosidic cleavage, an important feature when studying sequence.

The application of ammonia DCI to glycolipid materials has proven equally successful by providing molecular weight, sequence and lipid ester detail in a single analysis (Carr and Reinhold, 1984). This is illustrated in Figure 4 for a permethylated and reduced ceramide trihexoside. Reduction of this amide fatty acid conjugate, previous to analysis (Karlsson *et al.*, 1974), enhances DCI elimination of the methylated alkyl amine. These fragments provide a separate evaluation of the fatty ester composition (i.e. *m/z* 342, 368, 370, 398, 400) which is frequently the major cause of $(M + H)^+$ heterogeneity. The opportunity to compare these two groups of ions is of considerable value especially for relating structural features to the sphingosine moiety. Carbohydrate sequence ions can be observed which are identical to those in Figure 3 (e.g. *m/z* 254, 458, 662) and represent the major fragments in the spectra. All of these fragments are ammonium adduct ions. Alterations in structure are readily observed by DCI as exemplified below with a related sample. Extension of the carbohydrate chain by an amino sugar moiety (globotetraglycosylceramide, GL-4) provides the same sequence information with the appropriate shift in mass (Figure 5). Because of the basicity of this amino sugar residue, these sequence ions *m/z* 266, 470, 674, 878) appear as protonated fragments and not as ammonium adduct ions as in Figure 3 and 4. However, the spectral information is not compromised and a complete structural assignment is still very easy (Figure 5). The spectrum suggests that the terminal amino sugar residue is protonated and the sequence ions are derived by cleavage or pyrolysis at glycosidic linkages remote from the charged site (for further discussion of point see Jensen *et al.*, 1985).

11.2.2.2. Fast atom bombardment (FAB) mass spectrometry

Because of the heating required to initiate DCI desorption, extension of this technique to very polar and high molecular weight glycoconjugates is not

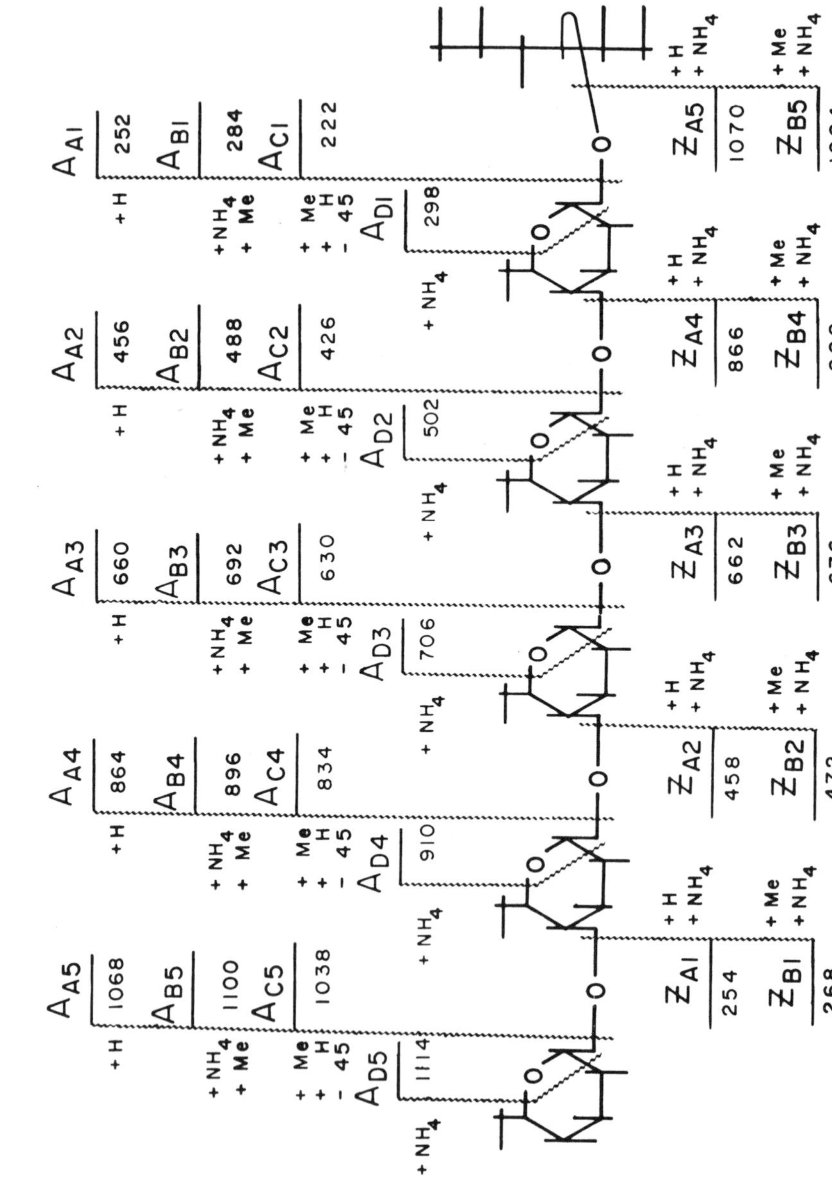

AA1 252
AB1 284
AC1 222
AD1 298
AA2 456
AB2 488
AC2 426
AD2 502
AA3 660
AB3 692
AC3 630
AD3 706
AA4 864
AB4 896
AC4 834
AD4 910
AA5 1068
AB5 1100
AC5 1038
AD5 1114
ZA1 254
ZB1 268
ZA2 458
ZB2 472
ZA3 662
ZB3 676
ZA4 866
ZB4 880
ZA5 1070
ZB5 1084
+H
+NH4
+Me
-45

Scheme 2

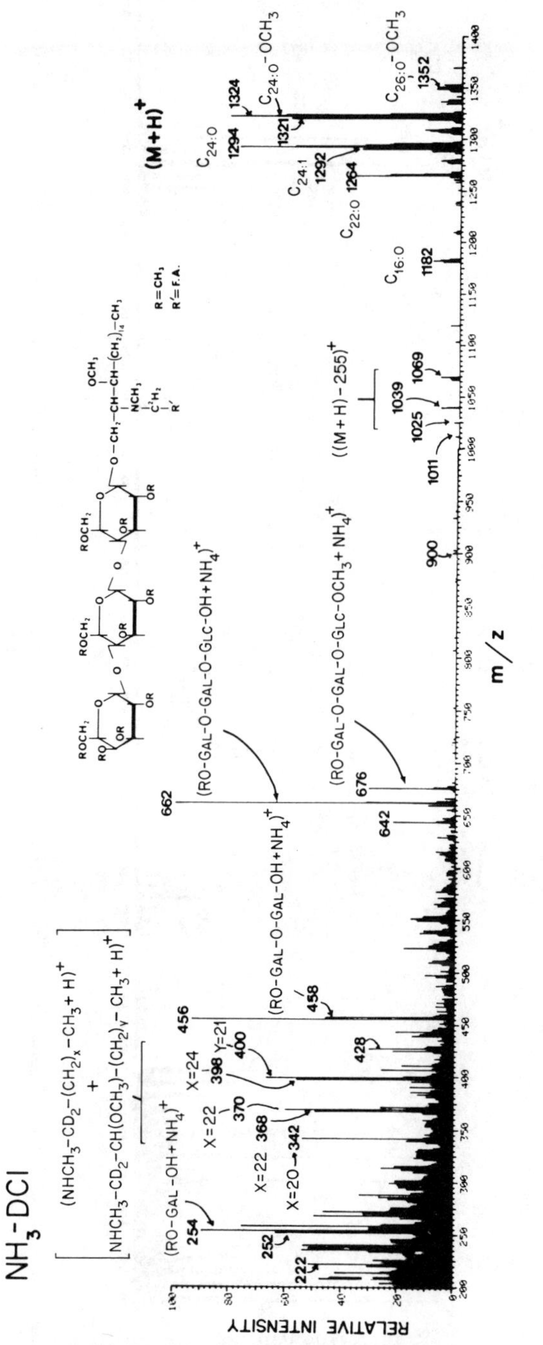

Figure 4: Direct chemical ionization MS of a permethylated and reduced cerebotrihexoside. Ammonia was used as the reagent gas. (Carr and Reinhold, 1984.)

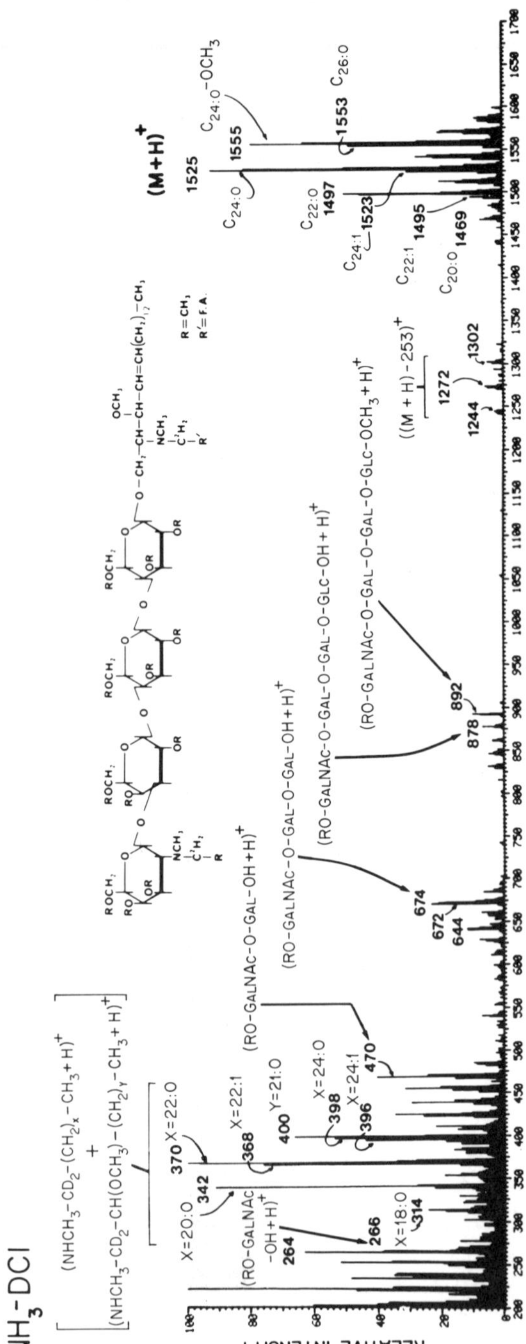

Figure 5: Direct chemical ionization MS of a permethylated and reduced GL-4. Ammonia was used as the reagent gas. (Carr and Reinhold, 1984.)

always successful. The energy needed to overcome sample–sample and sample–substrate interactions exceeds the molecular stability and the resultant spectra exhibit an increasing amount of sample pyrolysis. Beam-assisted desorption from a liquid matrix, using photons, ions or atoms, reduces this problem, and the technique of ionization by FAB (Barber *et al.*, 1981) has seen wide application in biopolymer research. An example is the molecular weight analysis of beta-glycans from Gram-negative bacteria (Dell and Ballou, 1983a,b). We have also investigated these beta-glycans for their possible role in osmotic adaptation (Miller *et al.*, 1986). As anticipated, permethylation and reductive cleavage (Section 11.2.1) followed by GC–MS showed only one component, that of 2-*O*-acetyl-1,5-anhydro-3,4,6-tri-*O*-methyl glucitol. Thus, the glycan was a 1-2 linked cyclic polysaccharide consisting only of glucose. The molecular weight distribution of the cyclic glycans from *Agrobacterium tumefaciens* indicated $(\text{glucose})_{19}$ to be in highest concentration with an envelope of homopolymers with a *dp* range from 17 through 23 (Figure 6). Weaker signals could be detected well below and above this *dp* range. It is difficult to determine the concentration of these materials because of the inability to find an appropriate internal standard. The relative distribution, however, may be more important for following the metabolic course of these periplasmic components.

An example of FAB desorption to provide mass spectral fragments indicative of oligomer sequence, in addition to molecular weight information, has been reported for a mycobacterial 6-*O*-methyl glucose polysaccharide (Dell and Ballou, 1983a,b). Two structural factors may have enhanced these sequence ion fragments; first, the sample was partially methylated (naturally occurring), a feature which seems to direct fragmentation to glycosidic rupture (see discussion above); second, the oligomer was terminated with an acidic group. Under such circumstances, charged terminating groups provide a series of fragments with incrementing glycosidic residues. (Note similar effects under DCI, Figure 5). In addition, both of these structural features may affect the distribution within the matrix for more favourable desorption.

Obtaining sequence information, however, is usually less predictable and more difficult. For oligosaccharides, the complications appear more in the selection of a suitable matrix and instrumental operation than from limitations imposed by the sample. This can be demonstrated with an amino glycoside antibiotic, viridopentose A (Harada *et al.*, 1983b). For greater sensitivity, positive ion extraction using a thioglycerol matrix provided abundant molecular weight-related ions (Figure 7; note that in this analysis the molecular weight-related ions are divided between the $(M + H)^+$ and $(M + Na)^+$ species), while greater structural detail was obtained with triethanolamine as the matrix and negative ion extraction (Figure 7; Santikarn and Reinhold, unpublished data). It should be noted that the matrix and ion extraction mode greatly modify the sequence ions and in some cases suppress them totally (e.g., see *m*/*z* 564 and 334, Figure 7b). Signal longevity, molecular ion intensity

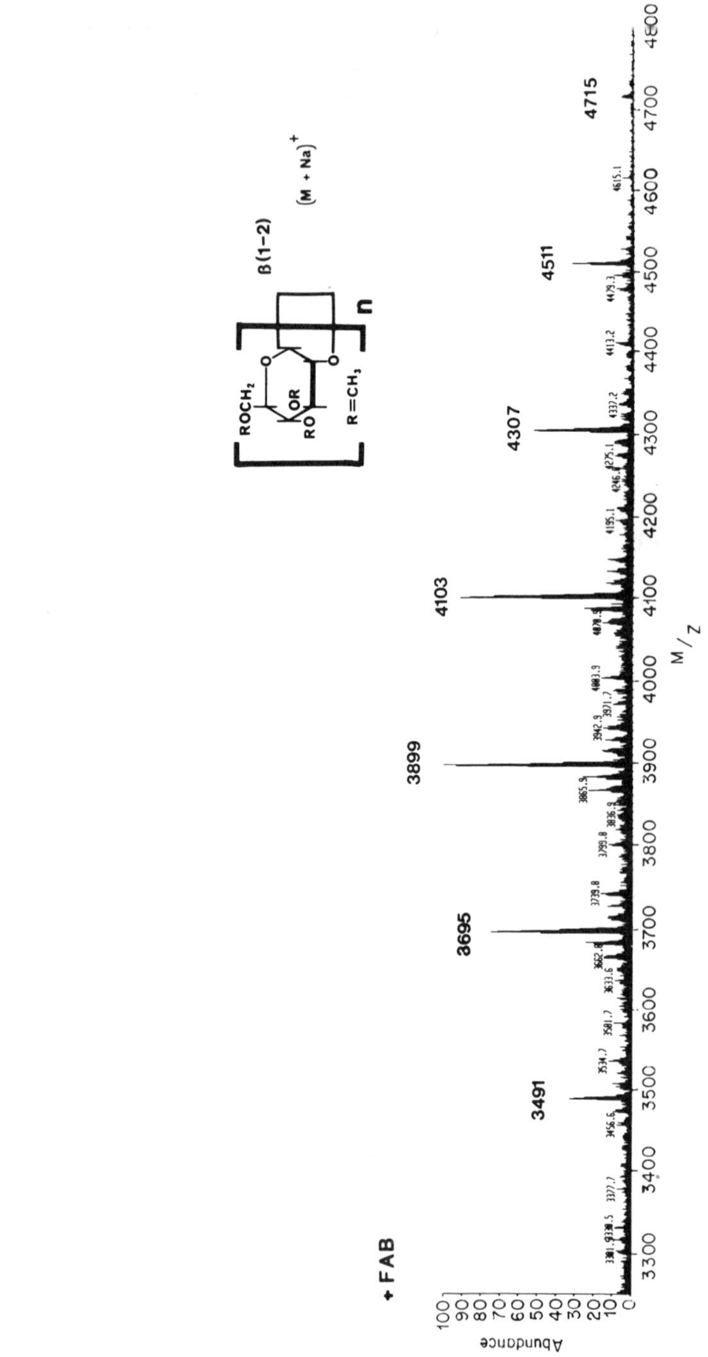

Figure 6: Fast atom bombardment MS of β(1–2) glycan isolated from *Agrobacterium tumefaciens*. (Miller *et al*., 1986)

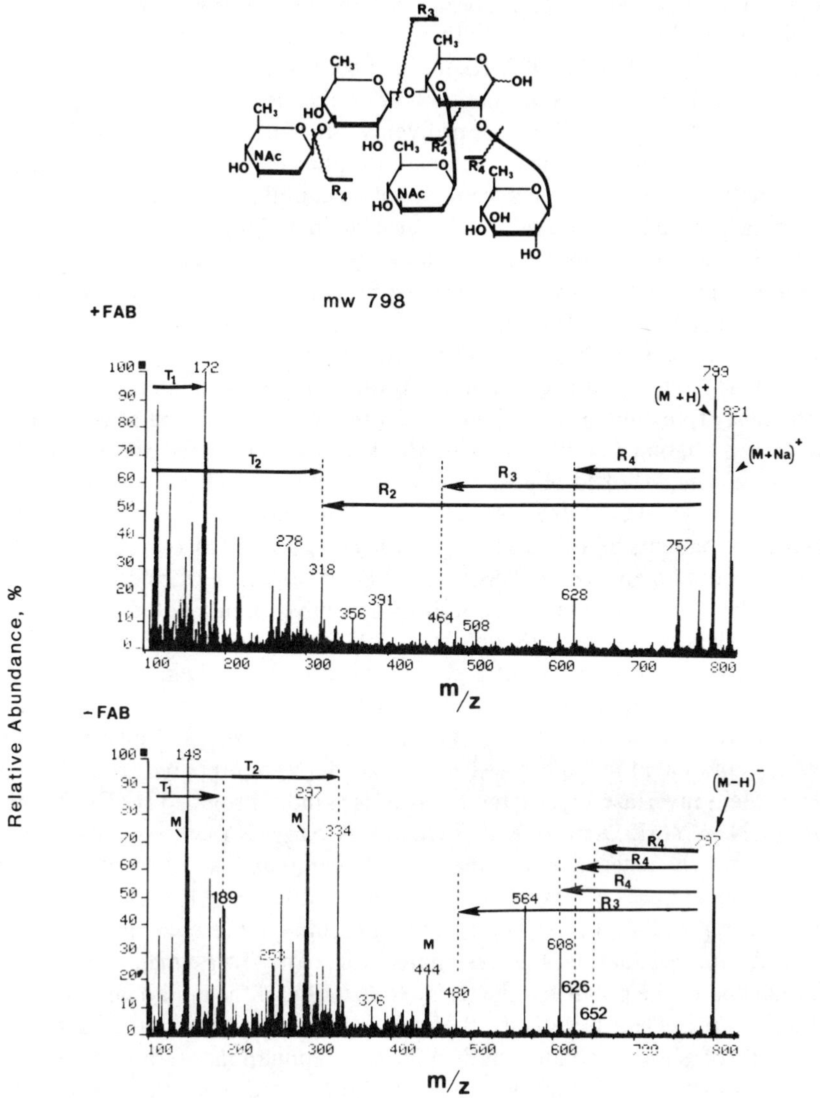

Figure 7: Fast atom bombardment MS of viridopentose A (Santikarn and Reinhold, unpublished data). The sample was kindly provided by Professor Suzuki, Meijo University, Nagoya, Japan. Upper spectrum: thioglycerol matrix with positive ion extraction. Lower spectrum: triethanolamine matrix with negative ion extraction.

sequence related information may not be optimized using the same liquid matrix, and the choice of matrix depends on the information desired and the sample to be analyzed. There has been much discussion on sample solubility and distribution within the matrix, and these variables greatly influence the resultant spectra (Chan and Cook, 1983; Keough, 1985; for a discussion of matrices see Gower, 1985). Trace sample analyses offer the usual difficulties because of contaminants; with FAB these problems are further complicated due to suppression by contaminants of sample ions and modification of the fragmentation pattern. Data systems capable of accumulating multiple scans and averaging out random noise can be crucial in pulling out weak signals. If adequate amounts of sample are available, three or four analyses using a cross section of matrices usually become most informative and are recommended.

As discussed above for viridopentose A, our preliminary studies with *N*-linked oligosaccharides have also provided spectra with sequence information when extracting negative ions. Additionally, these materials provide an opportunity to evaluate the influence of a polar group located at the terminal end of the oligomer combined with the spectral variations caused by chain branching. The *N*-linked oligosaccharides are prepared chemically by hydrazinolysis or enzymatically from glycoproteins and have also been isolated from the urine of patients or animals foraging on a plant called locoweed. The disease involved is due to an enzyme deficiency of acidic alpha mannosidase (Kjellman *et al.*, 1969) and the animal complications have been related to the plant alkaloid called swainsonine (Sadeh *et al.*, 1983). The urine oligosaccharides are characterized by having *N*-acetyl glucosamine at the reducing termini with a varying degree of highly branched mannose residues. An example of the spectral detail provided by FAB MS for a group of isomeric *N*-linked glycans (Man_8-G*N*Ac) is presented in Figure 8. These materials were prepared from the yeast glycoprotein invertase by enzyme hydrolysis (kindly provided by Dr. Trimble, State of New York Department of Health, Albany, NY). From the spectrum one can clearly determine the molecular weight and a series of fragments showing single losses of hexose units, R_8 through R_1. The R fragments include the reducing end of the structure and the suffix represents the number of carbohydrate residues in the designated ion. The T fragments include the terminal end of the polymer. The terminal location of the amino hexose residue appears to provide a negatively charged 'handle' for extraction of a series of oligomers of shorter chain length. Closer examination of this material for specific branching information would not be fruitful because this sample is known to be contaminated with isomeric structures.

The determination of oligosaccharide branching poses a unique analytical problem for carbohydrate researchers. As described above, fragment ion abundance is usually enhanced when extracting negative ions; however, the specificity of fragmentation usually appears more pronounced with permethylated samples and when the instrument is operated in the positive ion extraction

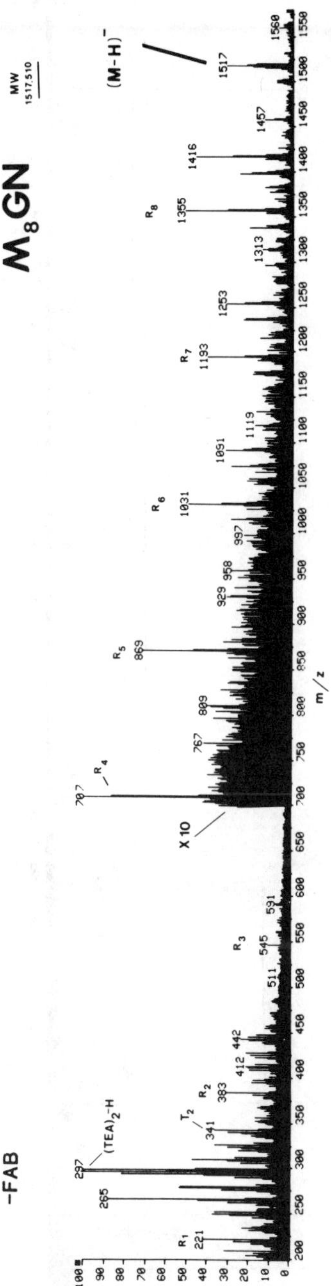

Figure 8: Fast atom bombardment MS of $Man_8GlcNAc$ using triethanolamine as a matrix and negative ion extraction (Santikarn, Trimble and Reinhold, unpublished data).

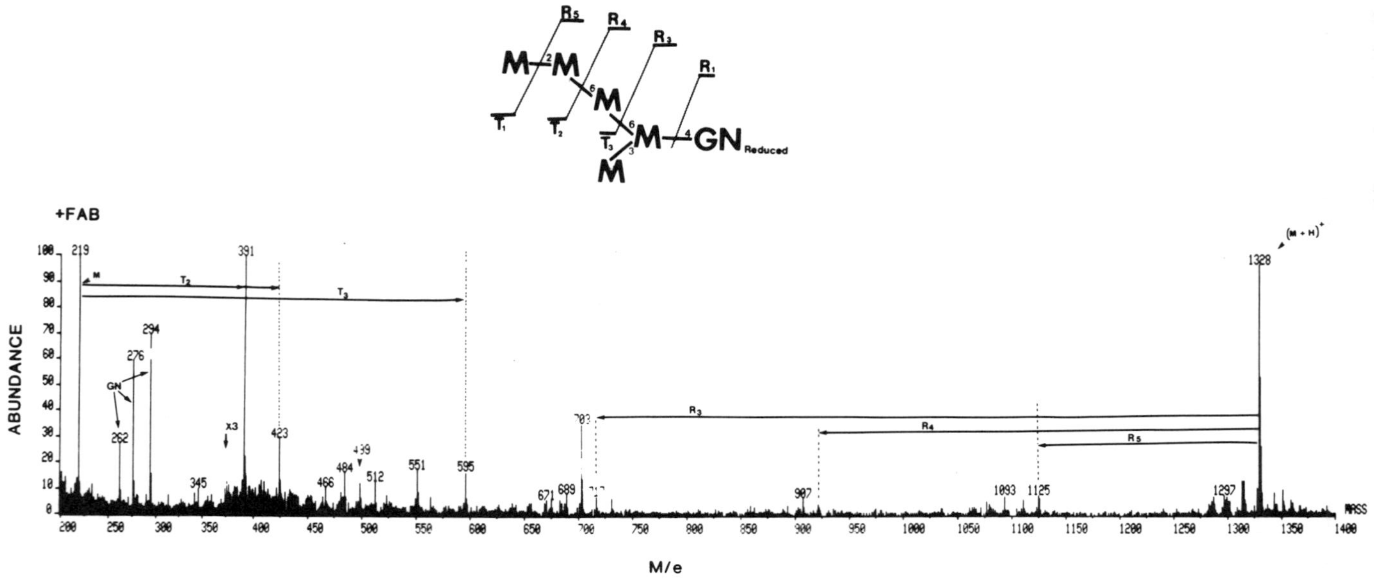

Figure 9: Fast atom bombardment MS of singly branched $Man_5GlcNAc$ using thioglycerol as the matrix and positive ion extraction (Santikarn and Reinhold, unpublished data).

mode. An isolated Man_5GN fraction was reduced, permethylated, and the sample analyzed by positive FAB MS (Figure 9). The molecular weight can be determined from the ion at *m/z* 1328, $(M + H)^+$. Closer analysis of this spectrum indicates that the major fragments can be accounted for by considering cleavage on either side of the glycosidic oxygen and a combination of hydrogen or methyl rearrangements (e.g. *m/z* 689, 717, see Scheme 3). Fragments which can provide sequence information (intervals in mass equal to monomer units) are observed from the terminal (T_1,T_2,T_3; Figure 9) and the reducing end (R_5 and R_3) of the oligomer.

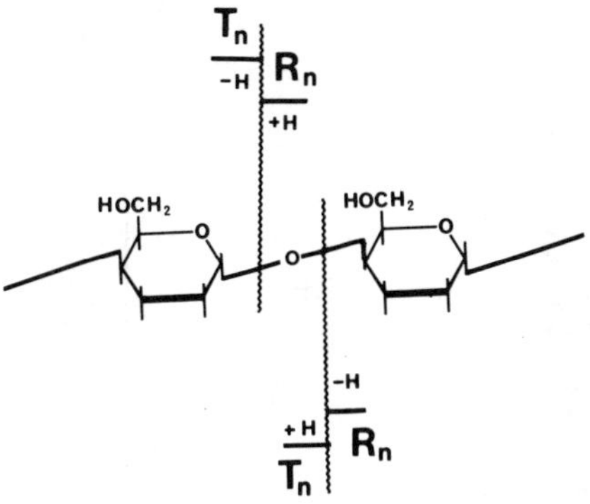

Scheme 3

To ascertain the ability of FAB desorption to differentiate different branching patterns within an isomeric structure, an HPLC fraction of Man_5GN was permethylated and analyzed (Figure 10). This *N*-linked glycan has a branched residue at the penultimate mannose on the non-reducing terminus. The spectrum indicates a $(M + H)^+$ ion at *m/z* 1312 with facile loss of methanol to give a major fragment, *m/z* 1280. The next grouping of ions (*m/z* 1062, 1076, 1090; combination of hydrogen or methyl rearrangments) indicates a loss of one hexose unit. No single cleavage of this glycan can occur that would result in the loss of two hexose residues and, if detected, double cleavage or contamination with an isomeric structure must be considered. Double cleavages would be expected to have a diminished probability and hence ion abundance. The spectra of this doubly branched isomer indicates R_5,R_3 and R_1 fragments, but no R_4 fragment. Unfortunately, some ion fragments appear in the spectrum which can be explained by isomer contamination (a singly branched isomer at a penultimate mannose residue) or by considering double cleavage; thus R_2 fragments are observed (*m/z* 482, loss of a mannose residue from *m/z* 700, Figure 10; *m/z* 499, loss of mannose residue from *m/z* 703, Figure 9). However, the absence of specific residues is very suggestive that branching information may be determined by FAB MS, but clearly the area requires further study.

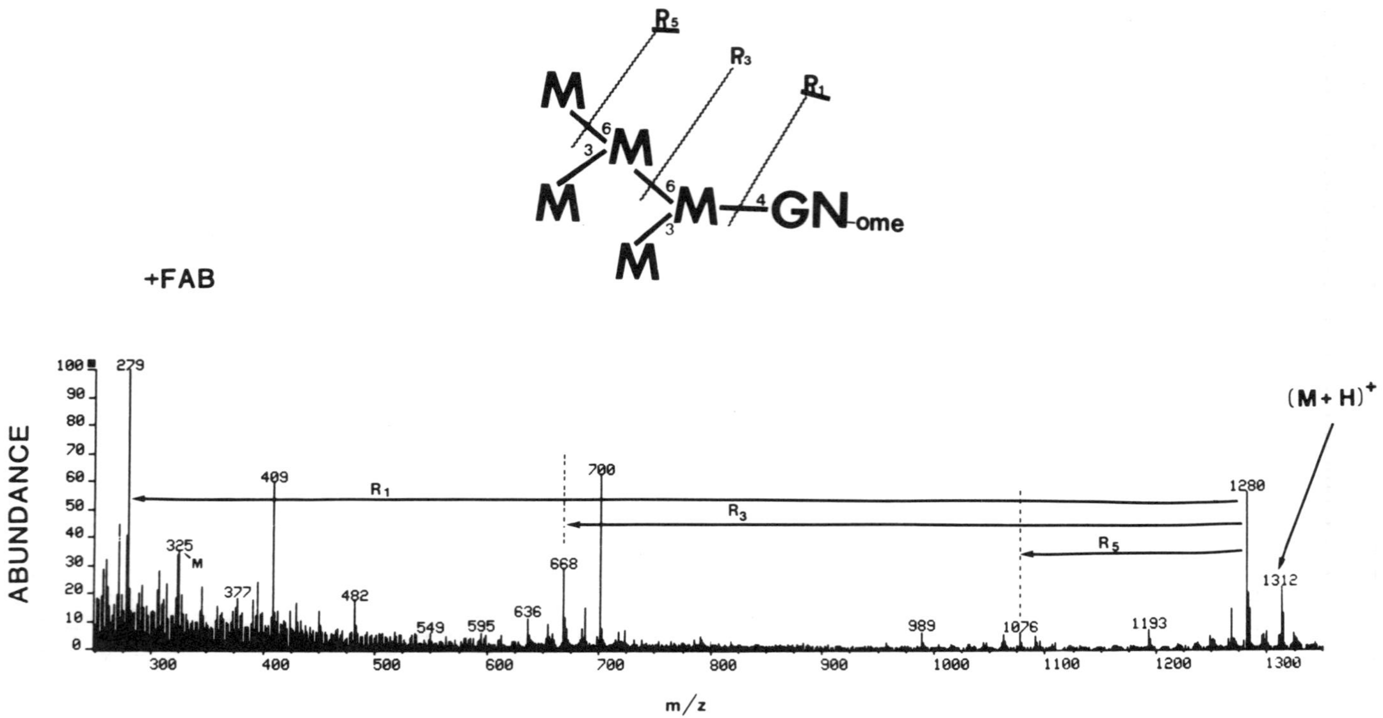

Figure 10: Fast atom bombardment MS of doubly branched Man_5Glc*N*Ac using thioglycerol as the matrix and positive ion extraction (Santikarn and Reinhold, unpublished data).

In summary, two questions are of considerable interest in the structural analysis of these FAB ionized materials. (i) Can the terminal location of a polar or charged residue provide greater sequence information? (ii) Can chain branching interrupt the sequence fragmentation pattern in such a way as to provide oligomer branch points? The answer to both questions is yes, if the charged oligomer undergoes primary cleavages only and randomly at glycosidic linkages remote from the charged site. Under these conditions the spectral pattern should represent a primary sequence with characteristic deletions at branching points. Unfortunately, glycosidic cleavage during FAB desorption is not generally random (Egge *et al.*, 1983; Hanfland *et al.*, 1983a,b; selectivity has also been observed under neutral gas collision, see Section 11.2.2.4) and our preliminary data suggest secondary cleavage.

Certainly the most troublesome aspect of the FAB ionization when considering sequence information is the spectral variations observed with changes in target matrix. In practice, biological isolates frequently bring their own contaminants, thereby adulterating the matrix and skewing the resultant spectra. These matrix effects become most evident with the analysis of trace components.

11.2.2.3. Desorption from a moving belt interface by FAB

In recent years, many different liquid chromatographic–mass spectrometric (LC–MS) interfaces have been constructed. From the time of its introduction (McFadden *et al.*, 1976) the moving belt interface has offered the most flexibility with respect to the choice of ionization. When using electron and chemical ionization, sample materials are vaporized from the belt at the heated tip projecting within the ionization block. Beam-assisted techniques have also proved successful for obtaining 'soft ionization' by using both charged (SIMS: Benninghoven *et al.*, 1980; Smith *et al.*, 1981) and neutral beams (FAB: Dobberstein *et al.*, 1983). In these cases samples are not thermally evaporated from the belt but are desorbed directly by surface bombardment. In a recent report FAB desorption from a moving belt interface has been applied to the study of a peptide hydrolysate of antiamoebin I (Stroh *et al.*, 1985). The HPLC–FAB MS belt interface in the antiamoebin study combines aerosol deposition (Smith and Johnson, 1981; Hayes *et al.*, 1983) of the LC solvent with subsequent drying and transport to the ion source for FAB ionization. Liquid matrix addition is unnecessary. On returning from the ion source the belt is washed with a fine brush scrubber which is saturated with solvent that is flowing in the opposite direction to belt transport. Chromatographic resolution of glucose oligomers (*dp* = 2–10; corn syrup) can be achieved on amino HPLC columns using an acetonitrile–water solvent. We have used this polyglucose mixture to evaluate the capability of HPLC–MS interfacing when combined with FAB ionization from the kapton belt (Reinhold and Lewis, unpublished

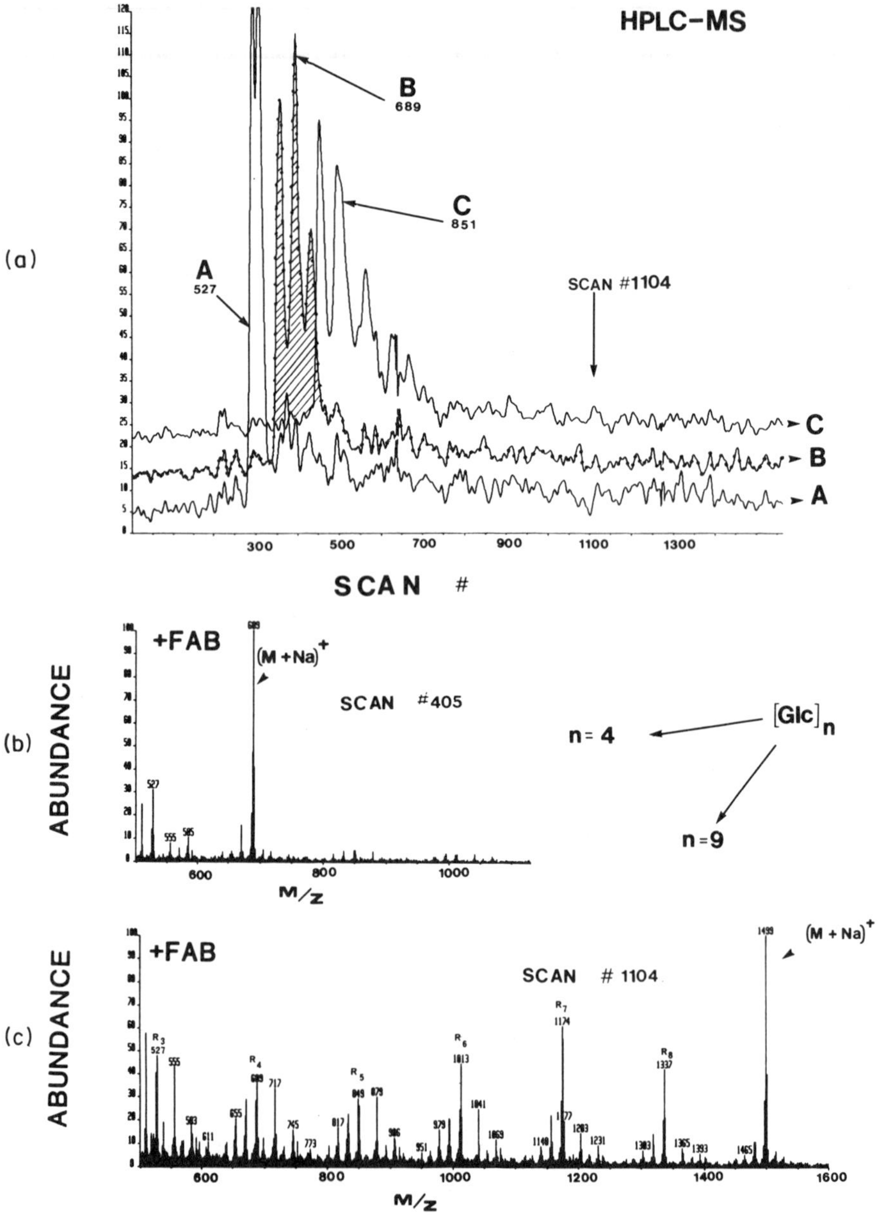

Figure 11: High performance liquid chromatography–fast atom bombardment mass spectrometry of corn syrup oligomers. No matrix was used. (Reinhold and Lewis, unpublished results). a) Selected ion chromatograms for trisaccharides, *m/z* 527, tetrasaccharides, *m/z* 689, and pentasaccharides, *m/z* 851. b) Mass spectrum of scan ♯405. c) Mass spectrum of scan ♯1104.

results). The chromatographic separation can be followed using appropriate selected ion chromatograms. Figure 11 shows plots for each of three oligomers (*dp* = 3,4 and 5; *m/z* 527, 689, 851, respectively). The ion chromatogram for *m/z* 527 $(M + Na)^+$ shows two peaks suggesting resolution of the anomeric forms of the trisaccharide. In addition to molecular weight data each of the components gave excellent oligomer sequence information. Presented in Figure 11 are the mass spectra taken at scans ♯405 and ♯1104 during the HPLC run. The ions of *m/z* 1337, 1174, 1013, 849, 689 and 527 represent fragmentation of the nonasaccharide at different glycosidic linkages (see Scheme 3) to provide the linear sequence information. Additional fragments can be observed which must represent fragmentation of the hexose ring. It is interesting that these fragments appear to have a structural counterpart when ionized by DCI and FAB with a liquid matrix.

The great advantage of FAB ionization in LC–MS of biopolymers is that it provides both molecular weight and sequence information. In addition, two important characteristics have been noted when this belt LC–MS interface has been utilized for oligosaccharide studies. First, when samples were analyzed and compared with conventional FAB, there was no loss in sensitivity. Second, there was a marked decrease in background noise resulting from the absence of matrix. The application of this HPLC-interface to larger molecular weight materials, with determination of oligomer branching, could prove most interesting.

11.2.2.4. Sequence studies using collision induced dissociation

In the discussions above, oligomer sequence information was more or less a by-product of molecular ionization. In this section gas phase collisions will be considered with respect to their ability to enhance fragmentation outside the ion source. The details of ion collisions are poorly understood for polyatomic systems but the fragmented products are known to vary with target gas and its pressure, scattering angle and ion kinetic energy. Collisions not only activate some of the ions for unimolecular decomposition but also induce charge neutralization, inversion, and stripping, as well as ion scattering (Cooks, 1978). The loss of sample signal is considerable following collision but part of these losses can be 'recovered' by signal amplification. The focusing of single ionic species for collision effectively filters out organic background ('chemical noise'); therefore, the daughter ions show good signal-to-noise ratio despite their low abundances.

The combination of ionization, fragmentation following collision (more frequently referred to as collision induced dissociation, CID), mass analysis and detection can be performed with a wide assortment of instruments having multiple analyzers (see Chapter 9).

Our interest in collision experiments were motivated, in large part, by the need to generate abundant fragment ions for the determination of oligomer sequence. Considering the current state-of-the-art for inducing glycosidic cleavage (see Sections 11.2.2.1 and 11.2.2.2), gas-phase collisions may not be the foremost technique to consider. However, other aspects of this preliminary study may turn out to have greater relevance and these points will be considered (Carr *et al.*, 1985). For these experiments, two mass spectrometers in tandem (BE–EB) were utilized, with an intermediate neutral gas collision cell (C) making the final configuration, BE–C–EB. Two features of MS/MS were evaluated. First, to assess the specificity of fragmentation by collision, iso-

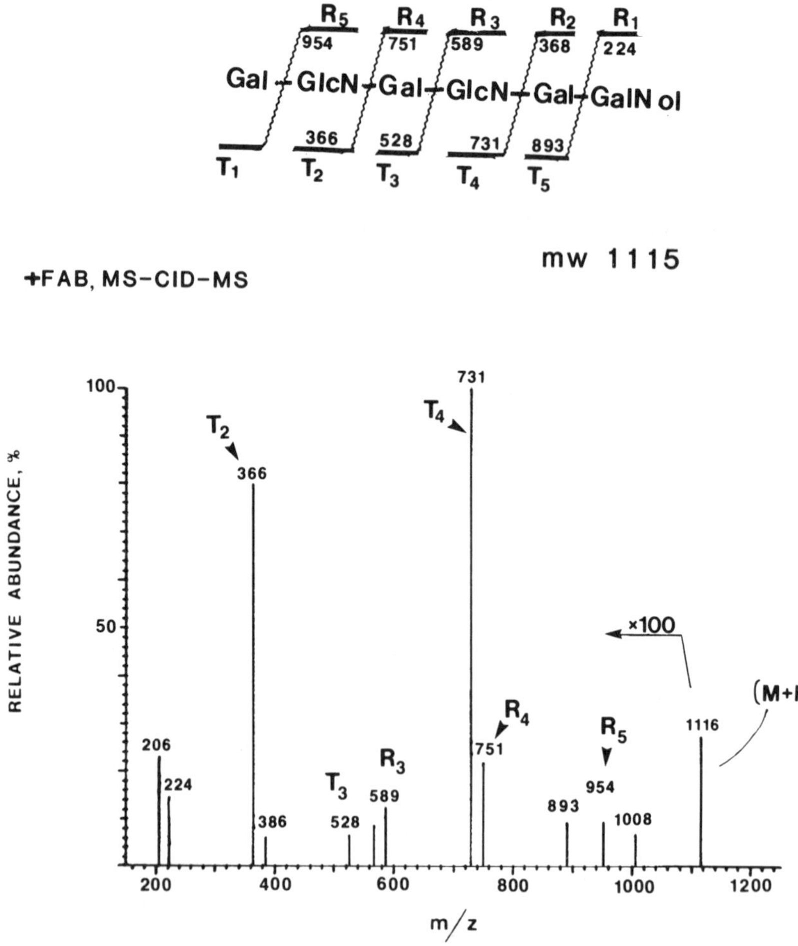

Figure 12: Spectrum of linear hexasaccharide obtained with MS/MS instrument: fast atom bombardment desorption, *BE* selection of *m*/*z* 1116, collision with helium gas and *EB* linked scan of daughter ion products. (Carr *et al.*, 1985.)

meric hexasaccharides were compared. Second, to study the separation and identification capability, $(M + H)^+$ ions in a complex mixture were isolated and characterized. The samples were ionized by FAB and selected by the double focusing combination, BE. Daughter ions produced in the collision cell (C) between BE and EB were recorded by a linked scan of the EB sectors such that the ratio E/B was maintained constant. Figures 12 and 13 show a comparison

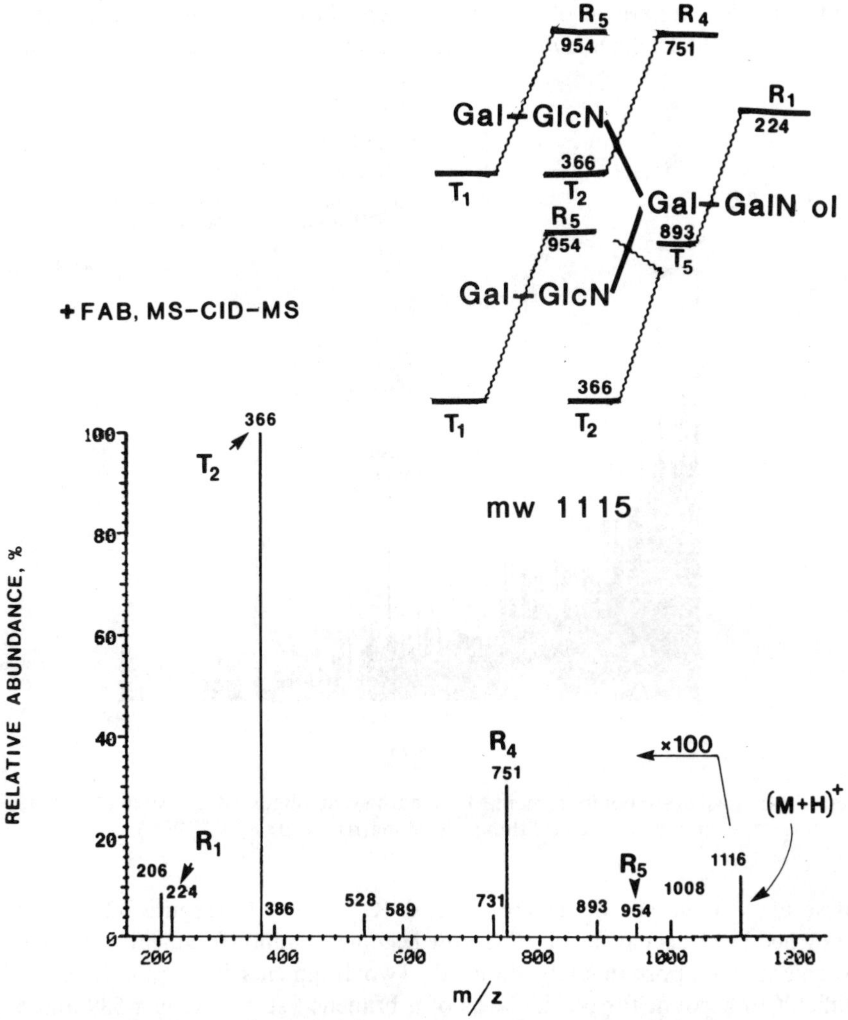

Figure 13: Spectrum of branched hexasaccharide obtained with MS/MS instrument. Fast atom bombardment desorption, *BE* selection of *m/z* 1116, collision with helium gas and *EB* linked scan of daughter ion products. (Carr *et al*., 1985.)

between a linear and branched hexasaccharide. The linear and branched saccharides each provided $(M + H)^+$ ion, *m/z* 1116, which was selected for collision. Helium was used as the neutral gas. The spectrum of the linear hexasaccharide (Figure 12) shows a series of fragments incrementing from the terminal end, T, as well as a series decreasing in mass for each carbohydrate residue from the reducing end, R. As noted above for FAB desorption, there is an enhanced selectivity for rupture at amino sugar linkages (e.g., compare relative abundances in Figure 12 of *m/z* 366 and 731, and in Figure 13 of *m/z* 366 and 751). A comparison of the two isomers shows a distinct ion abundance difference in the fragments *m/z* 731 and *m/z* 751. For the linear hexasaccharide

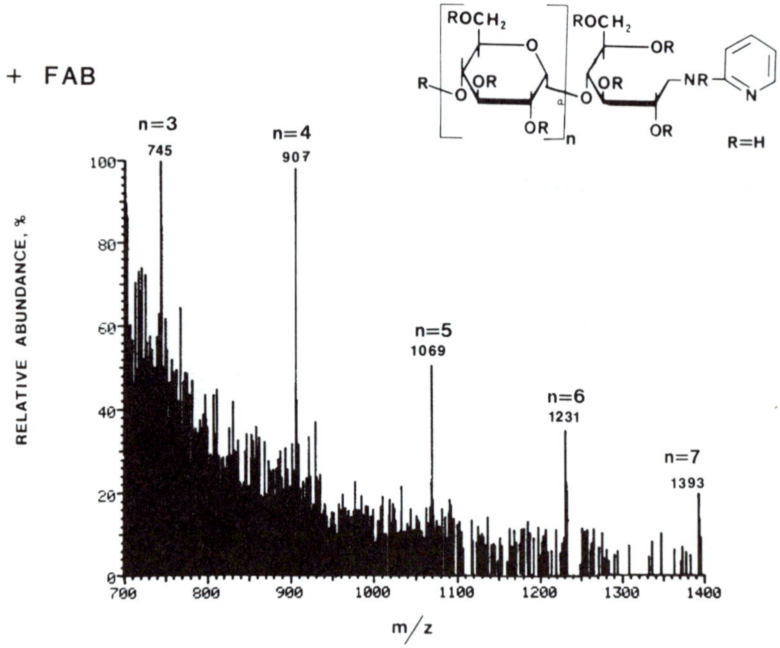

Figure 14: Fast atom bombardment MS of corn syrup oligomers prepared as pyridinylamine derivatives; thioglycerol matrix. (Carr *et al.*, 1985.)

these are accounted for as the T_4 and R_4 fragments (Figure 12). For the branched hexasaccharide, *m/z* 731 now becomes a minor fragment while the T_4 fragment has approximately doubled. Two fragments in Figure 13 are still difficult to account for on the basis of a branched structure, *m/z* 589 and *m/z* 731. These fragments are the R_3 and T_4 cleavages which can be readily accounted for only in the linear structure and their presence in the branched hexasaccharide is unexpected. An explanation may be that multiple collisions occur or that the sample is contaminated with the linear hexasaccharide. The samples were isolated from the same organism and separated by HPLC where incomplete separation could be a possibility (Dr. A. Wu, Texas A & M University, personal communication). The differences between the collision spectra indicate that sequence, including branching, information can be obtained by CID. An additional feature that MS/MS instrumentation brings to oligosaccharide characterization is the ability to select specific parent ions from complex mixtures and determine the structure of the isolated component. This capability is demonstrated in Figure 14 and 15. A mixture of oligosaccharides was labelled at the reducing end with a fluorescent group. The complex mixture was directly analyzed by FAB MS using thioglycerol as the matrix (Figure 14). To determine the structure of one of these components, the $(M + H)^+$ ion, m/z 1069, was selected and collided with helium. The daughter ion spectrum (Figure 15) shows a series of pyridinylamine fragments differing by a hexose unit.

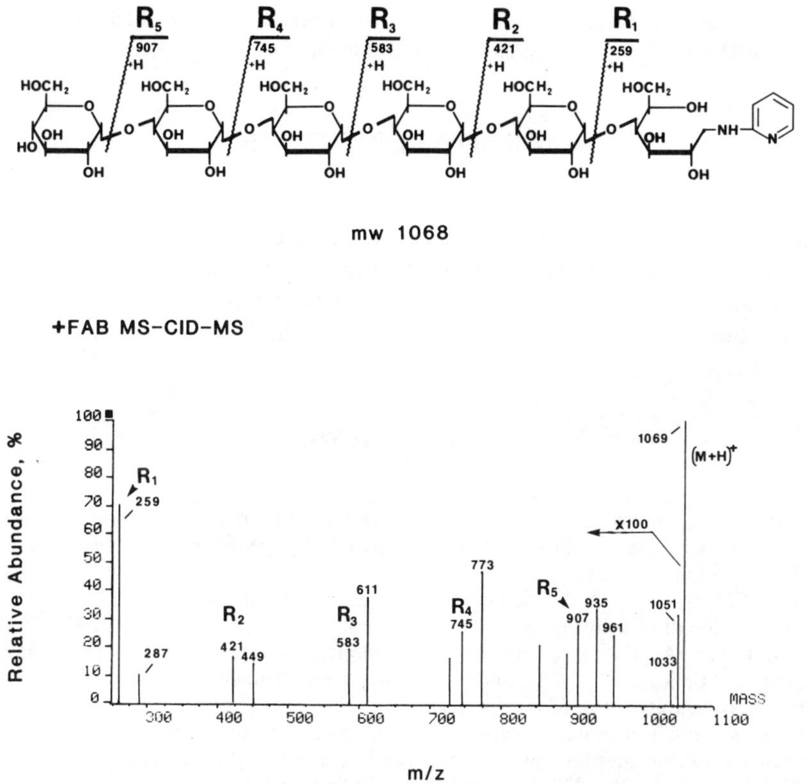

Figure 15: Spectrum of linear hexasaccharide obtained with MS/MS instrument. Fast atom bombardment desorption, *BE* selection of m/z 1069, collision with helium gas and *EB* linked scan of daughter ion products. (Carr *et al.*, 1985.)

11.3. SUMMARY

Many current problems in physiology and biochemistry depend on a detailed understanding of oligosaccharide and glycoconjugate structure. In contrast to other biopolymers, oligosaccharides do not have well established and sensitive analytical procedures for the determination of structure. Additionally, because of branching, linkage, and structural isomerism, oligosaccharides pose specific complications unique to this biopolymer. This chapter has discussed a new chemical procedure for the determination of linkage which utilizes the sensitivity and specificity of GC–MS for the assignment of structure. For the problems of oligomer sequence, earlier MS data have been reviewed, utilizing DCI and FAB ionization, and new information has been presented that indicates that FAB MS can provide a wealth of structural information including sequence, branching and molecular weight. In the final sections FAB ionization has been discussed in conjunction with HPLC interfacing and collision induced dissociation, two adjunct techniques that amplify the analytical value of MS instrumentation in the pursuit of oligosaccharide structure.

ACKNOWLEDGMENTS

The author would like to express his appreciation to Dr. Santikarn for sample analysis and generous time spent discussing aspects of this chapter. His further assistance, along with Dr. Her, in correcting the manuscript is gratefully appreciated. Financial support has been provided by the National Science Foundation and the National Institutes of Health.

REFERENCES

Baldwin, M. A., and McLafferty, F. W., (1973). Direct chemical ionization of relatively involatile samples. Application to underivatized oligopeptides. *Org. Mass Spectrom.*, **7**, 1353–1356.

Barber, M., Bordoli, R. S., Sedgwick, R. D., and Tyler, A. N. (1981). Fast atom bombardment of solids as an ion source in mass spectrometry, *Nature*, **293**, 270–275.

Benninghoven, A., Eicke, A., Junack, M., Sichtermann, W., Krizek, J., and Peters, H. (1980). Application of a secondary ion mass spectrometer as a detector in liquid chromatography. *Org. Mass Spectrom.*, **15**, 459–462.

Carr, S. A., and Reinhold, V. N. (1982). Sequence analysis of polypeptides by direct chemical ionization mass spectrometry', in *Methods in Protein Sequence Analysis* (Ed. M. Elzinga), pp. 263–270, Humana Press, Clifton, NJ.

Carr, S. A., and Reinhold, V. N. (1984). Structural characterization of glycosphingolipids by direct chemical ionization mass spectrometry, *Biomed. Mass Spectrom.*, **11**, 633–642.

Carr, S. A., Reinhold, V. N., Green, B. N., and Hass, J. R. (1985). Enhancement of structural information in FAB ionized carbohydrate samples by neutral gas collision, *Biomed. Mass Spectrom.*, **12**, 288–295.

Chan, K. W. S., and Cook, K. D. (1983). Factors affecting mass spectral sensitivity for ions sampled by field evaporation from a liquid matrix, *Anal. Chem.*, **55**, 1306–1309.

Cooks, R. G., Ed. (1978). *Collision Spectroscopy*, p. 458, Plenum Press, New York.

Constantin, E., Nakatani, Y., Ourisson, G., Hueber, R., and Teller, G. (1980). Spectre de masse de phospholipids et polypeptides non-protegés. Une technique simple d'obtention du spectre complet. *Tetrahedon Lett.*, **21**, 4745–4746.

Cotter, R. J. (1980). Mass spectrometry of nonvolatile compounds. *Anal. Chem.*, **52**, 1589A–1606A.

Cotter, R. J., and Yergey, A. L. (1981). Thermally produced ions in desorption mass spectrometry, *Anal. Chem.*, **53**, 1306–1307.

de Jong, E. G., Heerma, W., and Sicherer, C. A. X. G. F. (1979). Discrimination of permethylated disaccharides by their chemical ionization spectra, *Biomed. Mass Spectrom.*, **6**, 242–248.

Dell, A., and Ballou, C. E. (1983a). Fast atom bombardment, negative-ion, mass spectrometry of the mycobacterial O-methyl-*D*-glucose polysaccharide and lipopolysaccharides, *Carbohydr. Res.*, **120**, 95–111.

Dell, A., and Ballou, C. E. (1983b). Fast atom bombardment mass spectrometry of 6-O-methylglucose polysaccharide, *Biomed. Mass Spectrom.*, **10**, 50–56.

Dell, A., Morris, H. R., Egge, H., von Nicolai, H., and Strecker, G., (1983). Fast atom bombardment mass spectrometry for carbohydrate structure determination, *Carbohydr. Res.*, **115**, 41–52.

Dessort, D., Bisseret, P., Nakatani, Y., Ourisson, G., and Kates, H. (1983). Ammonia desorption chemical ionization mass spectrometry of phospholipids, *Chem. Phys. Lipids*, **33**, 323–330.

Dobberstein, P., Korte, E., Meyerhoff, G., and Pesch, R. (1983). Investigation of an LC/MS interface for EI, CI and FAB ionization, *Int. J. Mass Spectrom. Ion Phys.*, **46**, 185–190.

Egge, H., Dell, A., and von Nicolai, H. (1983). Fucose containing oligosaccharides from human milk. I. Separation and identification of new constituents, *Arch. Biochem. Biophys.*, **224**, 235–253.

Fournet, B., Strecker, G., Leroy, Y., and Montreuil, J. (1981). Gas-liquid chromatography and mass spectrometry of methylated and acetylated methyl glycosides. Application to the structural analysis of glycoprotein glycans, *Anal. Biochem.*, **116**, 489–502.

Gower, J. L. (1985). Matrix compounds for fast atom bombardment mass spectrometry, *Biomed. Mass Spectrom.*, **12**, 191–196.

Hanfland, P., Dabrowski, U., Peter-Katalinic, J., Kordowicz, M., Niermann, H., Egge, H., and Dabrowski, J. (1983a). Isolation and characterization of three different blood group B active complex glycosphingolipids from human erythrocyte membranes, in *Proc. 7th Int. Symp. on Glycoconjugates, Lund, Sweden.* (Eds. M. A. Chester, D. Heinegard, A. Lundblad, and S. Svensson), pp. 411–412.

Hanfland, P., Dabrowski, U., Kordowicz, M., Peter-Katalinic, J., Dabrowski, J., and Egge, H. (1983b). Purification and structural elucidation of three different blood group B and I or i glycosphingolipids from rabbit erythrocyte membranes, in *Proc. 7th Int. Symp. on Glycoconjugates, Lund, Sweden.* (Eds. M. A. Chester, D. Heinegard, A. Lundblad, and S. Svensson), pp. 413–414.

Hansen, G., and Munson, B. (1978). Surface chemical ionization mass spectrometry, *Anal. Chem.*, **50**, 1130–1134.

Harada, K., Ito, S., Takeda, N., Suzuki, M., and Tatematsu, A. (1983a). Sequence determination of permethylated oligosaccharides by chemical ionization mass spectrometry, *Biomed. Mass Spectrom.*, **10**, 5–12.

Harada, K., Ito, S., Suzuki, M., and Iwashita, T. (1983b). Structural investigation of the antibiotic sporaviridin. VII. Structural studies on the constituent pentasaccharides, viridopentoses, *Chem. Pharm. Bull. (Japan)*, **31**, 3829–3843.

Harrison, A. G. (1983). *Chemical Ionization Mass Spectrometry*, CRC Press, Boca Raton, FL.

Hayes, M. J., Lankmayer, E. P., Vouros, P., and Karger, B. L. (1983). Moving belt interface with spray deposition for liquid chromatography/mass spectrometry, *Anal. Chem.*, **55**, 1745–1752.

Hunt, D. F., Shabanowitz, J., Botz, F. K., and Brent, D. (1977). Chemical ionization mass spectrometry of salts, thermally labile organics with field desorption emitters as solids probes, *Anal. Chem.*, **49**, 1160–1163.

Jensen, N. J., Tomer, K. B., and Gross, M. L. (1985). Gas-phase ion decomposition occurring remote to a charged site, *J. Am. Chem. Soc.*, **107**, 1863–1868.

Karlsson, K.-A., Pascher, I., Pimlott, W., and Samuelsson, B. E. (1974). Use of mass spectrometry for the carbohydrate composition and sequence analysis of glycosphingolipids, *Biomed. Mass Spectrom.*, **1**, 49–55.

Keough, T. (1985). Cationization of organic molecules using fast atom bombardment mass spectrometry, *Anal. Chem.*, **57**, 2027–2034.

Kjellman, B., Gamstorp, I., and Brun, A. (1969). Mannosidosis: a clinical and histopathological study, *J. Pediatr.*, **75**, 366–373.

Lindberg, B. (1972). Methylation analysis of polysaccharides, in *Methods in Enzymology* (Ed. V. Ginsburg), Vol. 28, pp. 178–201, Academic Press, Orlando, FL.

McFadden, W. H., Schwartz, H. L., and Evans, S. (1976). Direct analysis of liquid chromatographic effluents, *J. Chromatogr.*, **122**, 389–396.

Miller, K. J., Kennedy, E., and Reinhold, V. N. (1986). Osmotic adaption by Gram-negative bacteria: possible role for periplasmic oligosaccharides, *Science*, **231**, 48–51.

Munson, B. (1977). Chemical ionization mass spectrometry: ten years later, *Anal. Chem.*, **49**, 772A–776A.

Ohashi, M., Barron, R. P., and Benson, W. R. (1981). Electron impact induced fragmentation of quaternary ammonium cations, *J. Am. Chem. Soc.*, **103**, 3943–3945.

Reed, R. I. (1958). Electron impact and molecular dissociation. Part I, *J. Chem. Soc.*, **1958**, 3432–3436.

Reinhold, V. N., and Carr, S. A. (1982). Direct chemical ionization mass spectrometry with polyimide-coated wires, *Anal. Chem.*, **54**, 499–503.

Reinhold, V. N., and Carr, S. A. (1983). New mass spectral approaches to complex carbohydrate structure, *Mass Spectrom. Rev.*, **2**, 153–221.

Reinhold, V. N., Coles, E., and Carr, S. A. (1982). New techniques for oligosaccharide sequencing, *J. Carbohydr. Chem.*, **2**, 1–18.

Rolf, D., and Gray, G. R. (1982). Reductive cleavage of glycosides, *J. Am. Chem. Soc.*, **104**, 3539–3542.

Rolf, D., and Gray, G. R. (1984). Analysis of the linkage positions in *D*-fructofuranosyl residues by the reductive cleavage method, *Carbohydr. Res.*, **131**, 17–28.

Rolf, D., Bennek, J. A., and Gray, G. R. (1985). Analysis of linkage positions in D-glucopyranosyl residues by the reductive-cleavage method, *Carbohydr. Res.*, **137**, 183–196.

Sadeh, S., Warren, C. D., Bugge, B., James, L. F., and Jeanloz, R. W. (1983). Characterization of oligosaccharides from the urine of loco-intoxicated sheep, *FEBS Letters*, **163**, 104–109.

Sisseret, P., Nakatani, Y., Ourisson, G., Hueber, R., and Teller, G. (1983). Ammonia chemical ionization mass spectrometry of lecithins on a gold support. *Chem. Phys. Lipids*, **33**, 383–392.

Smith, R. D., and Johnson, A. L. (1981). Deposition method for moving ribbon liquid chromatograph–mass spectrometry interfaces, *Anal. Chem.*, **53**, 739–740.

Smith, R. D., Burger, J. E., and Johnson, A. L. (1981). Liquid chromatography–mass spectrometry and electron impact and secondary ion mass spectrometry with ribbon storage interface, *Anal. Chem.*, **53**, 1603–1611.

Stillwell, R. N., Carroll, D. I., Nowlin, J. G., and Horning, E. C. (1983). Formation of trimethylsilyl molecular adduct ions in desorption chemical ionization mass spectrometry of nonvolatile organic compounds, *Anal. Chem.*, **55**, 1313–1318.

Stoll, R., and Rollgen, F. W. (1980). Thermal evaporation of intact positive ions of quaternary ammonium and phosphonium salts, *J. Chem. Soc. Chem. Commun.*, **103**, 3943–3948.

Stoll, R., and Rollgen, F. W. (1981) Thermal desorption of quasimolecular ions, *Org. Mass Spectrom.*, **16**, 72–75.

Stroh, J. G., Cook, J. C., Milberg, R. M., Brayton, L., Kihara, T., Huang, Z., and Rinehart, K. L. (1985). On-line liquid chromatography/fast atom bombardment mass spectrometry, *Anal. Chem.*, **57**, 985–991.

Takeda, N., Harada, K., Suzuki, M., Tatematsu, A., and Sakata, I. (1983). Structural characterization of underivatized menthyl glycosides using chemical ionization mass spectrometry, *Biomed. Mass Spectrom.*, **10**, 608–613.

Takeda, N., Harada, K., Suzuki, M., Tatematsu, A., and Sakata, I. (1985). Sensitivity enhancement in desorption chemical ionization mass spectrometry using silicon-activated wire emitter, *Mass Spectroscopy*, **33**, 59–63.

Thenot, J. P., Nowlin, J., Carroll, D. I., Montgomery, E., and Horning, E. (1979). Coated glass probe tips for the determination of thermally labile compounds by mass spectrometry, *Anal. Chem.*, **51**, 1101–1104.

van der Peyl, G. J. Q., Isa, K., Haverkamp, J., and Kistemaker, P. G. (1981). Gas phase ion/molecule reactions in laser desorption mass spectrometry, *Org. Mass Spectrom.*, **16**, 416–420.

Van Langenhove, A., and Reinhold, V. N. (1985). Determination of polysaccharide linkage and branching by reductive depolymerization: GC and GC–MS reference data, *Carbohydr. Res.*, **140**, 1–20.

Vestal, M. L. (1983). Ionization techniques for nonvolatile molecules, *Mass Spectrom. Rev.*, **2**, 447–480.

Virelizier, H., Hageman, R., and Jankowski, K. (1983). Systematic analysis of desorption chemical ionization (DCI) mass spectra of nucleic acids—7. *Biomed. Mass Spectrom.*, **10**, 559–566.

Mass Spectrometry in Biomedical Research
Edited by S. J. Gaskell
© 1986 John Wiley & Sons Ltd.

Chapter 12

FAST ATOM BOMBARDMENT MASS SPECTROMETRY OF OLIGONUCLEOTIDES

Lutz Grotjahn

12.1. INTRODUCTION

During the past few years the chemical synthesis of DNA fragments has become progressively more important for the modern biosciences. This has been made possible by major progress in synthesis technology, in particular the use of better coupling and deprotection methods, the employment of suitable solid supports, and refined isolation and characterization techniques.

Such molecules serve as 'Linkers and Adaptors' in recombinant DNA techniques and allow a vast number of new combinations of gene material to be produced. Other synthetic DNA fragments are used to perform site-directed mutagenesis of natural DNA or to achieve total synthesis of genes.

The synthesis of these small DNA fragments can be performed in a matter of days and as a consequence the longer part of the work is now the careful analysis of the deprotected oligonucleotides. Previously these analyses, which involve checking the sequence of the synthetic material, have been performed by the often laborious methods of chemical degradation (Maxam and Gilbert, 1977) or wandering spot techniques (Brownlee and Sanger, 1969) involving radioactive labelling. This chapter describes the determination and sequencing of oligonucleotides using fast atom bombardment mass spectrometry.

12.2. METHODS

All FAB mass spectra reported here were obtained on a Kratos MS 50S mass spectrometer equipped with a high field magnet and a Kratos FAB source. The atom gun used xenon and produced a beam of neutral atoms of 8–9 keV energy. The spectra were recorded at a magnet scan rate of 300 s/dec or in the ESA scan mode (3s sweep with a peak switch width of 200 to 10,000 ppm) by accumulation with a multichannel analyzer (Tracor TN 1500).

The oligodeoxyribonucleotides were synthesized following, in general, the established phosphate triester procedure (Grotjahn *et al.*, 1982; Conolly *et al.*, 1984). Ribonucleotides were of commercial origin (P-L Biochemicals, Milwaukee, USA). UGUU was prepared enzymatically and generously provided by Reinhard Lührmann, Max Planck Institut für Molekulare Genetik, D-1000 Berlin-Dahlem.

Samples were prepared for analysis in various ways. A water solution of the triethylammonium salt of each free oligonucleotide (*ca.* 4 µl, containing *ca.* 1 OD_{260} (*ca.* 10 nmol)) was injected into the glycerol matrix (*ca.* 2 µl). The negatively charged protected oligonucleotides (triethylammonium salts) were dissolved in methanol (dimers) or directly introduced into a 1 : 1 mixture of glycerol and triethylene glycol (higher oligomers).

2.3. RESULTS

Negative ion FAB mass spectrometry has been shown to be a suitable method for characterizing mononucleotides and for sequencing the oligomeric building blocks of the phosphotriester synthesis, as well as for sequencing the final free oligonucleotides (Sindona *et al.*, 1982; Grotjahn *et al.*, 1982, 1983, 1984, 1985; Panico *et al.*, 1983; Tazawa *et al.*, 1983; Conolly *et al.*, 1984).

Protected oligonucleotides with a negative charge at the terminal phosphate, produced either by initial chemical degradation of the terminal phosphate protecting group (tribromoethyl or cyanoethyl) or by reaction with the matrix (for example ethanolamines), generally yield the sequence information by consideration of the 3′- and 5′-phosphate sequence ions, formed by cleavage of the bond between the carbon of the sugar moiety and the phosphate oxygen. We have investigated all possible dinucleotides of a particular type, in which the terminal phosphate protecting group was removed. In all cases we could identify and sequence these by negative ion FAB mass spectrometry.

Figure 1 shows the negative ion FAB mass spectrum and the formula of the protected deoxyribonucleotide AG^-. The $(M - H)^-$ ion is found at m/z 1325, the 3′-phosphate sequence ion at m/z 816 and the 5′-phosphate sequence ion at m/z 716. The $(M - H)^-$ ion and the sequence ions are accompanied by loss of protecting groups, as described in the formula.

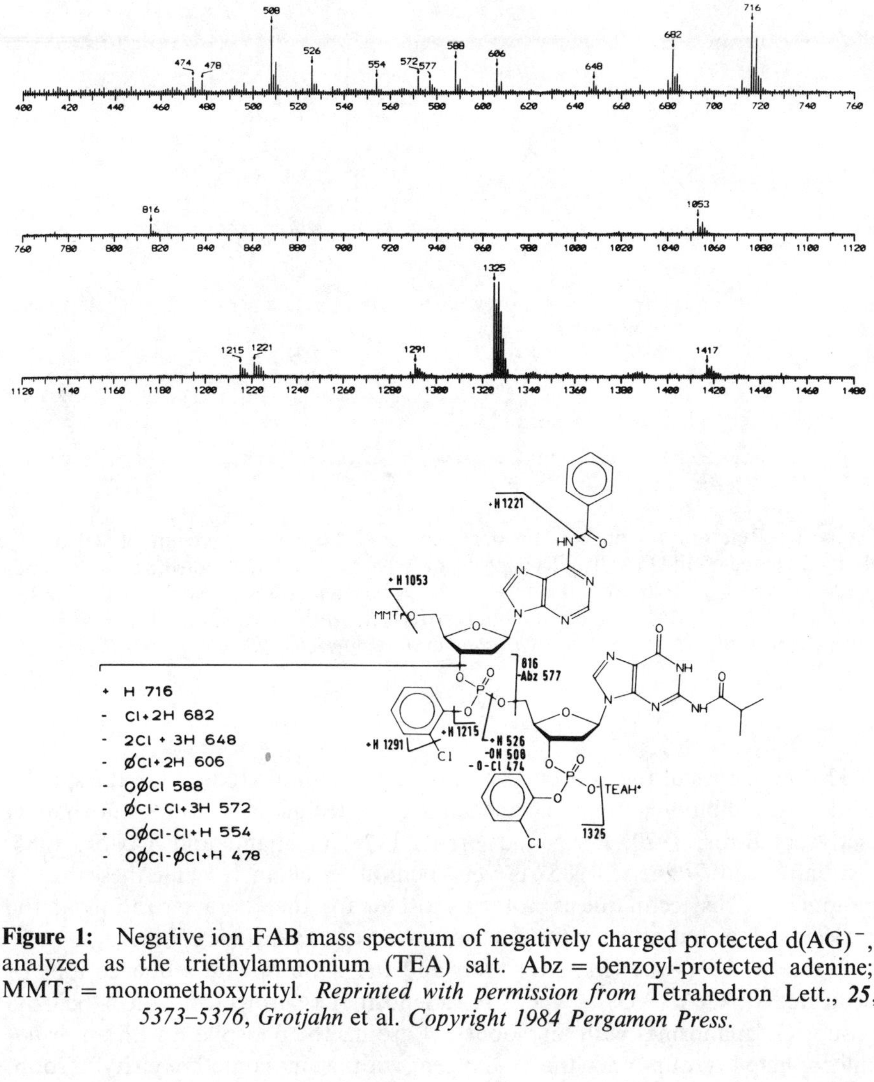

Figure 1: Negative ion FAB mass spectrum of negatively charged protected d(AG)$^-$, analyzed as the triethylammonium (TEA) salt. Abz = benzoyl-protected adenine; MMTr = monomethoxytrityl. *Reprinted with permission from* Tetrahedron Lett., ***25***, *5373–5376*, *Grotjahn* et al. *Copyright 1984 Pergamon Press.*

Figure 2 shows the relevant regions of the negative ion FAB mass spectrum of the protected deoxyribonucleotide TGAC$^-$. In comparison with the dinucleotides, the protected higher oligomers show a less complex fragmentation behaviour, with the monomethoxytrityl-protected 3′-phosphate sequence ions giving the most prominent peaks. From the set of these ions the base sequence of the oligomer is easily deduced.

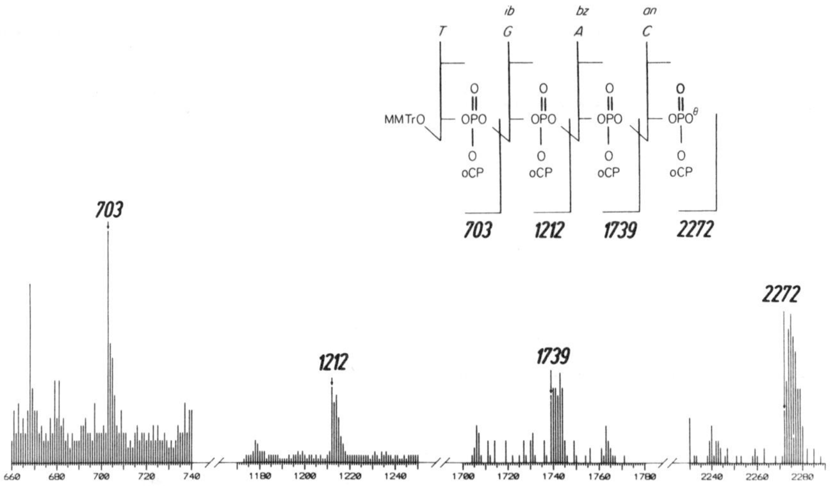

Figure 2: Relevant regions of the negative ion FAB mass spectrum of negatively charged protected d(TGAC)$^-$. The spectrum is background-subtracted and magnified, $\times$ 10, above m/z 1700. ib = isobutyryl, bz = benzoyl, an = anisoyl, oCP = *ortho*-chlorophenyl, and MMTr = monomethoxytrityl. *Reprinted with permission from* Tetrahedron Lett., ***25***, *5373–5376*, *Grotjahn* et al. *Copyright 1984 Pergamon Press.*

The intensities of the molecular and sequence ions are reduced in the spectra of the higher biooligomers. The accumulation technique with a multichannel analyzer (Biros, 1970; Peele and Brent, 1978; Grotjahn and Taylor, 1985; Grotjahn and Steinert, 1985) is very helpful in characterizing these higher oligomers. This technique is not only useful for the characterization of the molecular ion, but also for the determination of the sequence ions.

Figure 3 shows the (M − H)$^-$ ion and the 3′-phosphate sequence ions of protected d(AAAAAG)$^-$, where A (adenosine) is protected with a benzoyl group, G (guanosine) with an isobutyryl group, the phosphate with an *ortho*-chlorophenyl group and the 5′-oxygen with a monomethoxytrityl group. In the single magnetic scan FAB mass spectrum it was difficult to determine the sequence ions.

The determination of the sequence ions is much easier if the 'dead-time' between sequence ion regions is used to obtain a better signal-to-noise ratio and isotopic distribution by signal averaging the mass windows of the sequence ions. At present we manually jump from sequence-ion region to sequence-ion region and reload the sample, if necessary. We plan to automate this manual process, in order to gain more time for accumulation or to decrease compound consumption.

The top trace in Figure 3 shows the experimental results for the 3′-phosphate sequence ions and the $(M - H)^-$ ion together with the number of accumulated scans. There is good agreement with the theoretical isotopic distribution, which is shown in the middle trace of Figure 3. The lowest trace shows the shorthand structure of each 3′-phosphate sequence ion. The isotopic distributions of the smaller sequence ions are more strongly influenced by the chlorine isotopic distribution, whereas the higher sequence ions are more affected by the carbon distribution.

As we know from other investigations (e.g. tetrameric building block $TGAC^-$), in the case of protected oligonucleotides, the 3′-phosphate sequence ions are, in general, more intense than the corresponding 5′-phosphate sequence ions (Grotjahn *et al.*, 1984). As we had enough sample we used the same technique to determine the 5′-phosphate sequence ions.

Figure 4 (top) shows the experimental results for the 5′-phosphate sequence ions of protected $d(AAAAG)^-$. Again there is good agreement with the theoretical isotopic distribution (Figure 4, middle). Only in one case do we see an accompanying isotope cluster, which is derived from a larger sequence ion, after loss of an *ortho*-chlorophenyl group from the fourth 3′-phosphate sequence ion to give the partner accompanying the fourth 5′-phosphate sequence ion.

This ‘hopping method’, going from mass region to mass region in order to determine sequence and molecular ions, will be particularly useful in the future for the determination of higher oligomers. At present we use this method on a Kratos MS 50S for sequencing protected oligonucleotides up to 8 base units in length (e.g. $TCCTAGGA^-$ with the $(M - H)^-$ ion at m/z 4255.7) at a reduced accelerating voltage of 4 kV. With an inhomogeneous field magnet we will be able to investigate protected oligonucleotides with more base units at full accelerating voltage.

When oligodeoxyribonucleotides are only protected with a monomethoxytrityl group at the 5′-end the fragmentation behaviour is similar to the fully protected ones. The 3′-phosphate sequence ions from the 5′-end are in general more abundant than the corresponding 5′-phosphate sequence ions. Of course, the 3′-phosphate sequence ions are higher in mass by 272 u.

Figure 5 shows the sequence ion regions of the negative ion FAB mass spectrum of the monomethoxytritylated oligodeoxyribonucleotide CGTCGACG. The $(M - H)^-$ ion appears at m/z 2680 and the generally more intense 3′-phosphate sequence ions at m/z 2431 (7 base units), 2142 (6 base units), 1829 (5 base units), 1500 (4 base units), 1211 (3 base units), 907 (2 base units) and 578 (1 base unit), while the 5′-phosphate sequence ions appear at m/z 2199 (7 base units), 1870 (6 base units), 1566 (5 base units), 1277 (4 base units), 948 (3 base units), 635 (2 base units) and 346 (1 base unit). From the spectrum and the shorthand formula the sequence can be read from both ends.

The fragmentation behaviour of fully deprotected oligonucleotides allows a bidirectional sequence analysis (Grotjahn *et al.*, 1982). Figure 6 shows the negative ion FAB mass spectrum of the deoxyribonucleotide GGAATTCC. The $(M - H)^-$ ion is found at m/z 2407. The shorthand formula indicates the masses, which were formed by cleavage of the bond between the carbon of the sugar moiety and the phosphate oxygen. The main fragmentations result in the formation of sequence ions with either 3′- or 5′-terminal phosphates (3′- or 5′-phosphate sequence ions). A significant distinction was found for the two different types of phosphate sequence ions. 5′-phosphate sequence ions are always more abundant than the corresponding 3′-phosphate sequence ions with the same number of nucleotide units. This simple fragmentation behaviour allows an independent sequencing of the oligonucleotide chain either from the 3′- or from the 5′-end (bidirectional sequence analysis).

'Satellite' ions, corresponding to −18 u (loss of water) and +80 u (perhaps addition of HPO_3) are often observed with the sequence and matrix ions, but these do not disturb the sequence analysis. The less intense 3′-phosphate

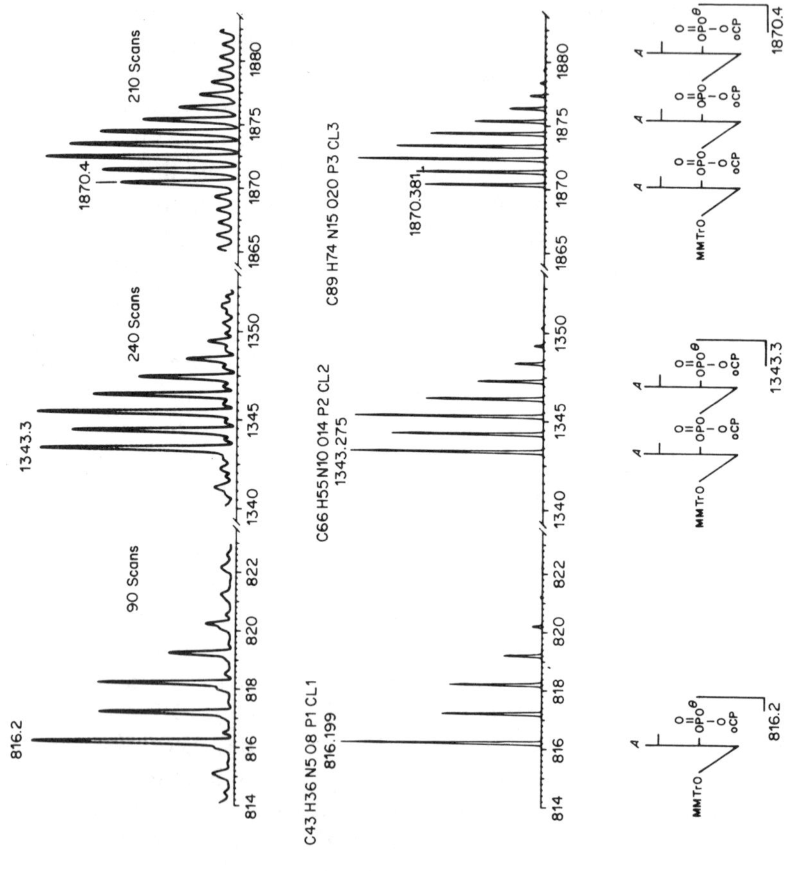

Figure 3: (part).

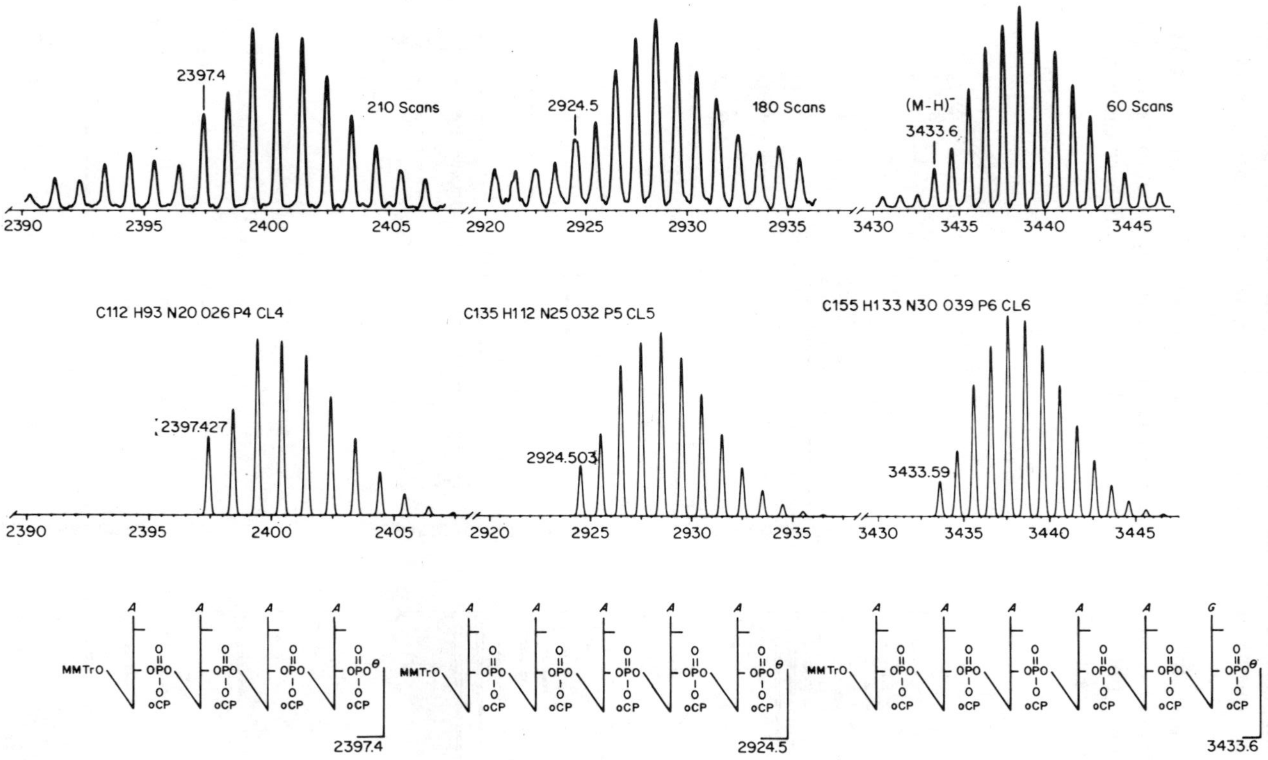

Figure 3: Molecular ion and 3′-phosphate sequence ions of negatively charged protected d(AAAAAG)$^-$. Top: experimental results; middle: simulated results; bottom: shorthand structures. ib = isobutyryl, an = anisoyl and oCP = *o*-chlorophenyl. The lower portion of the mass scale is shown separately on p. 220. *Reproduced from* Mass Spectrometry in the Health and Life Sciences (*Eds. A. L. Burlingame and N. Castagnoli*), *pp. 597–615, by permission of Elsevier Scientific Publishers.*

sequence ions (from the 5′-end) permit determination of the sequence information in the correct direction GGAATTCC, whereas the more intense 5′-phosphate sequence ions (from the 3′-end) permit determination from the opposite end.

Figure 7 shows the shorthand formula of the deoxyribonucleotide ACTCGATG. In this case, all 5′-phosphate sequence ions are higher in mass than the corresponding 3′-phosphate sequence ions. This does not influence the bidirectional sequence analysis.

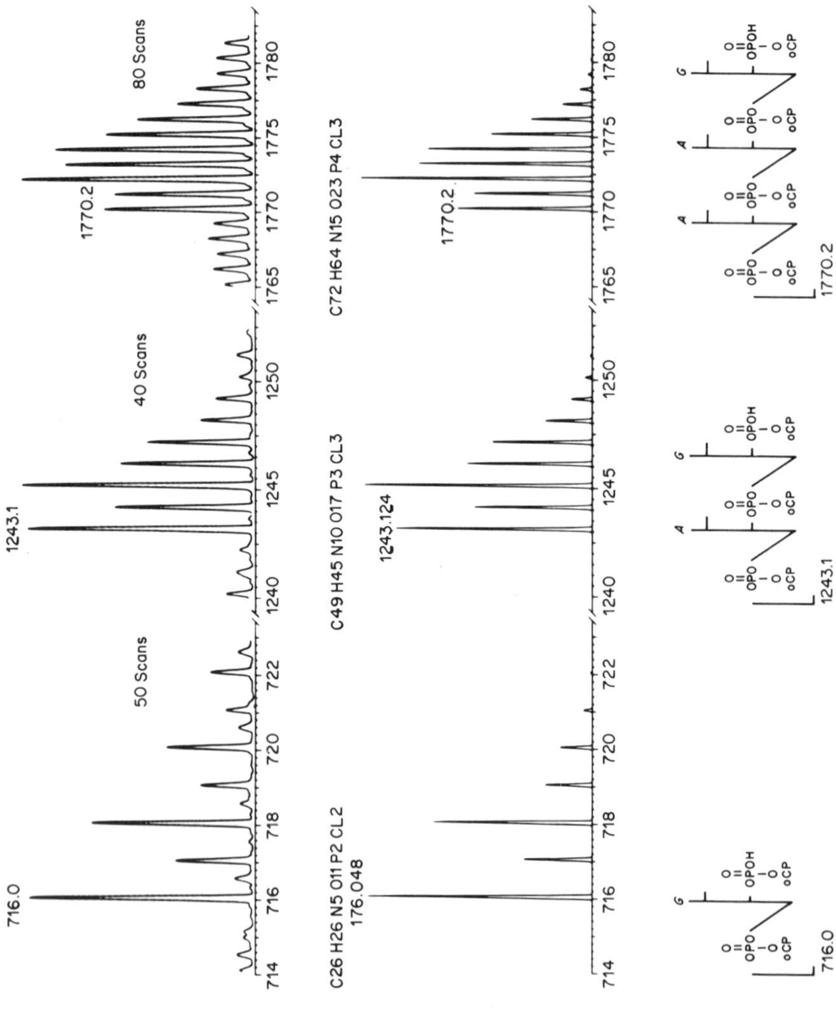

Figure 4: (Part).

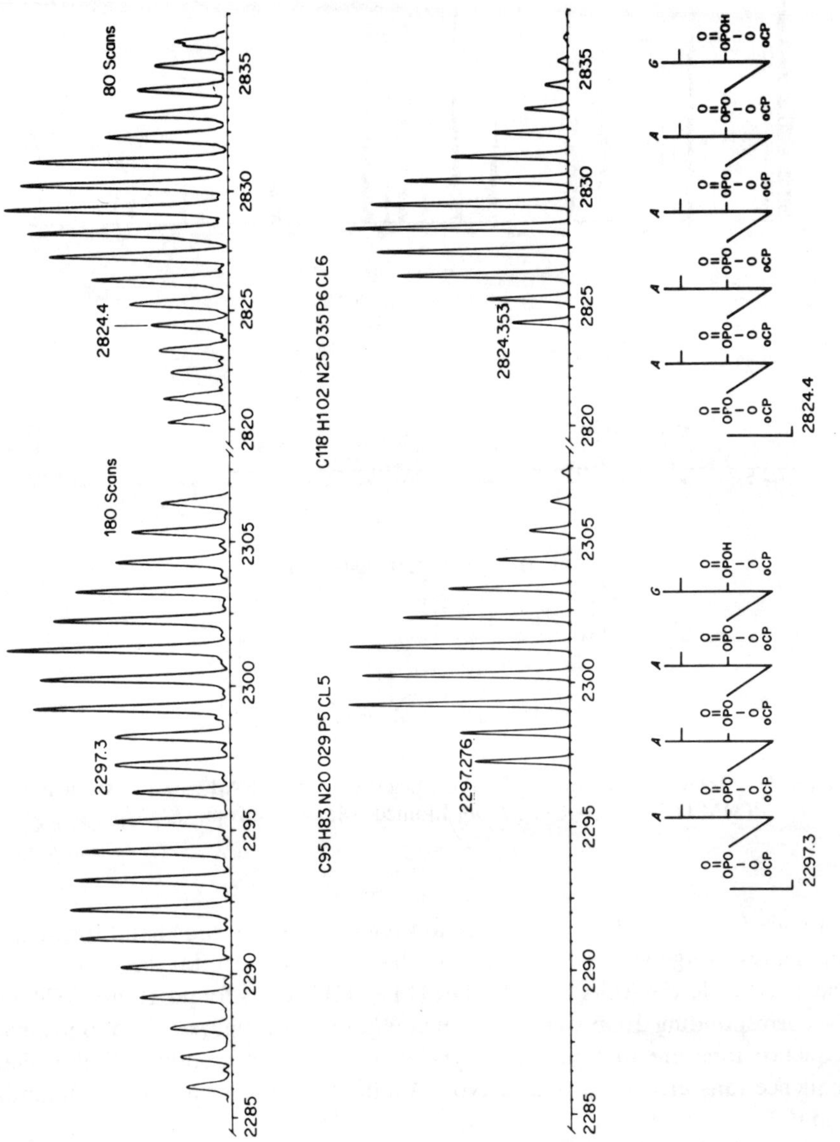

Figure 4: 5′-phosphate sequence ions of negatively charged protected d(AAAAAG)⁻. Top: experimental results; middle: simulated results; bottom: shorthand structures. ib = isobutyryl, an = anisoyl and oCP = o-chlorophenyl. The lower portion of the mass scale is shown separately on p. 222. *Reproduced from* Mass Spectrometry in the Health and Life Sciences (*Eds. A. L. Burlingame and N. Castagnoli*), *pp. 597–615, by permission of Elsevier Scientific Publishers.*

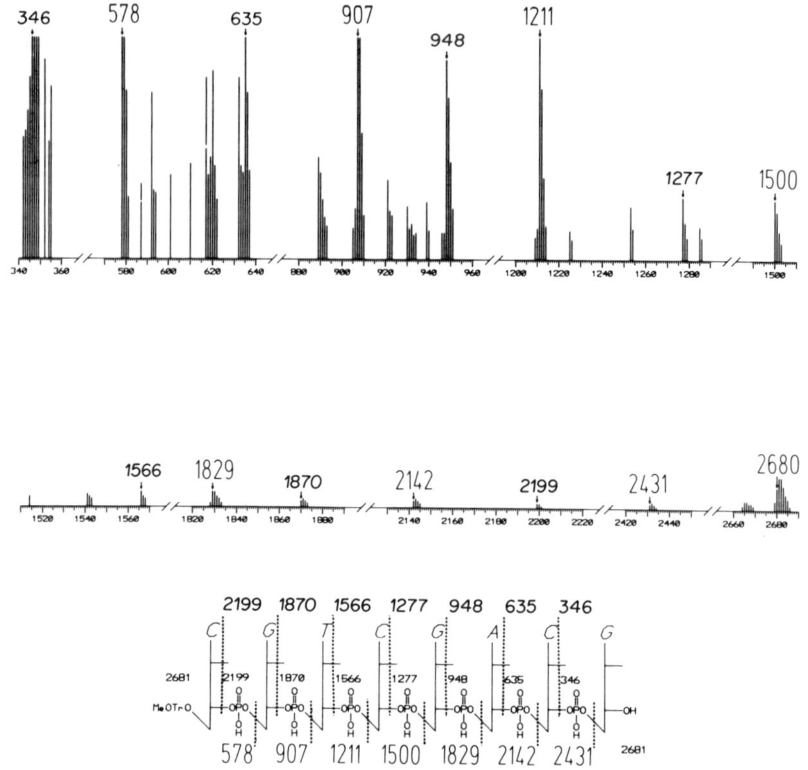

Figure 5: Relevant regions of the negative ion FAB mass spectrum of d(MMTrCGTCGACG). *From* Biomed. Mass Spectrom., *12*, *514–524*.

Figure 8 shows only the sequence ion regions and the shorthand formula of the largest oligodeoxyribonucleotide that we have so far sequenced, the decanucleotide GAAGATCTTC. The $(M - H)^-$ ion is found at m/z 3024 and the corresponding doubly charged ion at m/z 1512. In Figure 8, all 5′-phosphate sequence ions are in bold-faced type while the corresponding 3′-phosphate sequence ions are in light-faced type. Again the sequence analysis is unambiguous.

Figure 9 shows the largest free oligonucleotide yet determined by high field FAB MS (Grotjahn and Taylor, 1985). In these investigations we have again used the multichannel analyzer technique to determine the elemental composition of the $(M - H)^-$ ion from the tridecadeoxyribonucleotide CCTTCTGGAACTG. The peak matching unit of the mass spectrometer was set to alternate between this peak and the next highest caesium iodide cluster

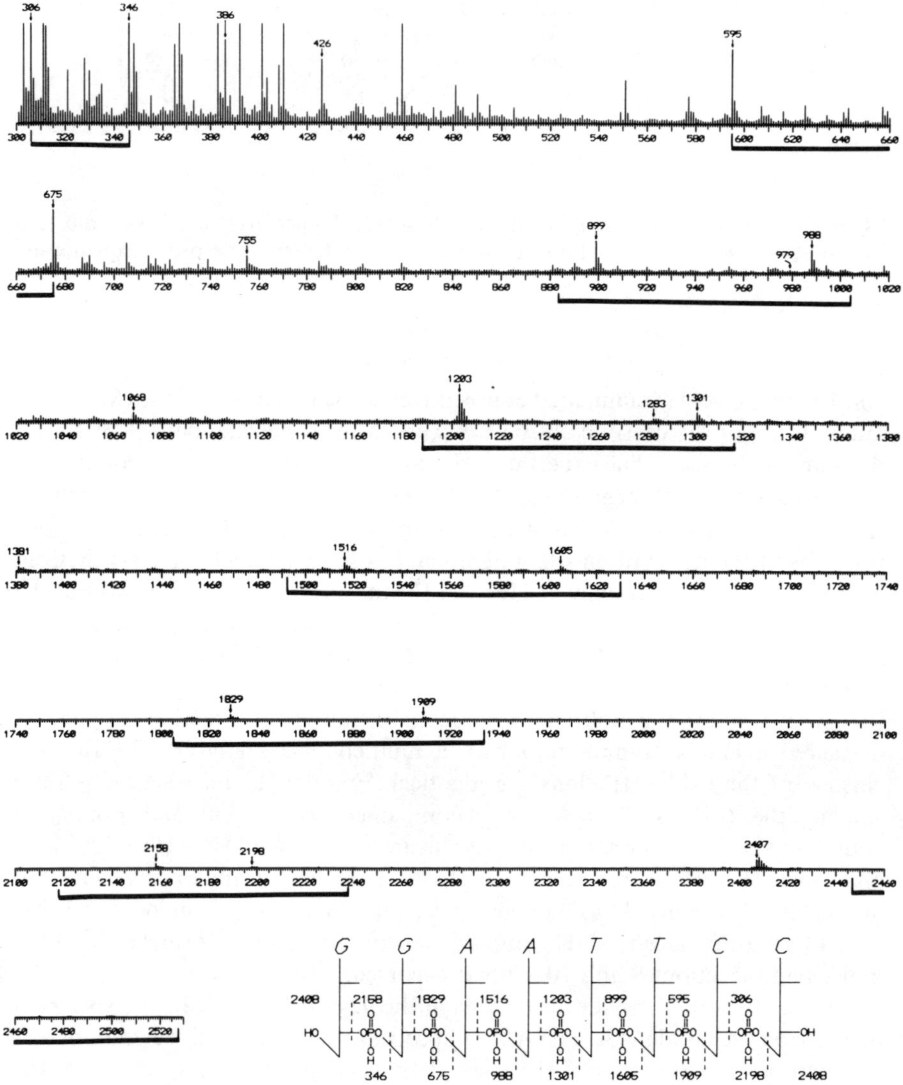

Figure 6: Negative ion FAB mass spectrum of d(GGAATTCC). The underlined regions indicate the sequence ion regions. *From* Biomed. Mass Spectrom., ***12***, *514–524*.

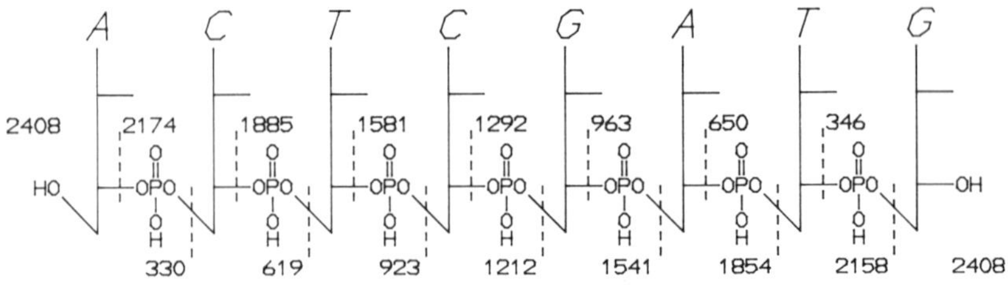

Figure 7: Shorthand formula of d(ACTCGATG). Upper mass numbers indicate the 5′-phosphate sequence ions, lower numbers the less intense 3′-phosphate sequence ions.

and 100 scans were accumulated consecutively. The maximum of the $(M - H)^-$ ion and that of the caesium cluster ion were in the same channel 1.9069 s after the start of the scan. The experimental and calculated results were equal.

Investigations with negative ion FAB MS allow the study of new synthetic strategies for small DNA fragments. As an example the investigation of four fractions of the deoxyribonucleotide nona T is shown, which were synthesized by a novel synthetic strategy at Hoechst (Frankfurt am Main). The aim of the mass spectroscopic research was to check the synthesis and to decide whether any replacement of $-NH_2$ by $-OH$ had occurred in the base part of the oligonucleotide.

Figure 10 shows the molecular ion region of four HPLC-fractions of nona T, measured again in combination with a multichannel analyzer. The isotopic clusters of the $(M - H)^-$ ions are identical. Fraction 1 and fraction 7–8 are similar; the $(M - H)^-$ ions are accompanied by sodium and potassium containing ions. Fraction 9 showed significant amounts of $(M - H + 42)^-$ ions and some $(M - H + 28)^-$ ions. Fraction 10–12 showed a similar spectrum with more $(M - H)^-$ ions. Thus the mass spectrometric investigation indicated that $-OH$ has not replaced $-NH_2$, and only acetylation and/or formylation of the molecule in fractions 9 and 10–12 had occurred.

Figure 11 shows the relevant regions of the negative ion FAB mass spectrum of the ribonucleotide AAAAAA; the spectrum (a) is that of the sodium salt while spectrum (b) is that of the triethylammonium salt. The spectrum of the sodium salt is complicated by the presence of sodium-containing sequence ions. The fourth and the fifth sodium-free sequence ions could not be detected. No such complication is observed for the triethylammonium salt. As reported previously for free oligodeoxyribonucleotides, all sequence ions were seen without sodium satellites; only the $(M - H)^-$ ions have a small sodium-containing partner.

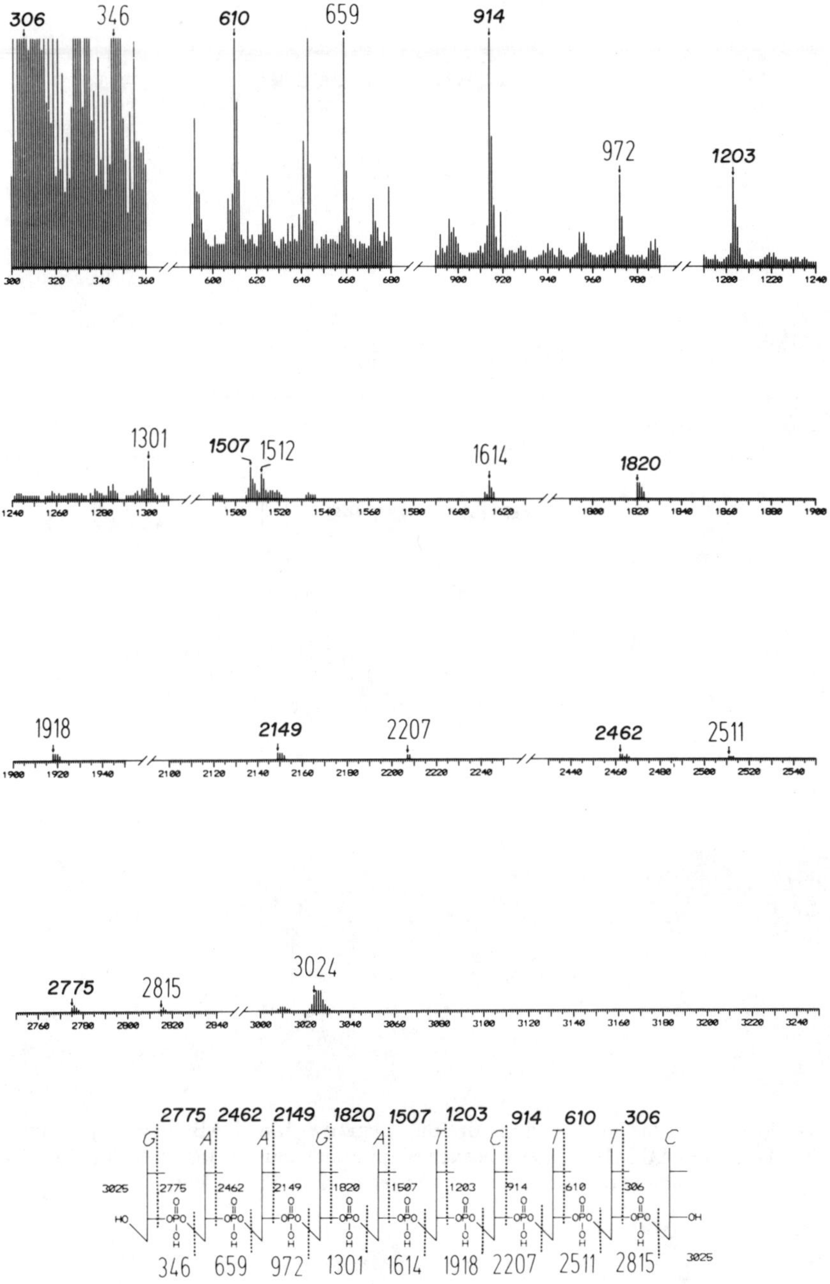

Figure 8: Sequence ion regions of the negative ion FAB mass spectrum of d(GAA-GATCTTC). *From* Biomed. Mass Spectrom., ***12***, *514–524*.

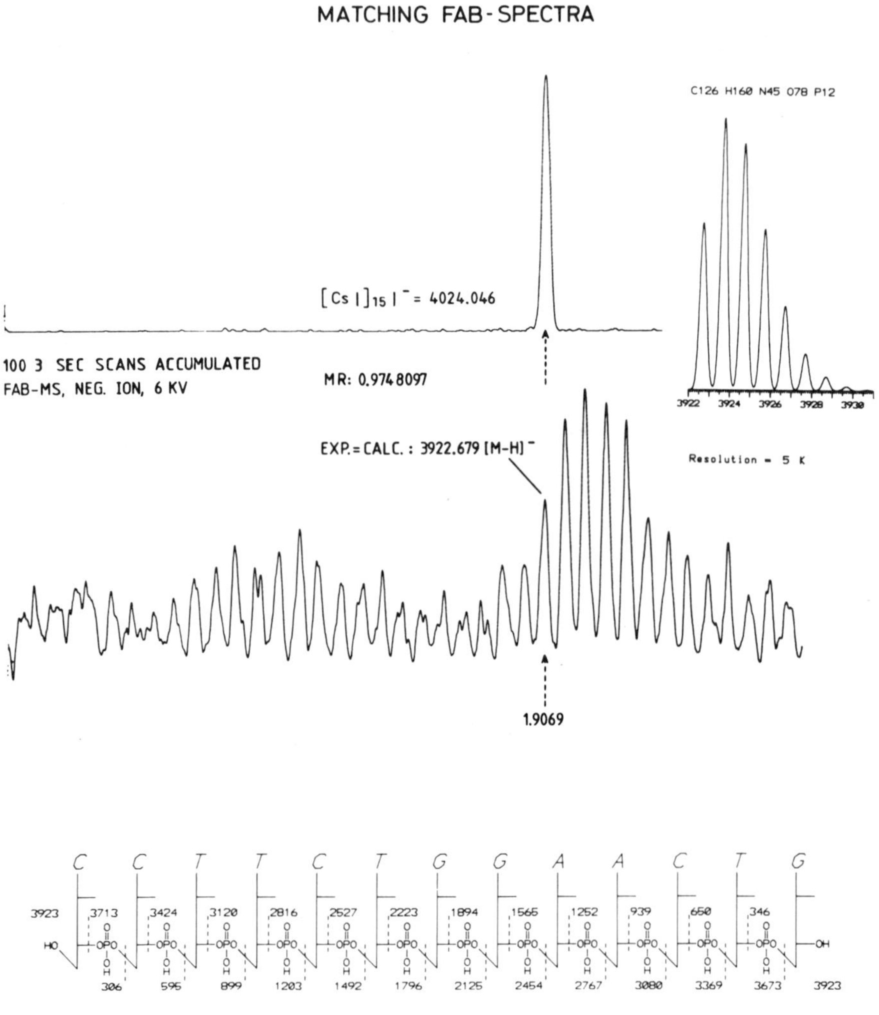

Figure 9: Molecular ion region of the negative ion FAB mass spectrum of d(CCTTCTGGAACTG). Upper trace: caesium iodide cluster. Insert on the right-hand side is the simulated ion cluster of the oligonucleotide.

Figure 12 shows the relevant regions of the negative ion FAB mass spectrum of the triethylammonium salt of ribonucleotide UGUU. The $(M - H)^-$ ion at m/z 1200 has a small sodium-containing satellite. The sequence ions are accompanied by ions, due to loss of water, and sometimes by +80 u ions, similar to those observed for deoxyribonucleotides. The first and third phosphate sequence ions coincide, while the 3′- and 5′-phosphate sequence ions for the dinucleotide confirm the sequence.

Thus we would expect to obtain sequence information for ribonucleotides similar to that obtained for oligodeoxyribonucleotides.

Figure 13 shows the relevant regions of the negative ion FAB mass spectrum of the modified deoxyribonucleotide GGsAATTCC (s indicates the exchange of a sulphur for one of the non-bonded oxygens on the phosphate linking adjacent nucleotides), containing the recognition sequence of the EcoR1 restriction endonuclease with a phosphorothioate internucleotidic linkage at the cleavage site. The background of the spectrum is higher than that for the same unmodified deoxyribonucleotide GGAATTCC (Figure 6) with the $(M - H)^-$ ion appearing at m/z 2423. The $(M - H)^-$ ion and the sequence ions are usually accompanied by relatively abundant ions 18 u lower (loss of water) and 22 u higher (addition of sodium). The 5′-phosphate sequence ions could be detected up to the fifth nucleotide and the 3′-phosphate sequence ions up to the third; of importance is the overlapping of the 3′- and 5′-phosphate sequence ions. The sequence analysis is fully compatible with the structure of the octamer (Conolly *et al.*, 1984).

For modified oligonucleotides, particularly for those where the phosphate group has been altered and rendered resistant to nucleases, this method can be successfully applied and might be the only method available, as all classical methods have so far failed.

12.4. CONCLUSIONS

Negative ion FAB MS is a well proven analytical method for characterizing synthetic DNA fragments which are used in genetic engineering. The method allows at present the determination of the molecular weight of protected and free oligonucleotides up to eight and 13 base units respectively, and sequence analysis up to eight and 10 base units. The mass spectrometric sequence analysis is less time consuming than the other sequencing procedures known to date. Mass spectrometric bidirectional sequence analysis of free oligonucleotides can be performed when *ca.* 1 OD_{260} or more material is available. Sequence analysis by classical methods would take approximately 2 days, while mass spectroscopic analysis requires approximately 1 hour.

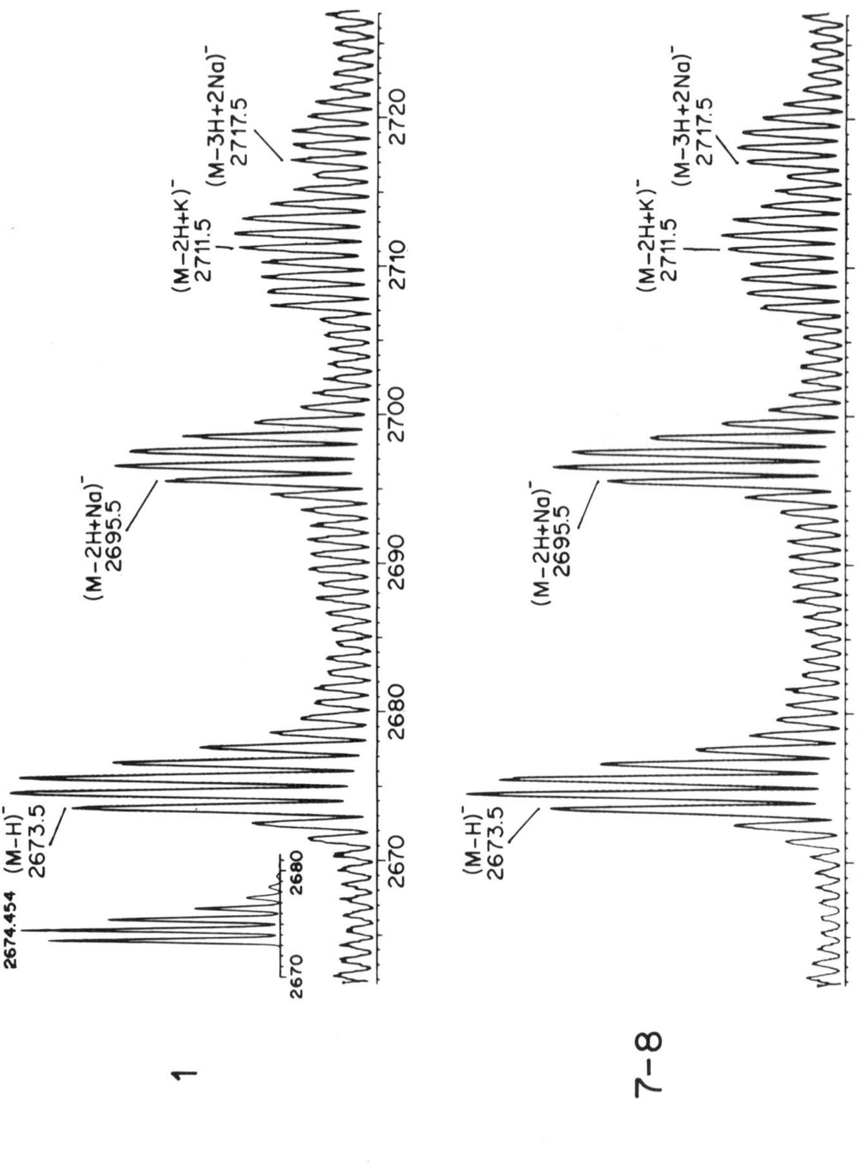
1
7-8
2674.454
2670
2680
(M-H)⁻
2673.5
(M-2H+Na)⁻
2695.5
(M-2H+K)⁻
2711.5
(M-3H+2Na)⁻
2717.5
2670
2680
2690
2700
2710
2720

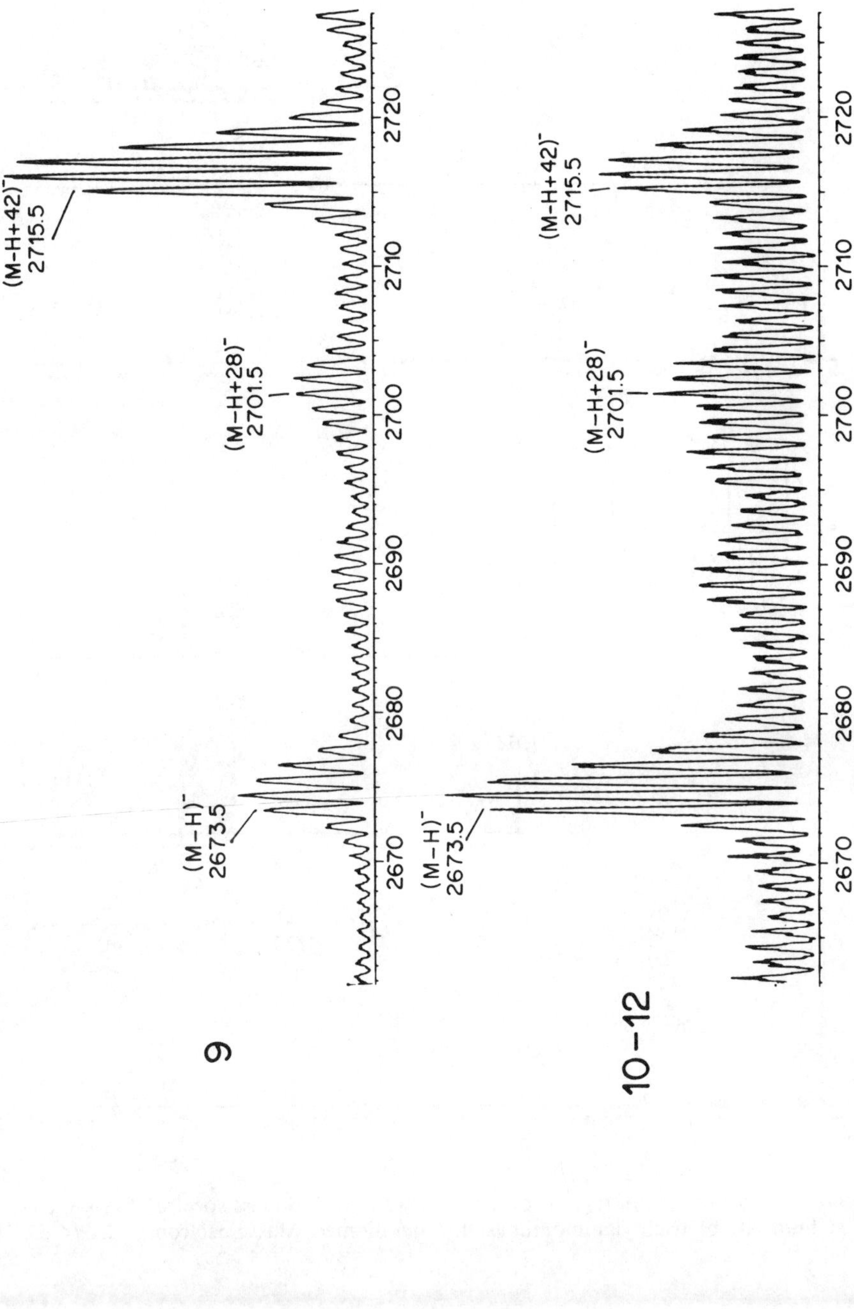

Figure 10: Molecular ion region of the negative ion FAB mass spectra of different HPLC fractions (1, 7–8, 9 and 10–12) of d(nonaT). Sample sizes: 1–4 OD_{260}. 40–100 scans were accumulated to give the data shown.

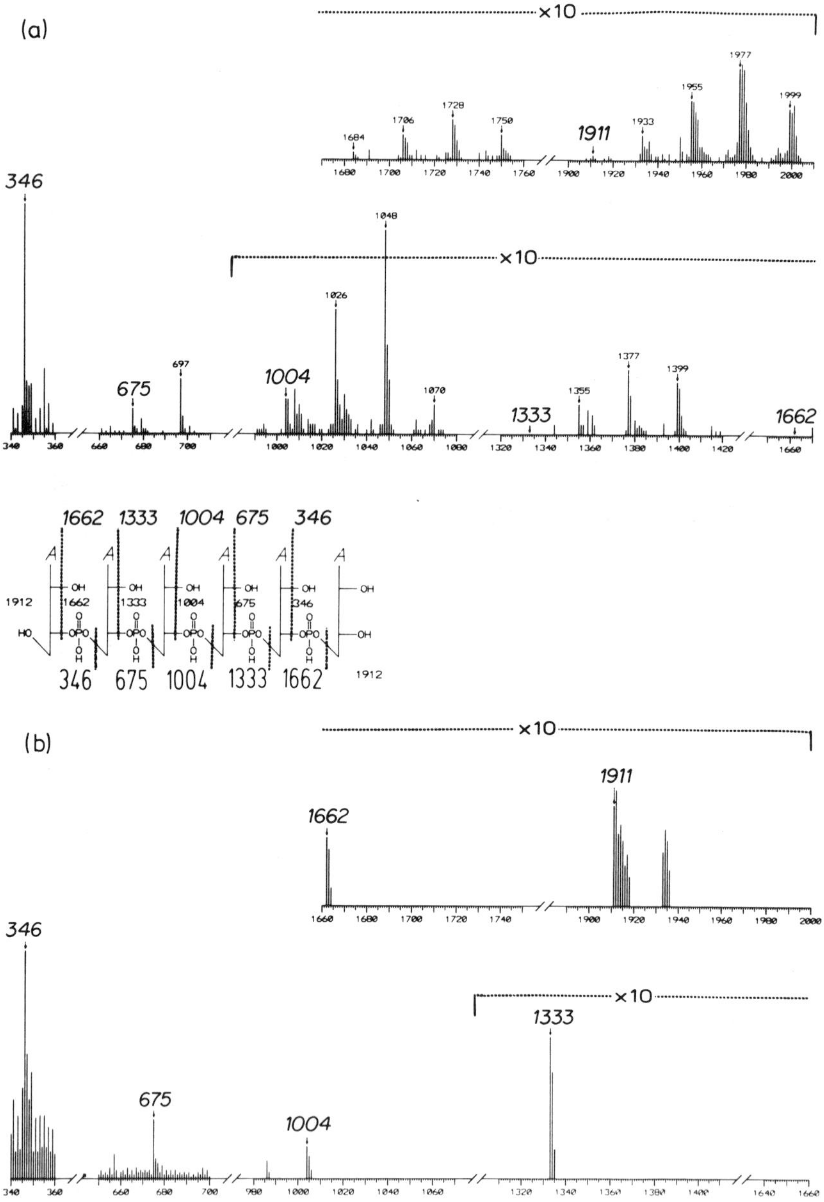

Figure 11: Sequence ion regions of the negative ion FAB mass spectra of AAAAAA. a) sodium salt; b) triethylammonium salt. *From* Biomed. Mass Spectrom., ***12***, *514–524*.

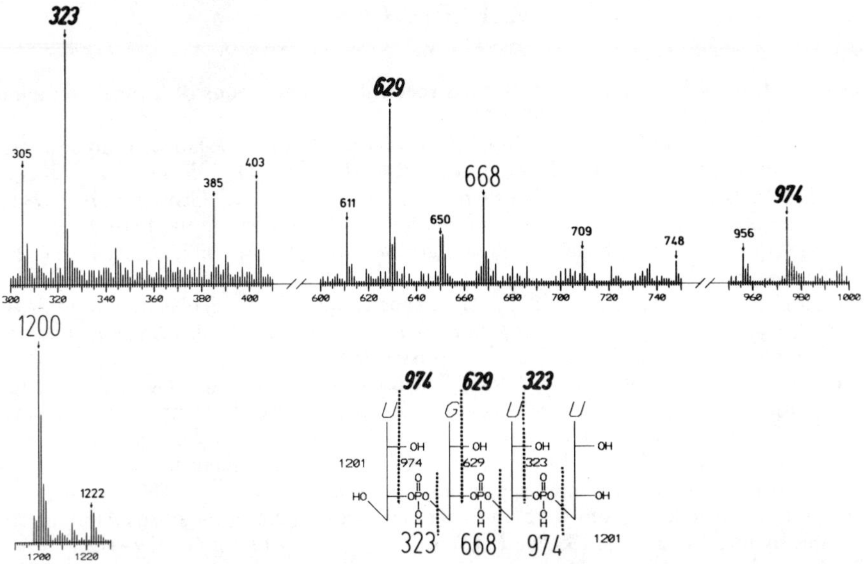

Figure 12: Sequence ion regions of the negative ion FAB mass spectrum of UGUU. *From* Biomed. Mass Spectrom., ***12***, *514–524*.

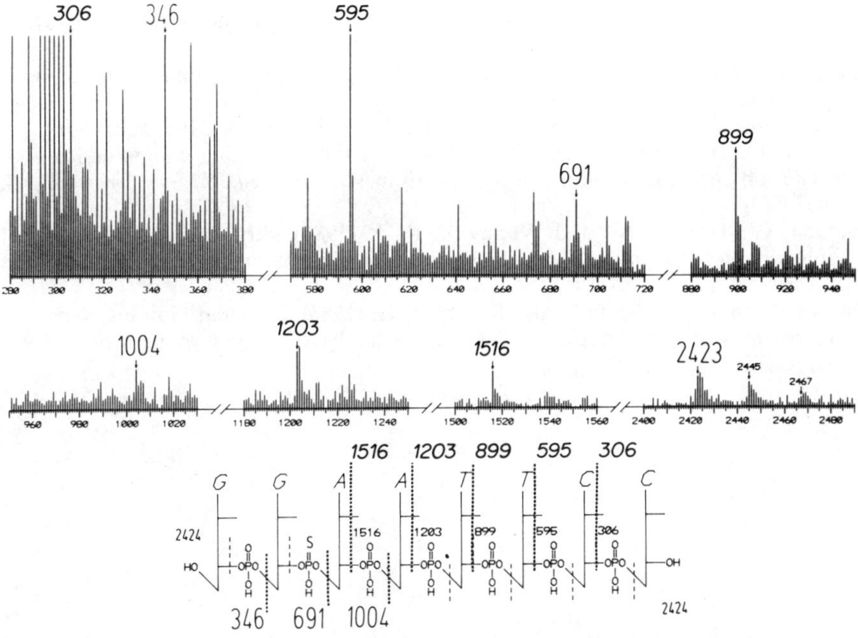

Figure 13: Sequence ion regions of the negative ion FAB mass spectrum of d(GGsAATTCC). *From* Biomed. Mass Spectrom., ***12***, *514–524*.

REFERENCES

Biros, F. J. (1970). Enhancement of mass spectral data by means of a time averaging computer, *Anal. Chem.*, **42**, 537–540.

Brownlee, G. G., and Sanger, F. (1969). Chromatography of ^{32}P-labelled oligonucleotides on thin layers of DEAE-cellulose, *Eur. J. Biochem.*, **11**, 395–399.

Conolly, B. A., Potter, B. V., Eckstein, F., Pingoud, A., and Grotjahn, L. (1984). Synthesis and characterization of an octanucleotide containing the EcoR1 recognition sequence with a phosphorothioate group at the cleavage site, *Biochemistry*, **23**, 3433–3453.

Grotjahn, L., and Steinert, H. (1985). Mass spectrometry in molecular design, in *Mass Spectrometry in the Health and Life Sciences* (Eds. A. L. Burlingame and N. Castagnoli jr.), pp. 597–615, Elsevier, Amsterdam.

Grotjahn, L., and Taylor, L. C. E. (1985). The use of signal averaging techniques for the quantitation and mass measurements of high molecular weight compounds using fast atom bombardment mass spectrometry, *Org. Mass Spectrom.*, **20**, 146–152.

Grotjahn, L., Frank, R., and Blöcker, H. (1982). Ultrafast sequencing of oligodeoxyribonucleotides by FAB–mass spectrometry, *Nucl. Acids Res.*, **10**, 4672–4678.

Grotjahn, L., Frank, R., and Blöcker, H. (1983). Sequencing of oligodeoxyribonucleotides by negative FAB–MS, *Int. J. Mass. Spectrom. Ion Phys.*, **46**, 439–442.

Grotjahn, L., Frank, R., Heisterberg-Moutsis, G., and Blöcker, H. (1984). Fast identification by FAB mass spectrometry of building blocks for oligonucleotide synthesis, *Tetrahedron Lett.*, **25**, 5373–5376.

Grotjahn, L., Blöcker, H., and Frank, R. (1985). Mass spectroscopic sequence analysis of oligonucleotides, *Biomed. Mass Spectrom.*, **12**, 514–524.

Maxam, A. M., and Gilbert, W. (1977). A new method for sequencing DNA, *Proc. Nat. Acad. Sci. U.S.A.*, **74**, 560–564.

Panico, M., Sindona, G., and Uccella, N. (1983). Bioorganic applications of mass spectrometry. 3. Fast-atom-bombardment-induced zwitterionic oligonucleotide quasimolecular ions sequenced by MS/MS. *J. Am. Chem. Soc.*, **105**, 5607–5610.

Peele, G. L., and Brent, D. A. (1978). Use of a time averaging computer for high resolution field desorption mass spectrometry, *Biomed. Mass Spectrom.*, **5**, 5607–5610.

Sindona, G., Uccella, N., and Weclawek, K. J. (1982). Structure determination of isomeric oligodeoxynucleotide salts by fast-atom-bombardment mass spectrometry, *J. Chem. Res. (S)*, 184–185.

Tazawa, I., Inoue, Y., Seki, S., and Kambara, H. (1983). Secondary ion mass spectrometry: the scope and limitations in structural analysis of oligoribonucleotides, *Nucl. Acids. Res.*, **12**, 205–208.

Mass Spectrometry in Biomedical Research
Edited by S. J. Gaskell
© 1986 John Wiley & Sons Ltd.

Chapter 13

HIGH MASS CAPABILITIES OF SECTOR MASS SPECTROMETERS

Brian N. Green and Robert S. Bordoli

13.1. INTRODUCTION

Until the beginning of the present decade, mass spectrometry was still largely confined to low molecular weight volatile materials. A few laboratories had specialized in field desorption (FD) ionization, gaining access to higher masses. However, the technical difficulties of FD had deterred many laboratories from adopting it as a routine method.

The introduction of the fast atom bombardment (FAB) ion source in 1981 (Barber *et al.*, 1981) undoubtedly revolutionized both low and high molecular weight mass spectrometry and has raised the expectations of mass spectrometrists considerably. FAB quickly demonstrated the ability to volatilize and ionize thermally fragile molecules including peptides up to about m/z 9500 molecular weight (Barber *et al.*, 1983). Furthermore, this new technique lacked the difficulties of using FD in that the preparation of samples and acquisition of mass spectra were relatively easy, and it has consequently been quickly adopted as an analytical technique. Today many laboratories are using FAB MS as a routine approach to both structural and quantitative applications.

The application of FAB MS to the solution of structural problems in biochemistry and medicine, particularly for 'big' molecules, for example oligopeptides, oligosaccharides and oligonucleotides, is an area of burgeoning interest.

The purpose of this chapter is to discuss the scope and limitations of sector mass spectrometers in the high molecular weight applications of mass spectrometry. High mass in this context is above m/z 3000.

13.2. INSTRUMENTATION

At about the time that FAB was gaining acceptance as a routine technique, the highest mass instrumentation available commercially had a mass range of 3000 daltons at 8 keV ion energy. It was soon realized that, although compounds of molecular weight around 10,000 daltons would produce ions of the intact molecule, a serious limitation was sensitivity. The amount of material and the number of ions that can be generated and detected determine not only the detection limit but also the mass measurement accuracy. Figure 1 indicates the way in which the sensitivity at the protonated molecular ion of a selection of peptides varies with molecular weight. Although the diagram is only given as a guide, it is immediately obvious that to work routinely at high mass requires the highest possible sensitivity.

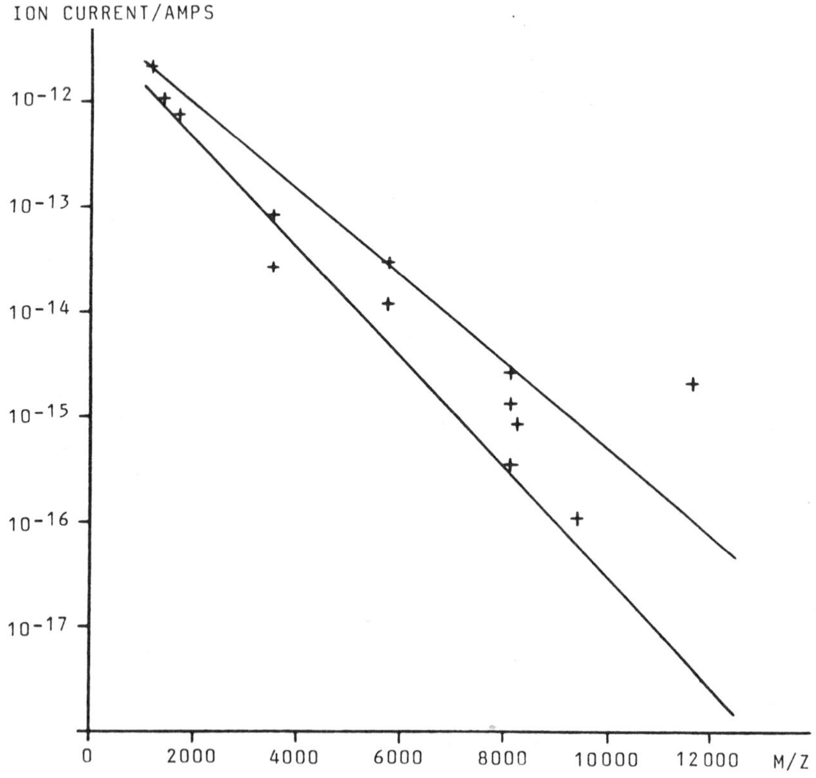

Figure 1: Variation of maximum parent ion current with mass for a series of peptides. Instrumental resolution 1000. The lines represent the range covered by the majority of data.

On earlier instruments one limitation was the necessity to reduce the ion accelerating voltage below the normal maximum of 8 kV to increase the mass range beyond m/z 3000. Since sensitivity, with a given instrument design, is found to be approximately proportional to ion energy, the spectrometer should be designed to transmit the highest mass ions at as high an energy as practicable.

The options available to increase the mass range at high sensitivity are to increase the magnetic field and the radius of the ion trajectory in the magnetic field. This can easily be seen from the basic mass spectrometer equation where the mass in daltons is given by:-

$$m/z = 4825\, B^2 R^2 / V$$

where B is the magnetic field (Tesla), R is the ion radius in the magnetic field (cm), and V is the ion energy (electron volts). Fortunately, to some extent, physics is on the side of the designer of a high mass instrument in that the mass is proportional to the square of both field and radius.

The maximum practically useful field produced by a scanning magnet having a ferromagnetic core and pole tips is about 2.4 Tesla. To achieve this field, a cobalt–iron alloy, which has the highest saturation induction of the materials currently available, is used. Because this alloy is expensive, it is usually incorporated only into the pole tips, the remainder of the core being constructed from low carbon steel laminations of the type used in transformers. A less expensive alternative to using the cobalt–iron alloy is to construct the magnet entirely from transformer steel, but the maximum magnetic field is reduced to about 1.9 Tesla, thus reducing the mass range to about two-thirds of the value on a comparable instrument using the alloy. Superconducting magnets can produce substantially higher magnetic fields, but are unsuitable for use in sector mass spectrometers at the present time, one major reason being their inability to scan the magnetic field quickly.

It should be emphasized that a commercially viable instrument for high mass applications also has to cover all the lower mass applications normally expected of a flexible instrument—and with no loss of performance.

The remaining option for achieving high mass is a large radius. The most direct approach for increasing the radius is to simply scale up a lower mass range instrument, but this leads to an overly large, heavy and expensive design. Broadly, two means have been used for increasing the radius without proportionately increasing the ion path length and physical size.

One of these involves the use of inhomogeneous (wedge shaped) magnetic fields (Taylor *et al.*, 1983) which are strongly focusing in the dispersion plane but are inevitably defocusing in the Z (magnetic field) direction. Z lenses are included to overcome these effects. In such a commercial instrument (Kratos, MS 50RF), the path length was increased only by about 29% when the radius was increased from 30.5 to 68.6 cm, increasing the mass range to 10,000 daltons at 8 keV ion energy with a 1.9 Tesla magnet.

Alternatively, the path length may be minimized by using non-normal angles between the magnetic field boundary and the ion beam at the entrance and exit of the magnetic sector (Hintenberger and Konig, 1959). Again lenses are beneficial in compensating for Z defocusing effects at the magnet boundaries. A schematic diagram of the ion optics of a commercial design using this latter approach (VG Analytical, ZAB-SE) is shown in Figure 2. Here, the magnet radius of 66 cm was chosen to give a mass range of 12,000 at an ion energy of 10 keV (15,000 at 8 keV ion energy) using cobalt–iron alloy pole tips. The mass range of this instrument is illustrated in Figure 3, where caesium iodide clusters up to $Cs_{60}I_{59}$ (m/z 15,462) are shown at an ion energy of 8 keV. Of course at 5 keV, with a loss of only a factor of two or so in sensitivity, it has a mass range of 24,000, sufficiently beyond the currently known highest mass limit for biomolecules ionized by FAB, to accomodate further developments in ion generation.

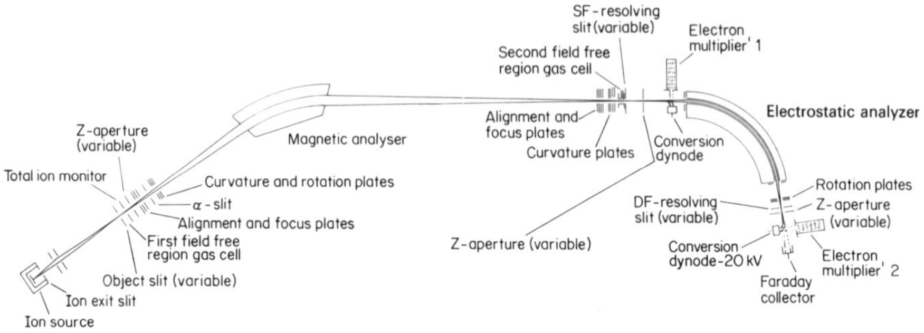

Figure 2: Ion optics of a mass spectrometer (VG ZAB-SE) using a homogeneous field magnet of 66 cm radius and 2.4 T, having a mass range of 12,000 daltons at 10 keV ion energy.

13.3 MASS ASSIGNMENT

The principal objective in obtaining a mass spectrum is to measure the masses of the recorded ions. There are two approaches which may be used to achieve this objective at high mass, which may be called high and low resolution approaches. High resolution refers to operation at a resolution sufficient to separate the isotope peaks in the ion cluster. Low resolution refers to operation at 1000–2000 resolution, such that there is no separation of the isotope peaks (Cotter *et al.*, 1984). Although partially resolved molecular ion clusters have been produced at m/z 8680 (Cottrell *et al.*, 1984) and bovine insulin at m/z 5730 resolved to 1 part in 10,000 (Figure 4), unresolved spectra at high mass often give sufficiently accurate data and are easier to obtain and interpret.

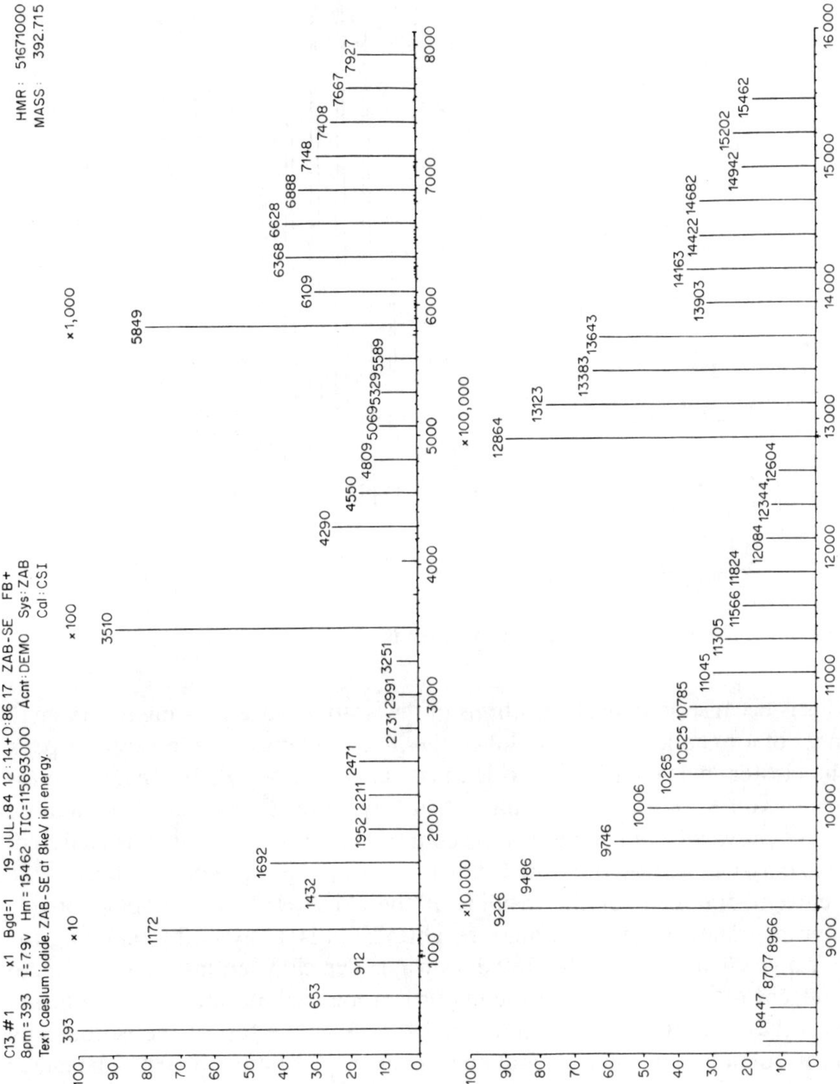

Figure 3: Caesium iodide spectrum showing mass range of over 15,000 daltons at 8 keV ion energy.

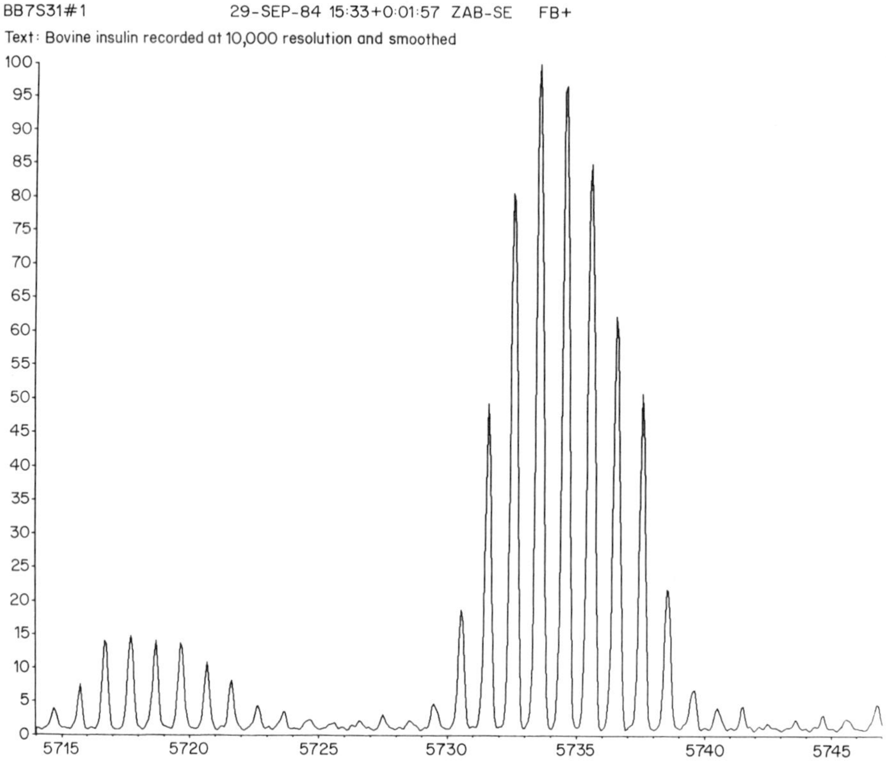

Figure 4: $(M + H)^+$ ion region of bovine insulin taken at a resolution of 10,000.

This is because statistical variations in the isotope pattern owing to the small number of ions collected lead to difficulties in identifying the monoisotopic peak in the cluster, particularly if a wide mass range is covered. If a narrow mass range is scanned in order to obtain statistically more reliable data, then a more accurate knowledge of the mass of the compound is needed before it is analyzed.

This may be illustrated by the following example using typical values of the ion current obtained for the insulins at the $(M + H)^+$ ion. Suppose bovine insulin of monoisotopic molecular weight 5729.6 is analyzed by scanning the mass range 7000–1000 in 60 s (100 daltons/s). Typically, an insulin will give an ion current of 5×10^{-16} A on the major isotope peak at unit mass resolution (10% valley definition) and about 1×10^{-14} A at the top of the isotopically unresolved molecular ion cluster at an instrument resolution of 1000. The factor of nearly 20 higher ion current in the unresolved case is the product of instrument transmission (a factor of 6) and a factor due to several of the isotope peaks in the cluster being partially summed together (a factor of 2.7). Figure 5 shows the calculated response for the $(M + H)^+$ ion cluster of bovine insulin assuming a mass spectrometer resolution of 1000. The theoretical isotope pattern is also given. It can be seen that the width of the unresolved cluster is mainly determined by the isotope pattern and not by the instrumental resolution. The width of the composite at 5% intensity is about 12 daltons, an effective resolution of 480.

Estimates of the reproducibilities in the measurement of mass and intensity to be expected on the basis of the number of ions collected (Campbell and Halliday, 1965) at instrumental resolutions of 6000 and 1000 are summarized in Table 1. In the resolved case, the reproducibilities refer to the largest peak in the cluster at *m/z* 5733.6. Since the determination of the isotope pattern involves measuring the intensity ratio between two peaks, variations in the patterns will be greater than the value given in the table.

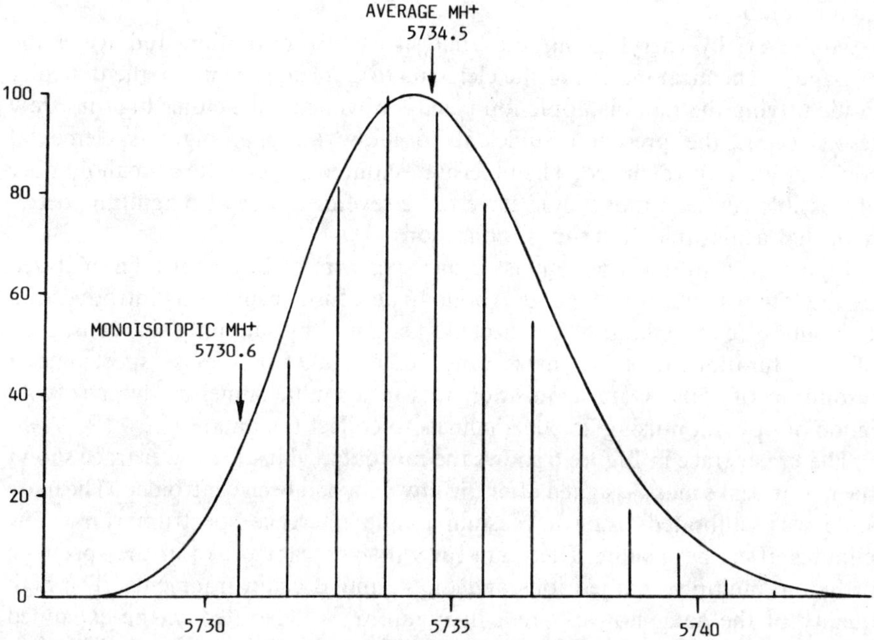

Figure 5: Calculated peak shape for the bovine insulin $(M + H)^+$ ion at a mass spectrometer resolution of 1000.

TABLE 1. Reproducibility in mass and intensity expected for bovine insulin at 6000 and 1000 resolution calculated on the basis of the number of ions collected.

Resolution	No. of ions collected	Standard Deviation: Mass (ppm)	Standard Deviation: Intensity (%)
6000	16	8.6	25
1000	3300	7.6	1.7

Two conclusions may be drawn. First, the expected accuracy in mass determination is not substantially different when the resolved and unresolved cases are compared. In fact, a single measurement on the largest peak in the resolved cluster is marginally worse than a measurement of the unresolved cluster. Averaging determinations on several peaks in the cluster (Barber *et al.*, 1984) would improve the accuracy of the mass determination on the resolved data, but not to make it dramatically better than on the unresolved data. Secondly, and possibly more importantly, the statistical variations in the intensity even of the major peaks in the cluster (25% standard deviation) make assignment of the monoisotopic ion unreliable.

Thus, with a given data collection time, over a given mass range, the expected accuracy in determining the mass of a molecule or fragment having normal isotopic composition is comparable whether analysis is carried out at low or high resolution.

Moreover, by carrying out the analysis at low resolution and using the average or chemical masses of the elements to calculate the mass, the difficulty in identifying the monoisotopic ion peak is obviated. At neither high nor low resolution is the precision sufficient to allow an unambiguous elemental composition to be deduced. The precision required to generate a small number of possible compositions would have to be several orders of magnitude better than that achievable under these conditions.

Figure 6 illustrates a wide mass range spectrum of bovine insulin analyzed by the low resolution technique. About 10 μg of the insulin was introduced in 1-thioglycerol. The data were collected over 2 min by summing six scans, each of 20 s duration, over the mass range 6200–1000, at a mass spectrometer resolution of 1500. Data acquisition was in a multichannel analyser (MCA) mode of operation using 25,000 channels to collect the data.

The upper trace in Figure 6 shows the raw data, whilst the lower trace shows the major peaks mass assigned after the raw data had been centroided. The mass scale was calibrated using a caesium iodide reference spectrum run a few minutes after the insulin. There are several significant peaks in the spectrum including multiply charged ions and the A and B chain fragments. The high quality of the data, not apparent in Figure 6, is illustrated in the expanded regions of the $(M + H)^+$, B chain and doubly charged ions shown in Figure 7.

The measured masses of the multiply charged ions and the A and B chain fragments are given in Table 2. With the exception of the A chain fragment ion, the measured values are within 0.25 dalton of the accepted values. For the $(M + H)^+$ ion the error is +0.18 dalton which is 31 ppm compared with a theoretical standard deviation of 8 ppm. Various contributions to the error may include a larger predicted error owing to ion statistical considerations with respect to the reference spectrum and the paucity of reference peaks used in this case. Measuring the centroid where the signal to background is only 4 : 1 and the short-term mass stability of the instrument may also add some error.

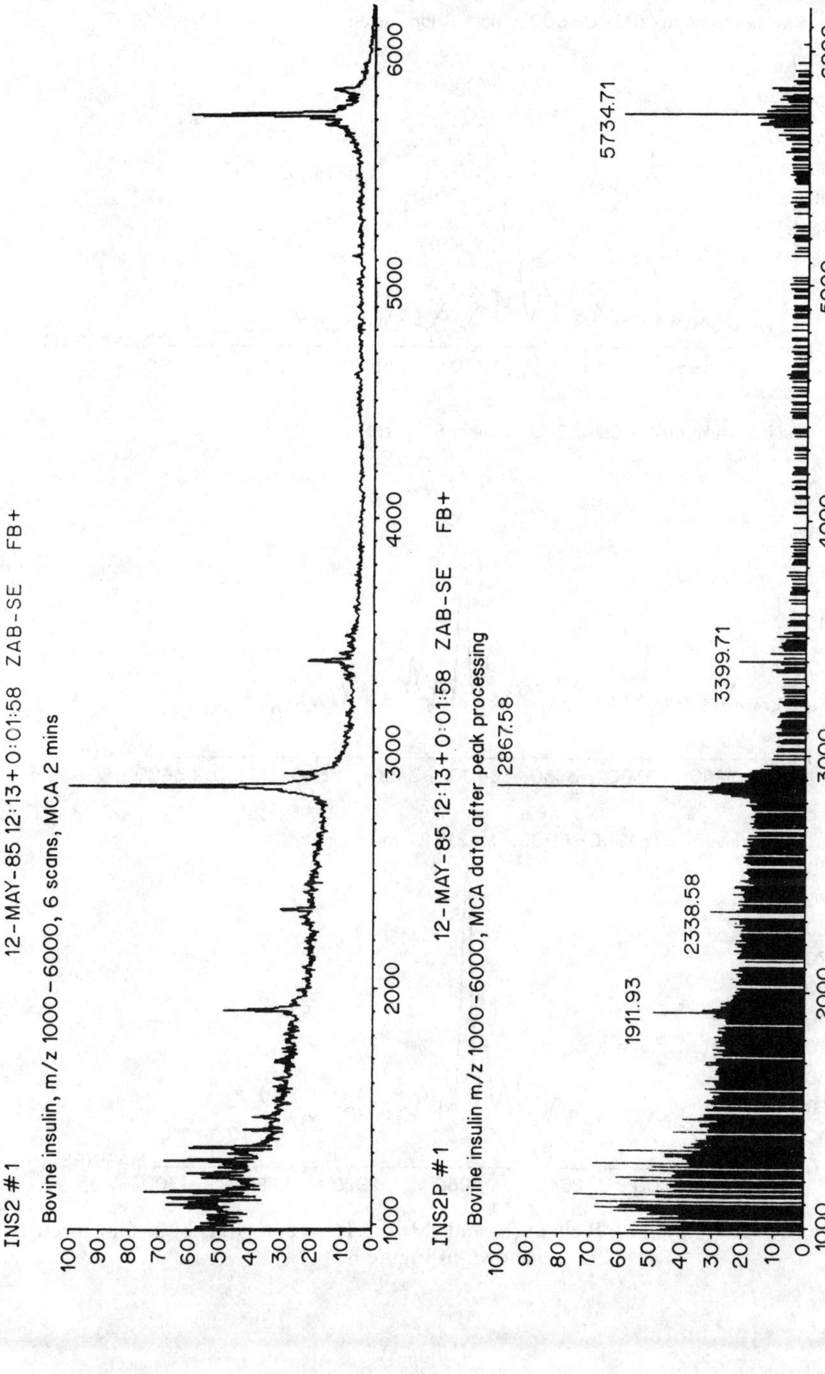

Figure 6: Spectrum of bovine insulin taken at 1500 resolution. Upper spectrum is raw data; the lower spectrum follows peak centroiding and mass assignment.

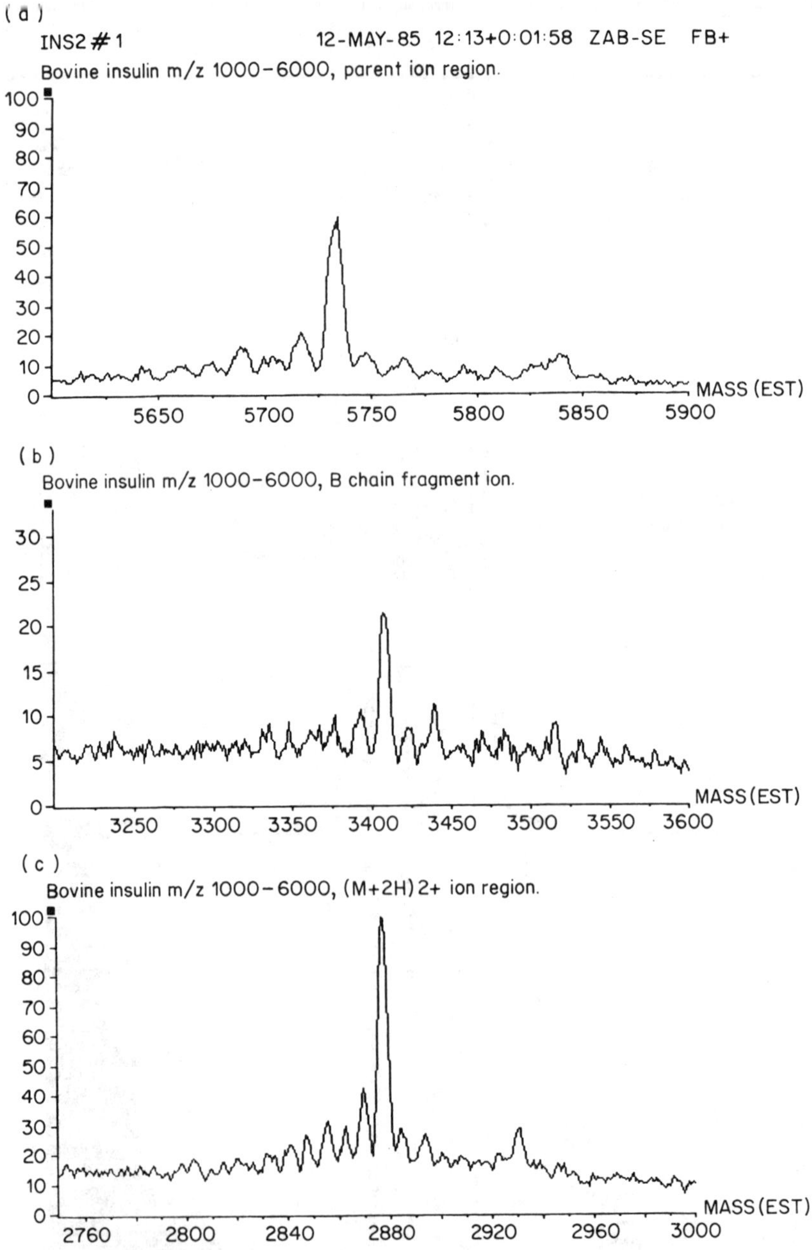

Figure 7: $(M + H)^+$ (a), B chain (b) and $(M + 2H)^{2+}$ (c) regions of insulin spectrum shown in Figure 6.

TABLE 2. Mass measurement errors from the low resolution insulin spectrum shown in Figure 6.

	Measured Mass	Calculated Mass[a]	Error (dalton)
$(M + H)^+$	5734.71	5734.53	0.18
B Chain	3399.71	3399.93	−0.22
$(M + H)^{2+}$	2867.58	2867.77	−0.19
A Chain	2338.58	2339.68	−1.10
$(M + 3H)^{3+}$	1911.93	1912.18	−0.25

[a] Calculated masses are based on chemical atomic weights, for comparison with measured masses derived from centroid calculations for the unresolved ion clusters.

Some of these effects would produce systematic errors whereas the others would introduce random errors. However, on the basis of a single measurement these cannot be separated and identified. Whilst the errors are somewhat larger than predicted, these measurements demonstrate that the data acquired using large mass range survey scans are sufficiently good to allow the mass assignment of both parent ions and fragments to well within 0.5 dalton.

Similar accuracies were obtained on the $(M + H)^+$ ions of two higher molecular weight compounds (eglin c and its *N*-acetyl derivative) where a narrower mass range (300 daltons) was scanned to compensate for the lower ion currents. The acquisition time per measurement was again 2 min. The results on six measurements are summarized in Table 3 (Richter *et al.*, 1985) and show a mean error of −0.32 dalton ±0.29 dalton standard deviation. Additional data from other compounds would be needed to establish the significance of the bias in the mean and whether it could be used to correct mass measurements made in this way.

TABLE 3. Results of 6 mass measurements made at low resolution on two related peptides (eglin c and its N-acetyl derivative).

Run N°.	Measured Mass: $(M + H)^+$	Calculated Mass: $(M + H)^{+a}$	Error (dalton)
1	8091.62	8092.08	−0.46
2	8092.12	8092.08	0.04
3	8091.50	8092.08	−0.58
	8133.46	8134.12	−0.66
4	8091.92	8092.08	−0.16
5	8134.03	8134.12	−0.09
	Mean error = −0.32 ± 0.29		

[a] See footnote to Table 2.

Of course, a limitation of the low resolution method is that mixtures of compounds whose molecular weights differ by only a few daltons will not be separated. A bias in the mass determination may also be caused, for example, by the partial reduction of disulphide bridges in peptides, which is known to occur in 1-thioglycerol commonly used as a FAB matrix. In order to reduce these potential sources of error, once the mass has been determined at low resolution, a very narrow scan (e.g. 30 daltons) over the appropriate mass range may be studied at higher resolution.

Figure 4 shows the $(M + H)^+$ region of bovine insulin acquired at 10,000 resolution by summing several scans over a period of 2 min. Mass measurement of several of the peaks in the isotope cluster against a suitable reference using the method described previously (Barber *et al.*, 1984) improves the mass determination, using state-of-the-art instrumentation, to typically better than ± 10 ppm (0.06 dalton) for insulin.

13.4. CALIBRATION COMPOUNDS

The most generally useful compounds for the calibration of FAB spectra are the alkali metal halides. Introduced in the solid form usually by evaporation from an aqueous solution, they produce intense, stable and readily interpretable spectra suitable for the calibration of the mass scale. The major peaks in the spectra occur at regular intervals of MX where M is the alkali metal and X is the halide. These cluster ions have the general formulae M_nX_{n-1} and $M_{n-1}X_n$ as positive and negative ions respectively, where $n = 1,2,3 \ldots$. Of these salts, caesium iodide is the most useful because it produces by far the most intense spectrum at the highest masses (Figure 3).

A disadvantage of caesium iodide, however, is that the peaks are separated by 260 daltons, which can be too large when narrow mass range, high resolution scans are to be acquired. If a rubidium iodide and caesium iodide solution in water, containing roughly equal molar proportions of the two salts, is applied to the sample holder mixed clusters are formed in addition to the rubidium iodide and caesium iodide clusters. The mixed clusters in the positive ion spectrum have the general formula $Cs_{n-m}{}^{85}Rb_{m-a}{}^{87}Rb_aI_{n-1}$ where m is the total number of rubidium atoms in the ion ($m = 0,1,2,3 \ldots n$) and a is the number of atoms of the rubidium-87 isotope in the ion ($a = 0,1,2$, etc).

The mass of any positive mixed ion may be calculated from the masses of the Cs_nI_{n-1} ions where Cs = 132.905433 and I = 126.904466 dalton using the expression:-

$$\text{Mass}\ (Cs_{n-m}{}^{85}Rb_{m-a}{}^{87}Rb_aI_{n-1}) = \text{Mass}\ (Cs_nI_{n-1}) - 47.993633\,m + 1.997384a.$$

Similarly the mass of any negative mixed ion may be calculated from the masses of the $Cs_{n-1}I_n$ ions using the expression:-

$$\text{Mass}\ (Cs_{n-m-1}{}^{85}Rb_{m-a}{}^{87}Rb_aI_n) = \text{Mass}\ (Cs_{n-1}I_n) - 47.993633m + 1.997384a.$$

The most intense peaks contain only a few rubidium atoms and are not therefore seriously reduced in intensity by being distributed over many isotopes. The spectrum can be considered as a caesium iodide spectrum in which increasing numbers of rubidium atoms replace caesium atoms. Hence the majority of the clusters are 48 daltons apart. The general characteristics of the spectrum from the mixture at low resolution are illustrated in the lower parts of Figures 8a and 8b in each of which only one of the repeating units of CsI is shown. At higher resolution, the rubidium containing peaks can be separated into multiplets, 2 daltons apart. These have been used successfully for calibrating narrow mass range scans at high resolution up to mass 6000. At low resolution, the spectrum is sufficiently intense to mass measure compounds as high as mass 12,000.

It is anticipated that as the number of investigators in high mass increases, other reference compounds, either organic, inorganic or organometallic, will be developed for more specific applications.

13.5. UPPER MASS LIMIT

The way in which the ion current for a series of oligopeptides falls as the mass increases has already been indicated in Figure 1. There is a paucity of data beyond m/z 10,000, but extrapolating from the Figure, a practical limit may be deduced to occur when the number of ions collected from one introduction of sample becomes too small to be usefully detected. At m/z 15,000, a current of roughly 10^{-18} A at the top of a peak 20 daltons wide whilst scanning over 200 daltons would allow 60 ions to be collected in 2 min, whilst only 24 ions would be collected if 500 daltons were scanned. Thus an upper limit in the region of m/z 15,000 is indicated on the basis of sensitivity.

Comparing the spectra of the $(M + H)^+$ regions in Figures 7(a) and 8 shows, however, that the upper limit is not only set by the number of ions collected, but also by the way in which the discrete $(M + H)^+$ peak seems to disappear into the background as the mass increases. Thus for insulin, the signal to background at an instrument resolution of 1000 is about 6 to 1, for eglin c, 3 to 1, for the dimer of insulin, 1 to 3 and finally for phospholipase A_2 there is no discrete peak corresponding to $(M + H)^+$ discernible above the background.

Whilst in the latter case this $(M + H)^+$ peak is absent, the background is composed of ions which must be related to the molecule in so far as the maximum occurs at about the expected mass (13,980), but because it is wide and statistically noisy the mass determination is very imprecise. Although the origin of the background in FAB has been the subject of some debate the picture is still unclear. The background below the molecular weight region may be composed

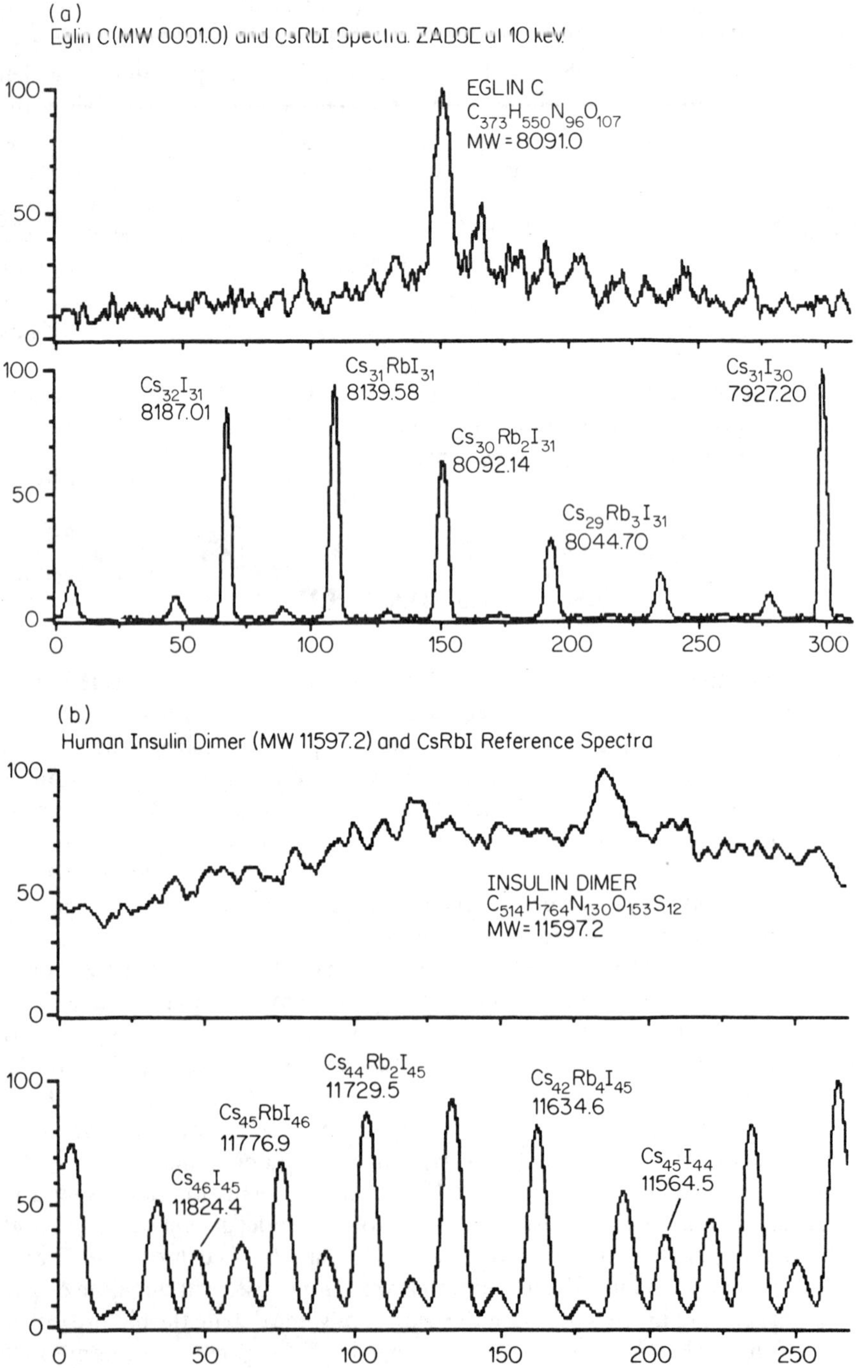
(a)
Eglin C(MW 8091.0) and CsRbI Spectra. ZABSE at 10 keV
EGLIN C
$C_{373}H_{550}N_{96}O_{107}$
MW = 8091.0
$Cs_{32}I_{31}$ 8187.01
$Cs_{31}RbI_{31}$ 8139.58
$Cs_{30}Rb_2I_{31}$ 8092.14
$Cs_{29}Rb_3I_{31}$ 8044.70
$Cs_{31}I_{30}$ 7927.20
(b)
Human Insulin Dimer (MW 11597.2) and CsRbI Reference Spectra
INSULIN DIMER
$C_{514}H_{764}N_{130}O_{153}S_{12}$
MW = 11597.2
$Cs_{46}I_{45}$ 11824.4
$Cs_{45}RbI_{46}$ 11776.9
$Cs_{44}Rb_2I_{45}$ 11729.5
$Cs_{42}Rb_4I_{45}$ 11634.6
$Cs_{45}I_{44}$ 11564.5

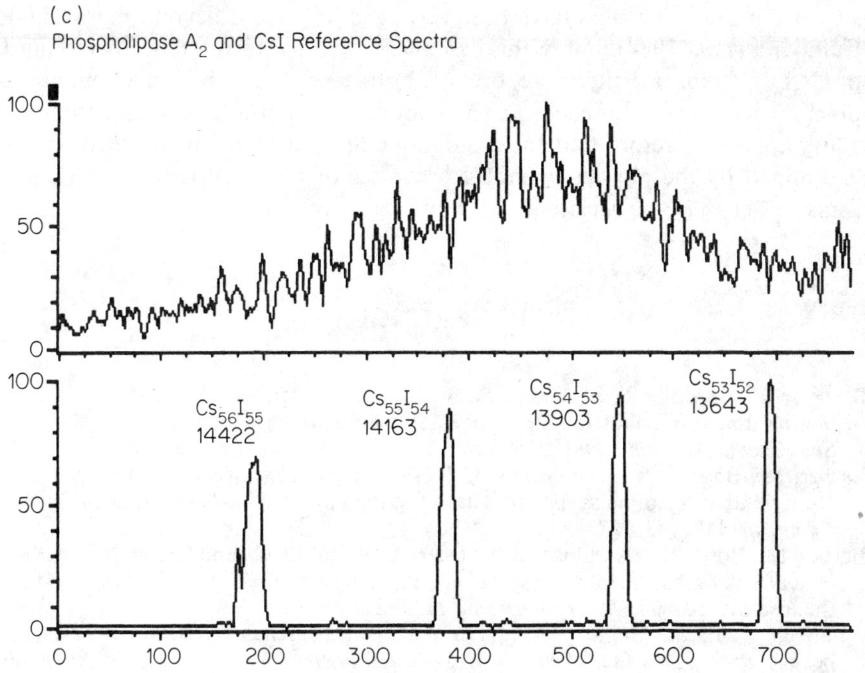

Figure 8: $(M + H)^+$ ion regions of three peptides of M_r (a) 8091, (b) 11,635 and (c) 13,980, showing progressive weakening of the $(M + H)^+$ peak with increasing mass.

of fragment ions derived from the molecule by decomposition of gas phase molecules and ions. Background ions above the molecular weight are more difficult to explain.

Adduct ions of the form $(MH + S)^+$ where S is a solvent matrix molecule are frequently observed in FAB spectra. Furthermore, the $(M + H)^+$ ions of simple peptides have been shown, using accelerating voltage scans, to have precursors which can contain several solvent molecules (Barber *et al.*, unpublished data). Keogh has also shown that mixed cluster ions of sucrose/glycerol/Na of the form $(M + Na + Gly_n)^+$ where $n = 1$–5 also decompose to form $(M + Na)^+$ species under collisional activation MS/MS conditions (Keogh, 1985). It is possible therefore to postulate mechanisms where the background ions are comprised of adducts to the sample molecule and fragments of these mixed cluster ions formed by decomposition in the selvedge region. Another source of the background ions may be beam damage products formed during the FAB process.

At the moment these rather speculative arguments can only be applied to oligopeptides, as no other classes of bio-organic compounds have been reported to yield molecular ions at masses greater than 10,000.

Caesium iodide clusters have been observed up to and beyond mass 25,000 (Campana *et al.*, 1984), but in this case there is no background of ions to mask the $Cs_nI_{n-1}^+$ ions. Furthermore, because both caesium and iodine are monoisotopic, the ion current for each cluster is not divided amongst several isotopes. It would appear therefore that for bio-organic molecules the upper mass limit is determined by the physico-chemical processes of the FAB ionization mechanisms rather than the sensitivity of the mass spectrometer.

REFERENCES

Barber, M., Bordoli, R. S., Sedgwick, R. D., and Tyler, A. N. (1981). Fast atom bombardment of solids (FAB): A new ion source for mass spectrometry, *J. Chem. Soc. Chem. Commun.*, **1981**, 325–327.

Barber, M., Bordoli, R. S., Elliot, G. J., Horoch, N. J., and Green, B. N. (1983). Fast atom bombardment mass spectrometry of human proinsulin, *Biochem. Biophys. Res. Commun.*, **110**, 753–757.

Barber, M., Bordoli, R. S., Elliot, G. J., Tyler, A. N., Bill, J. C., and Green, B. N. (1984). Fast atom bombardment mass spectrometry: A mass spectral investigation of some of the insulins, *Biomed. Mass Spectrom.*, **11**, 182–186.

Campbell, A. J., and Halliday, J. S. (1965). Mass measurement from low intensity ion beams, *Proc. 13th Ann. Conf. on Mass Spectrometry and Allied Topics, St. Louis*; American Society for Mass Spectrometry; pp. 200–203.

Campana, J. E., Colton, R. J., Wyatt, J. R., Bateman, R. H., and Green, B. N. (1984). Ultra-high mass spectrometry, *Appl. Spectrosc.* **38**, 430–432.

Cotter, R. J., Larsen, B. S., Heller, D. N., Campana, J. E., and Fenselau, C. (1984). Wide mass range scanning for the fast atom bombardment mass spectrometry of very large compounds, *Anal. Chem.*, **57**, 1479–1480.

Cotterell, J. S., Taylor, L. C. E., and Williams, D. H. (1984). 'A double focusing mass spectrometer for the study of high molecular weight bio-molecules', *Proc. 32nd Ann. Conf. on Mass Spectrometry and Allied Topics, San Antonio*; American Society for Mass Spectrometry; pp. 382–383.

Hintenberger, H., and Konig, L. A. (1959). Mass spectrometers and mass spectrographs corrected for image defects, in *Advances in Mass Spectrometry*, vol. 1 (Ed. J. D. Waldron), pp. 16–35, Pergamon Press, London.

Keogh, T. (1985). Cationisation of organic molecules using fast atom bombardment mass spectrometry, *Anal. Chem.*, **57**, 2027–2034.

Richter, W. J., Rashdorf, F., and Maerki, W. (1985). A two dimensional approach to the structural analysis of large peptides, in *Mass Spectrometry in the Health and Life Sciences*, Analytical Symposia Series, vol. 24 (Eds. A. L. Burlingame and N. Castagnoli), pp. 193–206, Elsevier, Amsterdam.

Taylor, L. C. E., Evans, S., Compson, K. R., and Ryan, P. A. (1983). The optimisation of mass spectrometer design to permit the analysis of high molecular weight molecules, *Proc. 31st Ann. Conf. on Mass Spectrometry and Allied Topics, Boston*; American Society for Mass Spectrometry, pp. 615–616.

Mass Spectrometry in Biomedical Research
Edited by S. J. Gaskell
© 1986 John Wiley & Sons Ltd.

Chapter 14

COLLISION-INDUCED DECOMPOSITION OF PEPTIDE IONS

Margaret M. Sheil and Peter J. Derrick

14.1. INTRODUCTION

Mass spectrometry has been applied to the problem of sequencing peptides for over 20 years (Arpino and McLafferty, 1976; Biemann, 1980). The technique offers the capability for precise determination of molecular weight, potential for obtaining full sequence information and the ability to analyze mixtures of peptides by gas chromatography–mass spectrometry (GC–MS) (Biemann, 1980; Carr *et al.*, 1981) and/or tandem mass spectrometry (MS/MS) (Hunt *et al.*, 1981a). Mass spectrometry may also be used to derive information concerning the intrinsic chemistry of peptide ions (Neumann and Derrick, 1984a). Early work on peptides using electron ionization (EI) and chemical ionization (CI) was limited by the low volatility and thermal instability of peptides, and necessitated the development of derivatization methods to enhance volatility (Arpino and McLafferty, 1976; Biemann, 1980; Carr *et al.*, 1981). One other limitation of electron ionization is that molecular ions may be absent or of very low intensity.

The problem of low volatility can be significantly alleviated by the use of ionization techniques such as field desorption (FD) (Beckey, 1969), fast atom bombardment (FAB) (Barber *et al.*, 1981a), secondary ion mass spectrometry (SIMS) (Benninghoven *et al.*, 1976), ^{252}Cf plasma desorption (McFarlane and Torgerson, 1976) and laser desorption (LD) (Posthumus *et al.*, 1978). The introduction of these techniques, which are capable of ionizing involatile, thermally labile compounds, has made it possible to study larger, underivatized peptides directly by mass spectrometry. Field desorption, for example, has been used to distinguish the different chains of haemoglobin and identify the mutant haemoglobin S implicated with the disease sickle-cell anaemia (Matsuo *et al.*,

1981b). Sundqvist *et al.* (1984) have used ^{252}Cf plasma desorption, in conjunction with time-of-flight mass spectrometry, to obtain molecular weight information for proteins up to 24,000 u. A recent study of over 25 peptides using secondary ion mass spectrometry has demonstrated rather convincingly that this technique provides sufficient information to enable the sequence of a small unknown peptide to be determined (Seki *et al.*, 1985). The large number of peptide studies employing fast atom bombardment are discussed in Chapters 10 and 16 by, respectively, Dell and Panico, and by Jardine.

14.2. COLLISION-INDUCED DECOMPOSITION

The potential for the study of peptides by these new ionization techniques is enhanced by combination with the technique of collision-induced decomposition (CID; alternatively termed CAD, collisionally-activated decomposition), whereby an ion of particular mass is separated and excited by collision with a neutral target gas in order to produce decomposition of that ion. CID can be readily studied in double-focusing mass spectrometers of reverse-geometry using the so-called mass analyzed ion kinetic energy spectroscopy (MIKES) technique (Beynon *et al.*, 1973), or with normal-geometry instruments employing linked-scanning techniques (Bruins *et al.*, 1978). Recent years have also seen rapid developments in the instrumentation for CID (McLafferty, 1983; Yost and Fetterolf, 1983; Cody and Freiser, 1982), with the introduction of multiple sector instruments, triple quadrupoles and Fourier transform ion cyclotron resonance (FT–ICR) spectrometers.

The utility of collision-induced decomposition for the study of peptides was first demonstrated by McLafferty and co-workers (Levsen *et al.*, 1974). They examined a number of di- and tripeptide derivatives using electron ionization and were able to use CID to distinguish between leucine and isoleucine residues. Despite this early success, there have been relatively few studies of CID of peptides reported in the open literature, and the majority of these have been published in the last 5 years, following developments in ionization techniques and, to a lesser extent, instrumentation. Because of the problem of volatility, there have been only a small number of examples of CID of peptides employing electron and chemical ionization (Hunt *et al.*, 1981b,c; Levsen *et al.*, 1974; Steinauer *et al.*, 1980; Bradley *et al.*, 1980, 1981; Steinauer and Schlunegger, 1982). Collision-induced decomposition appears to have the greatest potential to contribute to the study of peptide chemistry when combined with field desorption, fast atom bombardment and laser desorption.

14.2.1. FIELD DESORPTION

Field desorption (FD) of oligopeptides typically produces abundant molecular or 'pseudomolecular' ions and relatively few sequence ions, i.e., fragments arising from direct cleavage of the peptide backbone (Schulten, 1979; Weber and Levsen, 1980). Field desorption (FD) combined with CID, however, can in favourable cases yield complete sequence information. This potential was demonstrated by Weber and Levsen (1980) for a number of di- and tripeptides. Using FD/CID, Desiderio and Sabbatini (1982) have investigated the peptides leucine enkephalin and diphenylalanine and Matsuo *et al.* (1981a) have studied a number of peptides in the mass range 295–755 u. Both these groups employed the linked-scanning technique in which the magnetic and electric sectors are scanned simultaneously. Matsuo *et al.*'s study includes the CID spectra of the

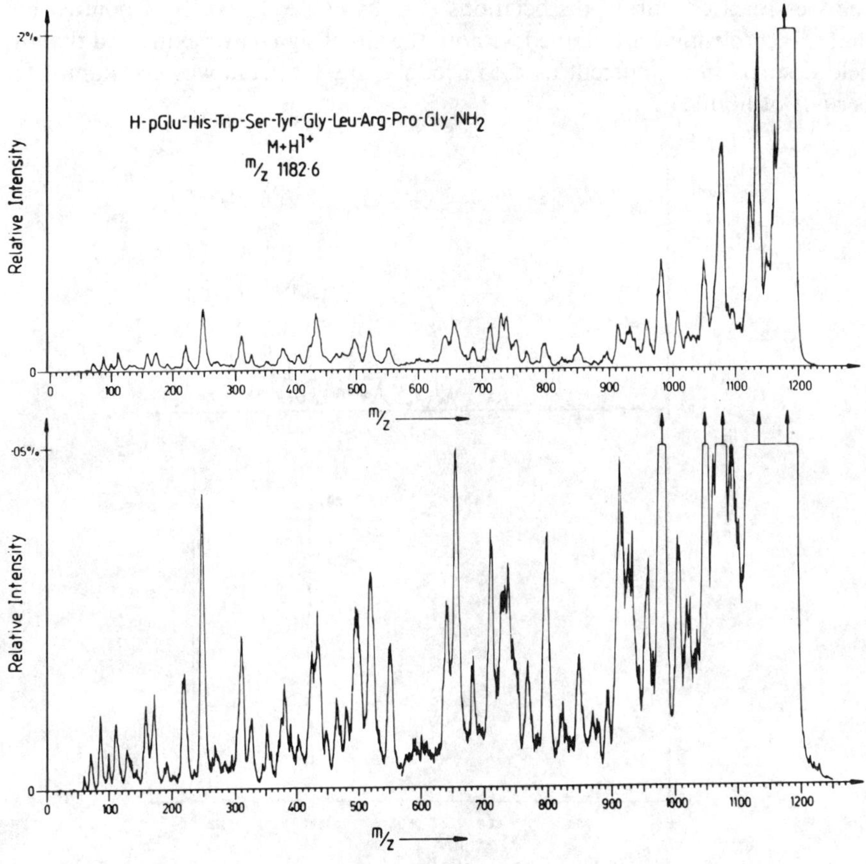

Figure 1: CID spectrum of the $(M + H)^+$ ion of luteinizing hormone releasing hormone (LHRH) (M_r 1181.6u).

$(M + Na)^+$ ions of eledoisin-related peptide (M_r 706.4 u) and two other peptides. These spectra show very intense $M^{+\cdot}$ peaks and the N-terminal sequence ions, $(B_n)^+$, (see below) are observed free of sodium. Such 'loss' of sodium has not been observed by Neumann and Derrick (unpublished data) for the $(M + Na)^+$ ion of eledoisin-related peptide. The CID spectrum of the $(M + H)^+$ ion of eledoisin-related peptide obtained by Matsuo *et al.* (1981a) is qualitatively very similar to that of Neumann and Derrick (1984b) for this ion.

Neumann and Derrick (1984b) have reported CID spectra of seven peptides selected as being representative of a larger number examined. The CID spectrum of the $(M + H)^+$ ion of luteinizing hormone releasing hormone (LHRH) (M_r 1181.6 u) is shown in Figure 1. This CID spectrum contains more sequence ions than would be typical for an oligopeptide of this size. The spectrum illustrates one of the shortcomings of the MIKES technique, namely low mass resolution because of peak broadening due to the translational energies released during dissociations (Cooks *et al.*, 1973). It is pointed out that this spectrum was obtained without the aid of signal averaging and that the field desorption ion current used to produce this spectrum was maintained for periods of hours.

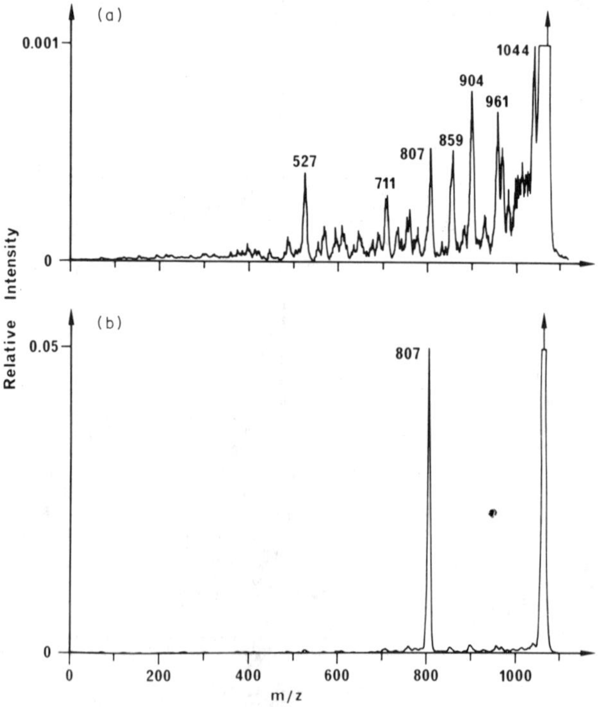

Figure 2: CID spectrum of the $(M + H)^+$ ion of bradykinin (M_r 1060.6u). The lower spectrum (b) was taken 20–30 minutes after spectrum (a).

Some insight into the intrinsic chemistry of peptide ions may also be gleaned from CID studies. The CID spectrum of the $(M + H)^+$ ion of bradykinin (M_r 1060.6 u) is characterized by one peak dominating the spectrum under certain conditions (Neumann and Derrick, 1984a). Figure 2 shows two spectra of the $(M + H)^+$ ion of bradykinin. Both spectra were recorded at the same emitter heating current (14 mA), but the lower spectrum (b) was taken 20–30 min after spectrum (a). The evolution of the form shown in (b) could be hastened by setting the emitter heating current at a slightly higher value. At 14.2 mA, for example, the change occurred in a few minutes. The difference between the two spectra lies essentially in the intensity of the *m/z* 807 peak which corresponds to rupture of the peptide bond between two prolyl groups accompanied by hydrogen transfer. The same phenomenon was also observed in the CID spectra of the *m/z* 572 fragment ion of bradykinin which also includes the two prolyl residues. In both cases the changes were irreversible, i.e., the earlier spectrum could not be obtained by again reducing the emitter heating current. The spectrum (b) was proposed to be associated with a novel structure of the peptide ion which is thermodynamically more stable than the conventional peptide structure. Collision properties such as attenuation and translational energy loss (Neumann and Derrick, 1984b; Neumann *et al.*, 1984) were unaffected by the changes, although the yield of fragment ions increased by over an order of magnitude.

14.2.2. FAST ATOM BOMBARDMENT

Collision-induced decomposition is particularly suited for the study of ions formed by fast-atom bombardent (FAB), since it is possible to focus on a particular ion and thus minimize interference stemming from the sample matrix and impurities. This was demonstrated by Hunt *et al.* (1981a) in a study of several peptides and peptide derivatives. Heerma *et al.* (1983) and Desiderio *et al.* (1983) have employed FAB and CID to study a heptapeptide and leucine enkephalin, respectively. Matsuo *et al.* (1983) have used both FAB and FD in conjunction with CID to study a mixture of three peptides. Heerma *et al.* (1983) point out differences between their observed masses and the masses of sequence ions calculated for straight cleavages. 'Hydrogen shifts' are proposed to account for the observed masses. Formation of C-sequence ions $(Y_n)^+$ (see below) is typically accompanied by hydrogen transfer; however, allowing for such transfers does not obviously account for the observed masses and in fact seems to increase the discrepancies. It seems likely that the problems are due to neglecting the translational energy lost as a result of collision, which 'shifts' peaks to lower mass in CID spectra obtained by scanning an electrostatic analyzer (Neumann and Derrick, 1984a,b). The limited mass resolution of such spectra would be another contributing factor.

A study which demonstrates the potential of FAB/CID for unequivocal sequence determination is that by Gross *et al.* (1982), in which they assign a sequence to the host specific toxin from *Helminthosporium carbonum* (HC-toxin). This is a cyclic peptide which had been incorrectly sequenced using electron ionization. In a subsequent study, workers from the same laboratory were able to sequence an analogue of that peptide, again using FAB/CID (Kim *et al.*, 1985). The peptides, human gastrin, bradykinin and two bradykinin analogues have been analyzed by Amster *et al.* (1983) using a triple sector instrument; again there seem to be errors in the mass assignments which stem from 'peak shifts' due to translational energy loss (Neumann and Derrick, 1984a). Following on from an earlier study, Katakuse and Desiderio (1983) have compared the information obtained from positive and negative FAB in CID spectra of leucine enkaphalin. Desiderio and Kai (1983) have used FAB/CID to locate ^{18}O atoms incorporated into leucine and methionine enkephalin samples for use as internal standards in quantification of these peptides in thalamus tissue. Desiderio and Katakuse (1983) have also employed FAB/CID to study the neuropeptide substance P. The FAB/CID spectrum of the $(M + H)^+$ ion of leucine enkephalin (M_r 555.3 u) obtained by Desiderio and co-workers (Desiderio *et al.*, 1983; Katakuse and Desiderio, 1983) differs from the FD/CID spectrum of this ion reported by the same group (Desiderio and Sabbatini, 1982). The FD/CID spectrum of this ion obtained in our laboratory (Neumann, G. M., and Derrick, P. J., unpublished data) is in better agreement with the FAB/CID spectrum of Desiderio and colleagues than with their earlier FD/CID spectrum.

Sequencing of cyclic peptides by mass spectrometry is inherently more difficult than sequencing of linear peptides because of the multiplicity of products of initial backbone cleavages. Gross and co-workers have reported a number of studies on the application of FAB and CID to the sequencing of cyclic peptides (Gross *et al.*, 1982; Kim *et al.*, 1985; Tomer *et al.*, 1983; 1984; Eckart *et al.*, 1985).

Using FAB/CID, Lippenstrau-Fischer and Gross (1985) have examined eight peptides containing from three to six amino acid residues and their corresponding polyamino-alcohols with a view to arriving at a general scheme for interpretation of the CID spectra of peptides. The scheme is exemplified using a pentapeptide.

14.2.3. LASER DESORPTION

At the time of writing there has to our knowledge been only one study reported of CID of peptides employing laser desorption ionization (Cody *et al.*, 1985). This study is particularly significant, because the authors used a Fourier transform ion cyclotron resonance (FT–ICR) instrument in this study of two peptides. This work demonstrates some advantages that FT–ICR instruments offer. Firstly, the measurements are relatively independent of the ion's translational energy. This alleviates the problem of broadening of peaks arising from the translational energy released during dissociation (Cooks *et al.*, 1973) and, further, the 'peak shifts' arising from translational energy lost during collision are not encountered (see below). These instruments afford simultaneous detection over a wide mass range, and pulsed ionization may be employed. These advantages, coupled with the high resolution of FT–ICR instruments with superconducting magnets, should see an increase in the use of these instruments for the study of CID of peptide ions.

14.3. CHARACTERISTIC FRAGMENT IONS IN CID OF PEPTIDES

Roepstorff and Fohlman (1984) have recently proposed a common nomenclature for the possible sequence ions of peptides which we have adopted here. The nomenclature is illustrated in Scheme 1. Fragment ions formed by cleavage of

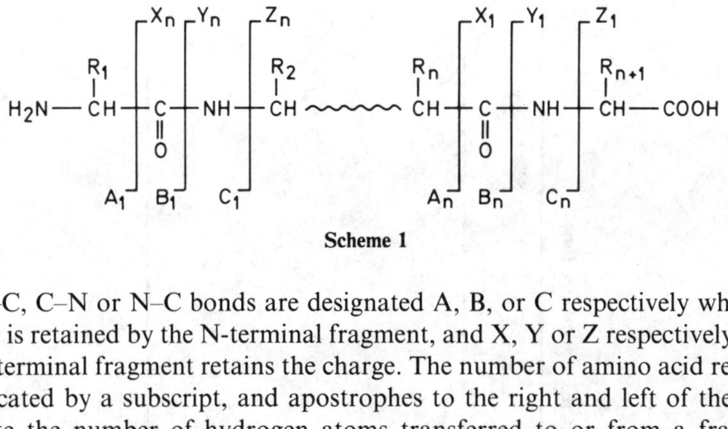

Scheme 1

the C–C, C–N or N–C bonds are designated A, B, or C respectively when the charge is retained by the N-terminal fragment, and X, Y or Z respectively when the C-terminal fragment retains the charge. The number of amino acid residues is indicated by a subscript, and apostrophes to the right and left of the letter indicate the number of hydrogen atoms transferred to or from a fragment respectively. An apostrophe is also used to indicate the retention of the additional proton by a fragment formed from a protonated peptide. Using this nomenclature, an amino acylium ion, for example, would be labelled $(B_n)^+$ and a C-terminal fragment formed by cleavage of the C–N bond with H-transfer and retention of the additional proton by the fragment ion would be designated $(Y''_n)^+$ (see below). The only obvious drawback with this nomenclature is that it is not immediately applicable to cyclic peptides, since they have no terminus.

TABLE 1. Summary of peptide sequence ions.

	Typical sequence ions		References	
Ionization	Mass spectra	CID spectra	Mass spectra	CID spectra
Electron ionization	$(A_n)^+$ $(B_n)^+$	$(A_n)^+$ $(B_n)^+$	Arpino and McLafferty (1976) Wipf *et al.*, (1973)	Levsen *et al.*, (1974) Steinauer and Schlunegger (1982)
Chemical ionization	$(B_n)^+$ $(Y''_n)^+$	$(A_n)^+$ $(B_n)^+$ $(Y''_n)^+$	Kiryushkin *et al.*, (1971) Mudgett *et al.*, (1977)	Hunt *et al.*, (1981b,c)
Field desorption	$(B_n)^+$ $(Y''_n)^+$	$(A_n)^+$ $(B_n)^+$ $(Y''_n)^+$	Neumann and Derrick (unpublished data)	Neumann and Derrick (1984b) Neumann *et al.*, (1984)
Fast atom bombardment	$(A_n)^+$ $(B_n)^+$ $(C''_n)^+$ $(Y''_n)^+$ $(Z''_n)^+$	$(A_n)^+$ $(B_n)^+$ $(Y''_n)^+$	Barber *et al.*, (1981b, 1982) Williams *et al.*, (1982) Buko *et al.*, (1983) Roepstorff *et al.*, (1985)	Lippenstrau-Fischer and Gross (1985)

Table 1 summarizes the typical sequence ions which have been observed in the positive ion mass spectra and the CID spectra of peptide ions following EI, CI, FD and FAB. The information has been gleaned from the work of a number of authors. There is some confusion with regard to the most common sequence ions occurring in fast atom bombardment. It may even be that it is not possible to generalize about the sequence ions found in FAB given the extensive nature of the fragmentation. Williams *et al.* (1982) have reported that ions of the types $(C''_n)^+$, $(Y''_n)^+$ and $(Z''_n)^+$ occur most frequently in FAB mass spectra. In a study of sixteen peptides, Buko *et al.* (1983) consider these three ion types together with $(A_n)^+$ and $(B_n)^+$ ions. Barber *et al.* have reported ions of the type $(A_n)^+$, $(Y_n)^+$, $(Y''_n)^+$ and $(Z''_n)^+$ in the FAB spectrum of bradykinin (Barber *et al.*, 1981b) and $(A_n)^+$, $(C''_n)^+$ and $(Y''_n)^+$ ions in the FAB mass spectra of the angiotensins (Barber *et al.*, 1982). Roepstorff *et al.* (1985) have found the dominating sequence ions to be $(B_n)^+$ and $(Y''_n)^+$ and that ions of the type $(A_n)^+$ are sometimes abundant in FAB spectra. The confusion is compounded because Roepstorff *et al.* (1985) and Buko *et al.* (1983) use different labels for the same ion, e.g. $(B_n)^+$ ions in the nomenclature used here are labelled $(A_n)^+$ by Buko *et al.*

CID of ions formed by EI differs from CID following the other ionization techniques in that with EI only sequence ions are normally sufficiently intense to study. With the other techniques 'pseudomolecular' ions can be employed as precursors for CID. In the CID spectra of amino acylium ions, $(B_n)^+$, formed by EI, the most abundant ions are lower mass ions of the same type ($(B_{n-1})^+$ etc.) (Levsen *et al.*, 1974). Similarly in CI/CID of amino acylium ions, lower mass acylium ions and acylium ions less small neutral molecules account for over 50% of the total fragment ion current (Hunt *et al.*, 1981c). As is clear from Table 1 the most commonly observed sequence ions in CID spectra of ions $(M + H)^+$ are the N-terminal fragments $(A_n)^+$ and $(B_n)^+$ and the C-terminal fragments $(Y''_n)^+$. The general mechanism shown in Scheme 2 describes the formation of acylium ions (Scheme 2a) and the C-sequence ions (Scheme 2b).

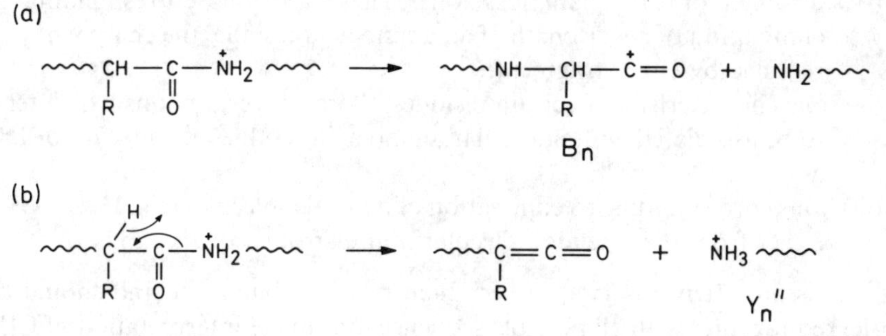

Scheme 2

Ions of the type $(A_n)^+$ may arise either from loss of CO from the corresponding acylium ion or from direct cleavage of the peptide backbone (Scheme 3). These ions $(A_n)^+$ are often more abundant than the corresponding acylium ions $(B_n)^+$ (Neumann and Derrick, 1984b).

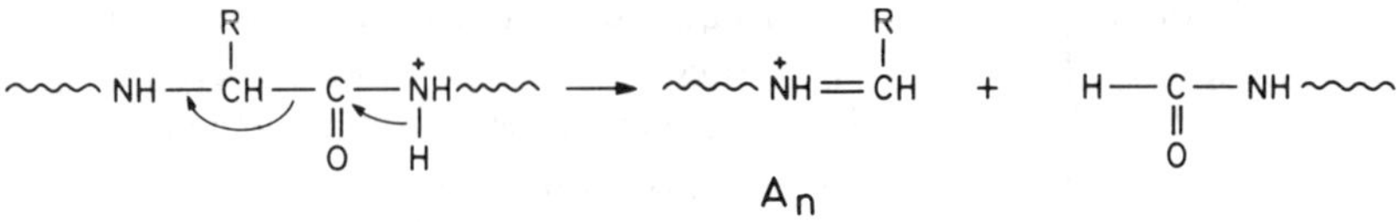

Scheme 3

It has been proposed that the site of protonation within a protonated peptide strongly influences which bonds within the peptide are cleaved (Neumann and Derrick, 1984a; Barber *et al.*, 1982). It should also be noted that for any given protonation site, the reactions shown in Scheme 2 are in competition. Beynon and co-workers have suggested that CID spectra of ions formed by negative chemical ionisation are influenced by both the site of deprotonation of the molecule and by the stability of the fragment ions formed (Bradley *et al.*, 1981).

Similarly, with cyclic peptides, Gross and co-workers have proposed that fragmentation can be understood in terms of a mechanism involving protonation of a ring nitrogen and cleavage of the adjacent peptide bond with subsequent fragmentation of the linear acylium ions so formed (Tomer *et al.*, 1984). They have also found that if one of the ring nitrogens is significantly more basic than the others, then fragmentation resulting from protonation of that site will tend to dominate the spectrum.

Whilst sequence ions clearly provide the most useful sequence information, there are several other types of ions which frequently occur in CID spectra of peptides (Neumann and Derrick, 1984b). These include:

(i) side-chain or side-chain loss ions. These are ions corresponding to elimination of a side chain of an amino acid residue; the charge may be retained by either fragment;

(ii) ions characteristic of certain residues; there are certain ions which tend to be associated with particular amino acid residues e.g. m/z 70 for Pro and m/z 120 for Phe;

(iii) ions corresponding to elimination of neutral molecules (e.g. H_2O, NH_3) either from the 'pseudomolecular' ion or from sequence ions.

These ions are often observed in very high relative abundance and should be considered together with all possible sequence ions in the interpretation of CID spectra of peptides.

14.4. DYNAMICS OF COLLISION-INDUCED DECOMPOSITION

Secure interpretation of CID spectra of peptide ions calls for some knowledge of the collision dynamics. The dynamics of collisions between massive ions with keV translational energies and thermal gas atoms is, however, a largely unexplored area. The situation is, if anything, worse at lower ion energies. One essential requirement for interpretation of CID spectra is a knowledge of the translational energies of fragment ions formed. In a CID experiment in which an electric sector is scanned (i.e. MIKES), the danger is that, if faulty assumptions are made concerning translational energies of fragment ions, mass assignments may be incorrect. In an experiment in which a linked scan of an electric and magnetic sector is employed, the danger is that some ions may not be transmitted when they ought to be, i.e. there may be 'peaks missing'. Other ions may be transmitted when they ought not be, i.e. spurious peaks may be present. The problems of 'peak-shifts' mentioned several times in earlier parts of this Chapter are a direct consequence of the translational energies of fragment ions being *lower* than was supposed. Consider an incident ion $(M_1)^+$ decomposing to a fragment ion $(M_2)^+$ following collision, and let the translational energy of $(M_1)^+$ prior to collision be E. The translational energy of $(M_2)^+$ will be equal to $(M_1/M_2)(E\text{-}\Delta E)$, where ΔE is the translational energy lost by $(M_1)^+$ in the collision. With light ions ΔE is generally small and tends to be neglected. With massive ions, ΔE can become large, and with the MIKES technique can easily 'shift' peaks by 5 mass units or more. The magnitude of the translational energy loss ΔE seems to bear a direct relationship to the incident ion mass, for a given collision gas and at a constant incident ion energy (Neumann and Derrick, 1984b; Neumann *et al.*, 1984). The problem of 'peak shifts' is particularly acute when using helium as the collision gas (Neumann *et al.*, 1984).

A second requirement in terms of collision dynamics is knowledge of the scattering. The effect of the energy released during fragmentation is equally important. While not much is known about either of these factors for massive ions, it can be assumed with some confidence that for a given parent ion small fragment ions are likely to have wider angular distributions than large fragment ions, and that the possible differences are likely to become greater as the mass of the parent ion considered becomes larger. Such differences in angular distributions may be one significant reason why CID spectra of large peptide ions (>1500 u) often contain only peaks of low intensities at low mass (Rumpf *et al.*, 1986).

The internal energies taken up by massive ions as a result of keV collisions with thermal gases can be large (10–20 eV), but may not always be sufficiently large to induce decomposition of those ions within the required time span (Neumann *et al.*, 1984). It has been concluded that the proportion of the available collision energy (centre-of-mass framework) taken up by an incident ion rises in proportion to the mass of the incident ion, for a given collision gas and a constant incident ion energy (Neumann *et al.*, 1984). At a certain mass (said to be 1500 u for helium and an incident ion energy of 8 keV), all the collision energy is taken up as internal energy. Above this particular mass, the internal energy uptake must fall as the collision energy will fall; however, it also appears that the *proportion* of the *lower* collision energy taken up as internal energy also falls (Neumann *et al.*, 1984). It has been suggested (Neumann *et al.*, 1984) that below this particular mass, the collision gas is forward-scattered, and that, above that mass the collision gas is back-scattered. The available evidence suggests that the mechanism of energy transfer to high energy peptide ions is vibrational excitation (Neumann and Derrick, 1984b; Neumann *et al.*, 1984; Gilbert *et al.*, 1985). The experimental results on internal energy uptake by massive ions largely concern helium as the collision gas, and strongly suggest that CID of peptide ions (at 8 keV) above about 1500 u will not be particularly successful, at least if helium is the collision gas. The ions will in most cases tend not to take up sufficient internal energy to induce extensive fragmentation.

The final aspect of the collision dynamics to be addressed is multiple collisions, and their potential usefulness for raising the amount of internal

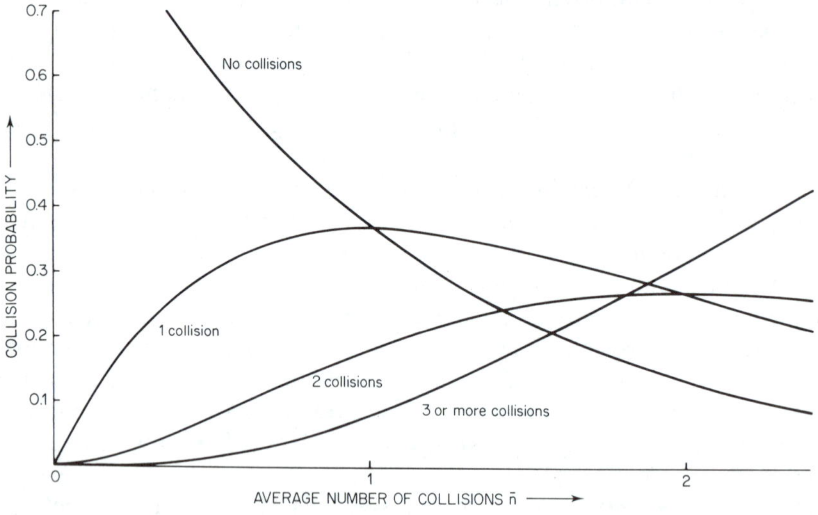

Figure 3: Collision probability for 0, 1, 2 and 3 collisions versus the average number of collisions $\bar{n}$. Collisions are assumed to conform to Poisson statistics. The average number of collisions $\bar{n}$ is directly related to the collision gas pressure for any given ion.

energy taken up by the incident ions. There are two opposing views, the difference between which hinges on the significances of the so-called 'optimum collision pressure' at which the yield of fragment ions (i.e. CID efficiency) is greatest. The one view (Holmes, 1985) would describe optimum collision pressures as multiple collision regimes; the other view (Neumann *et al.*, 1984) sees optimum collision pressures as effectively single collision regimes. Probabilities of different numbers of collisions are plotted against the average number of collisions in Figure 3 assuming that the collisions conform to Poisson statistics. The average number of collisions is related directly to the collision gas pressure and, at a particular pressure, for a given ion depends on the ion mass (Kim, 1983). It is generally agreed that the optimum collision pressure corresponds to roughly 60% attenuation. Considering 63% attenuation, i.e. the point on the graph (Figure 3) at which the average number of collisions is one, 37% of incident ions do not collide, 37% undergo 1 collision, 18% 2 collisions and 6% 3 collisions. These figures assume that an incident ion does not decompose before it has time to undergo a second or third collision. The multiple collision viewpoint stems from taking these figures at face value. Where it can be criticised is in ignoring loss processes. Suppose that the probability of ion loss is the same at the second or third collision as at the first, and suppose that for any collision, the probability of that collision leading to a detectable fragment ion is $\frac{1}{5}$ that of an ion being lost (scattering, charge exchange, etc.). Then of the 37% ions undergoing one collision, 7% give rise to detected fragment ions. Of the 18% undergoing two collisions, only 0.7% give detected fragment ions (14% are 'lost' at the first collision and $\frac{4}{5}$ of the remaining 4% are lost at the second collision). Detected fragment ions from ions undergoing three collisions are negligible. Of the fragment ions detected at the attenuation considered of 63%, 88% stem from single-collision events and 12% from double-collision events. If attenuation by 20% is considered, the comparable proportion of fragment ions stemming from single collisions is 92%, and for attenuation by 80% the proportion 86%. In other words, if 'loss processes' outweigh detection of fragment ions by a factor of 4, attenuations up to about 80% are effectively in the single-collision regime as far as detection of fragment ions is considered. That CID spectra can change as pressure and hence attenuation are raised may be attributed in part to the fragment ions formed in one collision, themselves undergoing collision and decomposing. Given that ion losses do tend to strongly outweigh formation of fragment ions (Todd and McLafferty, 1981), tuning the pressure in a CID experiment can reasonably be assumed to correspond to maximising the probability of one collision. This being the case, multiple collisions are not likely to be of significant advantage for raising the internal energy uptake of massive ions because the multiple collisions will merely enhance ion losses. There might be rather more hope of exploiting multiple collisions at very high ion energies (>25 keV), where loss processes might be relatively less important.

14.5. CONCLUSION

The literature on CID of peptide ions is still relatively small, and mainly concerns ions with mass <1000 u (for recent reports, see Todd (1986) in addition to the work cited above). In the few cases where comparison of CID spectra from different laboratories for a given peptide ion is possible, there is more often than not disagreement. The spectra may simply be quite different, or, if the spectra look the same, mass assignments may differ slightly. These disagreements are possibly a direct or indirect consequence of the peak shift problem, and will undoubtedly be clarified.

Below about 1000 u, CID spectra of peptide ions ionized by FAB, FD or LD provide a great deal of structural information, and if CID spectra of both 'pseudomolecular' and sequence ions are exploited, constitute bases for full and reliable sequence information. Although precise evidence is scant, the general impression is that the CID spectra of peptide ions are not strongly dependent on their mode of formation. Thus, for example, the CID spectrum of an $(M + H)^+$ ion of a given peptide formed by FAB can be reasonably expected to be similar to that of the same ion formed by FD. This generalization requires qualification, however, if a given peptide ion has two or more distinctly different structures which may be formed to different extents using different ionization techniques.

Above mass 1000 u, the CID spectra of peptide ions may contain too few peaks for determination of the full sequences. It seems from the limited experimental evidence that the problem becomes worse as ion mass increases beyond 1500 u. The cause of the problem is probably insufficient internal excitation. The solution to the problem is not clear; however, the use of higher incident ion energies merits further investigation.

ACKNOWLEDGMENTS

We are pleased to acknowledge the support of the Australian Research Grants Scheme. We are grateful to Professors M. L. Gross and F. W. McLafferty for communication of results prior to publications.

REFERENCES

Amster, I. J., Baldwin, M. A., Cheng, M. T., Proctor, C. J., and McLafferty, F. W. (1983). Tandem mass spectrometry of higher molecular weight compounds, *J. Am. Chem. Soc.*, **105**, 1654–1655.

Arpino, P. J., and McLafferty, F. W. (1976), in *Determination of Organic Structures by Physical Methods*, vol. 6 (Eds. F. C. Nachod, J. J. Zuckermann, and E. W. Randall), Academic Press, New York.

Barber, M., Bordoli, R. S., Sedgwick, R. D., and Tyler, A. N. (1981a). Fast atom bombardment of solids (F.A.B.): a new ion source for mass spectrometry, *J. Chem. Soc. Chem. Commun.*, **1981**, 325–327.

Barber, M., Bordoli, R. S., Sedgwick, R. D., and Tyler, A. N. (1981b). Fast atom bombardment mass spectrometry of bradykinin and related oligopeptides, *Biomed. Mass Spectrom.*, **8**, 337–342.

Barber, M., Bordoli, R. S., Sedgwick, R. D., and Tyler, A. N. (1982). Fast atom bombardment mass spectrometry of the angiotensin peptides, *Biomed. Mass Spectrom.*, **9**, 208–214.

Beckey, H. D. (1969). Field desorption mass spectrometry: a technique for the study of thermally unstable substances of low volatility, *Int. J. Mass Spectrom. Ion Phys.*, **2**, 500–503.

Benninghoven, A., Jaspers, D., and Sichtermann, W. (1976). Secondary-ion emission of amino acids, *Appl. Phys.*, **11**, 35–39.

Beynon, J. H., Cooks, R. G., Amy, J. W., Baitinger, W. E., and Ridley, T. Y. (1973). Design and performance of a mass analyzed ion kinetic energy (MIKE) spectrometer, *Anal. Chem.*, **45**, 1023A–1031A.

Biemann, K. (1980). Amino acid sequence in oligopeptides and proteins, in *Biochemical Applications of Mass Spectrometry*, First Supplementary Volume (Eds. G. R. Waller and O. C. Dermer), pp. 469–525, Wiley, New York.

Bradley, C. V., Howe, I., and Beynon, J. H. (1980). Analysis of underivatised peptide mixtures by collision-induced dissociation of negative ions, *J. Chem. Soc. Chem. Commun.*, **1980**, 562–564.

Bradley, C. V., Howe, I., and Beynon, J. H. (1981). Sequence analysis of underivatized peptides by negative ion chemical ionization and collision induced dissociation, *Biomed. Mass Spectrom.*, **8**, 85–89.

Bruins, A. P., Jennings, K. R., and Evans, S. (1978). Observations of metastable transitions in a double-focusing mass spectrometer using a linked scan of electric sector and magnetic sector fields, *Int. J. Mass Spectrom. Ion. Phys.*, **26**, 395–404.

Buko, A. M., Phillips, L. R., and Fraser, B. A. (1983). Peptide studies using a fast atom bombardment high field mass spectrometer and data system: 2—Characteristics of positive ionization spectra of peptides, *m/z* 858 to *m/z* 5729, *Biomed. Mass Spectrom.*, **10**, 408–419.

Carr, S. A., Herlihy, W. C., and Biemann, K. (1981). Advances in gas chromatographic mass spectrometric protein sequencing, *Biomed. Mass Spectrom.*, **8**, 51–61.

Cody, R. B., and Freiser, B. S. (1982). Collision-induced dissociation in a Fourier-transform mass spectrometer, *Int. J. Mass Spectrom. Ion Phys.*, **41**, 199–204.

Cody, R. B., Amster, I. J., and McLafferty, F. W. (1985). Peptide sequencing by tandem Fourier transform mass spectrometry, *Proc. Nat. Acad. Sci. (USA)*, **82**, 6367–6370.

Cooks, R. G., Beynon, J. H., Caprioli, R. M., and Lester, G. R. (1973). *Metastable Ions*, Elsevier, Amsterdam.

Desiderio, D. M., and Kai, M. (1983). Preparation of stable isotope-incorporated peptide internal standards for field desorption mass spectrometry quantification of peptides in biologic tissue, *Biomed. Mass Spectrom.*, **10**, 471–479.

Desiderio, D. M., and Katakuse, I. (1983). Fast atom bombardment-collision activated dissociation-linked field scanning mass spectrometry of the neuropeptide Substance P, *Anal. Biochem.*, **129**, 425–429.

Desiderio, D. M., and Sabbatini, J. Z. (1982). Field desorption collision activation linked scanning mass spectrometry of underivatized oligopeptides, *Biomed. Mass Spectrom.*, **8**, 565–568.

Desiderio, D. M., Katakuse, I., and Kai, M. (1983). Field desorption, fast atom bombardment–collisional activation mass spectrometry with accumulated linked scan technique for peptide sequence elucidation, *Biomed. Mass Spectrom.*, **10**, 423–426.

Eckart, K., Schwarz, H., Tomer, K. B., and Gross, M. L. (1985). Tandem mass spectrometry methodology for the sequence determination of cyclic peptides, *J. Am. Chem. Soc.*, **107**, 6765–6769.

Gilbert, R. G., Sheil, M. M., and Derrick, P. J. (1985). Energy transfer in the collision-induced decomposition of multiatomic ions, *Org. Mass Spectrom.*, **20**, 430–431.

Gross, M. L., McCrery, D., Crow, F., Tomer, K. B., Pope, M. R., Ciufetti, L. M., Knoche, H. W., Daly, J. M., and Dunkle, L. D. (1982). The structure of the toxin from *Helminthosporium carbonum*, *Tetrahedron Lett.*, **23**, 5381–5384.

Heerma, W., Kamerling, J. P., Slotbon, A. J., van Scharrenburg, G. J. M., Green, B. N., and Lewis, I. A. S. (1983). Fast atom bombardment and collisional activation mass spectrometry of a heptapeptide, *Biomed. Mass Spectrom.*, **10**, 13–16.

Holmes, J. L. (1985). Assigning structures to ions in the gas phase, *Org. Mass Spectrom.*, **20**, 169–183.

Hunt, D. F., Bone, W. M., Shabanowitz, J., Rhodes, J., and Ballard, J. M. (1981a). Sequence analysis of oligopeptides by secondary ion/collision activated dissociation mass spectrometry, *Anal. Chem.*, **53**, 1704–1706.

Hunt, D. F., Buko, A. M., Ballard, J. M., Shabanowitz, J., and Giordani, A. B. (1981b). Sequence analysis of polypeptides by collision activated dissociation on a triple quadrupole mass spectrometer, in *Soft Ionisation Biological Mass Spectrometry* (Ed. H. R. Morris), pp. 85–109, Heyden, London.

Hunt, D. F., Buko, A. M., Ballard, J. M., Shabanowitz, J., and Giordani, A. (1981c). Sequence analysis of polypeptides by collision activated dissociation on a triple quadrupole mass spectrometer, *Biomed. Mass Spectrom.*, **8**, 397–408.

Katakuse, I., and Desiderio, D. M. (1983). Positive and negative fast atom bombardment–collision activated dissociation-linked field scanned mass spectra of leucine enkephalin, *Int. J. Mass Spectrom. Ion Processes.*, **54**, 1–15.

Kim, M. S. (1983). Fragmentation probability in collisional-activation mass spectrometry (CA/MS). 1. Effects of sequential collisional excitation, *Int. J. Mass Spectrom. Ion Phys.*, **50**, 189–203.

Kim, S.-D., Knoche, H. W., Dunkle, L. D., McCrery, D. A., and Tomer, K. B. (1985). Structure of an amino acid analog of the host-specific toxin from *Helminthosporium carbonum*, *Tetrahedron Lett.*, **26**, 969–972.

Kiryushkin, A. A., Fales, H. M., Axenrod, T., Gilbert, E. J., and Milne, G. W. A. (1971), Chemical ionization mass spectrometry of complex molecules. VI, *Org. Mass Spectrom.*, **5**, 19–31.

Levsen, K., Wipf, H. -K., and McLafferty, F. W. (1974). Mass spectrometric studies of peptides. VI, *Org. Mass Spectrom.*, **8**, 117–128.

Lippenstrau-Fischer, D. L., and Gross, M. L. (1985), Fast atom bombardment and tandem mass spectrometry for sequencing peptides and polyamino alcohols, *Anal. Chem.*, **57**, 1174–1180.

McFarlane, R. D., and Torgerson, D. F. (1976). Californium-252 plasma desorption mass spectroscopy, *Science*, **191**, 920–925.

McLafferty, F. W. (Ed.) (1983). *Tandem Mass Spectrometry*, Wiley, New York.

Matsuo, T., Matsuda, H., Katakuse, I., Shimonishi, Y., Marayama, Y., Higuchi, T., and Kubota, E. (1981a). Field desorption–collisional activation mass spectrometry with accumulated linked-scan technique for peptide structure elucidation, *Anal. Chem.*, **53**, 416–421.

Matsuo, T., Matsuda, H., Katakuse, I., Wada, Y., Fujita, T., and Hayashi, A. (1981b). Field desorption mass spectra of tryptic peptides of human haemoglobin chains, *Biomed. Mass Spectrom.*, **8**, 25–30.

Matsuo, T., Matsuda, H., Aimota, S., Shimonishi, Y., Higuchi, T., and Maruyama, Y. (1983). Field desorption, fast atom bombardment collisional activation mass spectrometry with accumulated linked scan technique for peptide sequence elucidation, *Int. J. Mass Spectrom. Ion Phys.*, **46**, 423–426.

Mudgett, M., Bowen, D. V., Field, F. H., and Kindt, T. J. (1977). Peptide sequencing: the utility of chemical ionization mass spectrometry, *Biomed. Mass Spectrom.*, **4**, 159–171.

Neumann, G. M., and Derrick, P. J. (1984a). Intrinsic chemistry of peptide ions: field desorption and collision-induced decomposition of bradykinin. *Aust. J. Chem.*, **37**, 2261–2277.

Neumann, G. M., and Derrick, P. J. (1984b). Energy transfer in the collision-induced decomposition of peptide ions formed by field desorption, *Org. Mass Spectrom.*, **19**, 165–170.

Neumann, G., Sheil, M. M., and Derrick, P. J. (1984). Collision-induced decomposition of multiatomic ions, *Z. Naturforsch. Teil A*, **39**, 584–592.

Posthumus, M. A., Kistemaker, P. G., Meuzelaar, H. L. C., and Ten Noever de Brauw, M. C. (1978). Laser desorption–mass spectrometry of polar non-volatile bio-organic molecules, *Anal. Chem.*, **50**, 985–991.

Roepstorff, P., and Fohlman, J. (1984). Proposal for a common nomenclature for sequence ions in mass spectra of peptides, *Biomed. Mass Spectrom.*, **11**, 601.

Roepstorff, P., Hojrup, P., and Moller, J. (1985). Evaluation of fast atom bombardment mass spectrometry for sequence determination of peptides, *Biomed. Mass Spectrom.*, **12**, 181–189.

Rumpf, B. A., Allison, C. E., and Derrick, P. J. (1986). Mass discrimination in collisionally activated decomposition (CAD) and mass analyzed ion kinetic energy (MIKE) spectra, *Org. Mass Spectrom.*, **21**, 295–299.

Schulten, H.-R. (1979). Biochemical, medical, and environmental applications of field-ionisation and field-desorption mass spectrometry. *Int. J. Mass Spectrom. Ion Phys.*, **32**, 97–283.

Seki, S., Kambara, H., and Naoki, H. (1985). Sequence analysis for an unknown peptide by molecular secondary ion mass spectrometry, *Org. Mass Spectrom.*, **20**, 18–24.

Steinauer, R., and Schlunegger, U. P. (1982). Detection of fragment genesis in the mass spectrometer: amino acid sequence analysis by DADI/MIKE, linked scan and collisional activation mass spectrometry, *Biomed. Mass Spectrom.*, **9**, 153–157.

Steinauer, R., Walter, H., and Schlunegger, U. P. (1980). Detection of fragment genesis in the mass spectrometer. Amino acid sequence analysis by DADI/MIKE, linked scan and collisional activation mass spectrometry. *Helv. Chim. Acta*, **63**, 610–612.

Sundqvist, B., Roepstorff, P., Fohlman, J., Hedin, A., Håkonsson, P., Kamensky, I., Lindberg, M., Salehpour, M., and Save, G. (1984). Molecular weight determinations by californium plasma desorption mass spectrometry, *Science*, **226**, 696–698.

Todd, J. F. J. (Ed.) (1986). *Advances in Mass Spectrometry 1985*, Wiley, Chichester.

Todd, P. J., and McLafferty, F. W. (1981). Collisionally activated decomposition of gaseous ions: the effect of multiple collisions. *Int. J. Mass Spectrom. Ion Phys.*, **38**, 371–378.

Tomer, K. B., Crow, F. W., and Gross, M. L. (1983). Analysis of mixtures of ionic compounds using fast atom bombardment and a triple analyzer mass spectrometer, *Int. J. Mass Spectrom. Ion Phys.*, **46**, 375–378.

Tomer, K. B., Crow, F. W., Gross, M. L., and Kopple, K. D. (1984). Fast atom bombardment combined with tandem mass spectrometry for the determination of cyclic peptides, *Anal. Chem.*, **56**, 880–886.

Weber, R., and Levsen, K. (1980). Collision induced dissociation of field desorbed di- and tripeptides, *Biomed. Mass Spectrom.*, **7**, 314–316.

Williams, D. H., Bradley, C. V., Santikarn, S., and Bojesen, G. (1982). Fast atom bombardment mass spectrometry. A new technique for the determination of molecular weights and amino acid sequences of peptides, *Biochem. J.*, **201**, 105–117.

Wipf, H.-K., Irving, P., McCamish, M., Venkataraghavan, R., and McLafferty, F. W. (1973). Mass spectrometric studies of peptides. V. Determination of amino acid sequences in peptide mixtures by mass spectrometry, *J. Am. Chem. Soc.*, **95**, 3369–3375.

Yost, R. A., and Fetterolf, D. D. (1983). Tandem mass spectrometry (MS/MS) instrumentation, *Mass Spectrom. Rev.*, **2**, 1–46.

Mass Spectrometry in Biomedical Research
Edited by S. J. Gaskell
© 1986 John Wiley & Sons Ltd.

Chapter 15

PLASMA DESORPTION MASS SPECTROMETRY OF HIGH MOLECULAR WEIGHT BIOMOLECULES

Peter Roepstorff and Bo Sundqvist

15.1. INTRODUCTION

Following the development of particle induced desorption–ionization methods in mass spectrometry, analysis of large biomolecules has attracted considerable interest. Two of the techniques have proved their capacity for analyzing large molecules, ^{252}Cf-plasma desorption time-of-flight mass spectrometry (PDMS), introduced by Macfarlane and co-workers (Torgerson *et al.*, 1974), and fast atom bombardment (FAB) in combination with magnetic sector instruments, first described by Barber and co-workers (1981). The latter method has been very quickly adopted in many laboratories because it can be installed in existing magnetic sector mass spectrometers. It very soon became clear that sector instruments, in terms of mass range, could not meet the possibilities offered by FAB, and the development of new techniques to produce high field magnets was undertaken.

The time-of-flight (TOF) mass analyzer used in PDMS is inexpensive compared to large sector instruments and does not have any limitations of mass range. One of the major limitations for the development of PDMS has been that only a few laboratories have used the technique with home built instruments, and commercial instruments have become available only recently (Bio-Ion Nordic AB, Uppsala, Sweden). Analysis of medium molecular weight molecules, (M_r 1000–5000) is today readily achieved with either of the two techniques.

Therefore, this chapter will only be concerned with molecules above M_r 5000. We feel that mass spectrometric analysis of such large biomolecules will be a major challenge to mass spectrometrists in the coming years and that such studies may contribute significantly to the development of biochemistry, molecular biology and biotechnology.

15.2. THE PRINCIPLES OF THE PLASMA DESORPTION MASS SPECTROMETER

One of the decay-routes of the radioactive nucleus californium-252 is spontaneous fission. The two almost colinear fission fragments are fast heavy ions, typically of mass 100 u and 100 MeV energy. In a PDMS-system (Figure 1) one of the fission fragments is used to desorb and ionize molecules when it passes through a sample surface layer. The associated fission fragment from the same fission event is registered in the start detector and used as a time-marker for the desorption event. Depending on the polarity of the voltage applied to the samples, either positive or negative secondary ions can be accelerated towards a grid at earth potential. The secondary ions then drift in a field free region before being detected in an array of channel electron multiplier plates. When a fragment or molecular ion hits the front side of the first channel electron multiplier plate, secondary electrons are produced if the velocity of the ion is high enough. Depending on the charge state of the secondary ion (q) it will be accelerated to an energy $q \times V$ (eV) where V is the applied sample voltage in volts. The signals from start and stop detectors are processed by fast electronics before they enter a time-to-digital converter (TDC). As a fission fragment produces many (10–100) secondary ions, the TDC must be able to register many flight times per start signal. The flight times are accumulated in a time-of-flight (TOF) spectrum with the aid of a computer. The TOF spectra can then be converted to mass spectra assuming one charge state of the secondary ions. As flight time is proportional to the square root of the mass over charge of an ion, a doubly charged molecular ion will appear at a flight-time $T/\sqrt{2}$ where T is the flight time of the singly charged molecular ion, or in a mass spectrum at half the mass of the molecular ion.

The mass analyzer in itself has no limitations for high mass analysis. The limitations are rather in the production and detection of massive ions. The primary fast ions (fission fragments) used in PDMS are more efficient to desorb and ionize large molecules than are low energy (keV) particles (Kamensky *et al.*, 1982). In PDMS, biomolecular cluster ions of mass higher than 50,000 u (Sundqvist *et al.*, 1983) have been observed.

As already mentioned, the detection efficiency is connected to the number of secondary electrons produced when the molecular ion hits the stop detector.

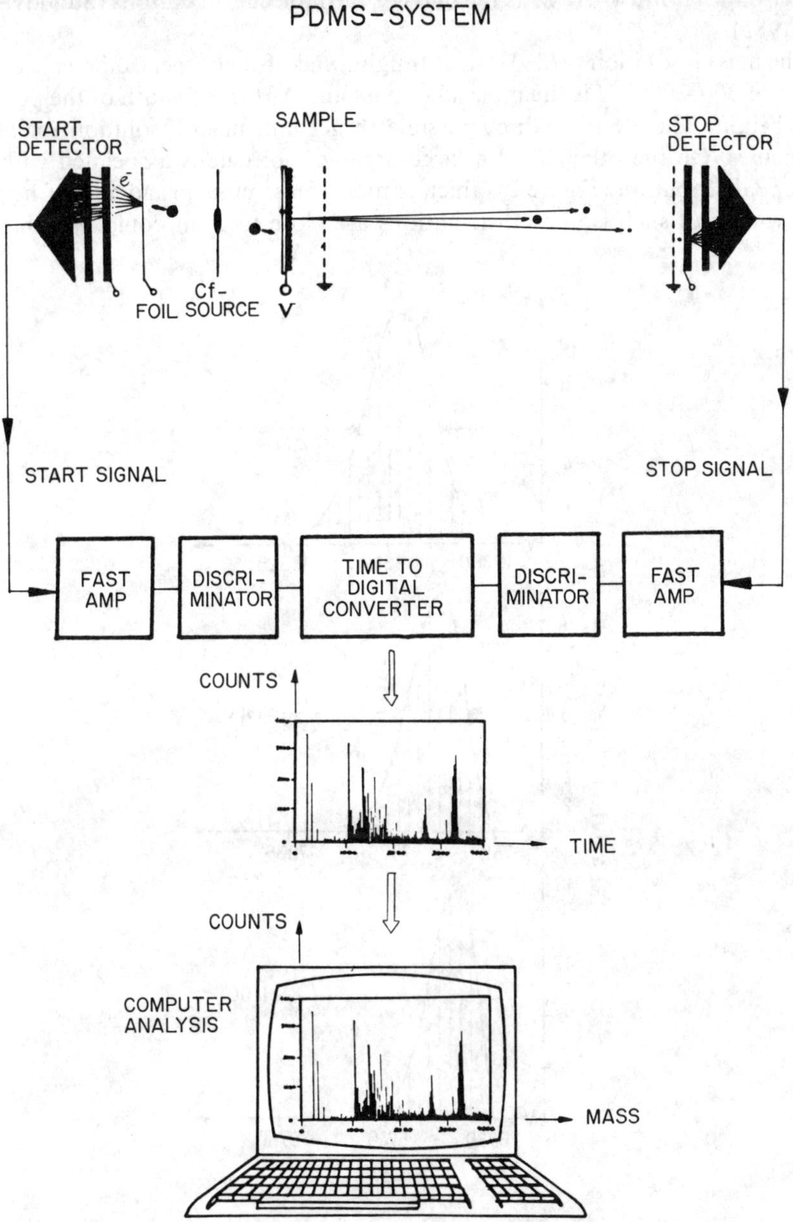

Figure 1: Principle of the plasma desorption time-of-flight mass spectrometer.

For efficient detection of large biomolecular ions the energy E(eV) should be larger than 1.7 M, where M is the mass of the molecule in daltons (Sundqvist *et al.*, 1985).

The mass resolution ($M/\Delta M$) in a straight time-of-flight spectrometer is of the order of 300–3000. M is the molecular mass and ΔM is the width of the peak at half height. Therefore it is not possible to get unit mass resolution for large molecules. On the other hand a large organic molecule is associated with an isotope distribution (Figure 2) which is many mass units broad at half height. The widths of such isotope distributions are close to what could be achieved

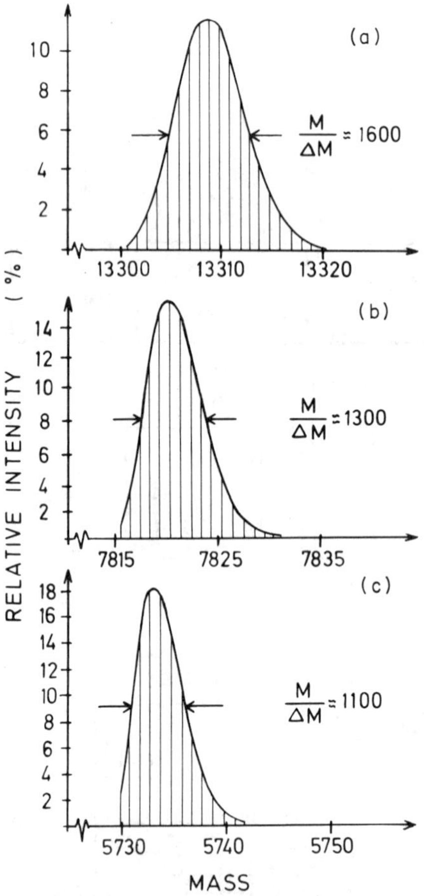

Figure 2: The theoretical isotope distribution for the molecular ion of tiger snake phospholipase homologue (a), siamensis neurotoxin (b) and bovine insulin c. The solid envelopes have inverted relative widths at half height ($M/\Delta M$) as indicated in the figure. (© *1985 Elsevier, reproduced with permission from Sundqvist* et al., *1985*).

with a high performance straight time-of-flight mass spectrometer (Sundqvist *et al.*, 1985). Such a mass analyzer should therefore be appropriate to use for determination of the isotopically averaged molecular weight of large molecules.

The precision in the mass determination with a PDMS–TOF spectrometer is of the order of $\pm(0.1–1)$ u in the mass range 1–5000 u. For larger molecules the relative contribution to the molecular ion peak from metastable decay products makes the peak shape more complex and the uncertainty in the centroid determination larger. This means that at present the precision is of the order of ± 10 u at 10,000 u and ± 100 u at mass 20,000 u. The technical details of a PDMS system have been described recently by Sundqvist and Macfarlane (1985).

15.3. PDMS OF LARGE BIOMOLECULES

The first indication of the possibility of obtaining molecular ions of molecules above M_r 5000 was obtained by PDMS studies of protected oligonucleotides by McNeal and MacFarlane (1981) and McNeal *et al.* (1982a,b). These studies showed that the positive ion spectra were dominated by molecular ion species $(M + H)^+$, $(M + Na)^+$ etc. The largest oligonucleotide studied was a 12-mer with molecular weight 6339 which produced a $(M + Na)^+$ peak at *m/z* 6352; a dimer ion was also observed at *m/z* 12637. In these studies it was further demonstrated that the negative PD-spectra contained sufficient fragment ions to allow sequence determination of the protected nucleotides (McNeal *et al.*, 1982a,b).

Analyses of small proteins were first carried out on the Tandem Accelerator at Uppsala University. Using 90 MeV ^{127}I ions as the primary beam and mass analysis by time-of-flight, it was possible to observe the $(M + H)^+$ ion of bovine insulin and also fragment ions corresponding to the A and B chains (Håkansson *et al.*, 1982). The insulin experiment was soon reproduced on a prototype ^{252}Cf-plasma-desorption time-of-flight mass spectrometer constructed in the same laboratory. This instrument was used in a series of studies of proteins ranging from insulin up to trypsin, with molecular weight 23,364 (Table 1) (Sundqvist *et al.*, 1983b, 1984a,b; Håkansson *et al.*, 1983; Kamensky *et al.*, 1983).

15.4. PD SPECTRA OF LARGE BIOMOLECULES

PD spectra look rather different from conventional mass spectra, as illustrated by the spectrum of insulin (Figure 3a). The spectrum is dominated by a rather

TABLE 1. Summary of molecular weight determinations of proteins by PDMS.

Molecule	Calculated mass (CM)	Measured mass (MM)	Δ(CM + 1–MM)	Relative precision (parts per thousand)	Reference (see footnote)
Bovine insulin	5733.5	5734.3 ± 2.0	+0.2	+0.03	1
Porcine insulin	5777.6	5778.0 ± 2.0	+0.6	+0.1	1
Human insulin	5807.6	5811.3 ± 2.0	−2.7	−0.5	1
Mono-I-A^{14} porcine insulin	5903.5	5927 ± 2.0	+0.6	+0.1	1
Mono-I-A^{19} porcine insulin	5903.5	5949 ± 5	−44.5[b]	0.0	1
Thai Cobra neurotoxin	7821	7819 ± 4	+3	+0.4	1
Bovine proinsulin	8681	8725 ± 10	−43[b]	+0.1	1,2
Porcine proinsulin	9082	9090 ± 10	−7	−0.8	1
PSP	11 711	11 794 ± 15	−82	−7.0	1
Cytochrome-C	12 384	12 335 ± 30	+50	+4.0	1
Tiger Snake II-1	13 309	13 285 ± 20	+25	+1.9	1,3
Bovine ribonuclease	13 682	13 529 ± 50	+162	+12.0	1
Porcine phospholipase A_2	13 980	13 968 ± 15	+13	+0.9	1
Porcine trypsin	23 463	23 406 ± 140	+58	+2.4	4

[a] Probably $(M + Na)^+$. Relative precision based on this assumption
[b] Probably $(M - H + 2Na)^+$. Relative precision based on this assumption

References to Table 1
1: Sundqvist *et al.*, (1984a). 2: Håkonsson *et al.*, (1983). 3: Kamensky *et al.*, (1983). 4: Sundqvist *et al.*, (1984b).

intense continuum or background, on top of which distinct peaks appear. To understand the origin of the continuum and the peak shape it must be borne in mind that the TOF mass spectrometer records all secondary ions accelerated. This includes ions obtained by prompt and delayed desorptions, ions from metastable decay both during acceleration and in the field free zone, and multiply charged ions. The large smooth background is most probably due to fragments of secondary ions which have decayed in the acceleration region before they have reached full energy (Chait, 1983). This is supported by the observation that approximately 20% of the background in the insulin spectrum

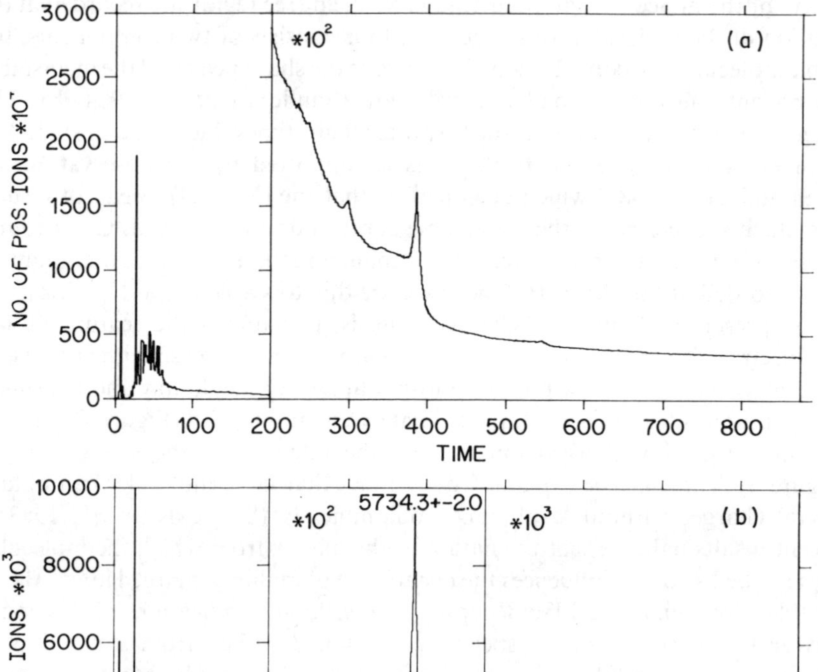

Figure 3: a) Raw PD-spectrum of bovine insulin. b) Background-subtracted PD-spectrum of bovine insulin. Time in arbitrary units. (*© 1983 Elsevier, reproduced with permission from Sundqvist* et al., *1983*).

is due to neutrals which, as they have been accelerated, must be the result of metastable decay (Sundqvist *et al.*, 1984a). The continuous background can be removed by a rather simple algorithm (Sundqvist *et al.*, 1984a) leading to the background subtracted spectrum shown in Figure 3b. The spectrum contains the $(M + H)^+$ peak, M^{2+} and peaks corresponding to the A and B chains. Furthermore, peaks corresponding to molecular clusters 2M, 3M and 4M, probably as $(nM + H)^+$ species, are observed. The molecular ion peak is broader, especially at the base, than expected from the mass resolution of the TOF spectrometer. This is partly because the instrument does not have energy focusing capabilities and partly because ions and neutrals due to metastable decays of the molecular ion occurring in the field free region are recorded at the position of the molecular ions. The peak thus consists of two components, the stable molecular ion contribution giving rise to the sharp peak and the metastable component causing a broader distribution (Sundqvist *et al.*, 1984a,b). The latter peak is shifted a little towards shorter flight times due to the postacceleration of light charged fragments. This is supported by the observations of Chait and Field (1985) which demonstrate that the $(M + H)^+$ peak of insulin gets much sharper when the metastable contribution is removed. Chait found in the case of bovine insulin that only a minor fraction of the total amount of ions recorded at the $(M + H)^+$ position are due to stable $(M + H)^+$ ions.

The presence of multiply charged ions is a characteristic feature of the PD-spectra. The intensities of multiply charged ions increase with increasing molecular weight; thus for porcine phospholipase A2 the doubly and the singly charged molecular ion are of approximately equal intensity (Figure 4) and for trypsin the doubly charged ion is twice the intensity of the singly charged (Figure 5). This may be explained by the fact that large molecules can obtain several charges without 'exploding' coulombically (Sundqvist *et al.*, 1983a). Recent results indicate that the nature of the matrix from which the molecules are desorbed strongly influences the occurrence of multiply charged ions (Alai *et al.*, 1986; Jonsson *et al.*, 1986; Roepstorff *et al.*, 1986). Indications of this could previously be seen from the spectra of insulin desorbed from a naphion (a polyfluorocarbon sulphonic acid polymer) backing shown by Macfarlane and co-workers (Jordan *et al.*, 1984).

In the high mass region of the spectra, peaks corresponding to the dimer, trimer, tetramer and so forth are often observed. It has at present not been demonstrated whether the multimers are desorbed as such or whether oligomerization takes place in the gas phase after desorption. The former, however, seems more probable. This is also supported by the extension of the continuum above the molecular ion region which demonstrates that many oligomers break down in the acceleration zone, i.e. within the first 10^{-6} s after desorption.

From the low mass section of the spectra it is obvious that extensive fragmentation takes place. However, most of these fragments are of too low mass to give useful structural information. For proteins, in contrast to

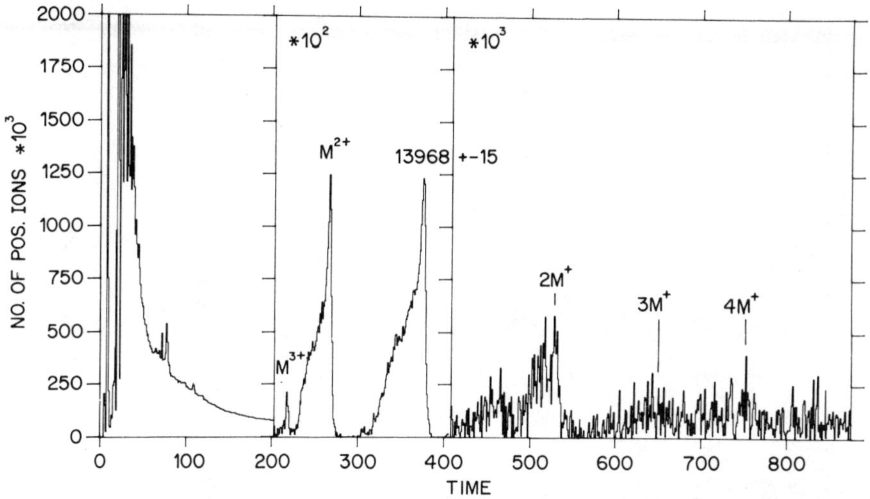

Figure 4: Background subtracted PD-spectrum of porcine phospholipase A2 (M_r 13,980). (© *1983 Elsevier reproduced with permission from Sundqvist* et al., *1983*).

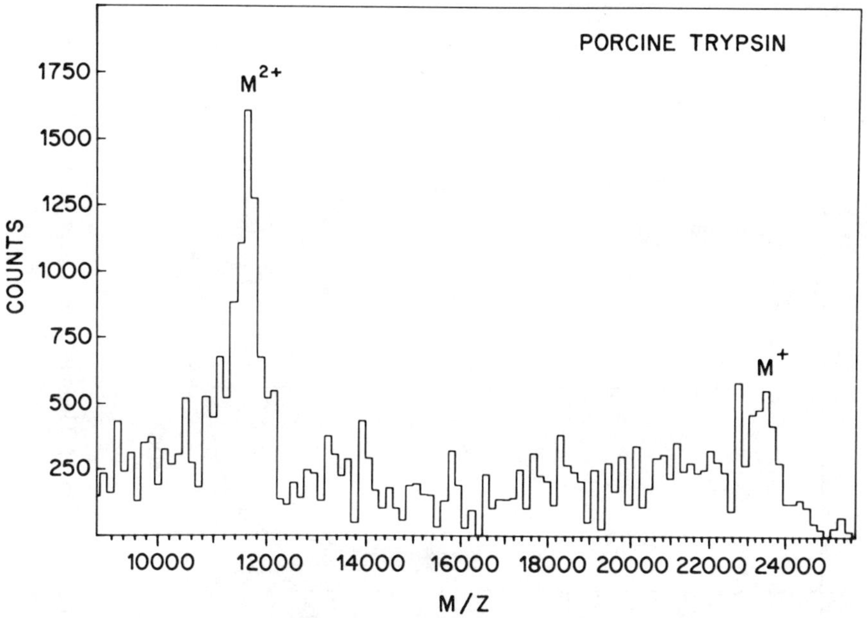

Figure 5: The high mass region of the PD-spectrum of porcine trypsin. The molecular weight determined from the centroids of the M^+ and M^{2+} peaks is 23 406 ± 140 (calculated M_r 23,463). (© *1984 American Association for the Advancement of Science, reproduced with permission from Sundqvist* et al., *1984b*).

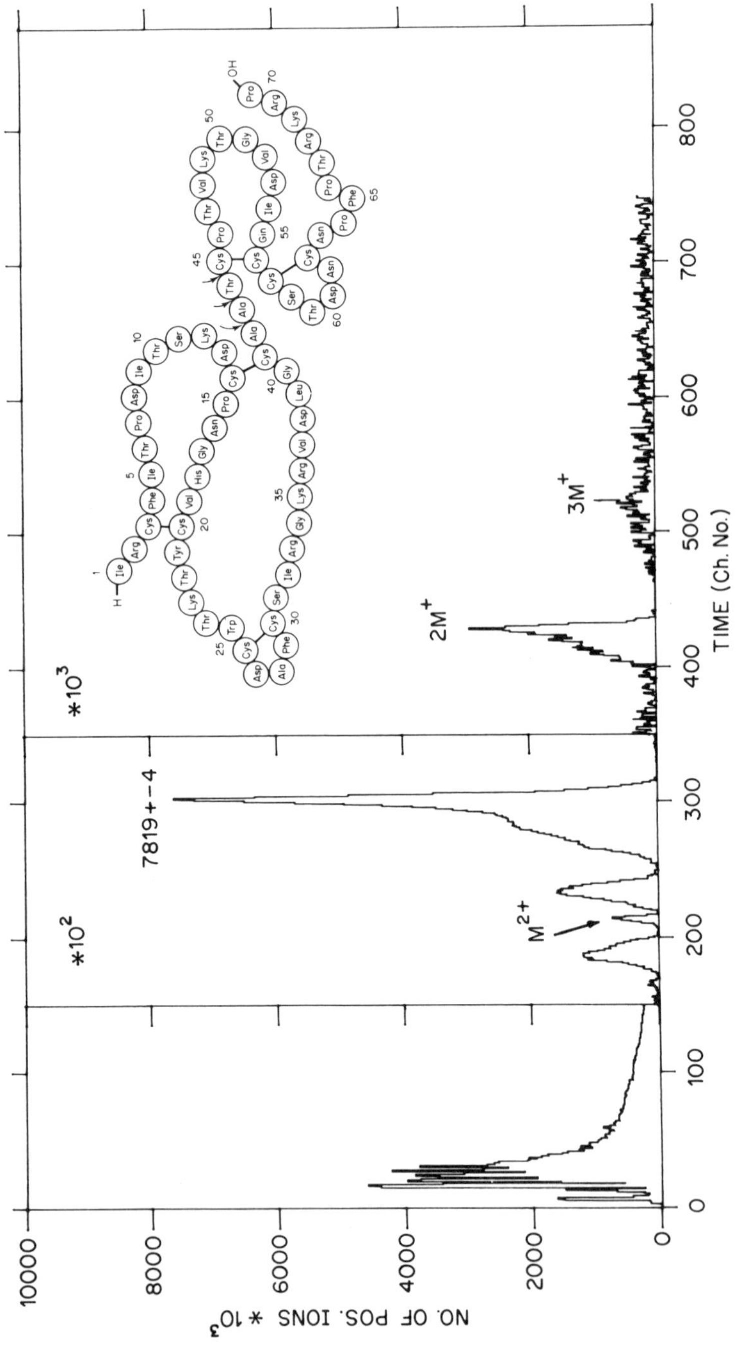

Figure 6: PD-spectrum of Thailand cobra neurotoxin. The arrows on the structure indicate possible cleavage points leading to fragment ions.

oligonucleotides, only a few examples of fragment ions at higher mass have been observed. For insulin, distinct peaks corresponding to the A and B chains are observed and, for a cobra snake neurotoxin with molecular weight 7820, fragmentation between two domains of the molecule is observed (Figure 6).

15.5. SAMPLE PREPARATION

In most plasma desorption instruments the primary ions enter the sample from behind. Here thin foils (0.5 mg cm^{-2}) of aluminium, aluminized Mylar or nickel are used as backing for the sample. For geometries where the primary ion enters from the front a thick sample backing may be used. For most studies so far the sample has been deposited on the backing by the electrospray method (Brunnix and Rudstam, 1981; Sundqvist *et al.*, 1984a). A drawback of the electrospray technique is that the solvents appropriate for spraying (i.e. organic solvents) are not always good solvents for large biomolecules, and for proteins nearly always cause denaturation. Another problem is that the quality of the spectra seems strongly affected by low molecular weight contaminants in the sample. The presence of alkali metal ions is very critical, and adduct ions containing up to several sodium or potassium atoms are nearly always observed for biological samples. Furthermore, even small amounts of alkali metal ions may entirely quench the molecular ion yields.

With the present technique it is very difficult to prepare biomolecules free of traces of salts. Therefore, other sample preparation techniques are under development. The first promising alternative was the use of an ion exchange surface (Jordan *et al.*, 1984). The aluminium foil was covered with a layer of a sulphonic acid-substituted fluorocarbon polymer, naphion, and insulin was ion-exchanged to the surface from a water solution. The development of other types of surfaces which allow adsorbtion of proteins from water solutions are in progress in the authors' laboratories; the most successful is nitrocellulose (Jonsson *et al.*, 1986). There are several advantages of such surfaces, such as the possibility of sample application from rather dilute water or buffer solutions, the possibility of selecting conditions under which the biomolecules are either in their native conformation or denatured and the use of solvents which involve no risk of reaction with or alteration of the biomolecules.

15.6. SENSITIVITY

In biochemical research high sensitivity is often a prerequisite for the practical application of a technique. Sundqvist *et al.* (1985) have calculated the theoretical sensitivity for detection of insulin by the PDMS technique to be approximately 10^{-13} mole for a sample area of 1 cm^2. Although such sensitivities have not been demonstrated in practice, detection limits of the order of 1–10 pmole for large peptides and small proteins have been demonstrated in the authors' laboratories and 1 pmole of insulin distributed on an area of 1 mm^2 was detected using an accelerator beam of primary ions (Sundqvist *et al.*, 1985). In routine analysis of pure samples, quantities of 0.1–0.5 nmole normally result in spectra with clearly detectable molecular ions. The sample purity, however, is very important and the presence of impurities will result in a splitting of the molecular ion intensity into several peaks due to adduct ions and eventually the whole spectrum is quenched. It is also our experience that the intensity of adduct ions relative to $(M + H)^+$ increases when the sample quantities are reduced. It has recently been demonstrated that the development of new sample application techniques can reduce these problems (Jonsson *et al.*, 1986).

An important feature of PDMS is that the flux of primary ions is very low. Thus, most of the sample is undamaged even after rather long exposures to primary ions and the sample may be recovered from the backing foil after analysis.

In the context of sensitivity it is important to realize that the detectors used so far are based on secondary electron production and such detectors have a threshold velocity for the production of secondary electrons when an ion hits the detector (Beuler and Friedman, 1980). With the present instruments, acceleration voltages of 15–20 kV and postacceleration of 3–7 kV are used. This should in principle be sufficient to detect molecules up to a molecular weight of approximately 15,000. In practice molecular ions up to M_r 24,000 and cluster ions well above 60,000 have been observed, but the detection efficiency for such large ions is probably extremely low. Increased sensitivity for analysis of large molecules will thus require development of new detectors.

15.7. APPLICATIONS

Thus far the number of applications of the PDMS technique to large biomolecules is rather limited. This is partly due to the limited number of instruments available for applied studies. Large molecules have only been studied in the last few years by a few groups, mainly interested in studies to evaluate the limits and possibilities of the technique. Applications of PDMS, including analyses of smaller molecules, have been reviewed recently (Sundqvist and Macfarlane, 1985).

A few recent exploratory examples will be discussed. The production of semisynthetic human insulin from porcine insulin by the Danish company Nordisk Gentofte was monitored by PDMS (Sundqvist *et al.*, 1984b). The only difference between human and porcine insulin is the C-terminal amino acid

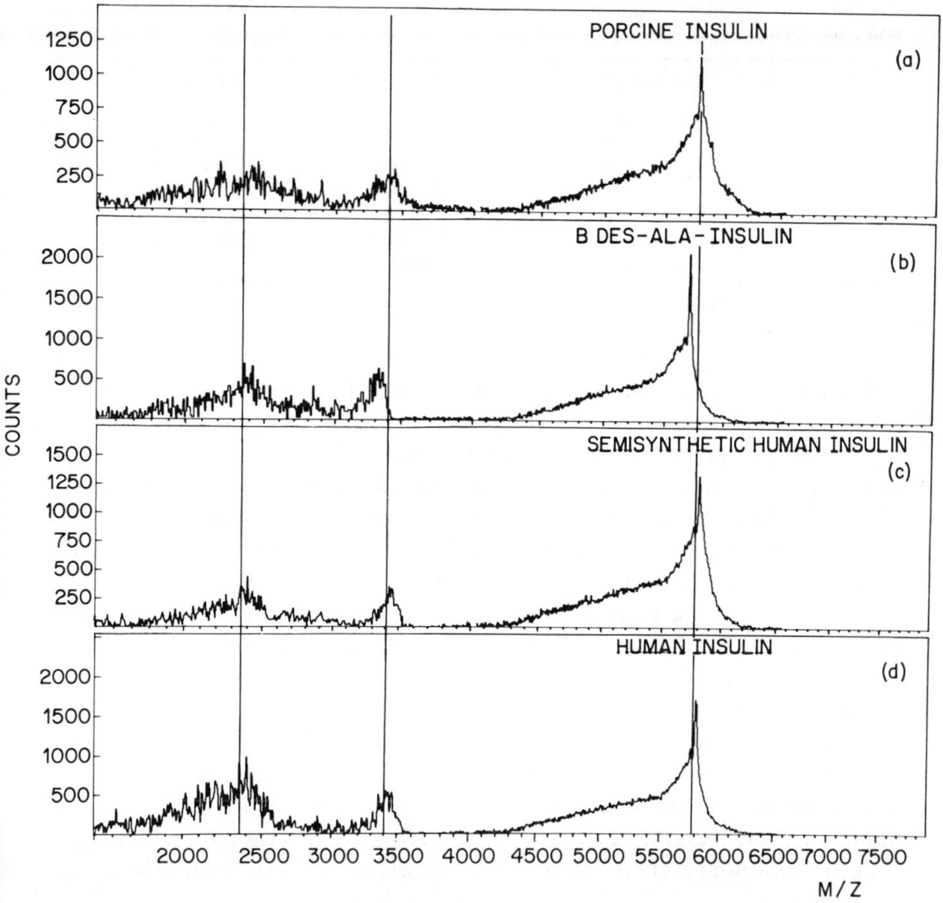

Figure 7: The high mass region PD-spectra of porcine insulin (a), B-des-ala insulin (b), semisynthetic human insulin (c) and human insulin (d). The mass differences observed for the molecular ion peaks relative to that of porcine insulin are -71.5 ± 3.0, $+31.9 \pm 3.0$ and 27.4 ± 3.0. The corresponding calculated values are -71, $+30$ and $+30$ respectively. (© *1985 Elsevier, reproduced with permission from Sundqvist* et al., *1985*).

residue in the B-chain. From the starting material, porcine insulin, the C-terminal alanine is removed, resulting in des-Ala insulin, to which threonine is coupled, resulting in semisynthetic human insulin, which can then be compared to authentic human insulin. The PDMS spectra are shown in Figure 7a-d respectively. The mass differences observed for $(M + H)^+$ as well as the B-chains correspond within the experimental error ($\pm$ 3 u) to those expected.

TABLE 2. Molecular weight determinations of the tryptic peptides from a cuticular protein from locust.

Peptide	Calculated mass	Measured mass	Molecular ion species assumed
T_2A2-2	280.3	302.8	$(M + Na)^+$
T_2A2-3	674.8	698.0	$(M + Na)^+$
T_2A3	2383.7	2428.7	$(M - H + 2Na)^+$
T_2A4	3828.4	3850.7	$(M + Na)^+$
T_2A5	3264.8	3311.4	$(M - H + 2Na)^+$
T_2A6	4324.0	4347.6	$(M + Na)^+$

Sequence determination of a cuticular protein from locust revealed that this protein is characterized by several repeated sequences (Højrup *et al.*, 1986). The molecular weight was determined by SDS-gel electrophoresis to be 21,600 whereas the calculated molecular weight based on the determined sequence is 15,323. Tryptic cleavage of the protein resulted in six peptides, the molecular weights of which were determined by PDMS (Table 2). These peptides add up to the calculated molecular weight, if the tryptic peptide T_2A2-3 occurs twice, as indicated by the sequence data. The discrepancy between the calculated molecular weight and the molecular weight based on SDS-gel electrophoresis could be explained if others of the tryptic peptides were repeated in the sequence, or if the protein contained post-translational modifications, which was not indicated. Molecular weight determination of the complete protein by PDMS showed the molecular weight to be 15,329 $\pm$ 50 Roepstorff *et al.*,1986). The combined results from PDMS analysis of the tryptic peptides and the complete protein confirmed the data from sequence determination and showed that the molecular weight determination based on SDS-gel electrophoresis was erroneous.

Molecular weight determinations of the components in peptide and protein mixtures by SDS-gel electrophoresis or gel-permeation chromatography are standard procedures in biochemistry, molecular biology and medicine. These techniques are rather imprecise ($\pm$ 5–15%) and strongly depend on characteristics of the molecules which are irrelevant to the molecular weight, such as conformation and hydrophobicity. Furthermore, SDS-gel electrophoresis is not well suited for proteins below M_r 10,000. Mixture analysis by PDMS was tested as shown in Figure 8. All four components were readily resolved and their molecular weights could be determined with a precision of 0.1% or better. A possible complication due to mixed cluster ions was not observed. The same mixture was analyzed by SDS-gel electrophoresis and gel permeation chromatography. In neither case were all components resolved and the electrophoretic procedure gave strongly misleading results about the composition and the molecular weight range of the mixture.

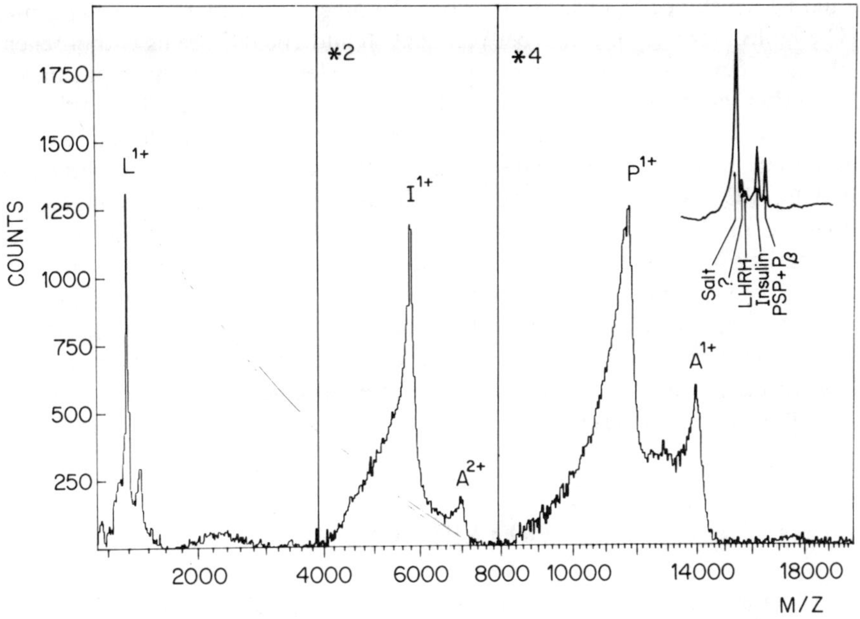

Figure 8: PD-spectrum of a 1 : 1 : 1 : 1 molar mixture of luteinizing hormone releasing hormone (L, M_r 1182), porcine insulin (I, M_r 5778) pancreatic spasmolytic peptide (P, M_r 11,711) and phospholipase A_2 (A, M_r 13,980). The insert shows the high performance gel permeation chromatogram of the same mixture. (*© 1985 Elsevier, reproduced with permission from Sundqvist* et al., *1985*).

15.8. FUTURE PROSPECTS

With the present state of the art, PDMS has been demonstrated to allow analysis of proteins up to a molecular weight of approximately 25,000; no data on high molecular weight nucleic acids and carbohydrates are presently available. For molecular weights above 10,000, PDMS is probably the only technique available today, whereas FAB with magnetic sector instruments offers an alternative below 10,000. In this latter mass range the ultimate sensitivity is best with PDMS whereas sector instruments offer better mass resolution. The major drawback of PDMS, in contrast to MS on sector instruments, is the truly integral nature of the technique because all ions, including ions due to metastable decompositions and delayed emission, are recorded, resulting in a high background level and broad peaks. This factor on the other hand may very well contribute to its high sensitivity for analysis of large molecules. Techniques which allow the option of elimination of metastable contributions in order to obtain a low background level and sharper peaks should be developed in order to improve the precision of mass determination.

The molecular weight range between 25,000 and 100,000 seems within reach in the next few years with improvements of the detection systems and sample preparation techniques. If these are achieved, most proteins will be within the analytical capacity of PDMS. Due to the high precision of the molecular weight determination compared to conventional techniques and the ease of obtaining the data, PDMS must be expected to become a very useful technique in bioscience in the future. Possible areas of application will include general characterization of biological macromolecules, studies and location of mutants, location of post-translational modifications, monitoring of protein modification reactions, control of sequence data and product control in biotechnology. The possibility of mixture analysis further increases the applicability because it allows rapid and precise peptide mapping after specific chemical or enzymatic cleavage of the molecules studied.

REFERENCES

Alai, M., Demirev. P., Fenselau, C., and Cotter, R. J. (1986). Glutathione as a matrix for plasma desorption mass spectrometry of large peptides, *Anal. Chem.*, **58**, 1303–1307.

Barber, M., Bordoli, R. S., Sedgwick, R. D., and Tyler, A. N. (1981). Fast atom bombardment of solids (FAB): a new ion source for mass spectrometry, *J. Chem. Soc., Chem. Commun.*, **1981**, 325–327.

Beuler, R. S., and Friedman, L. (1980). Threshold studies of secondary electron emission induced by macro-ion impact on solid surfaces, *Nucl. Instrum. Methods*, **170**, 309–315.

Brunnix, E., and Rudstam, G. (1981). Electrospraying, a method of making samples for β-counting allowing accurate correction for self scattering and self absorbtion, *Nucl. Instrum. Methods*, **131**, 131–140.

Chait, B. T. (1983). Prompt and metastable decompositions in ^{252}Cf-fission fragment ionization mass spectrometry, *Int. J. Mass Spectrom. Ion Phys.*, **53**, 227–242.

Chait, B. T., and Field, F. H. (1985). A study of the metastable fragments of ions produced by ^{252}Cf-fission fragment bombardment of bovine insulin, *Int. J. Mass Spectrom. Ion Processes*, **65**, 169–180.

Højrup, P., Roepstorff, P., and Andersen, S. O. (1986). Primary structure of a structural protein from the cuticle of the migratory locust, *Locusta migratoria, Biochem. J.*, in press.

Håkansson, P., Kamensky, I., Sundqvist, B., Fohlman, J., Peterson, P., McNeal, C. J., and Macfarlane, R. D. (1982). ^{127}I-Plasma desorption mass spectrometry of insulin, *J. Am. Chem. Soc.*, **104**, 2948–2949.

Håkansson, P., Kamensky, I., Kjellberg, J., Sundqvist, B., Fohlman, J., and Peterson, P. (1983). The observation of quasi molecular ions from a protein neurotoxin (mw 7821) using ^{127}I-plasma desorption mass spectrometry, *Biochem. Biophys. Res. Commun.*, **110**, 519–524.

Jonsson, G., Hedin, A., Håkansson, P., Sundqvist, B., Säve, G., Nielsen, P. F., Roepstorff, P., Johansson, K. E., Kamensky I., and Lindberg, M. (1986). Plasma desorption mass spectrometry of peptides and proteins adsorbed in nitrocellulose, *Anal. Chem.*, **58**, 1084–1087.

Jordan, E. A., Martin, C. R., McNeal, C. S., and Macfarlane, R. D. (1984). Desorption of insulin molecular ions from polymeric surfaces, *Proc. 32nd Ann. Conf. on Mass Spectrometry and Allied Topics, San Antonio*; American Society for Mass Spectrometry, pp. 26–27.

Kamensky, I., Håkansson, P., Sundqvist, B., McNeal, C. J., and Macfarlane, R. D. (1982). Comparison of biomolecule desorption yields for low and high energy primary ions, *Nucl. Instrum. Methods*, **198**, 65–68.

Kamensky, I., Håkansson, P., Kjellberg, J., Sundqvist, B., Fohlman, J., and Peterson, P. (1983). The observation of quasi molecular ions from a tiger snake venom component (mw 13,304) using ^{252}Cf-plasma desorption mass spectrometry, *FEBS Letters*, **155**, 113–116.

McNeal, C. J., and Macfarlane, R. D. (1981). Observation of a fully protected oligonucleotide dimer at m/z 12,637 by ^{252}Cf-plasma desorption mass spectrometry, *J. Am. Chem. Soc.*, **103**, 1609–1610.

McNeal, C. J., Ogilvie, K. K., Theriault, N. Y., and Nemer, M. J. (1982a). A new method for sequencing fully protected oligonucleotides using ^{252}Cf-plasma desorption mass spectrometry. 1. Negative ions of dinucleoside monophosphates, *J. Am. Chem. Soc.*, **104**, 972–975.

McNeal, C. J., Ogilvie, K. K., Theriault, N. Y., and Nemer, M. J. (1982b). A new method for sequencing fully protected oligonucleotides using ^{252}Cf-plasma desorption mass spectrometry. 2. Negative ions of subunits in the stepwise synthesis of a heptanucleotide, *J. Am. Chem. Soc.*, **104**, 976–980.

McNeal, C. J., Ogilvie, K. K., Theriault, N. Y., and Nemer, M. J. (1982c). A new method for sequencing fully protected oligonucleotides using ^{252}Cf-plasma desorption mass spectrometry. 3. Positive Ions, *J. Am. Chem. Soc.*, **104**, 981–984.

Roepstorff, P., Sundqvist, B., Højrup, P., Jonsson, G., Håkansson, P., Andersen, S. O., and Johansson, K. E. (1986). Application of plasma desorption mass spectrometry to molecular weight determination of a structural protein from insect cuticle, *Biomed. Environ. Mass Spectrom.*, in press.

Sundqvist, B., and Macfarlane, R. D. (1985). Plasma desorption mass spectrometry, *Mass Spec. Rev.*, **4**, 421–460.

Sundqvist, B., Hedin, A., Håkansson, P., Kamensky, I., Kjellberg, J., Salehpour, M., Säve, G., and Widdiyasekera, S. (1983a). Protein ion production by fast heavy ion induced desorption, *Int. J. Mass Spectrom. Ion Phys.*, **53**, 167–183.

Sundqvist, B., Håkansson, P., Kamensky, I., and Kjellberg, J. (1983b). Fast heavy ion induced desorption of molecular ions from small proteins, *Proc. 2nd Int. Workshop on Ion Formation from Organic Solids, Münster 1982* (Ed. A. Benninghoven), Springer Series in Chemical Physics, pp. 52–57, Springer, Berlin.

Sundqvist, B., Håkansson, P., Kamensky, I., Kjellberg, J., Salehpour, M., Widdiyasekera, S., Fohlman, J., Peterson, P., and Roepstorff, P. (1984a). ^{252}Cf-Plasma desorption time-of-flight mass spectrometry of proteins, *Biomed. Mass Spectrom.*, **11**, 242–257.

Sundqvist, B., Roepstorff, P., Fohlman, J., Hedin, A., Håkansson, P., Kamensky, I., Lindberg, M., Salehpour, M., and Säve, G. (1984b). Molecular weight determination by californium plasma desorption mass spectrometry, *Science*, **226**, 696–698.

Sundqvist, B., Hedin, A., Håkansson, P., Kamensky, I., Salehpour, M., and Säve, G. (1985). Plasma desorption mass spectrometry (PDMS) limitations and possibilities, *Int. J. Mass Spectrom. Ion Processes*, **65**, 69–89.

Torgerson, D. F., Skowronski, R. P., and Macfarlane, R. D. (1974). New approach to the mass spectroscopy of non-volatile compounds, *Biochem. Biophys. Res. Commun.*, **60**, 616–621.

Mass Spectrometry in Biomedical Research
Edited by S. G. Gaskell
© 1986 John Wiley & Sons Ltd.

Chapter 16

MASS SPECTROMETRIC ANALYSIS OF BIOLOGICAL OLIGOMERS

Ian Jardine

16.1. INTRODUCTION

The field of mass spectrometry has developed dramatically in the past few years. Major advances have been made in capillary gas chromatography–(accurate mass) mass spectrometry (GC–MS); high performance liquid chromatography–mass spectrometry (LC–MS); new ionization techniques such as fast atom bombardment (FAB MS) and ^{252}Cf-plasma desorption (^{252}Cf PDMS); extended mass range instruments; tandem mass spectrometry (MS/MS); and computer instrument control, data acquisition and data processing. With the availability of such a wealth of complementary mass spectrometry techniques, it is now possible to attempt the mass spectrometric analysis of important large biological molecules such as peptide and carbohydrate oligomers. Some examples will be described which serve to illustrate the application of diverse mass spectrometric techniques to the solution of biological structural problems.

16.2. CHOLECYSTOKININS

Cholecystokinin (CCK) is a biologically important polypeptide hormone which is present in gastrointestinal and neural tissue and which possesses a wide range of physiological functions (Mutt, 1980). Multiple molecular forms of immunoreactive CCK have been identified in many species. Porcine CCK peptides, for example, represent a family of linear peptides ranging in size from 58 amino acid residues down to four amino acid residues. Characterization of human forms of CCK, however, has been a much more difficult task.

Using human brain as a source of material, the first *chemical* characterization of a *human* form of CCK was recently accomplished (Miller *et al.*, 1984). This form was found to be identical to porcine CCK octapeptide, with the sequence Asp-Tyr(SO_3H)-Met-Gly-Trp-Met-Asp-Phe(NH_2). The peptide was characterized by amino acid analysis, automated peptide sequencing, and FAB MS on a total of 3 μg of isolated material (as determined by radio-immunoassay). The FAB MS analysis was carried out on half of the isolated material (1.5 μg) and was particularly important in characterizing the size and covalent modifications of this peptide which are not recognized by classical peptide sequencing methods.

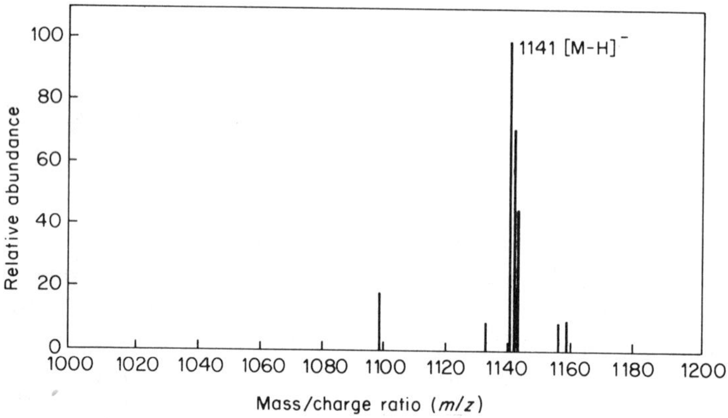

Figure 1: Negative ion fast atom bombardment mass spectrum (partial) of 1.5 μg of cholecystokinin octapeptide sulphate [Asp-Tyr(SO_3H)-Met-Gly-Trp-Met-Asp-Phe(NH_2); M_r 1142] isolated from human brain for the first time. *Reproduced from Miller* et al. (*1984*), *by permission of the International Society for Neurochemistry.*

The negative ion FAB mass spectrum of the isolated CCK-8 (Figure 1) was identical to standard porcine CCK-8 and consisted of an intense $(M - H)^-$ ion at *m*/*z* 1141 with a few weak fragment ions (not shown). Minimal fragmentation occurs with this molecule under these conditions because of the sulphate moiety, which produces a stable anionic species. The main fragment occurs at *m*/*z* 505 and corresponds to the fragment Asp-Tyr(SO_3)-Met. A minor fragment ion was also tentatively identified which corresponds to Asp-Tyr(SO_3)Met-Gly-Trp-Met. The unambiguous molecular weight determination of the molecule (1142 dalton) with supporting fragmentation information, in conjunction with the sequence data, confirms the covalent modifications of tyrosine sulphation and phenylalanine amidation.

As larger forms of human CCK are isolated, high mass mass spectrometry will be particularly valuable for the analysis of these materials mainly with respect to the confirmation of peptide size and sequence and to the determination of post-translational modifications, now that the DNA for the precursor of human CCK molecules has been sequenced (Kato *et al.*, 1984).

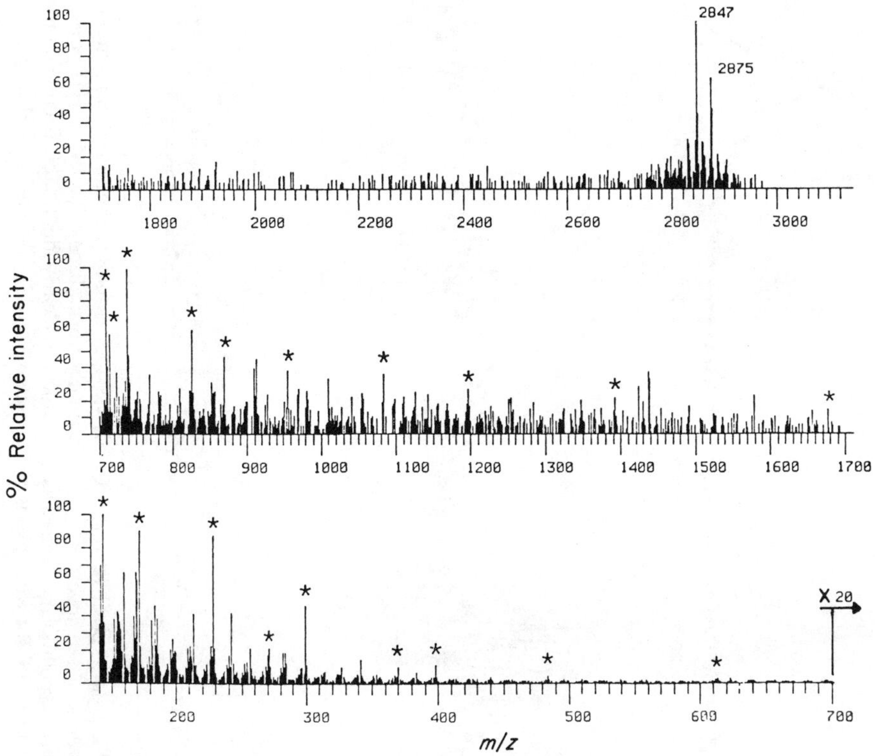

Figure 2: Computer recorded single scan positive ion fast atom bombardment mass spectrum of a mixture of melittin and N-formyl melittin. Computer calibration was effected using a CsI/RbI mixture. Note that the scaling of the mass axis has been adjusted in order that a more direct comparison with Figure 3 can be made. The intensities of all masses above m/z 700 have been multiplied by 20. The $(M + H)^+$ species are apparent at m/z 2847 and 2875. The $(M + H)^+$ ion for the ^{12}C isotope ion is at 2845.7 u. The $(M + 1)H^+$ or ^{13}C isotope ion is more intense, however, because of the large number of carbons in the molecule and the computer has chosen the most intense ion of the cluster (m/z 2846.7) to label. In doing so it has rounded 2846.7 up to 2847, although the more accurate mass is still available from file. Some clearly identifiable fragment ions are indicated by an asterix. The fragments are mainly from the N-terminus. The ion at m/z 738, for example, is the Gly-Ile-Gly-Ala-Val-Leu-Lys-Val fragment.

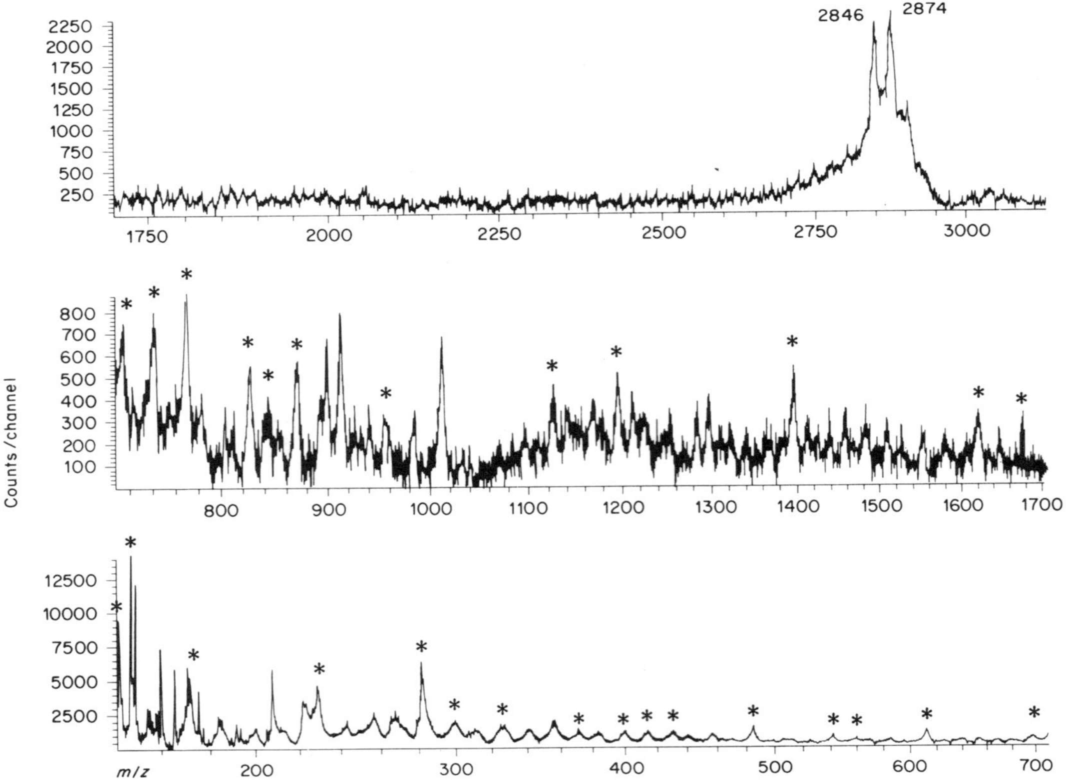

Figure 3: Low resolution (short flight path) positive ion ^{252}Cf PDMS spectrum of melittin and N-formyl melittin acquired in 10 000 seconds. Notice that all three segments of the spectrum are individually normalized. The $(M + H)^+$ ions at m/z 2846.5 and 2874.5 (average isotopic or chemical molecular weights) are intense, as are numerous clearly identifiable fragment ions indicated with an asterix. The ion at m/z 739, for example, is the Gly-Ile-Gly-Ala-Val-Leu-Lys-Val fragment.

16.3. CALMODULIN-BINDING AMPHIPHILIC PEPTIDES

Several amphiphilic, cytolytic peptides have been found recently to have a very high affinity for, and to inhibit expression of function in, calmodulin, an important calcium-dependent regulator protein. Melittin (26 amino acids; M_r 2844), δ-haemolysin (27 amino acids; M_r 3004) and mastoparans (a group of three tetradecapeptides; M_r approximately 1500) have the highest affinities of the peptides studied to date. In an ongoing investigation into the structural requirements for binding of these peptides to calmodulin, various derivatives and analogues are being synthesized to modify or maintain charge state, or to incorporate spectroscopic probes (e.g. EPR labels such as spin-imidate; NMR labels such as ^{13}C and ^{19}F tryptophan). Mass spectrometry is essential for the examination and confirmation of the structure of such synthetic analogues, since conventional peptide sequencing methods will not give information on modified amino acids.

The most powerful MS techniques now available for such analyses are FAB MS, ^{252}Cf PDMS and partial hydrolysis GC–MS. A computer acquired single FAB MS scan of isolated natural melittin (from bee venom) is shown in Figure 2. The structure of melittin is: Gly-Ile-Gly-Ala-Val-Leu-Lys-Val-Leu-Thr-Thr-Gly-Leu-Pro-Ala-Leu-Ile-Ser-Trp-Ile-Lys-Arg-Lys-Arg-Gln-Gln(NH_2). Molecular weight information is readily obtained including, as in this case, the presence of analogues (N-formylated melittin). A number of useful fragments are readily identifiable, particularly from the N-terminal end of the molecule.

^{252}Cf PDMS analysis of the same mixture of melittin and N-formyl melittin also readily generates molecular weight information (Figure 3) along with considerable fragmentation information (McNeal *et al.*, 1985). Contrary to the generally held impression that ^{252}Cf PDMS does not yield much useful fragmentation information, a number of the same fragment ions are present in both the FAB MS and ^{252}Cf PDMS spectra. Similar conclusions have been reached in recent publications on the ^{252}Cf PDMS analysis of smaller peptides (Fales *et al.*, 1985; Fohlman *et al.*, 1985). At present the resolution of the FAB mass spectrum is much higher than that of the ^{252}Cf PDMS spectrum. However, significant improvements in the resolution of the ^{252}Cf PDMS spectra have been achieved by metastable ion filtering (Chait and Field, 1984) and by the use of a reflectron which compensates for varying ion kinetic energies (Negra and LeBeyec, 1984). A ^{252}Cf PDMS experiment on the other hand is much easier to perform since calibration is very straightforward compared to FAB MS. The ultimately obtainable sensitivity for ^{252}Cf PDMS may also exceed that of FAB MS since an integrated spectrum of all generated ions is obtained in contrast to the scanned FAB MS magnetic system (Sundqvist *et al.*, 1985). A full ^{252}C fPDMS spectrum on 10 pmoles of insulin has been reported, for example (Fohlman *et al.*, 1984). Although the ^{252}Cf PDMS system may never compete with magnetic FAB MS with respect to resolution and therefore determination of isotopic distribution and accurate mass of molecular weight associated ions, the ^{252}Cf PDMS is much less expensive and much easier to operate and maintain. More detailed future comparisons of FAB MS and ^{252}Cf PDMS for the analysis of biological oligomers should clarify to what extent these powerful techniques are similar or complementary.

An elegant GC–MS technique for protein sequencing was recently developed in the laboratory of K. Biemann (Biemann, 1980). In this technique, the protein is partially hydrolyzed by acid or enzyme into a complex mixture of overlapping di- to hexa-peptides. This mixture of peptides is then derivatized to volatile *N*- α,(ω)-trifluoroethyl-*O*-trimethylsilyl polyamino alcohols and analyzed by GC–MS. There have been two major limitations to the more general application of this technique.

First, the procedure produces a very large amount of mass spectral data in the course of sequencing even a relatively small protein. Therefore, a computer program, PEPALG, was developed in Biemann's laboratory to interpret automatically the GC–MS data and identify the peptide derivatives (Herlihy and Biemann, 1981). This program was not readily available to other laboratories. The program, however, has now been implemented on an IBM-XT personal computer in both FORTRAN IV and PASCAL for off-line data analysis (Erickson and Jardine, 1986). Most GC–MS peptide sequencing runs can be processed in less than 2 hours with this approach. An example of the analysis of an enzymatic digest of melittin is shown in Figure 4. The mass spectrum of a computer-identified pentapeptide is shown in Figure 5 with the associated computer identification print-out. The availability of PEPALG in a microcomputer based version of both FORTRAN and PASCAL should now allow ready

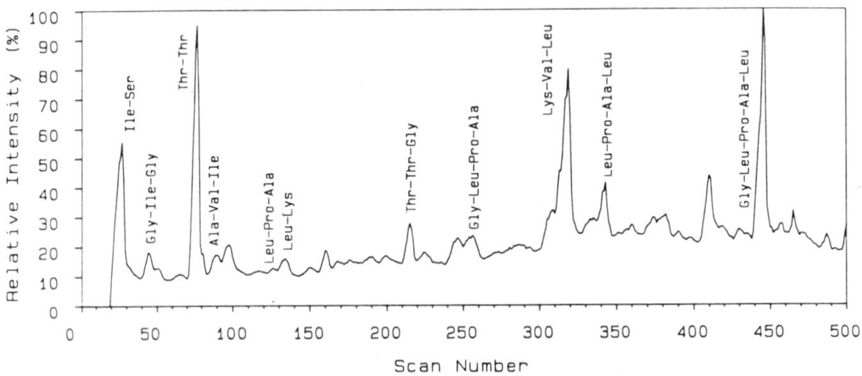

Figure 4: Total ion current chromatogram of a peptide mixture obtained by enzymatic hydrolysis of melittin using subtilisin and chymotrypsin and derivatized as the N-α,(ω)-trifluoroethyl-*O*-trimethylsilyl polyamino alcohols. PEPALG computer-identified peptides are indicated.

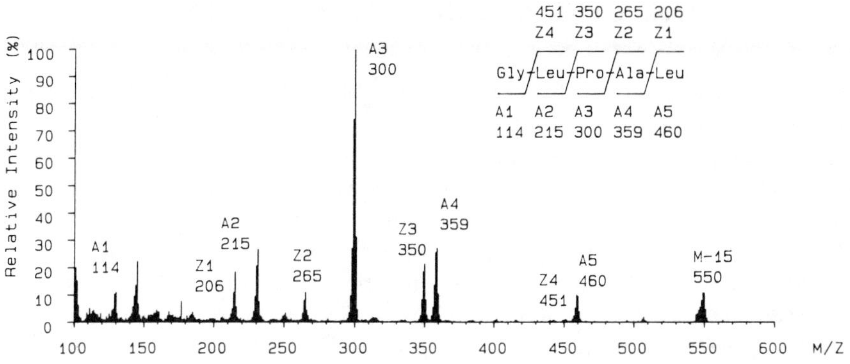

Scan # 443 Calculated Retention Index = 2738

	A1	A2	A3	A4	A5	Z1	Z2	Z3	Z4	*K	*F	*Z	*Y	*O	*C	M-15	Scr
GLPAL	6	25	100	47	10	3	11	36	0	0	0	0	0	0	0	18	338
Mass:	114	215	300	359	460	206	265	350	451							550	

Figure 5: EI mass spectrum of scan number 443 of Figure 4 and associated PEPALG output identifying the peptide as Gly-Leu-Pro-Ala-Leu from the unique sequence fragments shown. Note that the intensities reported on the PEPALG output have been weighted above m/z 300 relative to the plotted spectrum to accentuate the contribution of higher mass ions. The isotope clusters at each mass result from the use of BD_3 in the derivatization reduction step, since the BD_3 was not 100% deuterated. The columns on the PEPALG output headed by *K, *F, *Z, *Y, *O and *C refer to specific diagnostic ions for the amino acids. None of these happens to be present in the spectrum of Gly-Leu-Pro-Ala-Leu.

access to this powerful protein sequencing technique by any interested laboratory.

A second problem with this procedure is that a relatively large amount (50–100 nmoles) of the original peptide is required for analysis. The reason for this lies not in the sensitivity of the GC–MS analysis of the derivatized peptides, but in the manipulation of small samples and in the contaminants present in the solvents and reagents. To overcome this limitation Biemann is attempting to develop a derivatization procedure which utilizes only gaseous reagents. Preliminary work suggests that, at least for small standard peptides, an increase of at least two orders of magnitude in sensitivity may be possible (Zollinger and Biemann, 1985). If sustained for real samples, such a development would propel this GC–MS technique into the indispensible category for any MS laboratory working on peptides, particularly since the mass spectrometer is still unique in its ability to examine peptides modified either synthetically or by post-translational modification.

Application of these various MS techniques, coupled perhaps with MS/MS and LC–MS analysis, to modified amphiphilic peptides, which are difficult to analyze by conventional techniques, should allow ready confirmation of structural modifications of these molecules.

16.4. CYCLOSPORINE METABOLITES

The cyclic undecapeptide drug cyclosporine (M_r 1202) (Figure 6) is a fungal metabolite with potent immunosuppressive activity (Kahan, 1984). The compound is neutral, rich in hydrophobic amino acids and contains a number of N-methylated peptide linkages. Cyclosporine (CsA) is becoming the agent of choice for the prevention of the rejection of kidney, liver and heart–lung allografts. The drug is extensively metabolized *in vivo* by the liver into a number of hydroxylated and N-demethylated metabolites (Maurer *et al.*, 1984). In patients the rate and extent of metabolism can drastically affect the efficacy and toxicity (mainly renal) of the drug. Studies of the hepatic metabolism of cyclosporine are underway using animal models such as the rabbit to understand more clearly how the drug is handled by the liver (Hartman and Jardine, 1986).

CYCLOSPORINE A

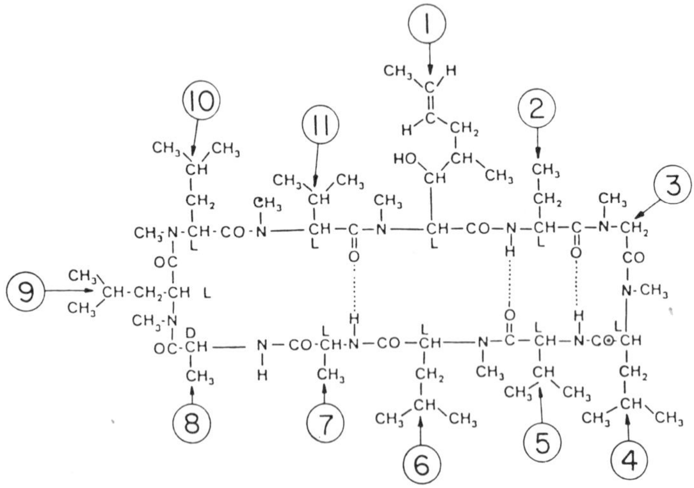

Figure 6: The structure of cyclosporine indicating the individual amino acids in the molecule.

After administration of radiolabelled cyclosporine to a rabbit, hepatic metabolites were collected via a biliary cannula over a 48 hour period. After extraction of the metabolites from the bile with ethyl acetate they were separated by HPLC on a C_{18} column with detection by UV absorption at 202 nm (Figure 7). Studies using [^{3}H]CsA and an on-line radioactivity flow detector indicated that virtually all UV-absorbing material was radioactive. Very little unmetabolized CsA could be detected in the rabbit bile.

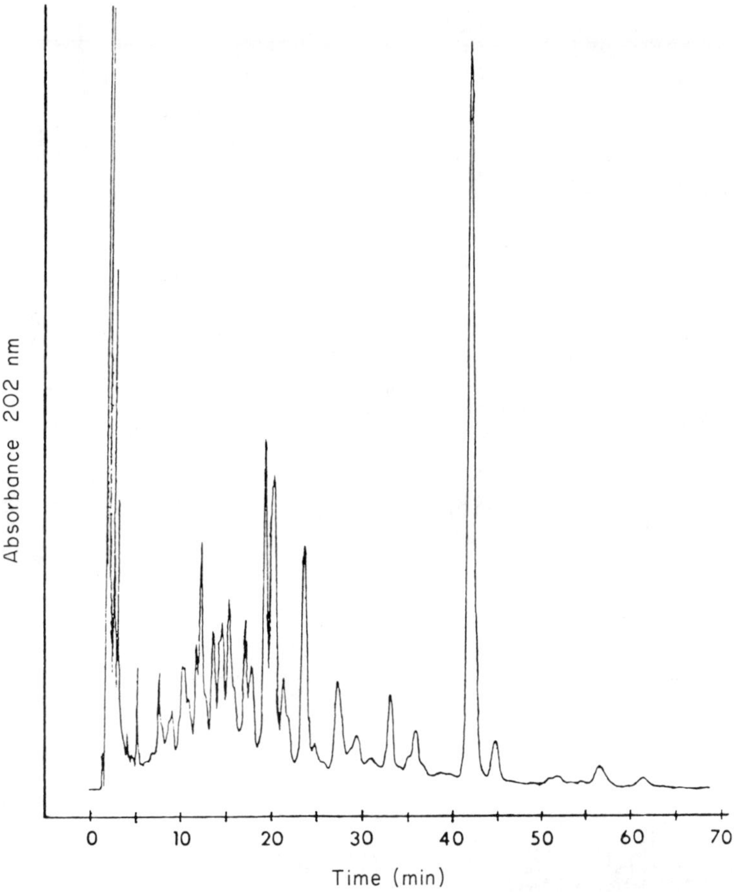

Figure 7: HPLC separation of rabbit biliary metabolites of cyclosporine.

LC–MS of the CsA rabbit metabolite mixture indicated that most of the metabolites were variously hydroxylated and N-demethylated species. This is illustrated in Figure 8 where mass chromatograms of mono-, di-, tri- and tetra-hydroxylated CsA's are shown. Numerous other permutations of hydroxylations and N-demethylations were observed from selected mass chromatograms. Hydroxylation on different carbons produces isomeric metabolites. In general, the retention time of the hydroxylated metabolites decreases with increasing hydroxylation in this reverse phase HPLC system, as expected. LC–MS can be seen to be a particularly powerful technique for the analysis of such complex metabolic profiles of compounds amenable to HPLC but not GC analysis.

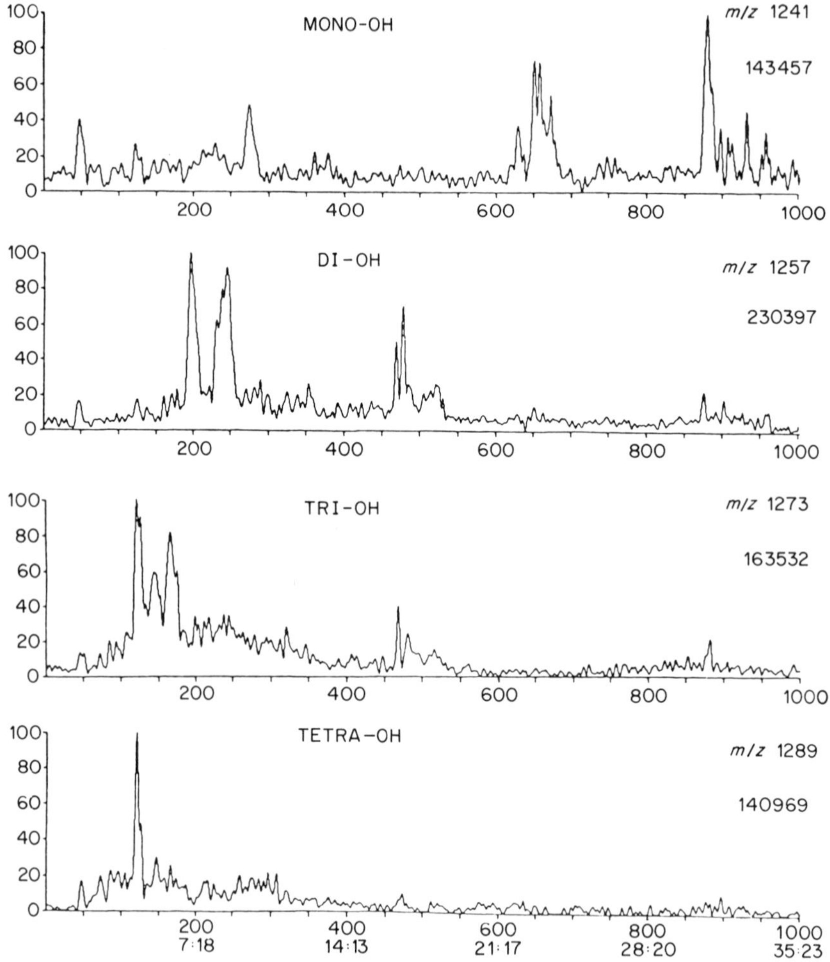

Figure 8: Mass chromatograms of hydroxylated cyclosporine metabolites obtained from LC–MS analysis (thermospray interface) of the mixture shown in Figure 7.

The major rabbit biliary metabolite (retention time 42 min in Figure 7) produced a $(M + H)^+$ ion at m/z 1233 (Figure 9) which is 30 u higher than that of the parent CsA molecule. This suggested that the metabolic conversion of a methyl group ($-CH_3$) to a carboxylic acid ($-COOH$) had taken place. The fragment ion at m/z 1090 is also present in the spectrum of standard cyclosporine and results from loss of the side chain of amino acid number 1 (Figure 6). The presence of this ion in the metabolite indicates that the metabolic transformation has taken place on amino acid No. 1.

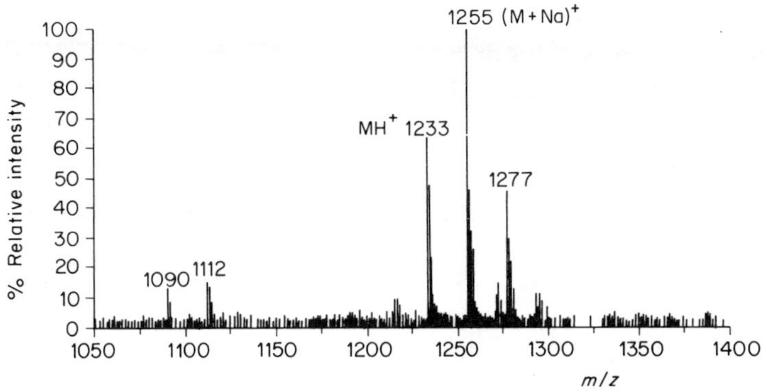

Figure 9: Positive ion thermospray LC–MS spectrum (partial) of the major metabolite of Figure 7. The fragment at *m/z* 1090 is present in the LC–MS spectrum of the parent drug and results from loss of the amino acid No. 1 side chain. The metabolic transformation in this component has, therefore, occurred on the side chain of amino acid No. 1.

To confirm which amino acid of the parent molecule had been converted to an acid, and to determine its structure, the metabolite was isolated by preparative HPLC and was Hydrolyzed in 6N HCl for 48 h at 110°C. The resulting amino acids were esterified with isobutanol/HCl, acylated with heptafluorobutyric anhydride, and identified by GC–MS. The peak normally eluting at a retention index of 1700 corresponding to the derivatized amino acid No. 1 in the parent molecule disappeared, and a new peak eluting at a retention index of 2146 was present. This peak produced a mass spectrum indicative of the derivatized amino acid No. 1 with a carboxylic acid group (Figure 10). In addition, the fragmentation pattern located the carboxylic acid to the η(eta) carbon of amino acid No. 1. The structure of the metabolite is shown in Figure 16.

Maurer *et al.* (1984) had previously described a metabolite hydroxylated on the terminal carbon of amino acid No. 1. Presumably alcohol dehydrogenase and aldehyde dehydrogenase act upon this hydroxylated metabolite to form the carboxylic acid metabolite. If this is the case, this metabolic transformation would involve one of the largest substrates ($M_r > 1200$) described for these enzymes.

A few of the most abundant metabolites were selected for further structure analysis. From FAB MS, one of the rabbit metabolites was found to be a dihydroxylated species with one hydroxyl group on amino acid No. 1 and the other hydroxyl group somewhere else on the ring. Total hydrolysis GC–MS revealed that amino acid No. 1 had been converted into the η(eta)-hydroxylated

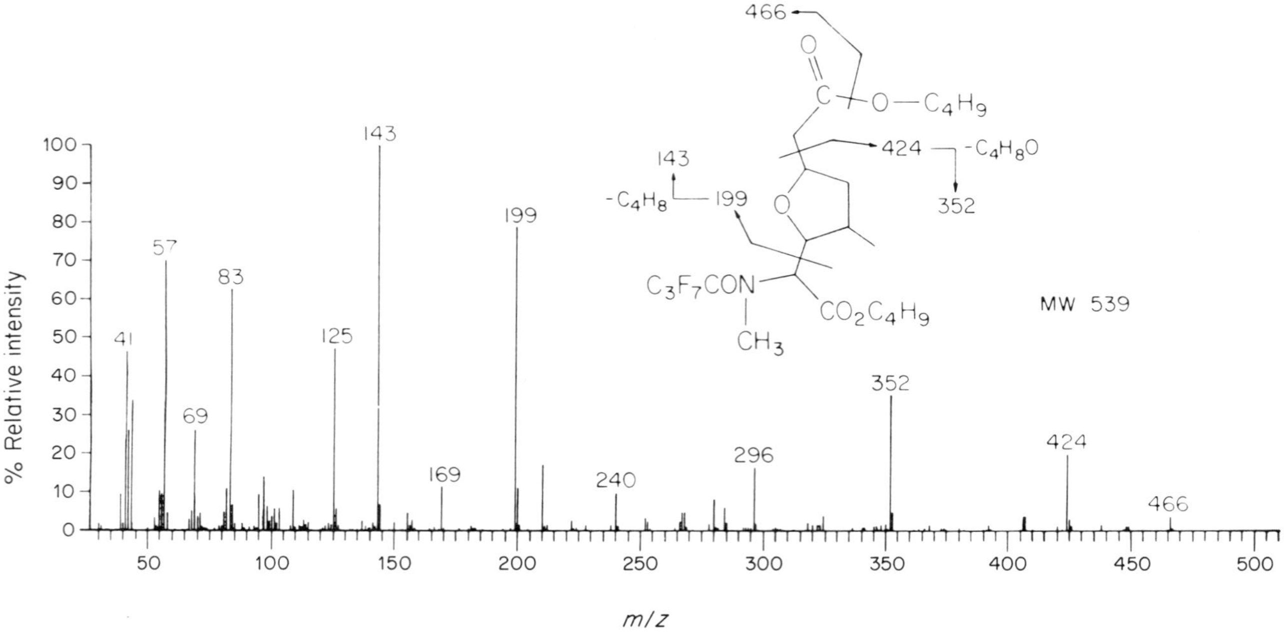

Figure 10: EI mass spectrum obtained for amino acid No. 1 of the major rabbit biliary metabolite indicating that the η (eta) carbon had been converted from a methyl to a carboxylic acid group. Note that as with the parent CsA molecule, the acid hydrolysis procedure promotes ring closure of the β-hydroxyl group with the double bond to form the cyclic ether.

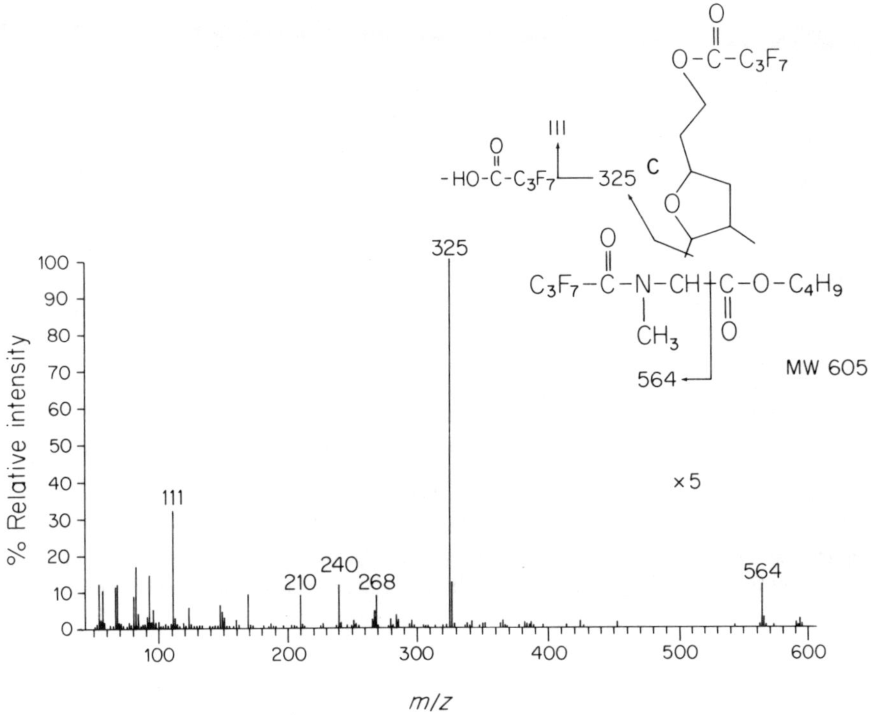

Figure 11: EI mass spectrum obtained for amino acid No. 1 of a purified dihydroxylated cyclosporine rabbit biliary metabolite indicating that metabolic hydroxylation had occurred on the η (eta) carbon.

species as deduced from the EI spectrum shown in Figure 11. The intensity of the amino acid GC peak corresponding to the *N*-methyl leucines 4,6,9 and 10 was reduced relative to CsA. The other amino acid peaks were the same and a few extra peaks appeared on the chromatogram. The EI spectrum of one of these peaks is shown in Figure 12 and is interpreted in terms of a hydroxylated *N*-methyl leucine. Under the conditions of acid hydrolysis this metabolite undergoes partial dehydration and partial lactone formation and this accounted for the other peaks in the chromatogram. The location of the hydroxyl group on the *N*-methyl leucine could not be deduced from this experiment. It was also not known which of the four *N*-methyl leucines in the CsA molecule had been hydroxylated.

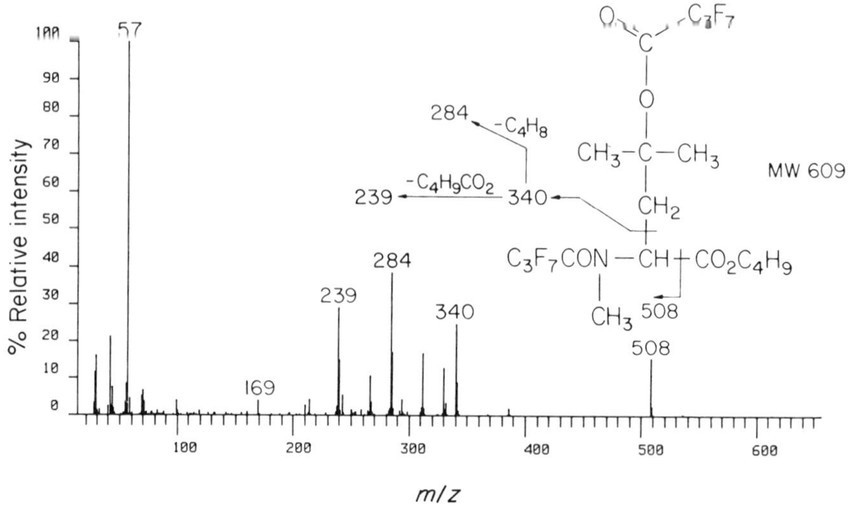

Figure 12: EI mass spectrum obtained from a hydroxylated *N*-methyl leucine after complete acid hydrolysis of a dihydroxylated cyclosporine metabolite.

When CsA was partially hydrolyzed in 6N HCl at 110°C for 45 minutes and the resulting peptides were converted into *O*-TMS polyamino alcohols (Figure 13) according to the procedure developed by Biemann (see above), GC–MS gave the total ion current chromatogram shown in Figure 14. The mass spectra of these derivatized peptides umambiguously determine their sequence. When this procedure was carried out on 300 μg of the isolated dihydroxylated CsA metabolite, the peaks for peptide 1–2 and for peptide 10–11 disappeared. The mass spectrum of a new 9–10 peptide is shown in Figure 15.

The spectrum of Figure 15 allows the location of the hydroxylation to be unambiguously assigned to the *N*-methyl leucine at amino acid No. 9. Furthermore, the intense ion at *m*/*z* 131 locates this hydroxylation to the γ-carbon of amino acid No. 9. The structure of this metabolite is therefore as shown in Figure 16. This structure had been determined previously (Maurer *et al.*, 1984) using NMR techniques but on milligram amounts of material.

A series of mass spectrometric methods, based on FAB MS, GC–MS and LC–MS have now, therefore, been developed for the complete structural elucidation of cyclosporine metabolites. As more peptide drugs are developed these techniques should prove useful for the solution of similar peptide structural problems encountered during metabolism studies.

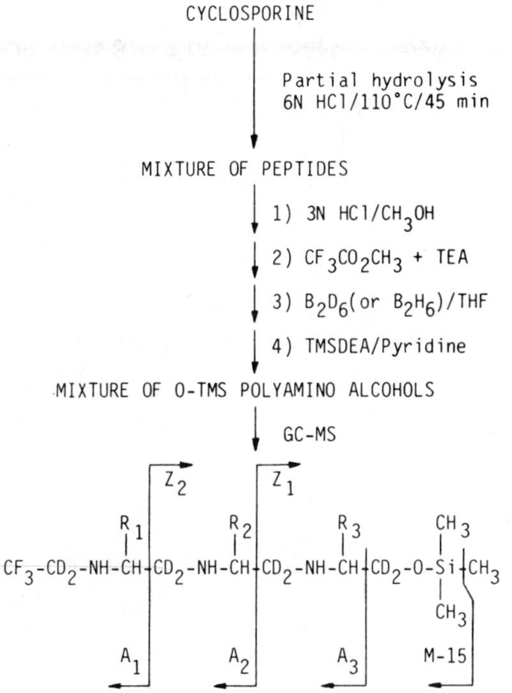

Figure 13: Partial acid hydrolysis and derivatization scheme for cyclosporine.

16.5. URINARY MANNOSE OLIGOSACCHARIDES

The chemical structures of the major urinary oligosaccharides of patients with mannosidosis have been characterized using monosaccharide composition analysis, methylation studies, acetolysis, Smith degradation and ^{13}C NMR (Matsuura *et al.*, 1981). Eight oligosaccharide fractions (tri- to decasaccharides) were obtained from the pooled urine of two patients with mannosidosis and the structures and relative amounts within each isomeric fraction of 17 oligosaccharides ($Man_xGlcNAc$; $x = 2$–9) were determined.

The EI mass spectra of each of the eight isolated oligosaccharide fractions, as their permethylated oligosaccharide alcohol mixtures, have been reported (Jardine *et al.*, 1984). It was found, in agreement with work from other laboratories (Egge *et al.*, 1982), that many of the oligosaccharide isomers can be distinguished on the basis of the EI mass spectra of these permethylated derivatives.

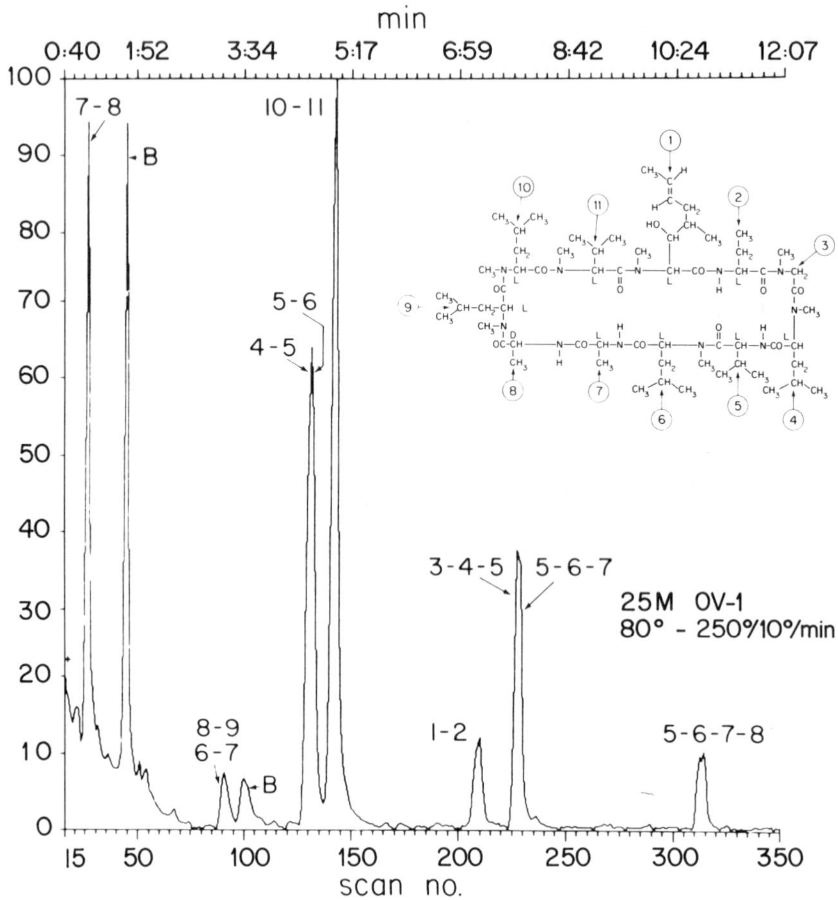

Figure 14: Total ion current GC–MS chromatogram of cyclosporine after partial hydrolysis and derivatization according to Figure 13. The peptides unambiguously identified from the resulting EI mass spectra are indicated. The peaks labelled 'B' are from background contaminants.

An example is shown in Figure 17. For the nonasaccharide mixture, isomer G_1 can be distinguished from isomers G_2 and G_3 on the basis of the ions at *m*/*z* 888 and 1034 for G_1, and *m*/*z* 595, 830, 1092 and 1266 for G_2 and G_3. The isomers G_2 and G_3 cannot be distinguished at present on this basis. Future MS/MS analysis on fragment ions from chromatographically separated G_2 and G_3 isomers may be able to accomplish such distinction. The MS/MS fragmentation of the *m*/*z* 830 ion from each isomer may differ, for example. In this study reasonable quantitative assessment of the relative amounts of isomers in the mixtures could also be made based on the ratio of diagnostic fragment ions (e.g. *m*/*z* 1092 and 888).

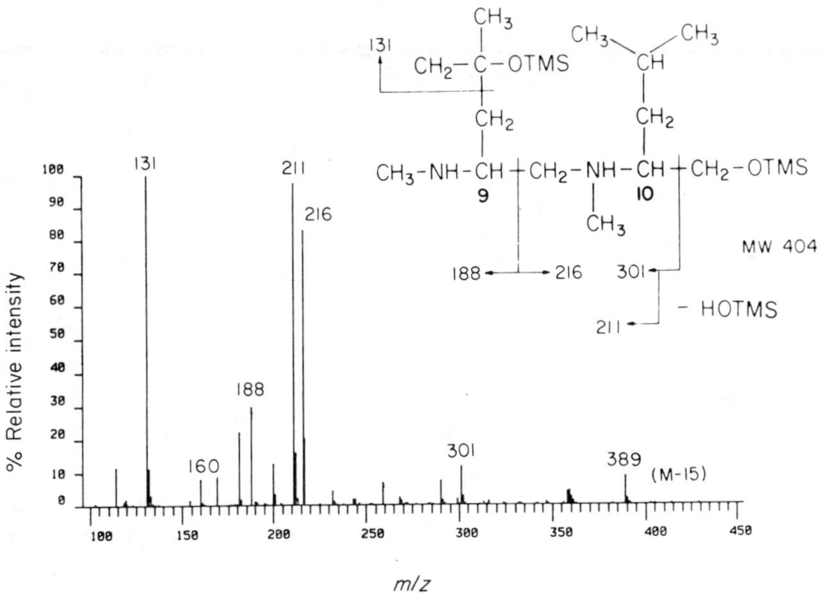

Figure 15: EI mass spectrum obtained from a partial hydrolysis digest of a dihydroxylated cyclosporine metabolite. The spectrum is specifically interpreted as resulting from a hydroxylated 9–10 peptide fragment where the hydroxylation is on the γ-carbon, as shown.

The power of EI fragmentation analysis for the examination of the structures of such complex oligosaccharides is apparent from this work. The technique, however, required a significant quantity of sample (approximately 100 μg). Developments in FAB mass spectrometry, particularly using permethylated oligosaccharides (see Dell and Panico, Chapter 10), may supersede such analysis although complementary information will presumably be available from techniques as different as EI MS and FAB MS.

16.6 BACTERIAL LIPOOLIGOSACCHARIDES

Lipopolysaccharides are biologically highly active substances from the outer membranes of Gram-negative bacterial cells (Anderson and Unger, 1983). They are of considerable compositional and structural diversity and pose challenging analytical problems. *Mycobacterium kansasii*, (the 'yellow bacillus'), for example, is characterized by the presence of seven species-specific neutral lipooligosaccharide antigens. Both the native glycolipids and their inherent oligosaccharides were purified and the major features of their unique structures were

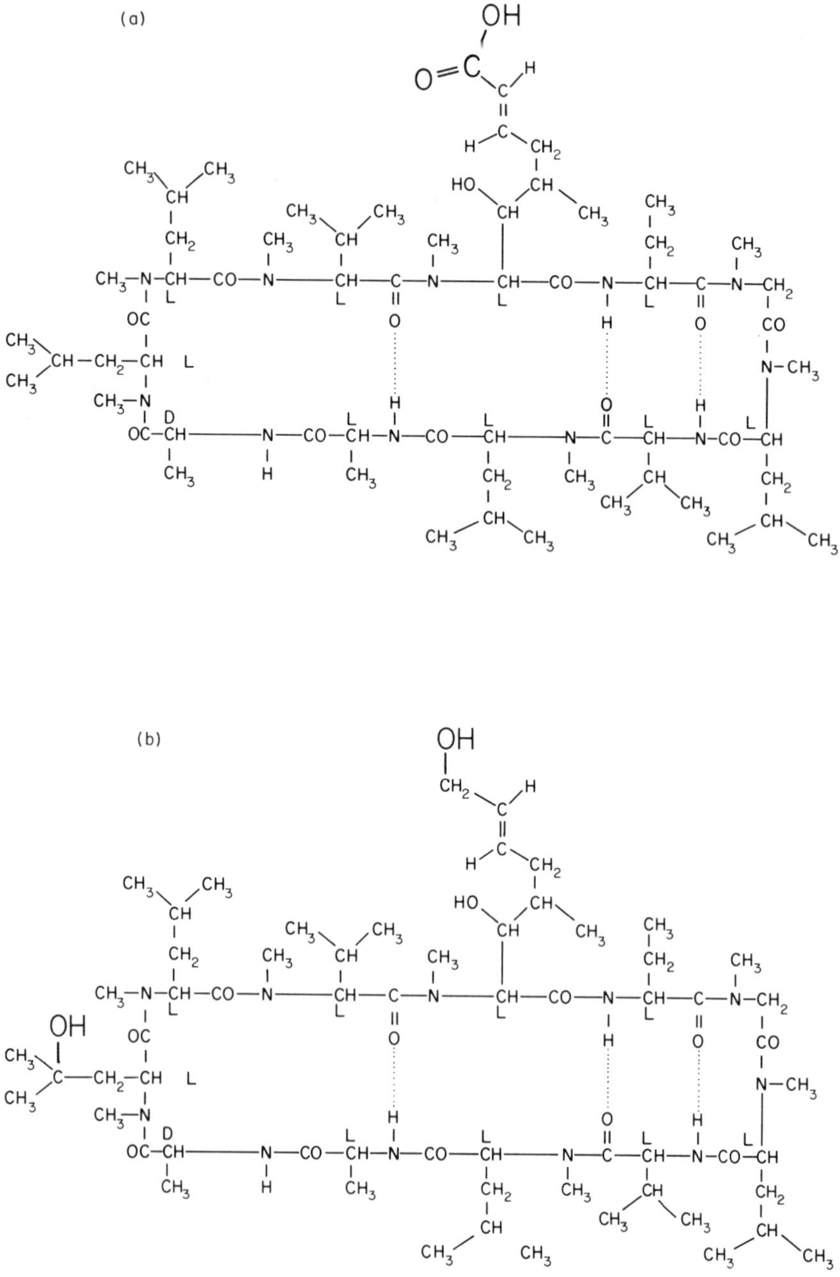

Figure 16: Structures of a carboxylic acid metabolite (a) and a dihydroxylated metabolite (b) of cyclosporine, as deduced entirely by mass spectrometric methods.

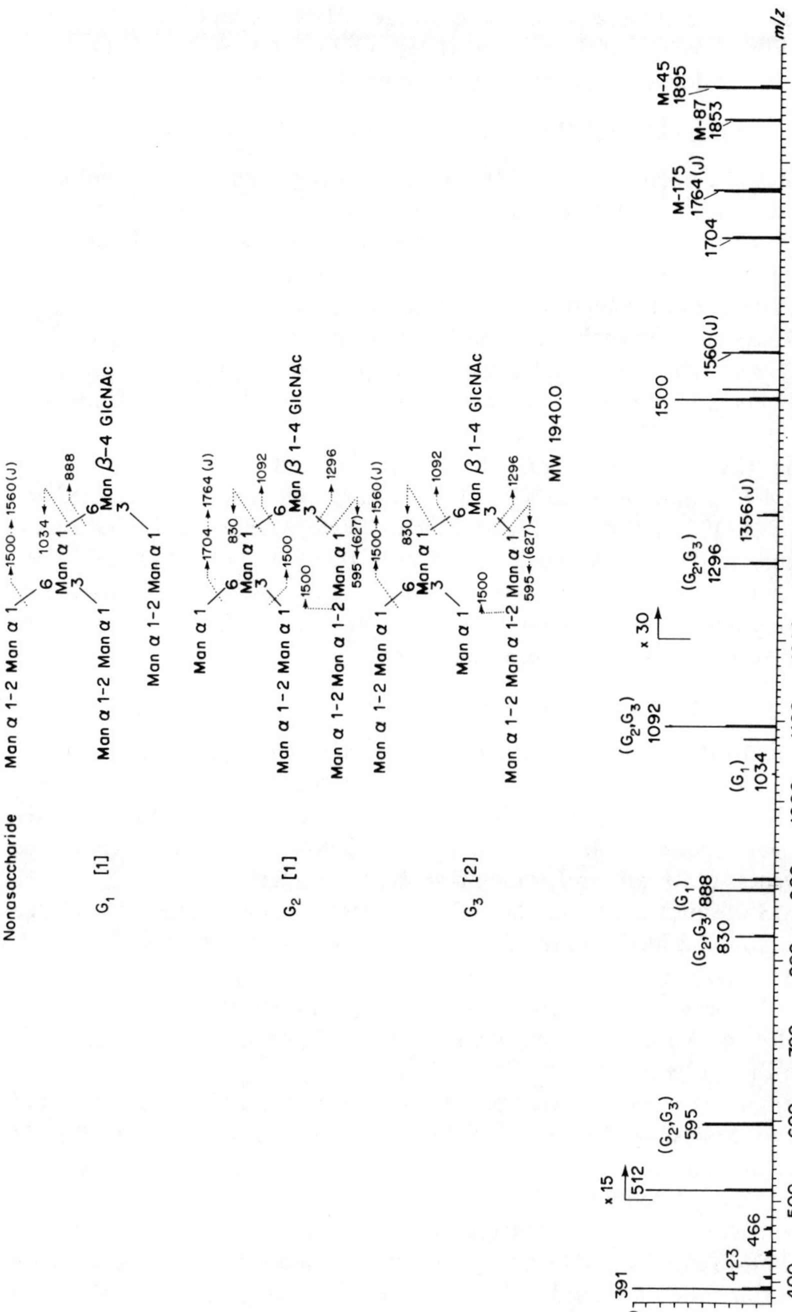

Figure 17: The 70 eV electron impact mass spectrum of a reduced and permethylated $Man_8GlcNAc$ nonasaccharide mixture isolated from the urine of a patient with mannosidosis. The relative concentrations of G_1, G_2 and G_3 as determined by other methods were approximately 1 : 1 : 2. (From Jardine *et al.* 1984.)

determined by acetolysis, partial acid cleavage, ^{1}H-NMR and ^{13}C-NMR, and CI and EI mass spectra of the permethylated products (Hunter *et al.*, 1983). In this way, it was deduced that they have in common a tetraglucose 'core',

β-*D*-Glc*p*-(1→3)-β-*D*-Glc*p*-(1→4)-α-*D*-Glc*p*-(1→1)-β-*D*-Glc*p*

where Glc*p* is glucopyranose. There is no reducing sugar in this molecule. Rather, the terminal glucose is linked in a (1→1) fashion to produce an α,α-trehalose substituent. Variable residues of xylose, 3-*O*-methylrhamnose, fucose and a novel *N*-acyl aminosugar are linked to the core, and the resulting oligosaccharides are acylated with 2,4-dimethyltetradecanoyl functions.

The actual structures of the oligosaccharide segments of the lipooligosaccharides of *M. kansasii* were then established by positive and negative fast atom bombardment mass spectrometry, acetolysis, partial acid hydrolysis, methylation analysis and NMR (Hunter *et al.*, 1985). Thus, the simplest lipooligosaccharide (LOS I′) was found to contain an additional α1→3 linked 3-*O*-methyl-*L*-rhamnopyranose (3-*O*-Me-*L*-Rha*p*) unit. Those of intermediate complexity (LOS I-III) contain an additional *D*-xylopyranose (*D*-Xyl*p*) residue or xylobiose in β1→4 linkage. Those of ultimate complexity (LOS IV–VIII) contain further *D*-Xyl*p* residues, and the distal *N*-acyl kansosamine (Kan*N*Acyl) and fucopyranosyl (Fuc*p*)-containing disaccharide, Kan*N*Acyl(1→3)Fuc*p*. Thus, for example, the structure of the oligosaccharide from LOS VII is:

Kan*N*Acyl(1→3)Fuc*p*(1→4)[-β-*D*-Xyl*p*(1→4)]$_6$-α-*L*-3-*O*-Me-Rha*p*(1→3)Glc$_4$

Polyclonal rabbit and murine monoclonal antibodies react only with the more complex Kan*N*Acyl-Fuc*p*-containing lipooligosaccharides, indicating that the Kan*N*Acyl distal end, not the trehalose end, contains the antibody binding site unique to *M. kansasii* and is responsible for the serological distinctiveness of *M. kansasii* among mycobacterial species.

FAB MS was particularly useful in this work since not only was the sequence of the oligosaccharides confirmed but the number of xylose residues in each molecule could be determined unambiguously. Classical oligosaccharide analysis left considerable doubt as to the exact number of xylose residues in each molecule. An example of the FAB MS of one of these molecules is shown in Figure 18.

The remaining structural question then was how many acyl residues are attached to these sugars in the intact LOS's and what is their location? FAB MS was unsuccessful in the analysis of the intact LOS's and so analysis by ^{252}Cf PDMS was attempted. ^{252}Cf PDMS analysis by C. J. McNeal and R. D. Macfarlane at Texas A & M University yielded interpretable mass spectra.

^{252}Cf PDMS turns out to be a very powerful technique for the direct analysis of complex lipooligosaccharides (Jardine *et al.*, 1986). In general, unambiguous molecular weight information was obtained in the positive ion mode (Figure 19). The heterogeneity within a sample, with respect to fatty acid composition,

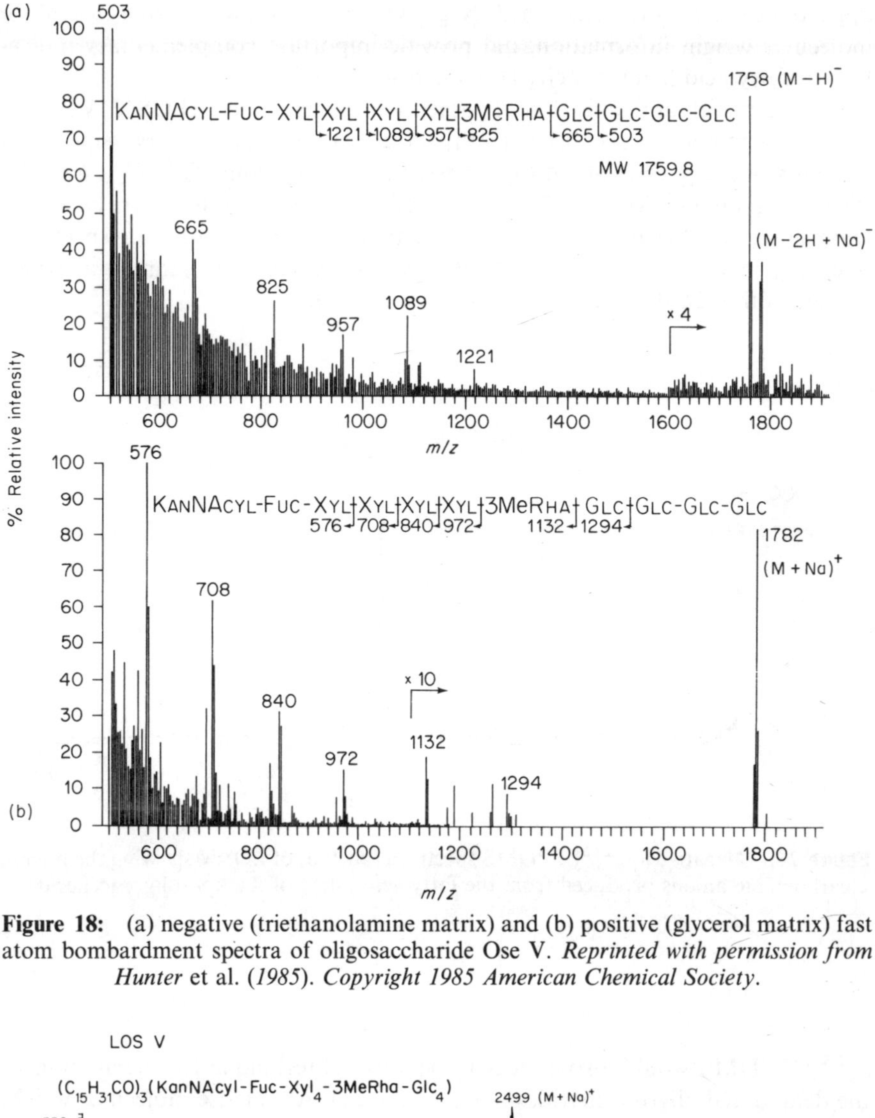

Figure 18: (a) negative (triethanolamine matrix) and (b) positive (glycerol matrix) fast atom bombardment spectra of oligosaccharide Ose V. *Reprinted with permission from Hunter* et al. (*1985*). *Copyright 1985 American Chemical Society.*

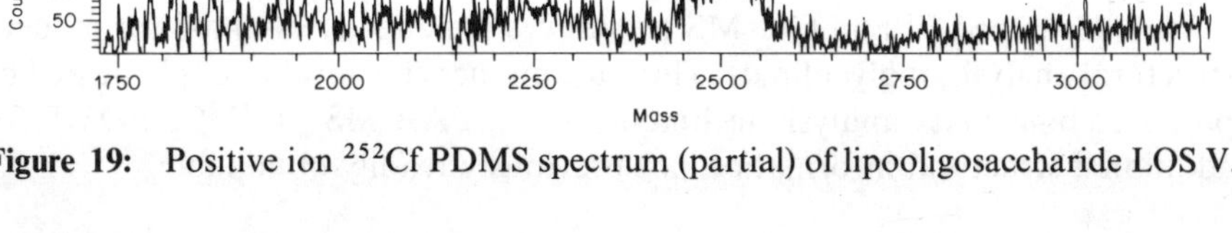

Figure 19: Positive ion ^{252}Cf PDMS spectrum (partial) of lipooligosaccharide LOS V.

appeared to be readily determined. Negative ion spectra, while not as useful for molecular weight information, did provide important complementary indication of fatty acid heterogeneity (Figure 20).

The bacterial antigenic lipooligosaccharides examined (LOS I–VIII) all appear to contain three 2,4,-dimethyltetradecanoyl (C_{16}) esters. Significant amounts of a C_{13} ester are always present along with some C_{14}, C_{15}, and C_{17} species. An intense $[M + 28]^+$ ion observed in all positive ion spectra *cannot* be from a C_{18} ester because no related anion appears in the negative ion spectra. It is most probable that the additional 28 u can be accounted for by a formyl residue, although this requires further clarification.

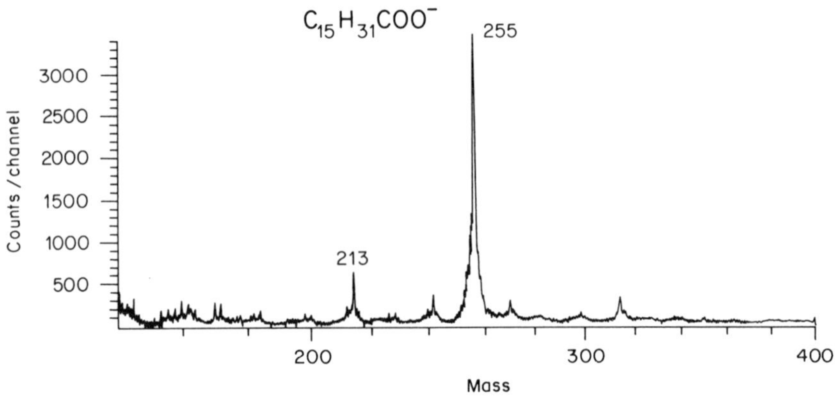

Figure 20: Negative ion ^{252}Cf PDMS spectrum (partial) of LOS V showing the intense carboxylate anions produced from the fatty acid esters of the lipo-oligosaccharide.

^{252}Cf PDMS would also appear to be a useful technique for the analysis of the delipidated oligosaccharides (e.g., Figure 21). An intense molecular weight associated ion is produced and it would be of considerable interest to compare the sensitivity of this analysis to that of the FAB MS experiment. Although numerous fragment ions were produced below *m/z* 1200, these will require further study for interpretation.

Well developed EI and GC–MS techniques are already essential for modern structural analysis of glycolipids. These techniques can now be supplemented by powerful high mass analysis techniques using FAB MS or ^{252}Cf PDMS for additional structural information not previously readily obtainable.

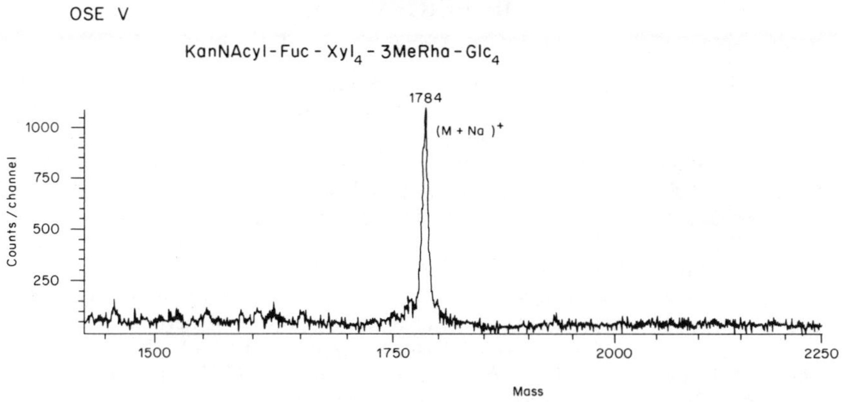

Figure 21: Partial positive ion ^{252}Cf PDMS spectrum of Ose V.

16.7. CONCLUSIONS

The application of mass spectrometry to the structural analysis of biological peptide and carbohydrate oligomers has entered a new era. As can be readily ascertained from the foregoing brief review and from other chapters in this volume, the structural problems in biomedical research which can be tackled by currently available mass spectrometry techniques are limited only by the amount and purity of sample available and by access to appropriate equipment.

ACKNOWLEDGEMENTS

The work described is currently being supported by the National Institutes of Health grant GM32928.

Each of the projects described is a collaborative effort with the following researchers and their essential contributions are gratefully acknowledged: Dr. L. J. Miller, Mayo Clinic (cholecystokinins); Dr. F. Prendergast, Mayo Clinic (amphiphilic peptides); Dr. N. Hartman, Mayo Clinic (cyclosporine metabolites); Dr. C. C. Sweeley, Michigan State University (urinary mannose oligosaccharides); Dr. P. J. Brennan, Colorado State University (bacterial lipooligosaccharides).

The author also gratefully acknowledges the ^{252}Cf PDMS contributions to this work from Drs. C. J. McNeal and R. D. Macfarlane of Texas A & M University.

REFERENCES

Anderson, L., and Unger, F. M. (1983). Bacterial Lipooligosaccharides. Structure, Synthesis and Biological Activities. *ACS Symposium Series* No. 231, American Chemical Society, Washington, D.C.

Biemann, K. (1980). Amino acid sequence in oligopeptides and proteins, *Biochemical Applications of Mass Spectrometry* (First Supplementary Volume) (Ed. Waller, G. R. and Dermer, O. C.), pp. 496–513, Wiley-Interscience, New York.

Chait, B. T., and Field, F. H. (1984). ^{252}Cf fission fragment ionization mass spectrometry of chlorophyll a, *J. Am. Chem. Soc.*, **106**, 1931–1938.

Egge, H., Michalski, J. C., and Strecker, G. (1982). Heterogeneity of urinary oligosaccharides from manosidosis: Mass spectrometric analysis of permethylated Man**9**, Man**8** and Man**7** derivatives, *Arch. Biochem. Biophys.*, **213**, 318–326.

Erickson, B. J., and Jardine, I. (1986) Implementation of the GCMS peptide sequencing program PEPALG on a personal computer for off-line analysis, *Biomed. Environ. Mass Spectrom.*, in press.

Fales, H. M., McNeal, C. J., Macfarlane, R. D., and Shimohigashi, Y. (1985). Californium-252 plasma-desorption mass spectrometry of polymethylenediamine-linked enkephalin peptides, *Anal. Chem.*, **57**, 1616–1621.

Fohlman, J., Sundqvist, B., Kamensky, I., Nyberg, F., Amberg, H., and Roepstorff, P. (1984). *Paper presented at Int. Symp. on Mass Spectrometry in the Health and Life Sciences, San Francisco, 1984.*

Fohlman, J., Peterson, P. A., Roepstorff, P., Höjrup, P., Kamensky, I., Säwe, G., Håkansson, P., and Sundqvist, B. (1985). Comparison of ^{252}Californium plasma desorption and fast atom bombardment mass spectrometry for analysis of small peptides, *Biomed. Mass Spectrom.*, **12**, 380–387.

Hartman, N. R., and Jardine, I. (1986). Mass spectrometric analysis of cyclosporine metabolites, *Biomed. Environ. Mass Spectrom.*, in press.

Herlihy, W. C., and Biemann, K. (1981). Advances in gas chromatographic mass spectrometric protein sequencing. (3) Automated interpretation of the mass spectra of the polyamino alcohol derivatives, *Biomed. Mass Spectrom.*, **8**, 70–77.

Hunter, S. W., Murphy, R. C., Clay, K., Goren, M. B., and Brennan, P. J. (1983). Trehalose-containing lipooligosaccharides. A new class of species-specific antigens from mycobacterium, *J. Biol. Chem.*, **258**, 10481–10487.

Hunter, S. W., Jardine, I., Yanagihara, D. L., and Brennan, P. J. (1985). The trehalose-containing lipooligosaccharides from mycobacteria; structures of the oligosaccharide segments and recognition of a unique N-acylkansosamine-containing epitope, *Biochemistry*, **24**, 2798–2805.

Jardine, I., Matsuura, F., and Sweeley, C. C. (1984). Electron ionization mass spectra of reduced and permethylated urinary oligosaccharides from patients with mannosidosis, *Biomed. Mass Spectrom.*, **11**, 562–567.

Jardine, I., Hunter, S. W., Brennan, P. J., McNeal, C. J., and Macfarlane, R. D. (1986). Heterogeneity of bacterial antigenic lipooligosaccharides determined by ^{252}Cf PDMS, *Biomed. Environ. Mass Spectrom.*, **13**, 273–276.

Kahan, B. D. (1984). *Cyclosporin, Biological Activity and Clinical Applications*, Grune and Stratton, Inc., Orlando, Florida.

Kato, K., Takahoshi, Y., and Matsubora, K. (1984). *Paper presented at 1st Int. Conf. on Neuronal Cholecystokinin, Brussels*, 1984.

Matsuura, F., Nunez, H. A., Grabowski, G. A., and Sweeley, C. C. (1981). Structural studies of urinary oligosaccharides from patients with mannosidosis, *Arch. Biochem. Biophys.*, **207**, 337–352.

Maurer, G., Loosli, H. R., Schreier, E., and Keller, B. (1984). Disposition of cyclosporine in several animal species and man. I. Structural elucidation of metabolites, *Drug Metab. Dispos.*, **12**, 120–126.

McNeal, C. J., Jardine, I., and Macfarlane, R. D. (1985). Unpublished results.

Miller, L. J., Jardine, I., Weissman, E., Go, V. L. W., and Speicher, D. (1984). Characterization of cholecystokinin from the human brain, *J. Neurochem.*, **43**, 835–840.

Mutt, V. (1980). Cholecystokinin: Isolation, structure, and functions, in *Gastrointestinal Hormones* (Ed. Glass, G. B. J.), pp. 169–221, Raven Press, New York.

Negra, S. D. and LeBeyec, Y. (1984). A ^{252}Cf time-of-flight mass spectrometer with improved mass resolution, *Int. J. Mass Spectrom. Ion Processes*, **61**, 21–29.

Sundqvist, B., Hedin, A., Håkansson, P., Kamensky, I., Salehpour, M., and Säwe, G. (1985). Plasma desorption mass spectrometry (PDMS). Limitations and possibilities. *Int. J. Mass Spectrom. Ion Processes*, **65**, 69–89.

Zollinger, M., and Biemann, K. (1985). Improvements of the reductive peptide sequencing method by the use of gaseous reagents, *Proc. 33rd Ann. Conf. Mass Spectrometry and Allied Topics, San Diego, California*; American Society for Mass Spectrometry; pp. 508–509.

Mass Spectrometry in Biomedical Research
Edited by S. J. Gaskell
© 1986 John Wiley & Sons Ltd.

Chapter 17

HIGH MASS FOURIER TRANSFORM MASS SPECTROMETRY: PRESENT CAPABILITIES AND FUTURE POTENTIAL

David H. Russell and Mauro E. Castro

17.1. INTRODUCTION

In the last decade there has been phenomenal growth in the field of mass spectrometry. A major reason for this growth has been the development of suitable ionization methods for large, non-volatile biomolecules (Beckey, 1977; Macfarlane and Torgerson, 1976; Benninghoven and Sichterman, 1978; Vastola *et al.*, 1976; Posthumus *et al.*, 1978; Baldwin and McLafferty, 1973; Cotter and Fenselau, 1979; Cotter, 1980; Murawski and Cook, 1984; Yergey and Cotter, 1982; Barber *et al.*, 1982). While the introduction of these ionization methods has been a virtual panacea to bioanalytical mass spectrometry, the ability to ionize and produce intact molecular ions of large biomolecules has increased the demand for instruments with extended mass range. Approaches for increasing the mass range of magnetic sector instruments are discussed by Green and Bordoli in Chapter 13. A second type of instrument with extended mass range is the time-of-flight (TOF) mass spectrometer (Macfarlane, 1983). The TOF instrument has better sensitivity than sector instruments, and due to simpler ion optics is better suited for the detection of large ions. However, the resolution of the TOF instrument is significantly less than that available with a magnetic sector instrument.

A third type of instrument with considerable potential for the analysis of large molecules is the Fourier transform (FT) mass spectrometer. The FT mass

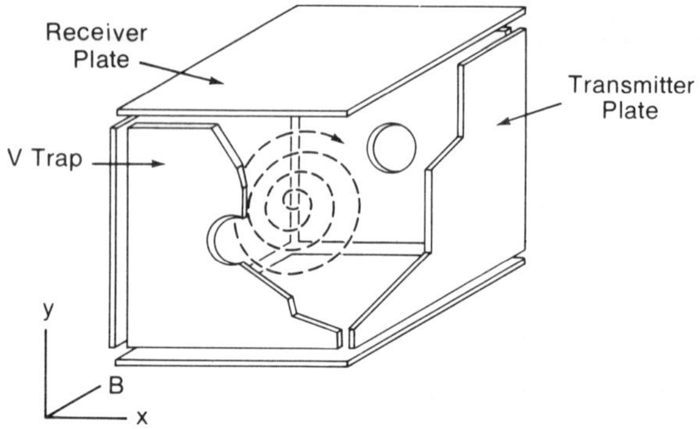

Figure 1: Schematic drawing of a cubic trapped ion cell commonly used for Fourier transform mass spectrometry experiments.

spectrometer is based on the principles of ion cyclotron resonance (Beauchamp, 1971). That is, an ion trapped within a strong magnetic field will acquire a circular motion in a plane that is perpendicular to the direction of the magnetic field (Figure 1). The angular frequency of the ion's circular motion, i.e., the cyclotron frequency (ω_c), is given by equation (1), where e is the charge on the ion, B is the magnetic field strength, and m is the mass of the ion. It is

$$\omega_c = eB/m \qquad (1)$$

important to note that ω_c is inversely proportional to mass and independent of the velocity of the ion (Cody *et al.*, 1982). The basic concepts of FTMS have been discussed in several reports (Comisarow and Marshall, 1974; Parisod and Comisarow, 1980; Marshall, 1979; Ghaderi *et al.*, 1981; Marshall *et al.*, 1979) and review articles (Wilkins and Gross, 1981; Wanczek, 1984; Johlman *et al.*, 1983; Gross and Rempel, 1984) and will not be covered in detail in this paper. Briefly, the coherent circular motion of the ions creates an image current in the signal receiver plates. The image current induced by the coherent ion packet will oscillate at the cyclotron frequency of the ions. The time-domain signal (Figure 2a) is then Fourier transformed to yield a frequency-domain signal which can be converted to a mass spectrum (Figure 2b) using equation (1).

The FTMS experiment is dramatically different from the conventional mass spectrometric experiment. An FTMS experimental sequence is shown in Figure 3. The first step in the experiment is to remove residual ions from the ion trap; this is accomplished by applying potentials (of opposite polarity) to the two electrostatic trapping plates of the ion cell (Figure 1). Following the ion quench step, the ionizing radiation (electron beam, photon beam, Cs^+ ion beam, etc.) is pulsed 'on' for a short period of time, typically 1–5 ms. Following the ionization step the ions can be detected promptly or additional steps can be introduced to

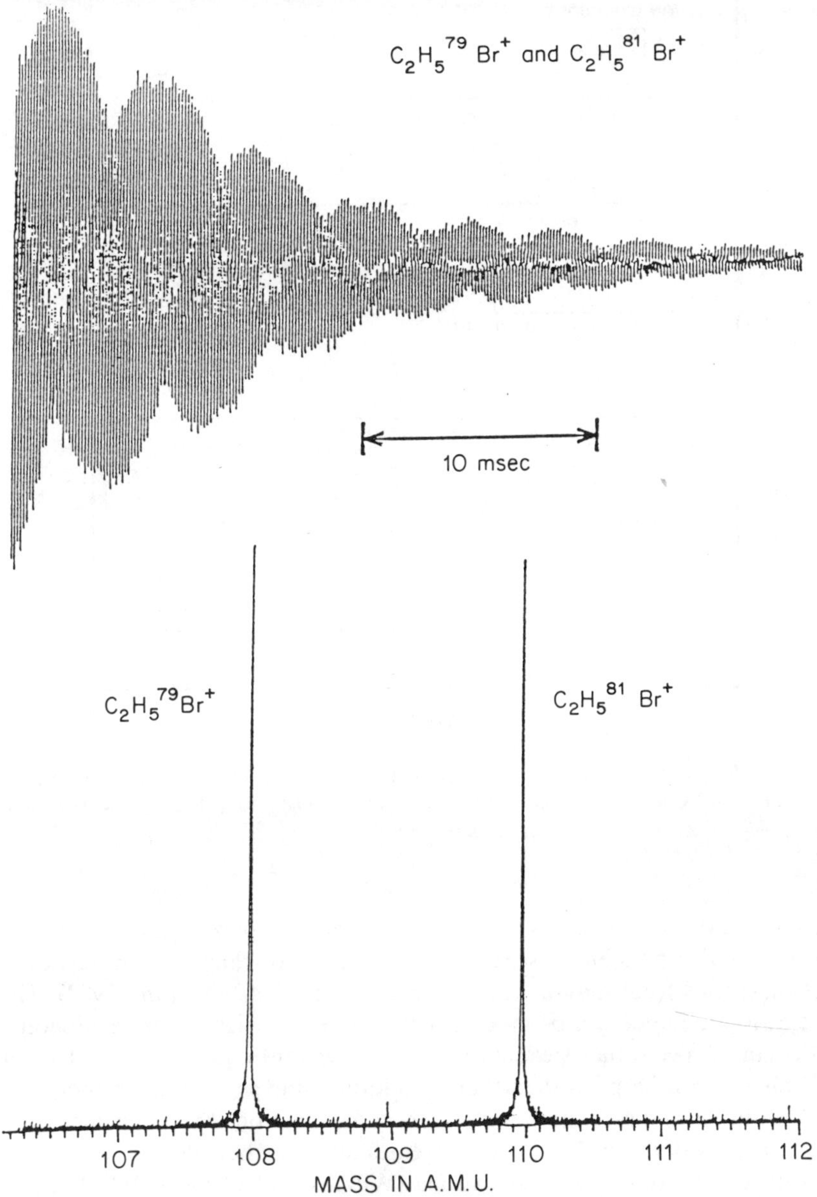

Figure 2: Time and frequency domain signals for two ions ($C_3H_5{}^{79}Br^+$ and $C_2H_5{}^{81}Br^+$) of different m/z ratio. *Reprinted with permission from Marshall (1985). Copyright 1985 American Chemical Society.*

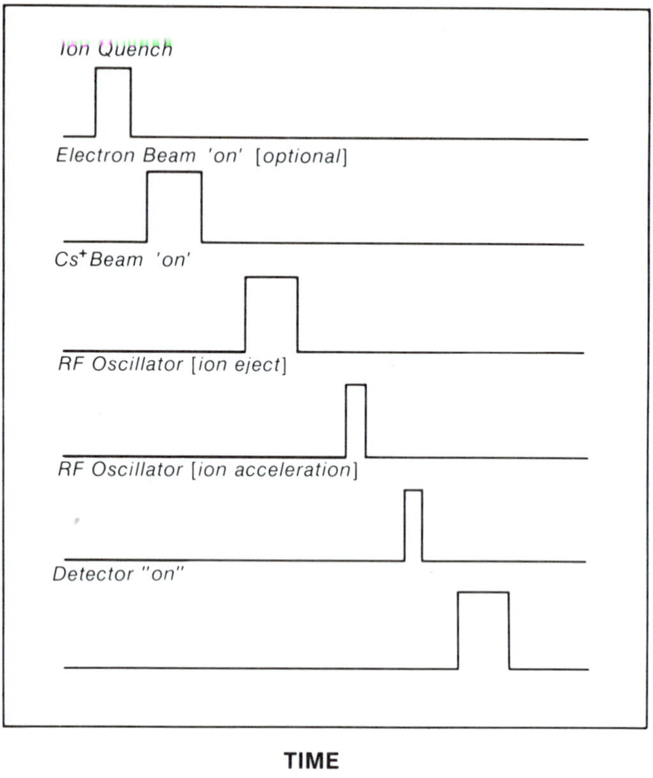

Figure 3: Sequence of pulses used to obtain an FT mass spectrum. Note that 'optional' pulses are controlled by the computer and can be used to extend the experimental capabilities.

allow ion–molecule reactions to occur, i.e., chemical ionization (Ghaderi *et al.*, 1981). It is also possible to select a particular ion or range of ion masses and eject these ions from the ion cell (Beauchamp, 1971; Cody *et al.*, 1982). Thus, ion ejection methods can be used to mass-select a particular ion, analogous to tandem mass spectrometry experiments with a sector or quadrupole instrument, and this ion can be probed further by additional experimental methods, e.g., collision-induced dissociation and/or photodissociation. Although such experiments are used routinely for fundamental studies, they also have particular advantages for analytical applications (Wilkins and Gross, 1981; Wanczek, 1984; Johlman *et al.*, 1983; Gross and Rempel, 1984). The total time required to complete the experimental sequence ranges from a few milliseconds to several seconds depending on whether time delays are introduced to allow ion–molecule reactions to occur.

Although the high/ultra-high resolution capabilities of FTMS are the most frequently cited advantages of the method, it is our view that these considerations are of secondary importance. Of greater importance is the versatility and adaptability of the method to a wide variety of experiments. Such experiments (photodissociation, laser desorption, multi-photon ionization, etc.) are all potentially of great analytical utility for structural characterization of large biomolecules. In this respect FTMS and time-of-flight methods share many similarities (Macfarlane, 1984; Holland *et al.*, 1983). For example, ionization and ion detection are separated in time and the ionization is performed by a pulsed (low duty cycle) excitation source. Also, a complete mass spectrum (e.g., m/z 50–5000) (broadband or low mass resolution mode) is obtained for each ionization event. FTMS has the added advantage of viewing a small mass range at high mass resolution. With both time-of-flight and FTMS the mass spectra can be signal-averaged to enhance the signal-to-noise ratio benefiting from the Fellgett advantage.

The basic concept of the time-of-flight mass spectrometer has distinct advantages over sector instruments, and these same considerations apply to the FTMS design (Holland *et al.*, 1983). For example, the systems are mechanically much less complex than sector instruments. Also, the instruments utilize pulsed (low duty cycle) ionization and the entire mass spectrum is viewed for each ionization event. The advantage of pulsed (low duty cycle) ionization is the low rate of sample consumption, which means that smaller sample sizes can be used. For example, a typical sample size needed for recording an entire mass spectrum with a sector instrument (at a scan rate of 10 s/decade) is *ca.* 1 μg. If we assume comparable sample consumption rates (e.g., 0.1 μg s^{-1}), a single-shot mass spectrum with pulsed ionization (pulse duration of 1 μs) corresponds to consumption of 1×10^{-13} g per pulse, a gain in sensitivity of *ca.* 10^7. This assessment of the sensitivity gain is conservative because we have not considered the ion transmission efficiency of the time-of-flight (> 10%) and FT instruments (near 100%) in comparison to the sector instruments (0.001–0.1%). With the pulsed ionization methods it is common practice to signal-average the mass spectrum for several ionization steps, e.g., 10–1000 spectra can be averaged. Even with signal averaging pulsed ionization can improve the detection levels by a factor of 10^2–10^5. For example, Niehuis *et al.* (1985) have recently demonstrated femtomole detection levels for small peptides such as gramicidin S.

There are two important points regarding the detection of large molecular ions by FTMS which should be noted (Marshall *et al.*, 1979; McIver *et al.*, 1980).

(i) The mass range of the FT mass spectrometer (given by equation (2)) is directly proportional to the magnetic field strength and inversely proportional to the minimum detectable frequency (f_{min}Hz). (As the mass of the analyte ion increases the cyclotron frequency of the ion decreases.) It is the ability to detect low frequency signals (in the 1–10 kHZ range) which ultimately limits the upper mass range of FTMS. For example, the FTMS system currently in use in our laboratory is equipped with a 3 Tesla magnet. If we assume that the detector system has a low frequency cut-off of 10 kHz, the upper mass limit is *ca.* m/z 4610. With modifications to the detector system it is possible to work in the 3–4 kHz frequency range, i.e., m/z 12,000–14,000 (Castro *et al.*, 1985). On the other hand, replacing the 3 Tesla magnet with an 8 Tesla system, and maintaining the working range of 3–4 kHz, increases the upper mass limit to *ca.* m/z 35,000.

$$m/z \text{ (max)} = 1.536 \times 10^7 \ B/f_{\min} \qquad (2)$$

(ii) The limiting mass resolution (given by equation (3) is determined by the observation time (t) of the time-domain signal, and this parameter is limited by the vacuum system (Marshall *et al.*, 1979; McIver *et al.*, 1980). That is, the time-domain signal decays by the ions losing coherence via collisions with the residual neutrals in the vacuum analyzer.

$$M/\Delta M = 1.733 \times 10^4 \ Bt/m \qquad (3)$$

The above discussion suggests that the ideal instrument for the mass spectrometry of large molecules should possess the following attributes: (1) pulsed (low duty cycle) ionization; (2) extended mass range, i.e., beyond 10,000 daltons; (3) ability to view the entire mass spectrum for each ionization event; (4) not be so complex as to preclude its use by the non-specialist. In the following sections of this chapter we will discuss instrument designs used for high mass studies, the present status of instrument development, and some future directions which should be evaluated.

17.2. DESIGN CONSIDERATIONS FOR HIGH MASS FTMS

The major consideration in designing an FTMS system for high mass work is the compatibility of the ionization source with the high vacuum requirements. The long signal observation times (t in equation (3)) required for ultra-high resolution have been achieved experimentally, but only when the analyzer vacuum is maintained below 10^{-8} Torr. Also, additional probe experiments (e.g., collision-induced dissociation and/or photodisssociation) require that the ions be trapped in the ion cell for long periods of time (10's of milliseconds to seconds), and the trapping efficiency of the ion cell decreases significantly at pressures above 10^{-7} Torr. Based on these considerations, there are two approaches to designing a high mass FTMS system: (i) use an ionization method which does not contribute neutral species to the vacuum analyzer, or (ii) produce the ions outside the FTMS analyzer and then transport the ions to the ion cell. Both of these general approaches have been used successfully.

Approach (i) is accomplished by using laser desorption ionization (McCrery *et al.*, 1982; Wilkins *et al.*, 1985; Shomo *et al.*, 1985) and kilovolt Cs^+ ion desorption ionization (Castro and Russell, 1984) and (ii) is performed by using either a tandem quadrupole–FTMS (QFT) apparatus (McIver *et al.*, 1985) or a two-section trapped ion cell (Ghaderi and Littlejohn, 1985). It is important to note that laser and Cs^+ desorption ionization are performed on solid samples. In contrast, the QFT and two-section trapped ion cell experiments can be performed by using either solid-state or liquid matrix ionization methods.

For high mass applications it is essential that superconducting high field magnet systems be used. Aspects of the construction of these systems (large, bulky cryogenic dewars and limited size of the room temperature bore) render the mechanical design of the system rather critical. Figure 4 shows three systems currently used for large molecule work by FTMS. The laser desorption ionization apparatus designed by Hein and Cody (1986) is modelled after the system used by McCrery *et al.* (1982) for the original laser desorption ionization–FTMS studies. The Hein and Cody system has been used extensively by both Wilkins *et al.* (1985) and Shomo *et al.* (1985) for a variety of studies. The laser system used for this work is a Tachisto TAC II model 215G CO_2 TEA gas laser. The laser delivers power densities of *ca.* 10^8 W cm^{-2} to the probe tip (0.3 J per 40 ns pulse at 10.6 μm with a repetition rate of 1 Hz.). The laser beam enters the vacuum system through a zinc selenide window and travels parallel to the magnetic field, striking the probe tip. The probe tip is positioned at the fringes of the ion cell; thus ions are formed external to the ion cell and only those ions which penetrate the cell are trapped and subsequently detected. The effect of positioning of the probe tip with respect to the ion cell has been studied, and increased secondary ion yields are obtained if the probe tip is within 1 mm of the ion cell (McCrery and Gross, 1985). Sufficient ions are produced to allow excellent signal-to-noise ratios for data collection from a single laser shot; thus, signal averaging is not required (McCrery *et al.*, 1985). The high power laser pulse generates abundant neutral species (analyzer pressure rises to greater than 10^{-5} Torr for a period of milliseconds) and to increase the mass resolution a delay of 3–10 seconds between laser pulse and detection is used to allow removal of the neutrals (Wilkins *et al.*, 1985).

The Cs^+ ion desorption ionization source adapted to FTMS by Castro and Russell (1984) is similar in design to the source used by Standing *et al.* (1982) and by Aberth *et al.* (1982). Cs^+ ions are generated by thermal heating (to *ca.* 1000°C) of a caesium alumino-silicate matrix. The Cs^+ ions are extracted from the solid matrix by applying a negative bias to the pulsing grid, and the ions are then accelerated and focused by the two-element lens assembly. Using this arrangement, Cs^+ ion beams of 10^{-9}–10^{-6} A cm^{-2} can be obtained, and the Cs^+ ion beam can be pulsed at a relatively high repetition rate (Hz–kHZ) with a low duty cycle (μs–ms). Typical operating conditions are Cs^+ beam energy of 5–10 keV, beam currents of 10^{-9}–10^{-8} A cm^{-}, pulse duration of 0.1–5 ms, and

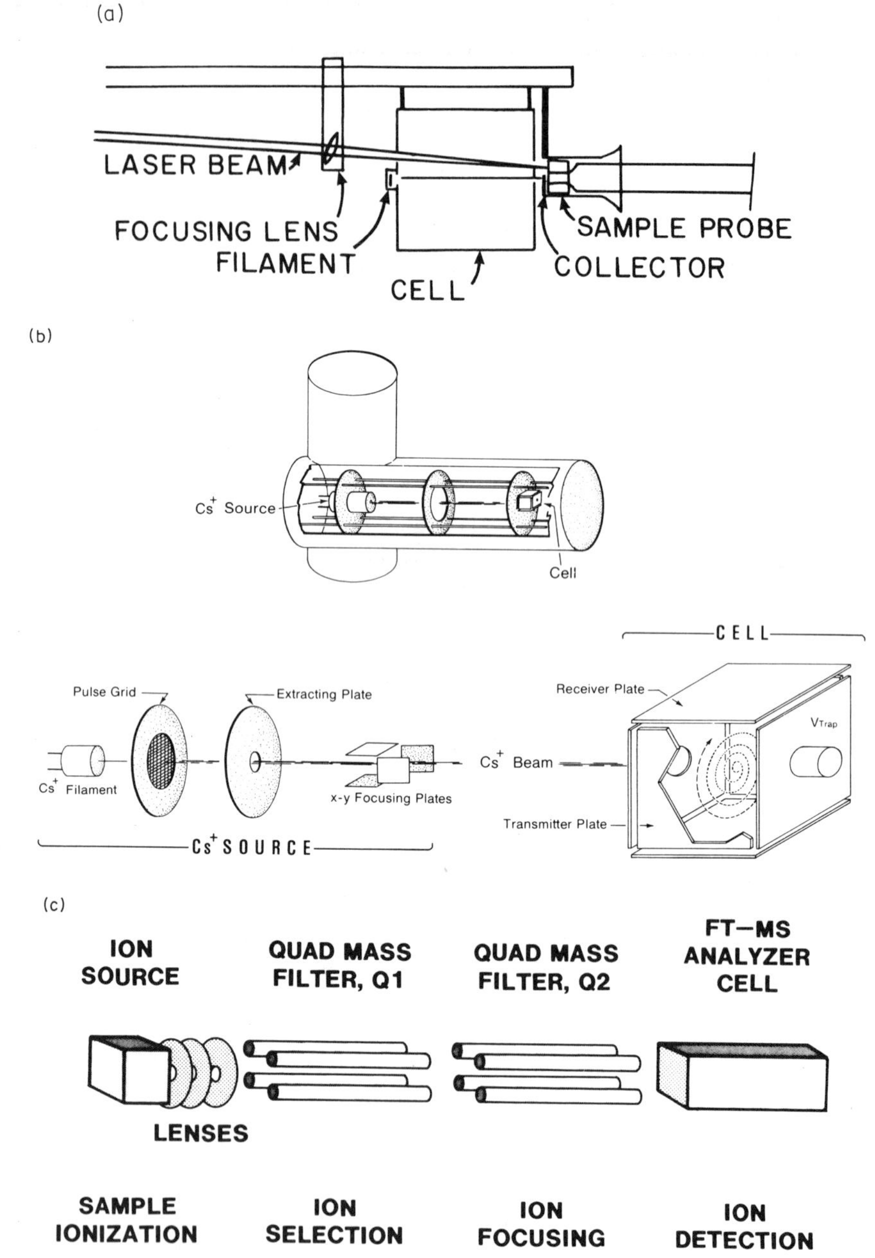

Figure 4: Schematic diagrams of the experimental systems used for (a) laser desorption (*Wilkins* et al. (*1985*), *with permission*); *(b)* Cs^+ desorption ionization-FTMS; and (c) tandem quadrupole-FTMS. *Reprinted with permission from Hunt* et al. (*1985*). *Copyright 1985 American Chemical Society.*

repetition rates of 0.1–10 Hz. Generally the secondary ion yield is sufficient for single scan data acquisition; however, it is common to signal-average the data obtained from 10–100 ionization events. In the Cs^+ desorption ionization experiment there is no detectable increase in the analyzer pressure when the ion gun is turned on; however, if the sample is irradiated repeatedly for a long period of time (several minutes) the analyzer vacuum slowly increases, generally from a baseline vacuum of 1×10^{-8} Torr to as much as 1×10^{-7} Torr.

In the tandem quadrupole–FTMS apparatus, Cs^+ desorption ionization a liquid matrix is used for ion production (Hunt *et al.*, 1985a,b). The novelty of the QFT apparatus is that ions are produced external to the FTMS ion cell; thus complications involving residual neutrals are avoided entirely. The quadrupole mass spectrometer system serves as an ion focusing system to inject the ions into the ion cell. This approach, devised by McIver *et al.* (1985), has produced very impressive results to date (Hunt *et al.*, 1985a,b).

Another method which should be comparable to the QFT apparatus is the newly designed two-section ion cell developed by Ghaderi and Littlejohn (1985). In this apparatus two cubic ion cells are positioned inside the magnetic field and the system is differentially pumped (Figure 5). The differential pumping arrangement permits one ion cell to be operated at relatively high pressure (e.g., 10^{-6} Torr) and the other ion cell to be maintained at ultra-high vacuum (e.g., $< 10^{-8}$ Torr). The ionizing radiation (electron beam, laser beam, Cs^+ beam, etc.) forms the ions over the length of both ion cells and the ion packet moves freely between the two ion cells. At the appropriate time a potential is applied to the centre plate between the two cells, and approximately one-half of the ion population is trapped in the high vacuum cell. The ions so contained are then

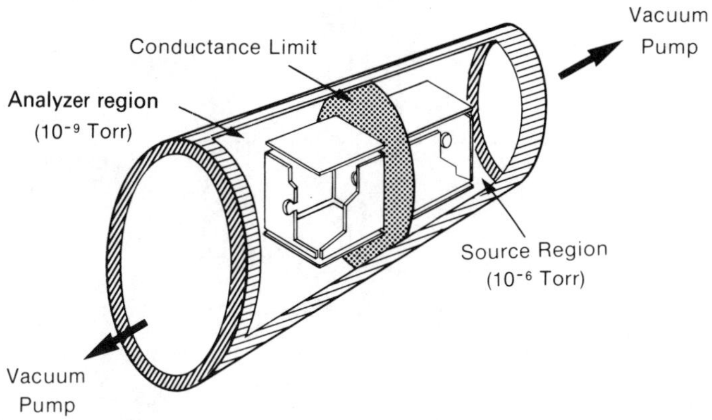

Figure 5: Schematic drawing of the two-section ion cell developed by Ghaderi and Littlejohn (1985).

detected under optimum vacuum conditions. Using this apparatus Cody *et al.* (1985) have demonstrated spectacular resolution of large molecular ions ($> m/z$ 1000) produced by laser desorption. It should also be possible to perform fast atom bombardment (FAB) ionization and liquid SIMS experiments using this arrangement; however, to date such experiments have not been performed.

17.3. DETECTION OF HIGH MASS IONS BY FTMS

There are two specific aspects concerning the detection of high mass ions by FTMS which we wish to emphasize. First, the physics of ion detection by FTMS is more favourable than for conventional ion detectors. Secondly, the long time scale of the experiment may have adverse effects on the sensitivity of FTMS as a high mass ion detector. In addition, the longer time scale of the FTMS experiment may influence the mass spectrum, i.e., the relative abundances of fragment ions arising from dissociation reactions associated with slow rate constants may be enhanced relative to those obtained with a sector mass spectrometer.

Examination of the fundamental considerations for detection of high mass ions suggests that the FTMS detection scheme should be unparalleled. For example, the detection of high mass ions by a method which requires the production of secondary electrons (electron multiplier or other charged particle

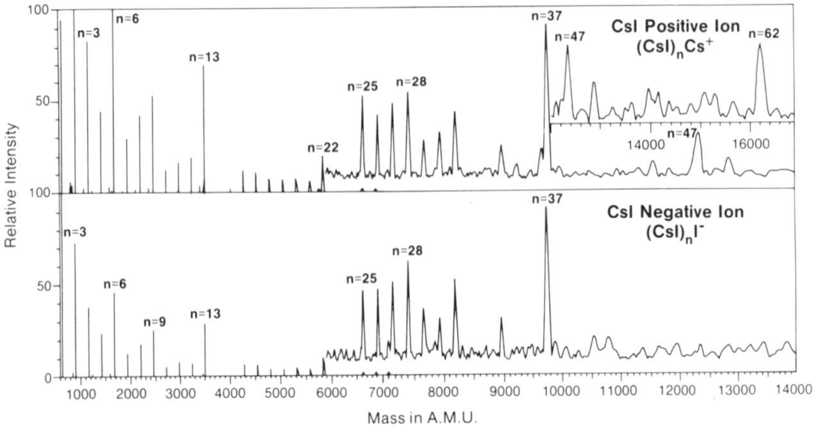

Figure 6: FT mass spectrum of CsI obtained by Cs^+ desorption ionization–FTMS. The highest mass ions detected correspond to $Cs(CsI)_n{}^+$ where $n = 62$. (Data were obtained by J. Amster, Cornell University).

detectors) is subject to mass discrimination (Beuhler and Friedman, 1980). The source of the discrimination is the reduction in the yield of secondary electrons as the velocity of the incident ion decreases (at constant kinetic energy the velocity of the ion is proportional to $m^{-1/2}$). As noted previously, the detection efficiency for FTMS is independent of the kinetic energy of the ion and depends only on the image current induced by the coherent ion packet, a quantity which is dependent only upon the density of charge particles. To a limited degree this principle is demonstrated by the results of Cs^+ desorption ionization of solid CsI (Castro *et al.*, 1985). In this study $Cs(CsI)_n{}^+$ ions were detected to m/z 16241, $n = 62$ (Figure 6). Comparison of the relative abundances of the cluster ions from the Cs^+ desorption ionization FTMS studies and SIMS studies of Ens *et al.* (1983) (Cs^+ ion SIMS with time-of-flight detection) lends credence to this idea. For example, Ens reports relative abundances for the $n = 7$–13 cluster ions as 2.3% (relative to $Cs(CsI)_1{}^+$), and the value obtained by FTMS is approximately 25%. The relative abundance for the $n = 14$–22 cluster ions is reported as 0.37%, as against 7% for FTMS. For the cluster ions $n =$ 23–37, Ens reports a value of 0.08%, and the value obtained by FTMS is ~1.4%. A similar enhancement in the relative abundances is observed for the negative cluster ions $(CsI)_nI^-$. These data suggest that the mass discrimination of the electron multiplier detector reduces the high mass sensitivity by a factor of approximately 20. Similar estimates in the detector discrimination have been made based on post-acceleration detector response (unpublished data).

A potential drawback to the detection of high mass ions by FTMS has to do with the time scale of the experiment. Recall that the time scale of the FTMS experiment ranges from milliseconds to seconds. That is, the ions are detected several milliseconds after ionization. Furthermore, the signal observation time ranges from several milliseconds to several seconds (dependent upon the desired resolution).

Chait and Field (1984) have recently reported data for the analysis of chlorophyll a which suggest that a substantial fraction (> 99%) of the molecular ions formed by desorption ionization (specifically ^{252}Cf fission fragment ionization or plasma desorption) undergo unimolecular dissociation with lifetimes of less than a few microseconds. This result has important implications for FTMS studies of large molecules: what fraction of the initially formed molecular ions will survive for the time period required for ion detection?

Although this question cannot be fully answered at this time, the $Cs(CsI)_n{}^+$ and $(CsI)_nI^-$ cluster ion data do shed some light on the general question. For example, Barlak was the first to report on the anomalous cluster ion intensities (measured with a magnetic sector mass spectrometer) for the $Cs(CsI)_n{}^+$ cluster ions (Barlak *et al.*, 1982; Campana *et al.*, 1984). Conversely, the intensities for the $Cs(CsI)_n{}^+$ cluster ions reported by Ens *et al.* (1983) (measured with a time-of-flight mass spectrometer) decreased monotonically with increasing mass. Ens attributed the anomalous cluster ion intensities to slow unimolecular dissociation reactions or collision-induced dissociation reactions during the flight time (several hundred microseconds) from ion source to detector in the instrument used by Barlak. Based on Ens *et al.* (1983) interpretation of the data and the much longer FTMS time-scale, one would expect the anomalous cluster ion intensities to be more pronounced, and indeed this is the case. For example, in the low mass region of the Cs^+ desorption ionization FT mass spectrum of CsI the $n = 10$–12 cluster ions are completely absent, as are the $n = 13$–15 cluster ions (Castro and Russell, 1985).

One additional point concerning the work on the $Cs(CsI)_n{}^+$ cluster ions concerns the mass resolution at high mass. Under the experimental conditions used for this work, the mass resolution (broadband mode) for the m/z 12,864 ion is unexpectedly low ($M/\Delta M = 130$). The mass resolution could be degraded if a substantial fraction of the ions underwent slow dissociation reactions. Recall that the mass resolution (given by equation (3)) is proportional to the observation time of the time-domain signal, and the time-domain signal decays by the ions undergoing collisions with residual neutrals. If the ion population is reduced by the occurrence of unimolecular or collision-induced dissociation, the time domain signal would decay at an accelerated rate, resulting in degradation of the mass resolution. Based on other studies it appears that, even at an analyzer vacuum of $<1 \times 10^{-8}$ Torr, the $Cs(CsI)_n{}^+$ ions readily undergo collision induced dissociation (Castro and Russell, 1985). This explanation is consistent with the exceptionally large CID cross-sections for $Cs(CsI)_n{}^+$ reported by Amster *et al.* (1983). Although there are no known examples of such behaviour for large organic molecules, more systematic studies must be performed, and are under way in our laboratory.

These studies can be summarized in the following manner. The data for the $Cs(CsI)_n{}^+$ cluster ions illustrate two important points: (i) detector response is dependent only upon the charge density of the ion packet and is independent of the mass of the ion being detected; and (ii) the longer time-scale of the FTMS experiment amplifies the anomalous cluster ion intensities. The latter consideration is important and illustrates that differences in relative abundances of molecular ions and/or fragment ions may occur for mass spectra obtained by FTMS *vis-à-vis* sector mass spectrometers. The magnitude of these differences will depend upon the dynamics of the dissociation reactions of large molecular ions formed by desorption ionization methods. At this time so little is known about the dissociation dynamics of large molecules that it is impossible to evaluate the importance of this factor. For example, the chlorophyll a data reported by Chait and Field (1984) may not be a general case but rather a result of the energy deposition function associated with the MeV particle-induced desorption process (Tabet *et al.*, 1985). Clearly, more definitive studies are needed to resolve this important issue.

17.4. ANALYSIS OF LARGE ORGANIC MOLECULES

Wilkins and co-workers (1985) have reported some impressive results from studies of relatively large organic molecules by laser desorption FTMS. This group has used the fluorinated polymer Krytox 16140 to demonstrate the mass range of the method (ions in excess of *m/z* 7000 have been detected), but more impressive are the data for small peptides, e.g., leucine enkephalin (M_r 555), angiotensin II (M_r 1045), and bradykinin (M_r 1059). These mass spectra (Figure 7), taken in the broadband, low resolution mode, contain abundant molecular ion signals, usually $(M + Na)^+$or $(M + K)^+$, and structurally significant fragment ions. For model tetra- and pentapeptides both acylium N-terminus sequence ions and C-terminus sequence ions were observed. However, in the case of angiotensin II, no C-terminus sequence ions (as reported for FAB MS; Konig *et al.*, 1983) were observed in the positive ion FT mass spectra, but the laser desorption FT mass spectra did contain abundant N-terminus sequence ions. Detailed comparison of the laser desorption FT mass spectra and FAB MS spectra (taken on a magnetic sector mass spectrometer) show significant differences. For example, more abundant molecular ions (expressed as a percentage of the total ion current) are observed by laser desorption FTMS

(a)

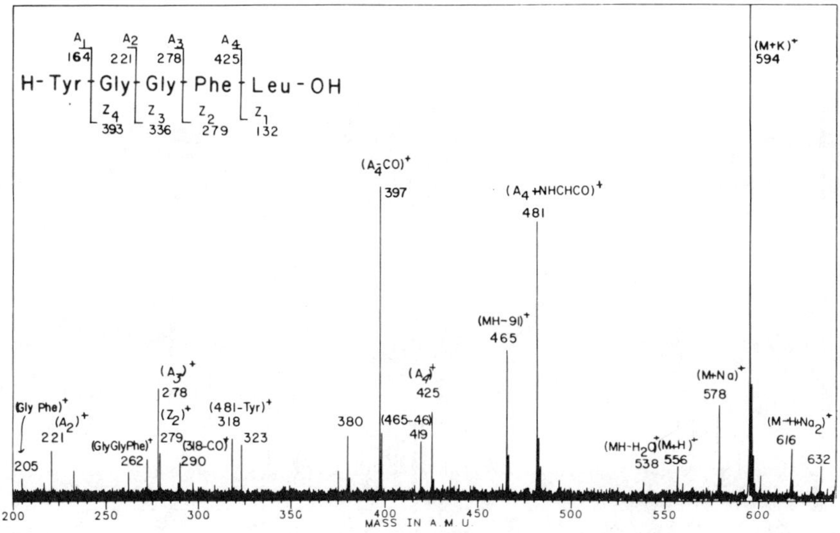

Figure 7: Laser desorption FT mass spectra for (a) leucine-enkephalin, (b): (p. 326) angiotensin II, and (c): (p. 326) bradykinin. *Reprinted with permission from Wilkins* et al. *(1985). Copyright 1985 American Chemical Society.*

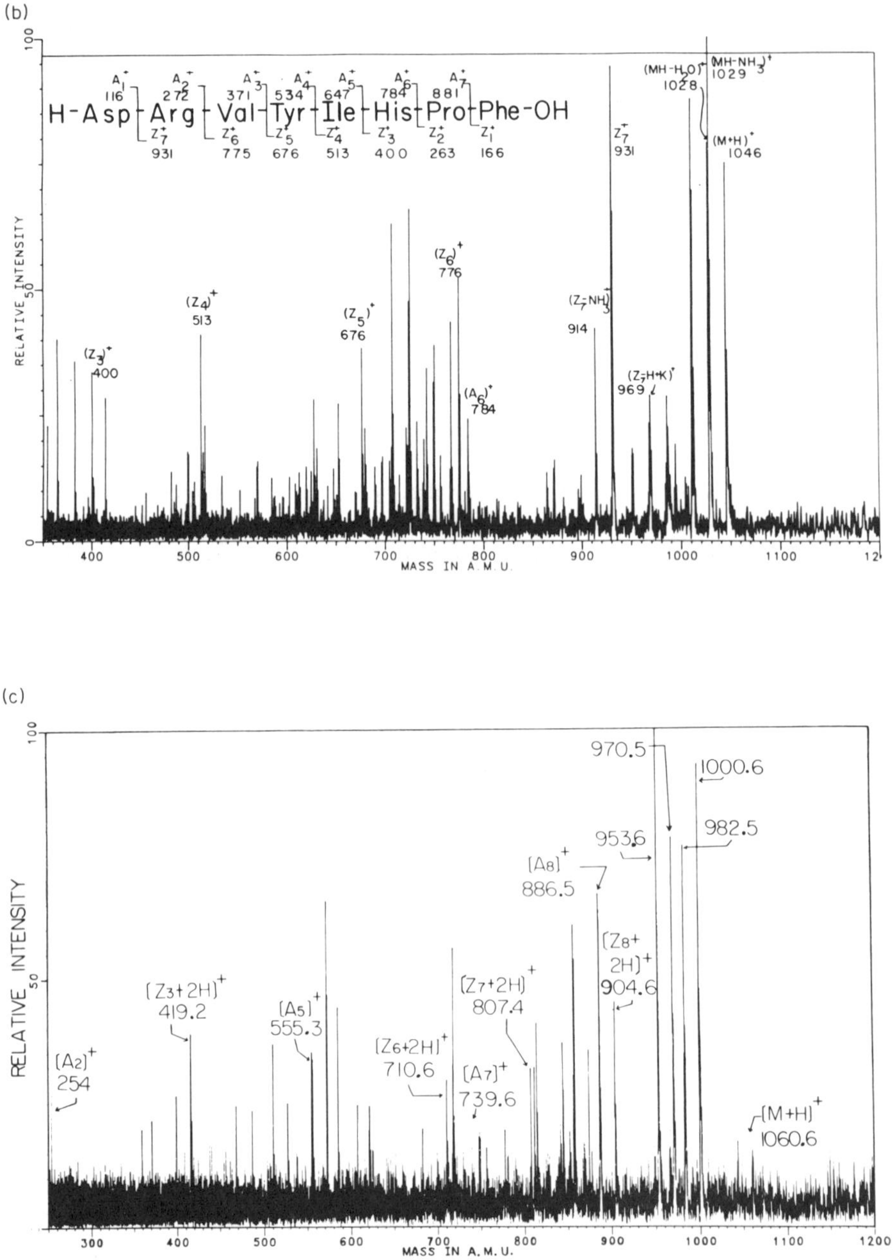

Figure 7: (continued). (b) angiotensin II; (c) bradykinin.

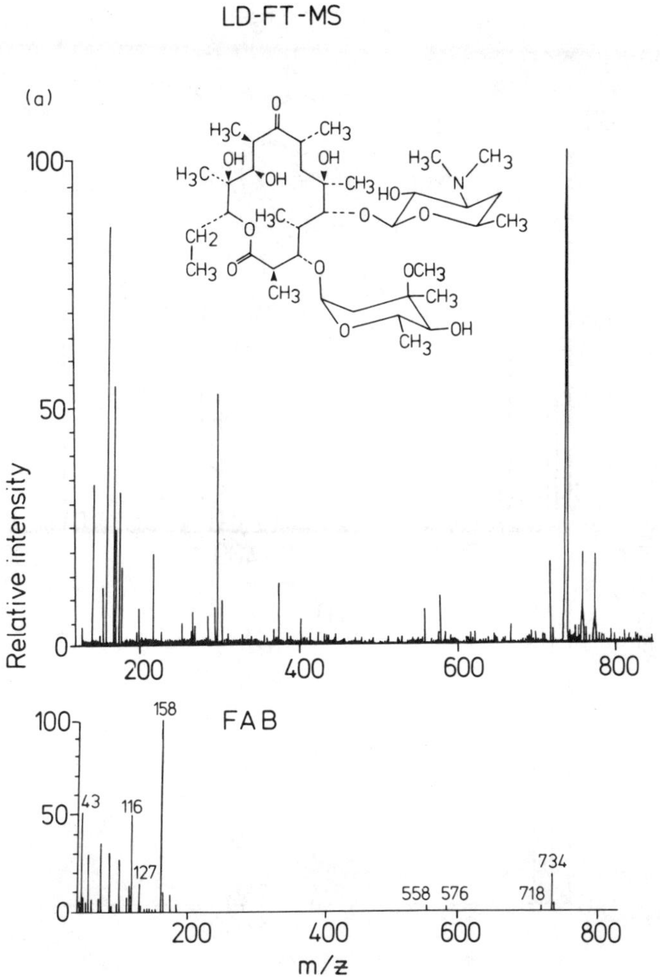

Figure 8: Laser desorption FT mass spectra for (a) erythromycin, (b): (p. 328) daunorubicin, (c): (p. 329) digoxin, and (d): (p. 330) amoxicillin. Positive ion fast-atom bombardment (FAB) ionization mass spectra are included also for comparisons of the qualitative aspects of the two methods. *Reprinted with permission from Shomo* et al. *(1985). Copyright 1985 American Chemical Society.*

than for FAB MS, and in the FAB MS spectra the major fragment ions are the N-terminus sequence ions. This result suggests a complementary value for laser desorption FTMS and FAB MS data.

Marshall's studies on laser desorption FTMS for the analysis of drugs demonstrate the quality of mass spectra that can be obtained (Shomo *et al.*,

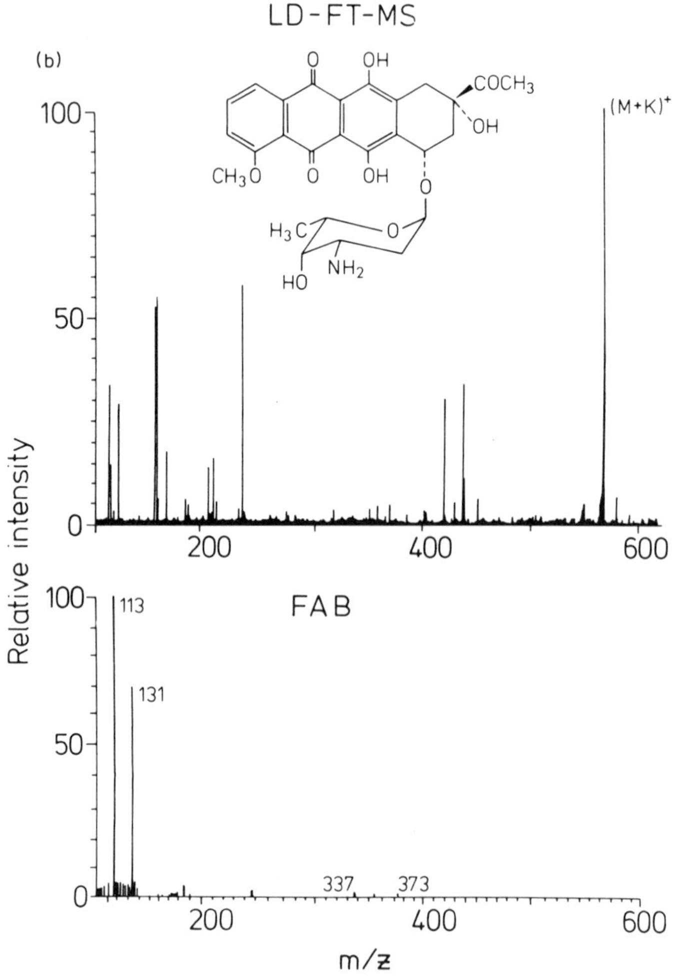

Figure 8: (continued). (b) daunorubicin.

1985). The laser desorption FTMS spectra for erythromycin, daunorubicin, digoxin and amoxicillin contain abundant molecular ions and important fragment ions. Although this class of molecules has been studied previously, the laser desorption FTMS spectra are obtained more easily and contain fewer fragment ions and background ions than do the FAB mass spectra (Figure 8). In addition, the low solubility of this class of molecules in the glycerol matrix, and the relatively high volatility of the thioglycerol matrix, make it more difficult to analyze these molecules by FAB MS. This study also suggests a complementary value for FAB MS and laser desorption FTMS, i.e., abundant

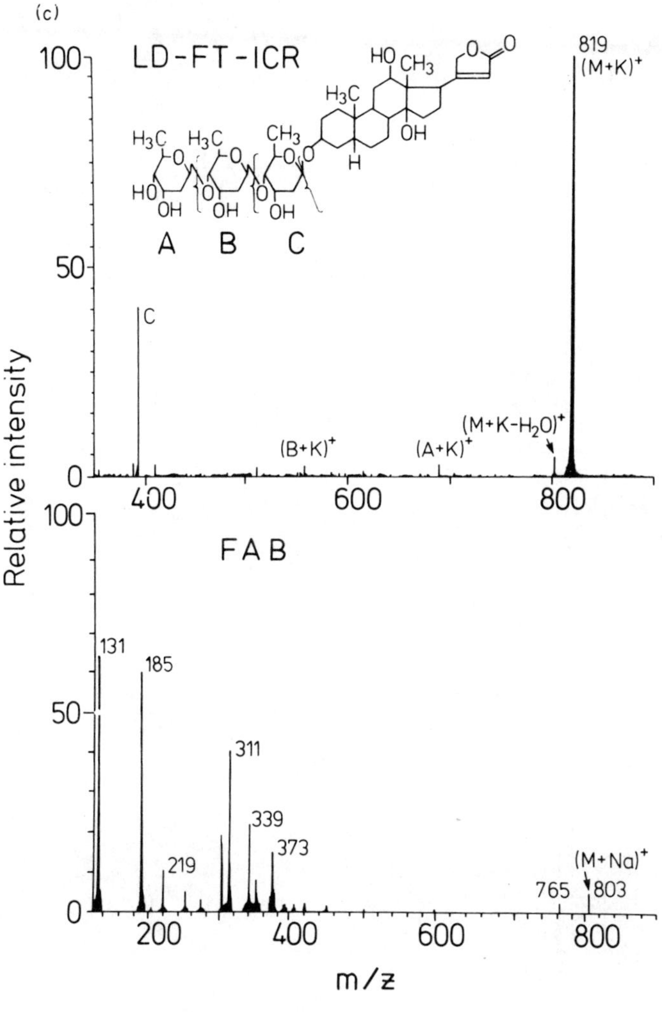

Figure 8: (continued). (c) digoxin.

molecular ions are observed by FTMS, whereas more structurally significant fragment ions are obtained with FAB MS.

Using laser desorption, Wilkins *et al.* (1985) have also acquired high resolution mass spectra of the molecular ion region of gramicidin S. A mass resolution in excess of 23,000 was obtained for the $(M + K)^+$ ion (m/z 1180) (Figure 9). It is important to note that these data were taken with a single section FTMS ion cell. More recently, Cody *et al.* (1985) have used a two-section FTMS ion cell and laser desorption for high resolution measurements. Cody

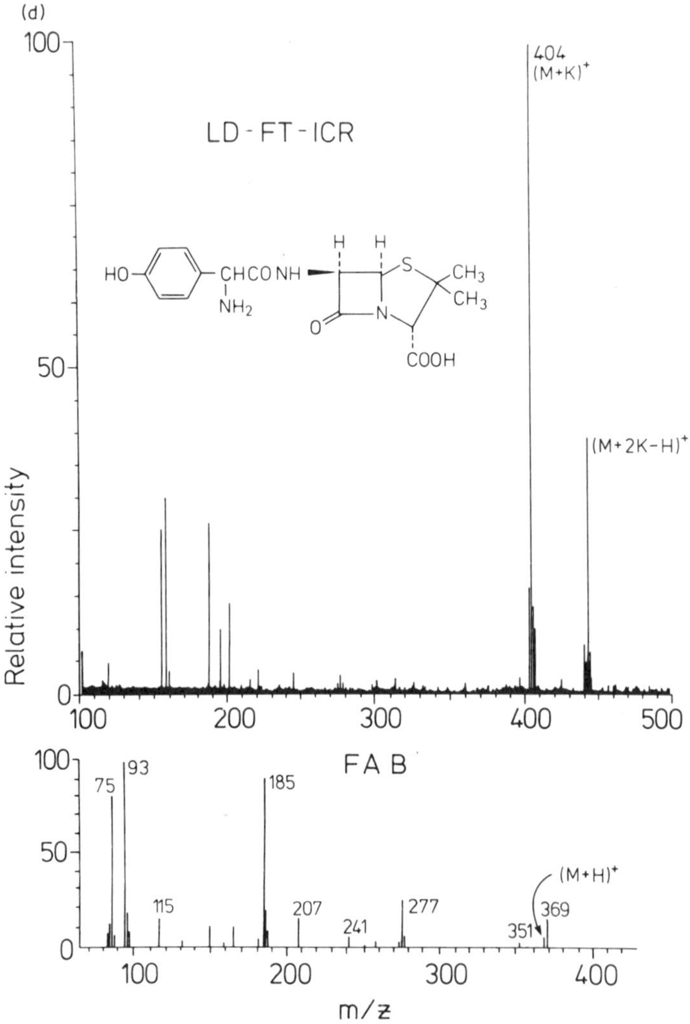

Figure 8: (continued). (d) amoxicillin.

reported a mass resolution of 150,000 (*m/z* 1180; gramicidin S) and 125,000 (*m/z* 1241; cyclosporin A) with this apparatus.

To date, the tandem quadrupole–FTMS data reported by Hunt and co-workers are the most impressive data for biomolecules. In recent work (Hunt *et al.*, 1985b), this group has obtained mass spectra (with Cs^+ desorption ionization from a liquid matrix) for vitamin B_{12} (M_r 1355), bradykinin (M_r 1141), gramicidin S (M_r 1141), melittin (M_r 2845), and renin substrate (M_r 1758) (Figure 10). The quality of the mass spectra is exceptional, and the reported

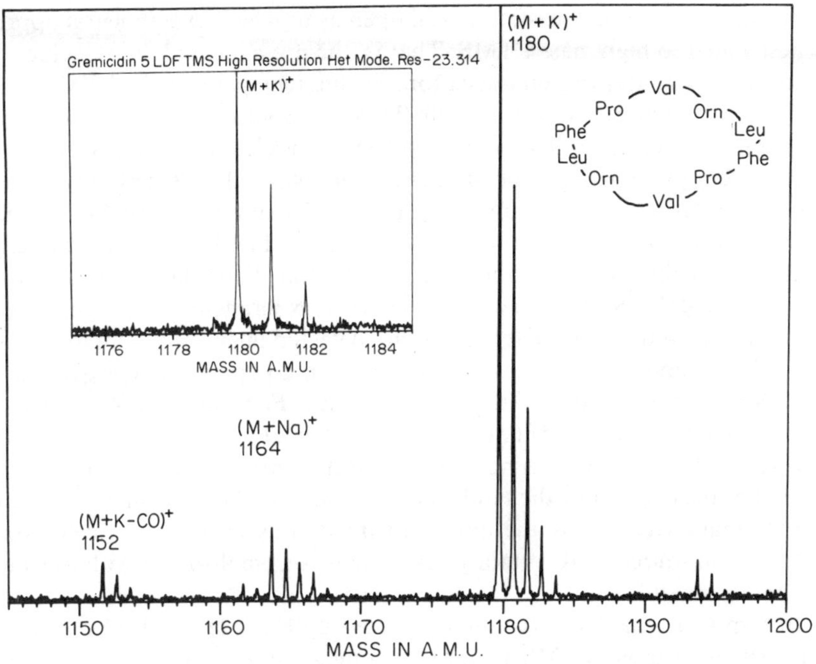

Figure 9: Laser desorption FT mass spectrum of gramicidin S. The inset is a scan of the molecular ion region showing a mass resolution of 23 000. *Reprinted with permission from Wilkins* et al. *(1985). Copyright 1985 American Chemical Society.*

mass resolution in the molecular ion region ranges from several thousand to in excess of 40,000. (The mass resolution reported for renin substrate was 20,000 and 40,000 for gramicidin S.) The quantity of sample used for each mass spectrum was in the 10 picomole range. In each case the spectrum is comparable to mass spectra obtained by using magnetic sector instruments.

17.5. FUTURE POTENTIAL FOR HIGH MASS FTMS

The studies discussed above are not truly novel in the sense that the molecules have been analyzed by other mass spectrometric methods. Of major importance is that such studies demonstrate the feasibility of high mass analysis with the FTMS. Also, where comparisons have been made, the quality of the FTMS mass spectra is certainly comparable to that obtained with other instruments. At this point the major limitations for high mass FTMS appear to be associated with the production of large molecular ions and the ability to detect such ions once they are formed. Certainly, the detection of cluster ions in excess of m/z 16,000 demonstrates the available mass range of FTMS.

At this juncture the question is still open as to which experimental approach is best suited to high mass FTMS. That is, the present capabilities include laser desorption, Cs^+ desorption ionization, the tandem quadrupole–FTMS system, and the two-section trapped ion cell. These capabilities span a broad range of cost and complexity. Based on our experience it is clear that the two-section cell is definitely preferable to the single-section cell, and it is certainly easier to produce intact molecular ions of large molecules with the liquid matrix ionization methods. In the final analysis one is faced with the simplicity of operation of the two-section ion cell, or the added complexity of the tandem quadrupole–FTMS. In principle these two experimental arrangements are comparable (in the sense that the ions are detected under optimum low vacuum conditions), and it is not clear that the more complex tandem quadrupole–FTMS system has any additional capabilities. For example, Wanczek *et al.* (1985) have demonstrated that ions can be formed external to the ion cell and injected into the ion cell using a simple electrostatic lens assembly. It is quite clear that the concept of the tandem quadrupole–FTMS system is a viable one, but its main attribute is the ability to inject ions into the ion cell from an external ion source. It is certainly desirable to accomplish this with as simple a system as possible.

In terms of sensitivity, the data reported by Hunt *et al.* (1985a,b) with the tandem quadrupole–FTMS required picomole amounts of sample, quantities comparable to the best results obtained to date with sector mass spectrometers

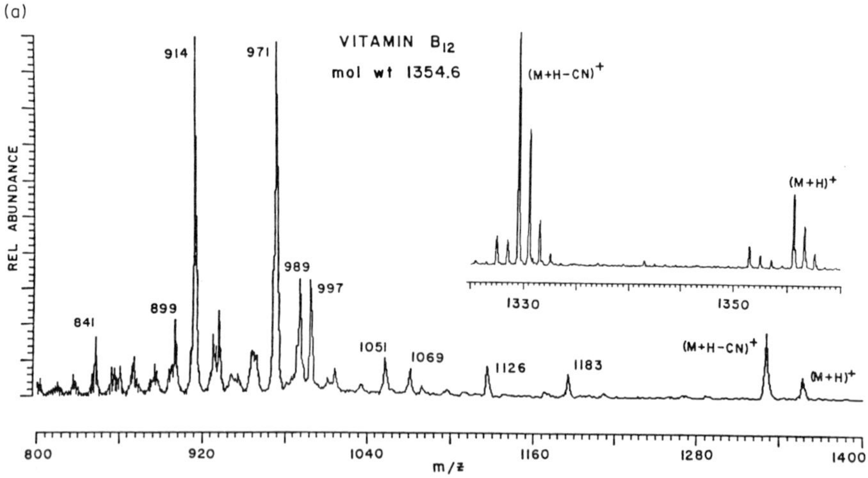

Figure 10: Mass spectra for (a) vitamin B_{12}, (b): (p. 333) bradykinin. (c): (p. 333) gramicidin S. (d): (p. 334) melittin, and (e): (p. 334) renin substrate, obtained with a tandem quadrupole-FTMS system. *Reprinted with permission from Hunt* et al. *(1985). Copyright 1985 American Chemical Society.*

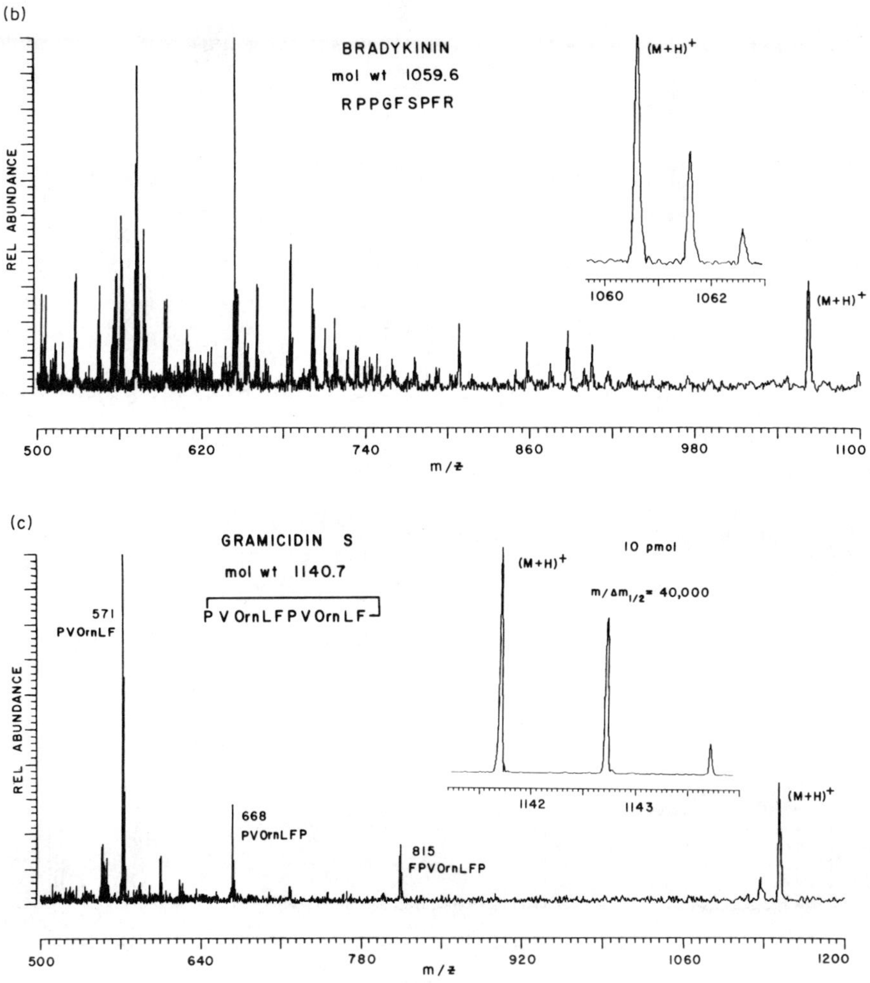

Figure 10: (continued). (b) bradykinin, (c) gramicidin S.

and FAB ionization. Similar sample quantities have been used for the laser desorption–FTMS and Cs^+ desorption ionization–FTMS work. In our laboratory we have concentrated a large amount of effort on sample preparation procedures and found that this variable can increase the sensitivity by orders of magnitude. For example, the Cs^+ desorption ionization FTMS spectrum of angiotensin I (Figure 11) was obtained on less than 1 picomole of sample. Based on such studies we estimate that reasonable spectra could be obtained on 50–100 femtomole samples. These sensitivity estimates are comparable to those reported recently by Niehuis *et al.* (1985) for TOF MS experiments.

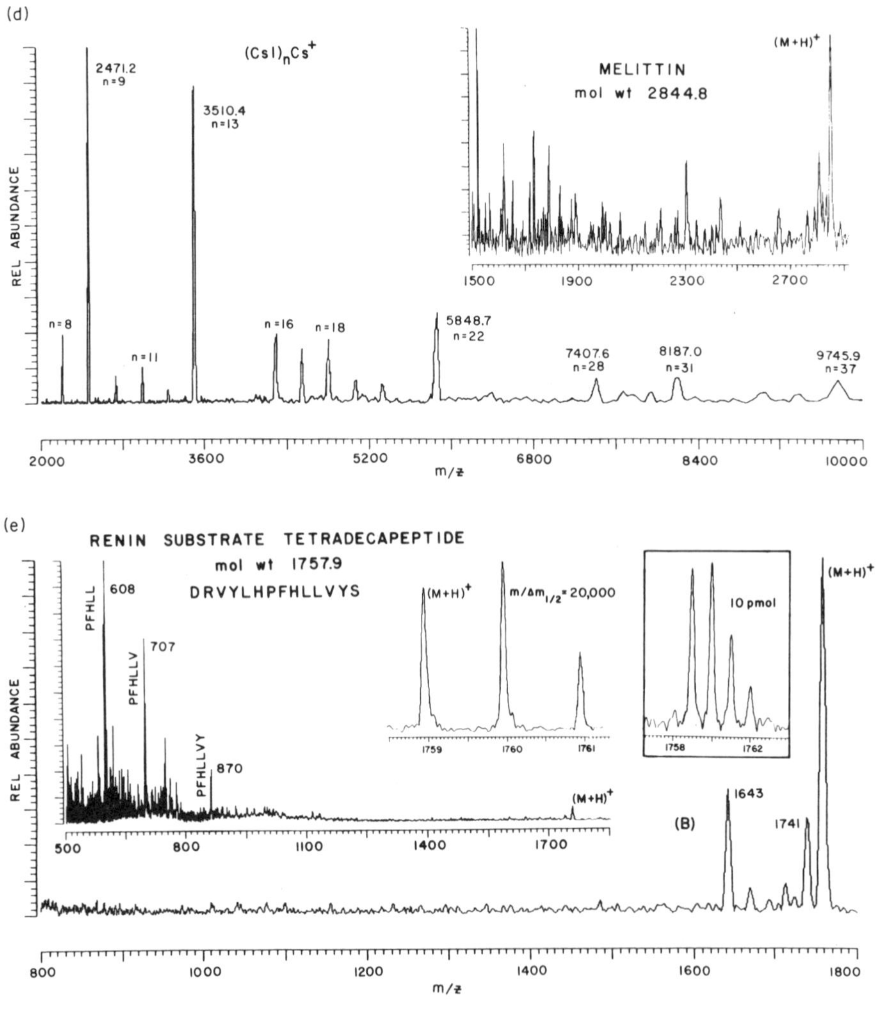

Figure 10: (continued). (d) melittin, (e) renin substrate.

In the last 3 years, enormous progress has been made in the development of ion cyclotron resonance based methods for the analysis of high mass molecules. The resolution achieved to date ($>100{,}000$ at masses greater than 1200), the demonstrated mass range ($>m/z$ 16,000 with a 3 Tesla magnet), and the sensitivity of the method (picomole levels) are quite impressive. All of these achievements have been realized by using ionization methods which are more difficult and less refined than FAB MS or by using prototype instruments, e.g., the tandem quadrupole–FTMS and two-section ion cells. There are many areas

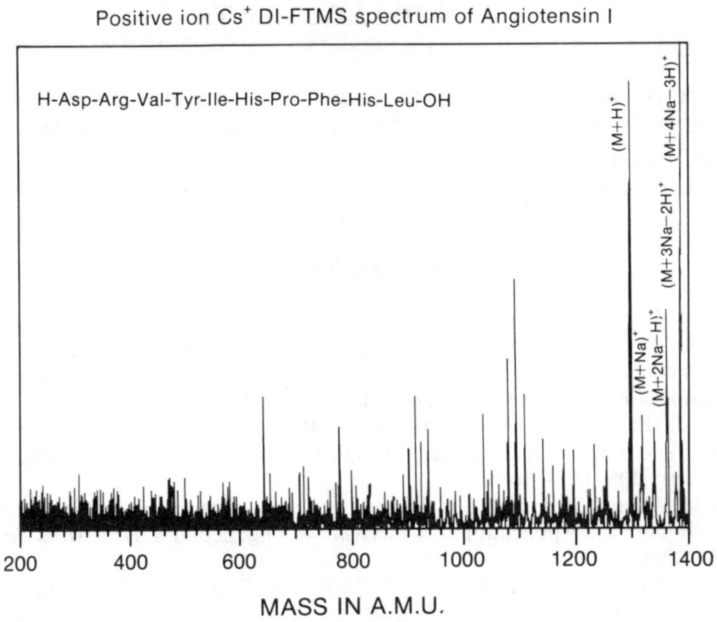

Figure 11: Cs^+ desorption ionization FT mass spectrum of angiotensin I. The spectrum was taken using 0.5 picomoles of sample and a Cs^+ ion beam density of 10×10^{-12} A cm^{-2}.

where improvements can be made, especially in the understanding and optimization of the ionization process, which could greatly improve the overall capabilities of high mass FTMS. For example, the ^{252}Cf ionization source is a natural for FTMS and to date only preliminary experiments have been performed (Castro, Russell, and Schweikert, unpublished results).

As pointed out in an earlier section of this chapter, the major importance of FTMS in analytical mass spectrometry is its versatility. The ability to produce ions by a variety of methods (e.g., electron impact ionization, various laser ionization methods, and particle-induced desorption ionization) for analysis by FTMS opens up a range of experiments for structural characterization of gas phase ions. These capabilities are extended even further with the introduction of FAB ionization capabilities using the quadrupole–FTMS system. Many of the structural probe methods were developed for fundamental studies, and only recently have these methods been utilized for purely analytical studies (Bowers *et al.*, 1984). These structural probes when combined with FTMS could solve many current problems, as well as opening up new areas of investigations.

ACKNOWLEDGMENTS

This work is supported by grants from the National Science Foundation (CHE-8418457) and the National Institutes of Health–General Medical Sciences (5R01-GM33780-02).

REFERENCES

Aberth, W., Straub, K. M., and Burlingame, A. L. (1982). Secondary ion mass spectrometry with cesium ion primary beam and liquid target matrix for analysis of bioorganic compounds, *Anal. Chem.*, **54**, 2029–2034.

Amster, I. J., Baldwin, M. A., Cheng, M. T., Proctor, C. J., and McLafferty, F. W. (1983). Tandem mass spectrometry of higher molecular weight compounds, *J. Am. Chem. Soc.*, **105**, 1654–1655.

Baldwin, M. A., and McLafferty, F. W. (1973). Direct chemical ionization of relatively involatile samples. Application to underivatized oligopeptides, *Org. Mass Spectrom.*, **7**, 1353–1356.

Barber, M., Bordoli, R. S., Elliott, G. J., Sedgwick, R. D., and Tyler, A. N. (1982). Fast-atom bombardment mass spectrometry, *Anal. Chem.*, **54**, 645A–657A.

Barlak, T. M., Wyatt, J. R., Colton, R. J., DeCorpo, J. J., and Campana, J. E. (1982). Secondary ion mass spectrometry of metal halides. Evidence for structure in alkali iodide clusters, *J. Am. Chem. Soc.*, **104**, 1212–1215.

Beauchamp, J. L. (1971). Ion cyclotron resonance spectroscopy, *Ann. Rev. Phys. Chem.*, **22**, 527–561.

Beckey, H. D. (1977). *Principles of Field Ionization and Field Desorption Mass Spectrometry*, Pergamon Press, Oxford.

Benninghoven, A., and Sichterman, W. (1978). Detection, identification, and structural investigation of biologically important compounds by secondary ion mass spectrometry, *Anal. Chem.*, **50**, 1180–1184.

Beuhler, R. J., and Friedman, L. (1980). Threshold studies of secondary electron emission induced by macro-ion impact on solid surfaces, *Nucl. Instrum. Methods*, **170**, 309–315.

Bowers, W. D., Delbert, S., Hunter, R. L., and McIver, R. T., Jr. (1984). Fragmentation of oligopeptide ions using ultraviolet laser irradiation and Fourier transform mass spectrometry, *J. Am. Chem. Soc.*, **106**, 7288–7289.

Campana, J. E., Colton, R. J., Wyatt, J. R., Bateman, R. H., and Green, B. N. (1984). Ultra-high mass spectrometry, *Appl. Spectrosc.*, **38**, 430–432.

Castro, M. E., and Russell, D. H. (1984). Cesium ion desorption ionization with Fourier transform mass spectrometry, *Anal. Chem.*, **56**, 578–581.

Castro, M. E., and Russell, D. H. (1985). Desorption ionization studies of cesium iodide by FTMS, *Anal. Chem.*, **57**, 2290–2293.

Castro, M. E., Russell, D. H., Ghaderi, S., Cody, R. B., Amster, I. J., and McLafferty, F. W. (1986). Detection of m/z 16,253 ions by Fourier transform mass spectrometry, *Anal. Chem.*, **58**, 483–485.

Chait, B. T., and Field, F. H. (1984). ^{252}Cf fission fragment ionization mass spectrometry of chlorophyll a, *J. Am. Chem. Soc.*, **106**, 1931–1938.

Cody, R. B., Burnier, R. C., and Freiser, B. S. (1982). Collision-induced dissociation with Fourier transform mass spectrometry, *Anal. Chem.*, **54**, 96–101.

Cody, R. B., Kinsinger, J. A., Ghaderi, S., and Hein, R. E. (1985). Soft ionization methods in Fourier transform mass spectrometry, *Proc. 33rd Ann. Conf. on Mass Spectrometry and Allied Topics, San Diego*; American Society for Mass Spectrometry, pp. 186–187.

Comisarow, M. B., and Marshall, A. G. (1974). Fourier transform ion cyclotron resonance spectrometry, *Chem. Phys. Lett.*, **25**, 282–283.

Cotter, R. J. (1980). Mass spectrometry of nonvolatile compounds. Desorption from extended probes, *Anal. Chem.*, **52**, 1589A–1597A.

Cotter, R. J., and Fenselau, C. (1979). The effects of heating rate and sample size on the direct exposure/chemical ionization mass spectra of some biological conjugates, *Biomed. Mass Spectrom.*, **6**, 287–293.

Ens, W., Beavis, R., and Standing, K. G. (1983). Time-of-flight measurements of cesium iodide cluster ions, *Phys. Rev. Lett.*, **50**, 27–30.

Ghaderi, S., Kulkarni, P. S., Ledford, E. B., Jr., Wilkins, C. L., and Gross, M. L. (1981). Chemical ionization in Fourier transform mass spectrometry, *Anal. Chem.*, **53**, 428–437.

Ghaderi, S., and Littlejohn, D. (1985). Fourier transform mass spectrometry using multiple-cell configurations, *Proceedings of the 33rd Ann. Conf. on Mass Spectrometry and Allied Topics, San Diego*; American Society for Mass Spectrometry, pp. 727–728.

Gross, M. L., and Rempel, D. L. (1984). Fourier transform mass spectrometry, *Science*, **226**, 261–268.

Hein, R. E., and Cody, R. B. (1986). Observation of cationized organic molecules and fragments by laser desorption Fourier transform mass spectrometry, *Anal. Chem.*, in press.

Holland, J. F., Enke, C. G., Allison, J., Stults, J. T., Pinkston, J. D., Newcome, B., and Watson, J. T. (1983). Mass spectrometry on the chromatographic timescale: realistic expectations, *Anal. Chem.*, **55**, 997A–1012A.

Hunt, D. F., Shabanowitz, J., McIver, R. T., Jr., Hunter, R. L., and Syka, J. E. P. (1985a). Ionization and mass analysis of nonvolatile compounds by particle bombardment tandem quadrupole–Fourier transform mass spectrometry, *Anal. Chem.*, **57**, 765–768.

Hunt, D. F., Shabanowitz, J., Yates, J. R., III, McIver, R. T. Jr., Hunter, R. L., Syka, J. E. P., and Amy, J. (1985b). Tandem quadrupole-Fourier transform mass spectrometry of oligopeptides, *Anal. Chem.*, **57**, 2728–2733.

Johlman, C. L., White, R. L., and Wilkins, C. L. (1983). Applications of Fourier transform mass spectrometry, *Mass Spectrom. Rev.*, **2**, 389–415

Konig, W. A., Aydin, M., Schulz, V., Rapp, V., Hohn, M., Pesch, R., and Kalikhevitch, V. N. (1983). Fast atom bombardment for peptide sequencing—a comparison with conventional ionization techniques, *Int. J. Mass Spectrom. Ion Processes*, **46**, 403–408.

Macfarlane, R. D. (1983). Inelastic surface collisions and the desorption of massive molecular ions, *Phys. Scripta*, **T6**, 110–116.

Macfarlane, R. D. (1984). ^{252}Cf plasma desorption mass spectrometry: large molecules, software, and the essence of time, *Anal. Chem.*, **55**, 1247A–1264A.

Macfarlane, R. D., and Torgerson, D. F. (1976). Californium-252 plasma desorption ionization, *Science*, **191**, 920–924.

Marshall, A. G. (1979). Theoretical signal-to-noise ratios and mass resolution in Fourier transform ion cyclotron resonance mass spectrometry, *Anal. Chem.*, **51**, 1710–1714.

Marshall, A. G. (1985). Fourier transform ion cyclotron resonance mass spectrometry, *Acc. Chem. Res.*, **18**, 316–322.

Marshall, A. G., Comisarow, M. B., and Parisod, G. (1979). Relaxation and spectral line shape in Fourier transform ion cyclotron resonance spectrometry, *J. Chem. Phys.*, **71**, 4434–4444.

McCrery, D. A., Ledford, E. B., Jr., and Gross, M. L. (1982). Laser desorption Fourier transform mass spectrometry, *Anal. Chem.*, **54**, 1435–1437.

McCrery, D. A., and Gross, M. L. (1985). Laser desorption Fourier transform mass spectrometry for study of nucleosides, oligosides, and glycosides, *Anal. Chem. Acta.*, **178**, 91–104.

McIver, R. T., Jr., Ledford, E. B., Jr., and Hunter, R. L. (1980). Theory for broadband detection of ion cyclotron resonance signals, *J. Chem. Phys.*, **72**, 2535–2547.

McIver, R. T., Jr., Hunter, R. L., and Bowers, W. D. (1985). Coupling a quadrupole mass spectrometer and a Fourier transform mass spectrometer, *Int. J. Mass Spectrom. Ion Processes*, **64**, 67–77.

Murawski, S. L., and Cook, K. D. (1984). Sampling of ions from volatile solutions by electrohydrodynamic mass spectrometry, *Anal. Chem.*, **56**, 1015–1020; and references therein.

Niehuis, E., Greifendorf, D., Friese, T., Benninghoven, A. (1985). Analytical applications of high performance TOF–SIMS, *Proc. 33rd Ann. Conf. on Mass Spectrometry and Allied Topics, San Diego*; American Society for Mass Spectrometry, pp. 565–566.

Parisod, G., and Comisarow, M. B. (1980). Fourier transform ion cyclotron resonance study of the ion molecule chemistry of $CpCr(CO)_2NS$ and related compounds, *Adv. Mass Spectrom.*, **8**, 216–223.

Posthumus, M. A., Kistemaker, P. G., Muezelaar, H. L. C., and Ten Noever de Brauw, M. C. (1978). Laser desorption mass spectrometry of polar nonvolatile bio-organic molecules, *Anal. Chem.*, **50**, 985–991.

Shomo, R. E., Marshall, A. G., and Weisenberger R. G. (1985). Laser desorption Fourier transform mass spectrometry *vs.* fast-atom bombardment magnetic sector mass spectrometry for drug analysis, *Anal. Chem.*, **57**, 2940–2944.

Standing, K. G., Chait, B. T., Ens, W., McIntosh, G., and Beavis, R. (1982). Time-of-flight measurements of secondary organic ions produced by 1 keV–16 keV primary ions, *Nucl. Instrum. Methods*, **198**, 33–38.

Tabet, J. C., Jablonski, M., and Cotter, R. J. (1985). Time-resolved laser desorption: III. The metastable decomposition of chlorophyll a and some derivatives, *Int. J. Mass Spectrom. Ion Processes*, **65**, 105–117.

Vastola, F. J., Mumma, R. O., and Pirone, A. J. (1976). Analysis of organic salts by laser ionization, *Org. Mass Spectrom.*, **3**, 101–104.

Wanczek, K. P. (1984). Ion cyclotron resonance spectrometry: a review, *Int. J. Mass Spectrom. Ion Processes*, **60**, 11–60.

Wanczek, K. P., Kofel, P., Allemann, M., and Kellerhals, Hp. (1985). External generation of ions in ICR spectrometry, *Int. J. Mass Spectrom. Ion Processes*, **65**, 97–103.

Wilkins, C. L., and Gross, M. L. (1981). Fourier transform mass spectrometry, *Anal. Chem.*, **53**, 1661A-1676A.

Wilkins, C. L., Weil, D. A., Yang, C. L. C., and Ijames, C. F. (1985). High mass analysis by laser desorption Fourier transform mass spectrometry, *Anal. Chem.*, **57**, 520–524.

Yergey, A. L., and Cotter, R. J. (1982). Thermal desorption mass spectrometry of organic salts, *Biomed. Mass Spectrom.*, **9**, 286–292.

PART III

TRACE ANALYSES

Mass Spectrometry in Biomedical Research
Edited by S. J. Gaskell
© 1986 John Wiley & Sons Ltd.

Chapter 18

OVERVIEW TO PART III

Simon J. Gaskell

In appropriate applications, mass spectrometry is a technique capable of exquisite sensitivity. The qualification is important, however, in that the lowest detection limits (in the femtogram range) are achieved only with certain classes of compound (generally of low molecular mass) and in applications where the identity of the analyte is known or suspected. Nevertheless, MS, usually in the form of combined GC–MS, has made major contributions to biomedical trace analyses (here defined as determinations of minor components of complex mixtures). Innovative techniques designed to improve the quality of trace analyses are the subject of this Part.

18.1. SENSITIVITY OF DETECTION

Optimal sensitivities of detection are achieved by single selected ion monitoring (SIM), that is, a static mode of operation in which the mass spectrometer transmits ions of a single *m/z* value. Minor sacrifices in sensitivity are made by monitoring a few additional ions by rapid switching between appropriate *m/z* values. SIM should be distinguished from the 'mass chromatography' technique in which selected ion current profiles are reconstructed from data obtained by repetitive scanning of a broad mass range. Detection limits using the latter technique are substantially poorer than those obtained by SIM, and quantitative precision is inferior.

Sensitivity of detection is also influenced by the method of sample introduction and the ionization technique employed. Combined chromatography–MS is generally favoured since this achieves a controlled and reproducible delivery of analyte to the ion source. By increasing the chromatographic resolution (for example, by using capillary columns in GC–MS), improved sample concentration is achieved leading to increased peak height to peak area ratios and lowered detection limits. Improved chromatographic resolution also permits the more precise definition of the retention time, improving selectivity of analysis. Approximate theoretical detection limits during a GC–MS/SIM experiment are readily calculated. Consider a GC–MS analysis where the chromatographic peak width for the analyte is of 10 s duration. If dual ion monitoring is performed, chromatographic integrity will be maintained using sampling intervals of 100 mS. A recognizable peak will be obtained during SIM if a minimum of about 20 ions is detected in the sampling interval corresponding to the GC peak maximum; this corresponds to a minimum total number of ions required for detection of about 1000. Basic sensitivities of detection on modern mass spectrometers, using electron impact ionization, are typically of the order of 10^{-8}Cμg^{-1}, corresponding to the detection of approximately 0.6×10^{11} ions. Thus, the 1000 ions required for detection during the GC–MS/SIM experiment described above correspond to less than 20 fg of analyte. Data presented in Chapter 19 indicate that limits of detection of this order may be achieved in practice.

It should be noted that the basic sensitivity of detection (Cμg^{-1}) for a particular analyte is strongly dependent on the proportion of the total ion current associated with the ionic species which is detected. Thus, when derivatization of an analyte is possible, or required for chromatographic reasons, sensitivity is improved by choosing a derivative with a mass spectrum in which ion current is concentrated in a single ionic species, ideally of high mass. The same objective may frequently be achieved using chemical ionization (CI). Proper comparison of relative sensitivities of EI and CI procedures, however, must take note of the relative efficiencies of ionization (Foltz, 1978).

The detection limit for a chosen analyte in a complex biological extract is frequently determined by the level of background 'chemical' noise generated by other components of the extract, rather than by absolute signal intensity. Thus, the *effective sensitivity* of detection is strongly influenced by the *selectivity* of detection.

18.2. SELECTIVITY OF DETECTION

The selectivity of a trace analysis method is determined by each stage of the procedure, as indicated in Table 1. Clearly, the various stages are of complementary value; isomers, for example, are frequently not distinguishable by MS but may be separated chromatographically.

Selectivity at the MS stage of analysis is determined by the method of ionization and by the approach to characterization of the ions formed (Table 1).

TABLE 1. Selectivity of trace analysis procedures.

Stage of procedure (with examples)	Fractionation/selection based on:
Sample extraction:	
— solvent or solid phase	polarity, acid/base properties
— immunoadsorption	stereochemistry, molecular size
Extract fractionation:	
— straight or reverse phase chromatography	polarity
— ion exchange chromatography	acid/base properties
Chromatography coupled to MS:	
— GC, LC	partition properties
MS ionization:	
— positive CI	proton affinity (for example)
— electron capture	electron affinity
MS detection:	
— SIM (low or high resolution)	molecular weight (elemental composition)
— selected reaction monitoring/tandem MS	MS fragmentation

The value of electron capture (EC) ionization (which is, strictly speaking, distinct from negative chemical ionization) is illustrated by the work described in Chapters 21, 22 and 25. Very low detection limits have been reported for compounds of high electron affinity (see, for example, Lewy and Markey, 1978), reflecting both an improved ionization efficiency (compared with EI and CI) and a reduced background, achieved by selective ionization.

The general lack of fragmentation observed in CI and EC analyses (compared with EI) may also improve selectivity of detection by reducing the background contribution from fragment ions derived from other mixture components.

Trace analyses of polar, labile or involatile compounds require ionization techniques such as FD or FAB/SIMS. Desiderio and Fridland (Chapter 24) report the analysis of neuropeptides using FD and FAB. Contributions to Part I of this book (Chapters 5–8) provide additional examples.

A further approach to enhance the selectivity of MS detection, in addition to selective ionization, is to improve the characterization of the ions formed. Two approaches may be followed:

(i) MS resolution may be increased, thus narrowing the range of possible elemental compositions corresponding to the ions detected. GC–high resolution MS/SIM analyses of prostaglandins are reported by Ishibashi *et al.* in Chapter 23.

(ii) Selected fragmentations may be monitored, where the fragmentations are defined in terms of the m/z ratios of both parent and daughter ions. For example, decompositions in the first field-free region of a double-focusing mass spectrometer may be detected, as in the GC–MS/metastable peak monitoring technique discussed by Harvey (Chapter 20) and in the FD and FAB analyses described by Desiderio and Fridland (Chapter 24). Using a reverse geometry (BE) instrument, fragmentations in the second field-free region may be monitored in a similar fashion (Kondrat and Cooks, 1978). Both these approaches to selected reaction monitoring, using a double-focusing mass spectrometer, are limited in terms of the resolution (and hence precision of determination) of the parent or daughter ions (Chapter 9, Section 9.2). Unit effective resolution, or better, of both parent and daughter ions, with consequent improvement in selectivity during selected reaction monitoring, is achieved using a multiple analyzer instrument. Examples of this approach are found in Chapters 19 (Section 19.2), 25 and 26.

Impressive examples have been reported of the use of tandem MS for the identification of trace components of complex mixtures introduced directly into the ion source (Hunt *et al.*, 1985). The approach is particularly valuable for rapid screening for a number of components or compound classes (Chapter 25, Section 25.6). For very sensitive analyses of targeted compounds, however, retention of a coupled chromatographic separation is generally desirable (Chapter 25, Section 25.5; Chapter 26).

18.3. SOME CONSIDERATIONS FOR QUANTITATIVE ANALYSES

Considerations of sensitivity and selectivity of detection apply equally, whether the objective of analysis is the characterization or the quantification of trace components. Certain aspects of quantitative analysis, however, require separate consideration and are discussed briefly here.

(i) Limit of quantification. The limit of detection is readily defined as, for example, that concentration of analyte in a biological sample which yields a signal to noise ratio of 3 during MS analysis. The limit of quantification, however, will be higher; its value will be determined by the required precision.

(ii) Accuracy (the agreement between the determined concentration of analyte and the true value). In general, the accuracy of quantification of endogenous components of a biological sample is difficult to establish. Accurate determination of analyte added to the sample is a necessary, but insufficient, criterion for accurate determination of the endogenous compound. Further in which a constant quantity of internal standard is added to various volumes of the biological sample. A linear relationship between quantity of analyte determined and volume of sample extracted confirms the validity of the internal standard employed. Additional confidence in quantitative accuracy is obtained if data obtained by different procedures (for example, using different derivatives or ionization techniques) are in agreement.

(iii) Precision (the variability of the experimental result, usually expressed as a standard deviation or coefficient of variation). Depending on the application, it may be necessary to determine both within- and between-assay variation. The importance of mass spectrometric quantitative procedures is in large measure due to the potential for high precision provided by internal standardization using isotopically labelled analogues. Losses are compensated throughout the procedure, and during combined chromatography–MS the coelution, or very closely separated elution, of analyte and internal standard compensates for temporal variations in MS sensitivity (Claeys *et al.*, 1977). Precise determination of response ratios (analyte/internal standard) is clearly important. Measurements of peak heights are frequently adequate but in GC–MS/SIM analyses close to the detection limit, optimal precision has been shown to be achieved by an alternative approach (Thorne *et al.*, 1984). Linear regression analysis of analyte signal intensity on isotopically labelled internal standard signal intensity yields a slope value equal to the response ratio. The improved precision derives from the fact that neither baseline levels nor peak tops need be defined and the data implicit in the total peak profiles are used.

The measurement of response ratios for authentic mixtures of analyte and internal standard permits the generation of a standard curve; this and other aspects of quantitative MS are discussed in the excellent review by Garland and Powell (1981).

REFERENCES

Claeys, M., Markey, S. P., and Maenhaut, W. (1977). Variance analysis of error in selected ion monitoring assays using various internal standards. A practical study case, *Biomed. Mass Spectrom.*, **4**, 122–128.

Foltz, R. L. (1978). Quantitative analysis of abused drugs in physiological fluids by gas chromatography/chemical ionization mass spectrometry, in *Quantitative Mass Spectrometry in Life Sciences II* (Eds. A. P. De Leenheer, R. R. Roncucci, and C. Van Peteghem), pp. 39–62, Elsevier, Amsterdam.

Garland, W. A., and Powell, M. L. (1981). Quantitative selected ion monitoring (QSIM) of drugs and/or metabolites in biological matrices, *J. Chromatog. Sci.*, **19**, 392–434.

Hunt, D. F., Shabanowitz, J., Harvey, T. M., and Coates, M. (1985). Scheme for the direct analysis of organics in the environment by tandem mass spectrometry, *Anal. Chem.*, **57**, 525–537.

Kondrat, R. W., and Cooks, R. G. (1978). Direct analysis of mixtures by mass spectrometry, *Anal. Chem.*, **50**, 81A–92A.

Lewy, A. J., and Markey, S. P. (1978). Analysis of melatonin in human plasma by gas chromatography negative chemical ionization mass spectrometry, *Science*, **201**, 741–743.

Thorne, G. C., Gaskell, S. J., and Payne, P. A. (1984). Approaches to the improvement of quantitative precision in selected ion monitoring: high resolution applications, *Biomed. Mass Spectrom.*, **11**, 415–420.

Mass Spectrometry in Biomedical Research
Edited by S. J. Gaskell
© 1986 John Wiley & Sons Ltd.

Chapter 19

HIGH SENSITIVITY DETERMINATIONS OF STEROIDS IN PHYSIOLOGICAL FLUIDS AND TISSUES

Simon J. Gaskell, Valerie J. Gould and Heather M. Leith

19.1. INTRODUCTION

The identification and quantification of steroids in physiological fluids and tissues are of importance to the understanding, diagnosis and monitoring of disorders of the endocrine system, including endocrine-related cancers. Major advances have followed the replacement of classical colorimetric assay techniques with competitive binding assays, notably the various forms of immunoassay which are now in routine use. The contribution of mass spectrometry to steroid endocrinology is in three areas:

(i) Study of metabolic processes, including the characterization of metabolites of endogenous and exogenous substrates. Stable isotope labelling techniques are important; their use has been discussed elsewhere (Gaskell, 1983).

(ii) Profiling of physiological fluids, especially urine, to establish the relative abundances of individual steroids and thereby assess disorders of metabolism. Generally this is achieved by capillary gas chromatography (reviewed by Shackleton, 1985), but combined GC–MS serves to confirm identifications or establish the presence of minor components. There are recent reports (Shackleton, 1983) of urinary steroid conjugate profiling using fast atom bombardment or related techniques of secondary ion MS. The approach is rapid and simple but suffers from the difficulty of distinguishing isomeric compounds.

(iii) Rigorous quantification of steroids using GC–MS with selected ion monitoring (SIM). The first report of multiple SIM (Sweeley *et al.*, 1966) included the example of separate detection of the trimethylsilyl ethers of epiandrosterone and dehydroepiandrosterone, which coeluted during GC. The use of isotopically-labelled internal standards and GC–MS/SIM has now been widely applied to the quantification of steroids in serum and other fluids, exemplified by the pioneering work of Siekmann (1979), Bjorkhem (Bjorkhem *et al.*, 1976) and Dehennin (Reiffsteck *et al.*, 1982), as well as by studies in this laboratory (Riad-Fahmy *et al.*, 1979; Gaskell *et al.*, 1983, 1984). When adequately validated, the GC–MS procedures may serve as reference methods for the assessment of routine assays, such as radio-immunoassay.

In this chapter, the analytical approach adopted in this laboratory for the trace analysis of steroids is described and illustrated by reference to analyses of fluids and tissues.

19.2. ANALYTICAL APPROACH

A general analytical procedure for the determination of unconjugated steroids (and other analytes) in biological samples consists of four stages—extraction, fractionation, derivatization and GC–MS. The fractionation stage may sometimes be omitted, even when solvent extraction is employed, if the analyte is present in high concentrations and if adequate selectivity of detection is achieved during instrumental analysis. GC–MS of a total solvent extract is generally undesirable, however, since the high loading of sample applied to the GC column may affect chromatographic performance and may induce fluctuations in MS response. Immunoadsorption extraction of steroids (Glencross *et al.*, 1981; Gaskell and Brownsey, 1983) provides a more rapid extraction technique and, more importantly, is much more selective than solvent extraction. Table 1 indicates the extraction efficiencies of a number of steroids, obtained using a microcellulose-coupled anti-oestradiol serum (Gaskell and Brownsey, 1983). The relative extraction efficiencies parallel the cross-reactions determined for the antiserum (at much greater dilution) using conventional

TABLE 1. Extraction efficiencies using cellulose-coupled anti-oestradiol-17β serum and cross-reactions of the liquid antiserum*.

Steroid	Extraction efficiency† (%)	Cross-reaction‡ (%)
Oestradiol-17β	81	100
Oestrone	65	24
Oestriol	53	9.5
Testosterone	3	<0.1
Progesterone	3	<0.1
Cortisol	<1	—

* Data from Gaskell and Brownsey (1983).
† Extraction of 0.5 ml plasma using 0.5 ml of a nominal 1/100 dilution of coupled antiserum.
‡ Determined according to the criteria of Abraham (1969); antiserum dilution 1/150,000.

procedures (Abraham, 1969). In general, constituents coextracted by immunoadsorption are readily distinguished from the intended analyte during GC–MS. Reliable determinations of steroid hormones may, therefore, in many instances (such as analyses of blood plasma) be made by GC–MS of total immunoadsorption extracts, following derivatization (Gaskell and Brownsey, 1983; Gaskell *et al.*, 1983, 1984). The satisfactory extraction efficiencies of several structurally-related steroids using a single coupled anti-serum may be turned to advantage in multicomponent analyses if appropriate corrections are made (ideally using isotopically-labelled internal standards) for differing recoveries.

For the most demanding problems in steroid trace analysis, such as the determinations of very low concentrations in saliva and tissue samples discussed below, fractionation of the total extract is required. Liquid chromatographic separations using Sephadex-based gels (Sjövall and Axelson, 1982) have been found effective, particularly where selection for a particular structural feature of the analyte may be made using ion exchange chromatography. Thus, for example, the weak acidity of oestrogens (associated with the phenolic moiety) permits their isolation using triethylammoniohydroxypropyl-Lipidex 5000, a lipophilic anion exchange derivative of Sephadex LH-20 (Axelson and Sjövall, 1977). Conversion of 3-keto-steroids to unsubstituted oximes confers a weakly basic property which allows separation on sulphoethyl-Sephadex LH-20, a cation exchange material (Axelson and Sjövall, 1976).

The choice of derivative is of key importance, the essential requirements being good GC properties and a high yield of a high mass ion, ideally one which retains the full structure of the parent analyte. (Additional considerations apply when detection is by selected reaction monitoring; see below.) Derivatives may usually be found which yield an abundant high mass ion following electron impact ionization and this mode is generally preferred to chemical ionization since a greater total ion yield is usually obtained. A preliminary assessment of fluorinated derivatives (pentafluorobenzyl oximes, heptafluorobutyrates) of steroids for SIM analysis using electron capture negative chemical ionization MS was disappointing (Gaskell and Brooks, 1983); with careful control of the ion source temperature, good sensitivities could be achieved but these remained inferior to those obtained with electron impact.

Extremely high sensitivities of detection during GC–electron impact MS/SIM may be achieved following appropriate choice of derivative. Figure 1 shows the analysis of 100 fg of 5-androstene-3β-17β-diol as the *bis-tert*-butyldimethylsilyl (TBDMS) ether, with monitoring of the $(M–C_4H_9)^+$ ion. Fluoroacyl derivatives are also of value for steroid trace analyses by GC–electron impact MS; molecular ions are frequently of high abundance (as in the example of enol heptafluorobutyrates) and the large mass increment accompanying derivatization (which is advantageous in that the monitoring of higher mass ions tends to be less susceptible to interference) is achieved without undue increase in GC retention.

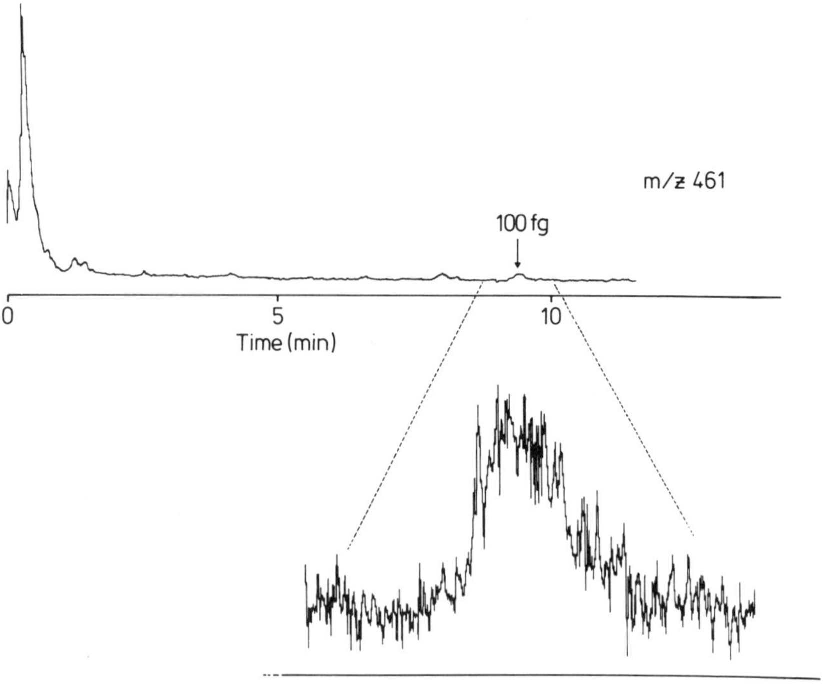

Figure 1: GC–MS analysis of 100 fg 5-androstene-3β,17β-diol *bis-tert*-butyldimethylsilyl ether with dual ion monitoring of *m/z* 461 ($(M - C_4H_9)^+$) and *m/z* 464. (Only the *m/z* 461 trace is shown). The GC column was fused silica, of the bonded phase type, operated at 250°. The MS resolution was 650. *Reproduced from Gaskell (1984), by permission of Kemia-Keni.*

During the analysis of biological extracts, the detectability of the intended analyte is influenced by the selectivity of detection as well as the absolute signal intensity achieved. In many instances, a reduction in signal intensity is acceptable if there is a greater concomitant improvement in the ratio of analyte signal to background ('chemical') noise. The use of high resolution SIM for the determination of steroids was pioneered by Millington (1975) and has been used extensively in this laboratory. 'High resolution' is ill-defined and, in practice, useful improvements in selectivity are achieved with resolutions from 1500 to 10,000 (10% valley definition). The resolution chosen is dependent on the complexity of the sample and the acceptable sacrifice in absolute signal intensity.

The observation that substantial proportions of high mass ions derived from some steroid derivatives (notably TBDMS ethers) spontaneously fragment in the first field-free region of a double-focusing mass spectrometer prompted the development of the metastable peak monitoring method for trace analysis (Gaskell and Millington, 1978; Gaskell *et al.*, 1980a; Thorne and Gaskell, 1985).

With regard to selectivity of detection, the principal merit of the method is that focusing conditions are dependent on the *m/z* ratios of both the ion detected and its parent so that an additional dimension of selectivity is introduced. There is the further advantage that increased specificity is achieved without increasing demands on the stability of focusing conditions, in contrast to requirements for high resolution SIM. Useful application of the metastable peak monitoring technique has been demonstrated in analyses for steroids in biological fluids (Gaskell *et al.*, 1980b,c; Finlay and Gaskell, 1981) and tissue (Gaskell *et al.*, 1979). A limitation to this approach is that the *m/z* ratio of the parent ion may be defined with only poor precision. (The opposite applies when second field-free region decompositions are studied using a reverse-geometry instrument. Here parent ion resolution is good but mass resolution of the daughter ion is very poor. Refer to Chapter 9 for an additional discussion.) The logical development of this approach is thus the use of a multiple analyzer instrument.

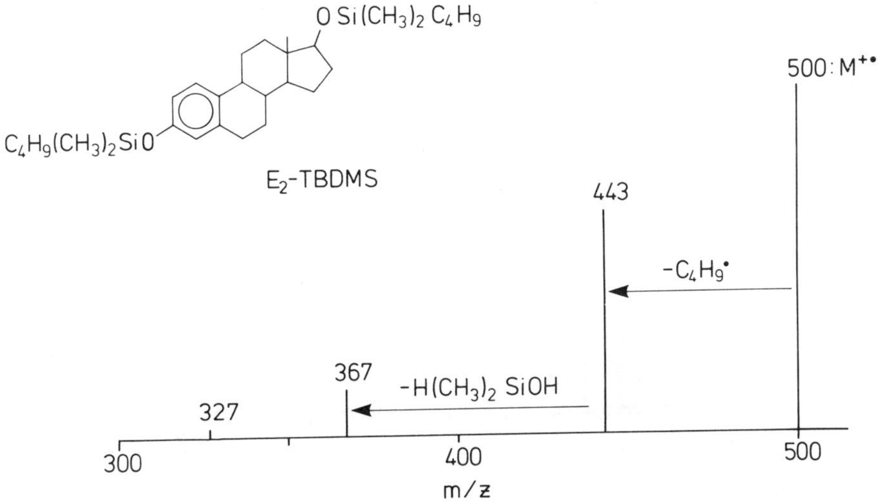

Figure 2: Scan of daughter ions formed following collisional activation of *m/z* 500 from oestradiol-17β (E_2) *bis-tert*-butyldimethylsilyl ether. The analysis was performed on a hybrid EBQQ instrument (VG 7070 EQ). Parent ions were selected using the double focusing instrument (*EB*); collisional activation (air; 35 eV) was effected in the first quadrupole and the spectrum shown resulted from a scan of the second quadrupole.

Figure 2 shows the daughter ion spectrum observed following collisionally-activated decomposition (CAD) of $M^{+\cdot}$ ions from oestradiol-17β *bis*-TBDMS ether; the spectrum was recorded using a hybrid (EBQQ) instrument, where Q1 functions as a high transmission collision cell. The $M^{+\cdot} \rightarrow (M - C_4H_9)^+$ transition was monitored during GC–MS to provide a highly selective trace analysis.

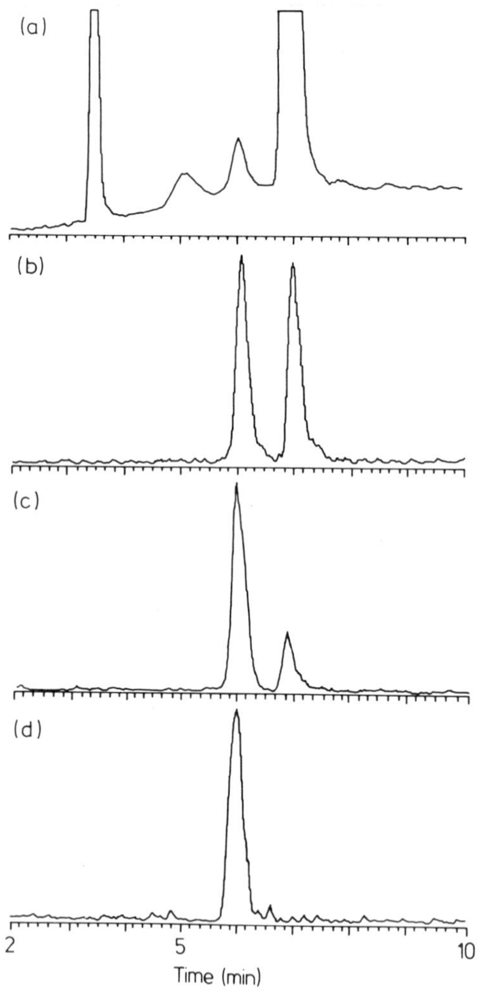

Figure 3: GC–MS analyses of a plasma extract for oestradiol-17β, as the *bis*-TBDMS ether. a) Low resolution (1000) selected ion monitoring of m/z 500. b) Selected reaction monitoring of m/z 500 → 443; parent ion resolution 1000. c) as (b), parent ion resolution 2500. d) as (b), parent ion resolution 5000. Reproduced from Gaskell *et al.* (1985).

Figure 3 illustrates analyses of a plasma extract for oestradiol-17β, as the TBDMS derivative, using several modes of operation (Gaskell *et al.*, 1985). Clearly the selected reaction monitoring mode is of superior selectivity, with optimal results achieved when the parent ion resolution is increased to 5000.

19.3. ANALYSES FOR STEROIDS IN SALIVA

The concentrations of steroids in saliva, in common with those of other analytes such as neutral drugs, are considered to reflect the non-protein bound levels in blood plasma (Riad-Fahmy *et al.*, 1982). Since it is generally assumed that it is the non-protein bound form which is physiologically active, much recent interest has been attached to the determination of salivary steroid concentrations (Read *et al.*, 1984). The analytical problems are substantial, however, since the concentrations of steroids in saliva are generally 1–5% of the total concentrations in plasma or serum. Various forms of immunoassay, usually radio-immunoassay (RIA), are the methods of choice for routine application

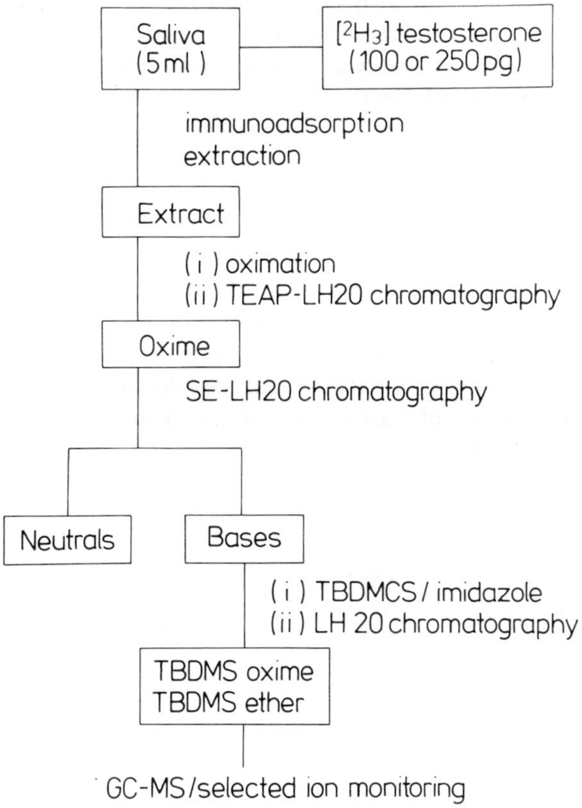

Figure 4: Experimental flow diagram for analyses of saliva for testosterone. (Abbreviations: SE, sulphoethyl; TEAP, triethylammoniohydroxypropyl; TBDMCS, *tert*-butyldimethyl chlorosilane).

but the uncertainty surrounding the specificity of such assays suggests the need for reference methods to assist in their validation. With care in assay design and scrupulous attention to their own validation, procedures based on stable isotope dilution and GC–MS can provide reference data.

Methods of analysis for a number of steroids in saliva, using mass spectrometry, have been developed in this laboratory (Gaskell *et al.*, 1980b,c; Gaskell and Finlay, 1984; Finlay *et al.*, 1982; Leith *et al.*, 1986a). The example discussed here is the determination of testosterone, with particular attention to saliva from female subjects, where the concentrations are low, and to saliva from prostatic cancer patients treated with diethylstilboestrol to suppress circulating testosterone levels. The requirement is for a specific and precise quantitative assay with a detection limit below *ca.* 2 pg/ml. Figure 4 shows the flow diagram of the procedure adopted (Gould *et al.*, 1986). High selectivity of extraction is obtained using a solid phase coupled antiserum (Gaskell and Brownsey, 1983). Conversion to the oxime derivative (Figure 5) confers mildly basic properties to the analyte, which may then be isolated using sulphoethyl-Sephadex LH–20, a cation exchange material (Axelson and Sjövall, 1976). Preparation of the *tert*-butyldimethylsilyl (TBDMS) oxime, TBDMS ether (Figure 5) provides a derivative with excellent GC–MS properties, yielding abundant $(M–C_4H_9)^+$ ions suitable for selected ion monitoring analyses. A modest increase in MS resolution, to 2500, gives a useful improvement in selectivity of detection while retaining high transmission of the ion beam and consequently adequate signal intensities. Two peaks are observed during GC–MS corresponding to the syn and anti oxime isomers (Figure 5). While the distribution of signal intensity increases the detection limit, the availability of two sets of response ratios (analyte/internal standard) for each determination provides an additional check on assay selectivity. The sensitivity of analysis was sufficient for the determination of concentrations of testosterone of approximately 2 pg/ml or higher.

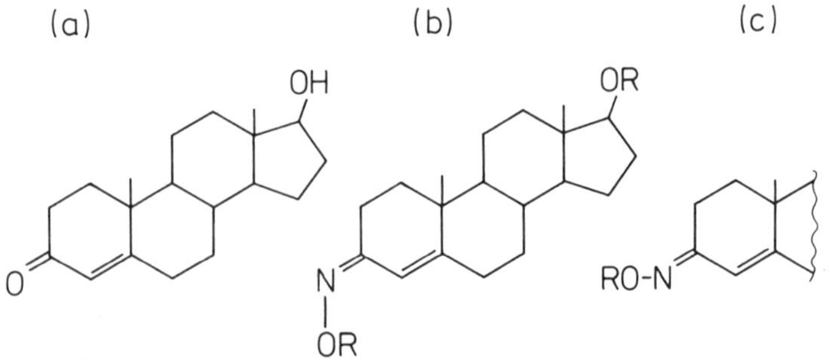

Figure 5: Structures of testosterone (a), the syn/Z (b) and anti/E (c) isomers of the oxime (R = H) and the *tert*-butyldimethylsilyl oxime, *tert*-butyldimethylsilyl ether ($R = (CH_3)_2C_4H_9Si$).

The coefficient of variation (within-assay) for analysis of a female saliva (7 pg/ml) was 11.8%. Measurement of testosterone in female saliva following addition of known quantities of the steroid indicated a recovery averaging 100.4%. Figure 6 shows the results of a series of analyses of various volumes of a single male saliva pool; the data indicate excellent linearity.

Application of the testosterone assay to the analysis of saliva from female subjects (18–27 years) indicated concentrations in the range 15–53 pmol/l (4–15 pg/ml). These data lie at the lower end of the range of previously reported RIA values suggesting that the earlier assays have been subject to significant interference. The concentrations of testosterone in diethylstilboestrol-treated patients, determined by GC–MS (in contrast to RIA values), indicated effective suppression of testosterone production (Gould *et al.*, 1986).

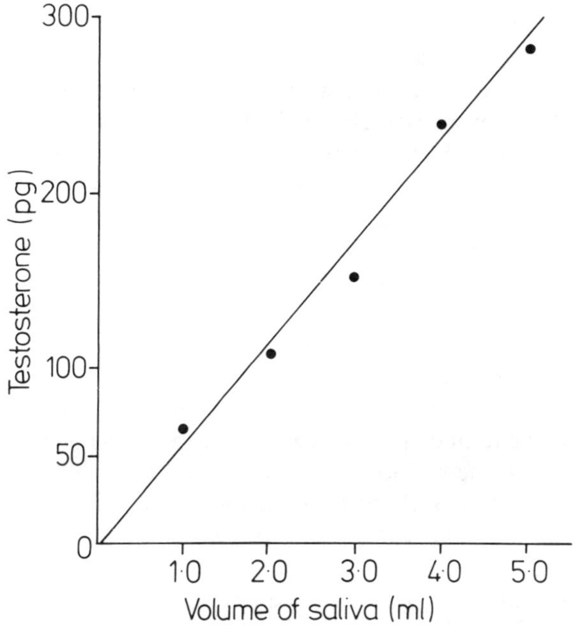

Figure 6: Quantification of testosterone in various volumes of a male saliva pool. Linear regression of quantity of testosterone (y pg) on volume of saliva (x ml) gave the equation $y = 56.6x - 0.82$ pg ($r = 0.991$).

19.4. ANALYSES FOR STEROIDS IN HUMAN BREAST TUMOUR TISSUE

Breast tumour tissue, removed at mastectomy, is generally analyzed for oestrogen receptor content to permit a prediction of the likely response of the patient to endocrine therapy, following recurrence of the disease. Thus, for example, 60% of patients with oestrogen receptor-positive primary tumours will respond to treatment with the antioestrogen, tamoxifen (Mourisden *et al.*, 1978). Oestrogen receptor content, however, represents an incomplete assessment of the endocrine status of the tumour cell and a knowledge of the concentrations of the oestrogens themselves, and of other steroids which bind to the oestrogen receptor, would be of complementary value. Furthermore, since the majority of assays for the oestrogen receptor detect only unoccupied cytosolic receptor, the presence of high concentrations of oestrogens might lead to significant under-estimation of receptor concentration. A number of studies of steroid concentrations in breast tissue homogenates and cytosol fractions have been reported (see, for example, Edery *et al.*, 1981). The purpose of the work summarized here was to use a sensitive, specific and well validated mass spectrometric procedure to establish relative concentrations of several steroids in breast tumour tissue cytosols.

Cytosol samples were supplemented with [^{13}C]- or [^{2}H]-labelled analogues as internal standards and were extracted with chloroform. The total extracts were fractionated on short columns of triethylammoniohydroxypropyl-Lipidex 5000, an anion exchange derivative of Sephadex LH-20 (Axelson and Sjövall, 1977). Elution with a reverse-phase solvent provided a fraction containing 5-androstene-3β,17β-diol (Figure 7). Non-polar lipids were then removed using a straight-phase solvent. Finally, the eluting solvent was acidified to permit recovery of the weakly acidic oestrogens (oestrone, oestradiol-17β and oestriol: Figure 7). Selected ion monitoring analysis of a cytosol extract for oestradiol, as the trimethylsilyl (TMS) ether, is shown in Figure 8. A detection limit for oestradiol of 2–4 pg/mg cytosol protein was established. Detection limits for oestrone (as the pentafluorobenzyl oxime, TMS ether) and 5-androstene-3β,17β-diol (as the *bis*-heptafluoro butyrate) were similar but that for oestriol (as the *tris*-TMS ether) was about 5-fold higher. The validation of each assay included the determination of steroid added to aliquots of cytosol. For the example of oestradiol, linear regression of amount of steroid determined (y) on amount added (x; 20–120 pg) gave the equation, $y = 1.11\,x + 19.1$ pg, with a

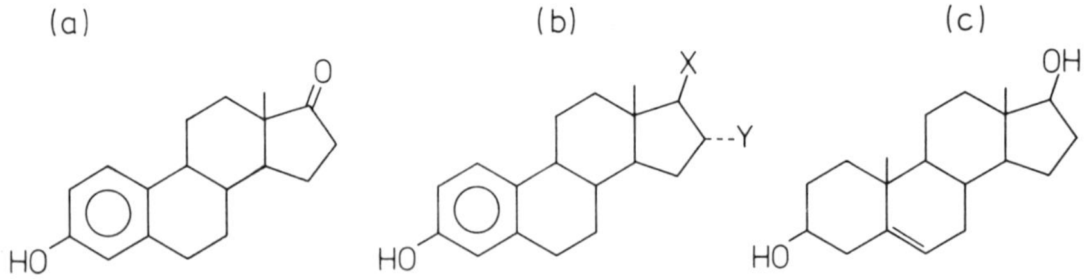

Figure 7: Structures of steroids determined in breast tumour tissue cytosol. a) oestrone; b) oestradiol (X = OH, Y = H), oestriol (X = Y = OH); c) 5-androstene-3β,17β-diol.

correlation coefficient of 0.996, indicating satisfactory precision and accuracy in the determination of added analyte. Similar data were obtained for oestrone and 5-androstene-3β,17β-diol, with a somewhat poorer correlation observed for oestriol.

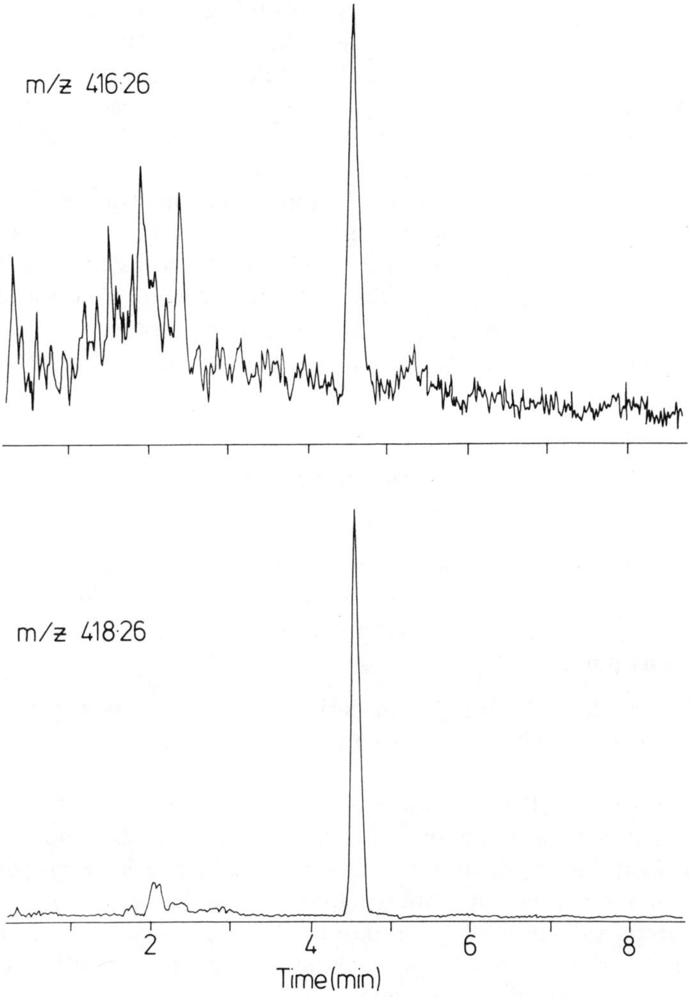

Figure 8: GC–MS/selected ion monitoring analyses for oestradiol-17β, as the *bis*-TMS ether, in a breast tumour cytosol extract. Upper trace = m/z 416.26 ($M^{+\cdot}$ for the analyte). Lower trace = m/z 418.26 ($M^{+\cdot}$ for the $[^{13}C]_2$-analogue). (Gas chromatography used a 7 m DB-1 bonded phase, fused silica column at 200°. The mass spectrometric resolution was 2500). The concentration of oestradiol-17β was determined to be 2.2 pg/mg cytosol protein.

The principal conclusions drawn from a study of forty-one breast tumours were as follows (Leith, H. M., Nicholson, R. I., Williams, M., and Gaskell, S. J., unpublished data):

(i) A wide range of oestradiol concentrations was observed (<2–316 pg/mg protein), with a tendency to higher concentrations in receptor-positive than receptor-negative tumours. The difference only reached statistical significance ($p = 0.05$), however, when concentrations were expressed relative to the DNA content of the tumour homogenate.

(ii) Oestriol was detected in only one sample. Oestrone was detected in two-thirds of the samples at a concentration of <10–950 pg/mg protein.

(iii) In some samples, the concentration of 5-androstene-3β,17β-diol was much higher (up to 70-fold) than that of oestradiol. The importance of this observation relates to the ability of 5-androstene-3β,17β-diol to displace oestradiol from the oestrogen receptor (Nicholson *et al.*, 1978). Relative concentrations such as those observed are sufficient to modify significantly the oestrogenic environment of the tumour cell.

19.5. CONCLUSIONS

Mass spectrometric techniques of great selectivity and sensitivity are now available for the analysis of steroids in a variety of physiological fluids and tissues. The examples given in this chapter were chosen particularly to illustrate the following points:

(i) An increase in MS resolution during conventional SIM and the use of selected reaction monitoring both significantly increase the selectivity of MS detection.

(ii) The quality of the analysis, however, is dependent on both instrumental and pre-instrumental techniques. In the steroid field, the introduction of immunoadsorption extraction for targeted analyses has been particularly beneficial in improving and simplifying procedures.

(iii) Careful validation of quantitative data is required. In particular, when the MS data are to be used to assess routine methods, validation at least as rigorous as that applied to the routine assays is essential.

In the area of analyses for non-conjugated steroids, particularly from tissue samples, it is likely that future improvements in methodology will focus on the isolation of 'cleaner' fractions for GC–MS determinations. The principal beneficial effect of such developments will be to maintain and improve GC performance with consequent improvements in both sensitivity and selectivity.

ACKNOWLEDGMENTS

We thank B. G. Brownsey, R. I. Nicholson, A. O. Turkes (all of the Tenovus Institute for Cancer Research, Cardiff) and M. Williams (City Hospital, Nottingham) for their collaboration and the Tenovus Organization for generous financial support.

REFERENCES

Abraham, G. E. (1969). Solid-phase radioimmunoassay of estradiol-17β, *J. Clin. Endocrinol. Metab.*, **29**, 866–870.

Axelson, M., and Sjövall, J. (1976). Selective liquid chromatographic isolation procedure for GC–MS analysis of 3-ketosteroids in biological materials, *J. Chromatog.*, **126**, 705–716.

Axelson, M., and Sjövall, J. (1977). Analysis of unconjugated steroids in plasma by liquid gel chromatography and glass capillary gas chromatography mass spectrometry, *J. Steroid Biochem.*, **8**, 683–692.

Björkhem, I., Blomstrand, R., Lantto, O., Svensson, L., and Ohman, G. (1976). Toward absolute methods in clinical chemistry: application of mass fragmentography to high-accuracy analyses, *Clin. Chem.*, **22**, 1789–1801.

Edery, M., Goussard, F., Dehennin, L., Scholler, R., Reiffsteck, J., and Drosdowsky, H. A. (1981). Endogenous oestradiol-17β concentration in breast tumours determined by mass fragmentography and by radioimmunoassay; relationship to receptor content. *Eur. J. Cancer*, **17**, 115–120.

Finlay, E. M. H., and Gaskell, S. J. (1981). Determination of testosterone in plasma from men by gas chromatography/mass spectrometry, with high-resolution selected-ion monitoring and metastable peak monitoring, *Clin. Chem.*, **27**, 1165–1170.

Finlay, E. M. H., Morton, M. S., and Gaskell, S. J. (1982). Identification and quantification of dehydroepiandrosterone sulphate in saliva, *Steroids*, **39**, 63–71.

Gaskell, S. J. (1983). Analysis of steroids by mass spectrometry, in *Methods of Biochemical Analysis*, vol. 29 (Ed. D. Glick), pp. 385–434, John Wiley, New York.

Gaskell, S. J. (1984). High resolution and metastable ions in biochemical trace analyses using mass spectrometry, *Kemia-kemi*, **11**, 12–14.

Gaskell, S. J., and Brooks, P. W. (1983). High sensitivity analyses of steroids using gas chromatography-negative chemical ionisation mass spectrometry, *Int. J. Mass Spectrom. Ion Phys.*, **48**, 241–244.

Gaskell, S. J., and Brownsey, B. G. (1983). Immunoadsorption to improve gas chromatography/high resolution mass spectrometry of estradiol-17β in plasma, *Clin. Chem.*, **29**, 677–680.

Gaskell, S. J., and Finlay, E. M. H. (1984). Identification and quantification of steroids in saliva by gas chromatography–mass spectrometry, in *Immunoassay of Steroids in Saliva; Proc. Ninth Tenovus Workshop* (Eds. G. F. Read, D. Riad-Fahmy, R. F. Walker, and K. Griffiths), pp. 33–42, Alpha Omega Publishing, Cardiff, U.K.

Gaskell, S. J., and Millington, D. S. (1978). Selected metastable peak monitoring; a new, specific technique in quantitative gas chromatography–mass spectrometry, *Biomed. Mass Spectrom.*, **5**, 557–558.

Gaskell, S. J., Finney, R. W., and Harper, M. E. (1979). The determination of testosterone in hamster prostate by gas chromatography mass spectrometry with selected metastable peak monitoring, *Biomed. Mass Spectrom.*, **6**, 113–116.

Gaskell, S. J., Finlay, E. M. H., and Millington, D. S. (1980a). The determination of testosterone in human blood plasma using multiple metastable peak monitoring, in *Advances in Mass Spectrometry 8* (Ed. A. Quayle), pp. 1908–1910, Heyden, London.

Gaskell, S. J., Finlay, E. M. H., and Pike, A. W. (1980b). Analyses of steroids in saliva using highly selective mass spectrometric techniques, *Biomed. Mass Spectrom.*, **7**, 500–504.

Gaskell, S. J., Pike, A. W., and Griffiths, K. (1980c). Analysis of testosterone and dehydroepiandrosterone in saliva by gas chromatography–mass spectrometry, *Steroids*, **36**, 219–228.

Gaskell, S. J., Collins, C. J., Thorne, G. C., and Groom, G. V. (1983). External quality assessment of assays for cortisol in plasma: use of target data obtained by gas chromatography/mass spectrometry, *Clin. Chem.*, **29**, 862–867.

Gaskell, S. J., Brownsey, B. G., and Groom, G. V. (1984). Analyses for progesterone in serum by gas chromatography/mass spectrometry: target data for external quality assessment of routine assays, *Clin. Chem.*, **30**, 1696–1700.

Gaskell, S. J., Porter, C. J., and Green, B. W. (1985). Trace analysis by GC/MS using a tandem double focusing/quadrupole instrument, *Biomed. Mass Spectrom.*, **12**, 139–141.

Glencross, R. G., Abeywardene, S. A., Corney, S. J., and Morris, H. S. (1981). The use of oestradiol-17β antiserum covalently coupled to sepharose to extract oestradiol-17β from biological fluids, *J. Chromatog. Biomed. Applic.*, **233**, 193–197.

Gould, V. J., Turkes, A. O., and Gaskell, S. J. (1986). Gas chromatography–mass spectrometric analysis of salivary testosterone with reference to diethylstilboestrol treated prostatic cancer patients, *J. Steroid Biochem.*, **24**, 563–567.

Leith, H. M., Truran, P. L., and Gaskell, S. J. (1986a). Quantification of progesterone in human saliva. *Biomed. Environ. Mass Spectrom.*, **13**, 257–261.

Millington, D. S. (1975). Determination of hormonal steroid concentrations in biological extracts by high resolution mass fragmentography, *J. Steroid Biochem.*, **6**, 239–245.

Mourisden, M., Palshof, T., Patterson, J., and Battersby, L. (1978). Tamoxifen in advanced breast cancer, *Cancer Treat. Rev.*, **5**, 131–141.

Nicholson, R. I., Davies, P., and Griffiths, K. (1978). Interaction of androgens with oestradiol-17β receptor proteins in DMBA-induced mammary tumours—a possible oncolytic mechanism, *Eur. J. Cancer*, **14**, 439–445.

Read, G. F., Riad-Fahmy, D., Walker, R. F., and Griffiths, K., Eds. (1984). *Immunoassays of Steroids in Saliva; Proc. Ninth Tenovus Workshop*, Alpha Omega Publishing, Cardiff, U.K.

Reiffsteck, A., Dehennin, L., and Scholler, R. (1982). Estrogens in seminal plasma of human and animal species: identification and quantitative estimation by gas chromatography–mass spectrometry associated with stable isotope dilution, *J. Steroid Biochem.*, **17**, 567–572.

Riad-Fahmy, D., Read, G. F., Gaskell, S. J., Dyas, J., and Hindawi, R. (1979). A simple, direct radioimmunoassay for plasma cortisol, featuring a ^{125}I-radioligand and a solid-phase separation technique, *Clin. Chem.*, **25**, 665–668.

Riad-Fahmy, D., Read, G. F., Walker, R. F., and Griffiths, K. (1982). Steroids in saliva for assessing endocrine function, *Endocrine Rev.*, **3**, 367–395.

Shackleton, C. H. L. (1983). Inborn errors of steroid biosynthesis: detection by a new mass-spectrometric method, *Clin. Chem.*, **29**, 246–249.

Shackleton, C. H. L. (1985). Analysis of steroids in *Glass Capillary Chromatography in Clinical Medicine and Pharmacology* (Ed. H. Jaeger), pp. 153–243, Marcel Dekker, New York.

Siekmann, L. (1979). Determination of steroid hormones by the use of isotope dilution–mass spectrometry; a definitive method in clinical chemistry, *J. Steroid Biochem.*, **11**, 117–123.

Sjövall, J., and Axelson, M. (1982). Newer approaches to the isolation, identification and quantitation of steroids in biological materials. *Vitamins Hormones*, **39**, 31–129.

Sweeley, C. C., Elliott, W. H., Fries, J., and Ryhage, R. (1966). Mass spectrometric determination of unresolved components in gas chromatographic effluents, *Anal. Chem.*, **38**, 1549–1553.

Thorne, G. C., and Gaskell, S. J. (1985). Dual metastable peak monitoring; application to the analysis of oestradiol-17β as the bis(tert-butyldimethylsilyl) ether, *Biomed. Mass Spectrom.*, **12**, 19–24.

Mass Spectrometry in Biomedical Research
Edited by S. J. Gaskell
© 1986 John Wiley & Sons Ltd.

Chapter 20

TRACE ANALYSIS OF THE CANNABINOIDS

David J. Harvey

20.1. INTRODUCTION

Cannabis, a drug derived from the plant *Cannabis sativa* L., is one of the oldest and most widely abused drugs known to man. Its use can be traced back to at least 3000 B.C. in China and today its use in Western countries is on the increase with as many as 50 million people in the USA estimated to have used the drug at least once. Use of the drug, which is usually by smoking, leads to a state of euphoria, well-being and hallucinations, with the effects of a single dose lasting for a few hours. Although the drug is commonly regarded as harmless, chronic use produces serious decrements in many body functions such as immune response, learning ability and reproduction, and the accumulation of drug in fat which accompanies repeated use can lead to a state of permanent cerebral depression resembling partial anaesthesia.

The major active component of this drug is delta-9-tetrahydrocannabinol (delta-9-THC, I, also known as delta-1-THC), one of over 60 related compounds termed cannabinoids, unique to *Cannabis* (Turner *et al.*, 1980). Other major cannabinoids include cannabidiol (CBD, II), a biochemical precursor of delta-9-THC, cannabinol (CBN, III), a chemical degradation product, and cannabichromene (CBC, IV). These compounds are largely devoid of the psychoactive effects of delta-9-THC but, like delta-9-THC, are present in *Cannabis sativa* in widely varying concentrations depending on the age, strain and geographical origin of the plant. Lower homologues of these and other cannabinoids, particularly homologues containing a propyl side chain, are common and in fresh plant material, most, if not all of the cannabinoids are found as their carboxylic acid derivatives of the type shown in structure V.

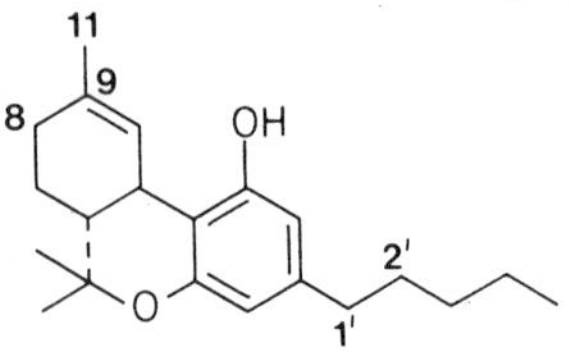

Δ9-THC (I)

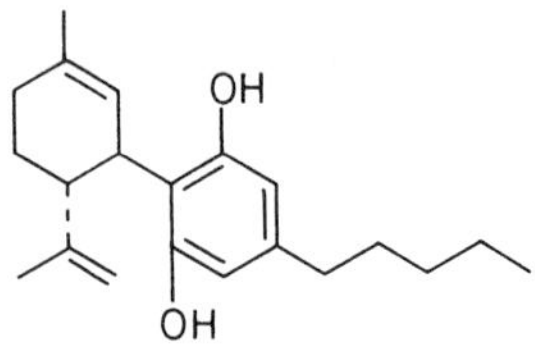

Cannabidiol (II)

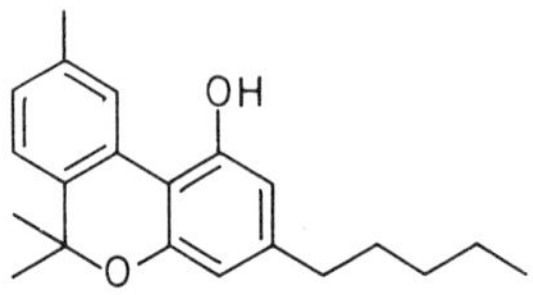

Cannabinol (III)

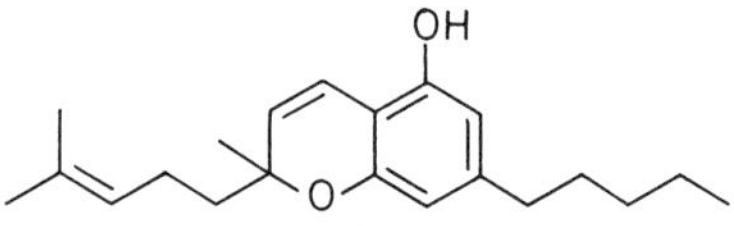

Cannabichromene (IV)

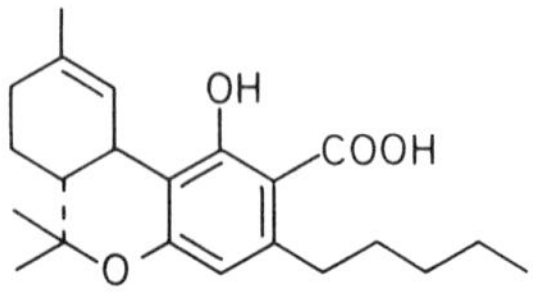

Δ9-THC-acid-A (V)

These acids are relatively unstable and rapidly decarboxylate when the drug is smoked.

Because of their high lipophilicity, the cannabinoids are good substrates for the enzyme cytochrome P-450 and metabolism is thus very extensive (Harvey and Paton, 1984); for example, over 100 metabolites have been reported from delta-9-THC alone. The metabolic routes shown by this compound are outlined in Figure 1 and are dominated in man and most animals by hydroxylation at C-11 and oxidation of the resulting pharmacologically active alcohol (VI) to the inactive delta-9-THC-11-oic acid (VII). This acid, in the form of its acid glucuronide, is the main urinary excretion product in man (Williams and Moffat, 1980). As little, if any, delta-9-THC itself is excreted in urine, identification and measurement of this acid is now of prime importance in the forensic determination of cannabis use (Law and Moffat, 1985). Another consequence of the drug's high lipophilicity is rapid uptake into tissues. This, coupled with its high activity, results in low maximum concentrations in blood which rapidly drop to values too low to be measured by most existing analytical techniques.

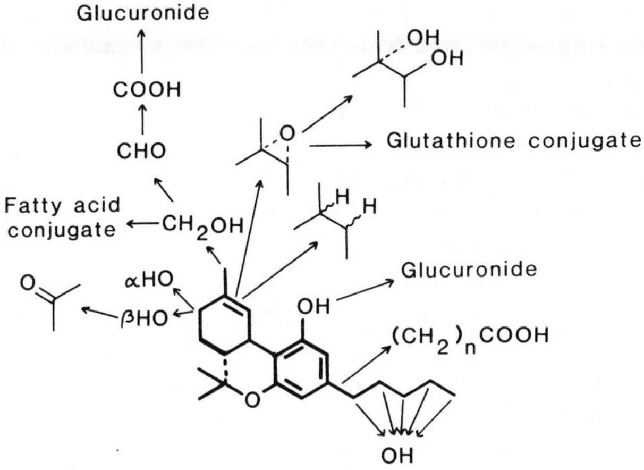

Figure 1: Metabolic routes shown by delta-9-THC.

For example, after a typical dose in man of 5–10 mg, maximum concentrations in blood are in the 100–500 ng/ml range and these fall to below 1 ng/ml by about 5 hours after administration (Agurell *et al.*, 1985). The most sensitive analytical techniques currently available have detection limits of about 0.1–1.0 ng/ml in plasma (Harvey, 1984) and this is not sufficient to enable the final phases of the pharmacokinetic curve to be determined in man. However, studies in animals, where higher doses and radiolabelled cannabinoids can more readily be given, have given values of about 30 hours to 8 days for the half-life of the final phase (Harvey and Paton, 1984; Garrett and Hunt, 1977). An interesting point to emerge from these studies is that the elimination phase does not appear in the pharmacokinetic curve until some 48 hours after dosing. Thus most published values of the half-life of the drug in humans, in the 30–60 hour range, are probably under-estimates.

The need for more sensitive techniques has prompted a number of studies in recent years and some very elegant methods have been developed (see Foltz, 1984; Harvey, 1984, 1985 for reviews). These are mainly based on electron-capture GLC, radio-immunoassay or GC–MS. Because of the cost involved with GC–MS, techniques based on radio-immunoassay are usually the methods of

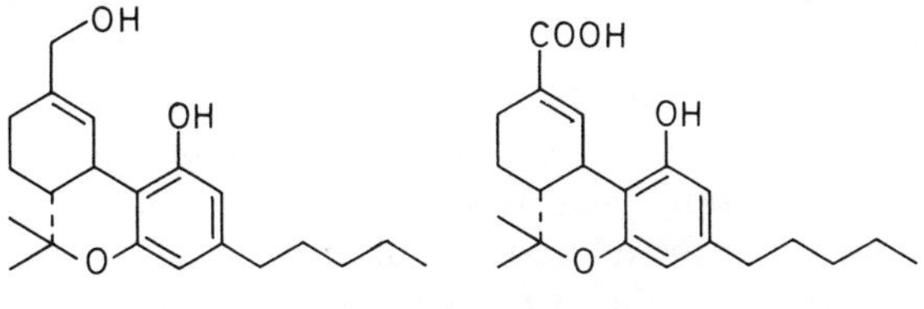

11-OH-Δ^9-THC (VI) 11-COOH-Δ^9-THC (VII)

choice for forensic work but, as this technique becomes less reliable at low concentrations because of cross-reactivity, GC–MS is usually used as a confirmatory method.

The first GC–MS method for the quantification of delta-9-THC in physiological fluids was published by Agurell *et al.*, (1973). The drug was extracted with 1.5% isoamyl alcohol in petroleum ether, concentrated by chromatography on Sephadex LH-20 and measured by GC–MS using a deuterated internal standard. The detection limit was about 1 ng/ml. The method has been improved considerably in recent years with, for example, the introduction of trimethylsilyl (TMS) derivatives but the basic technique of extraction, purification and measurement remains the same. Rosenthal *et al.* (1978) later developed a similar technique for the quantification of delta-9-THC and its two major metabolites, 11-hydroxy-delta-9-THC and delta-9-THC-11-oic acid with detection limits in the 0.5 ng/ml range for delta-9-THC. Experiments with hexane rather than with 1.5% amyl alcohol in petroleum ether showed that delta-9-THC could be efficiently extracted from plasma with this solvent; this did not produce an increase in sensitivity but did allow the clean-up stage to be dispensed with.

Several methods based on chemical ionization have been developed by Foltz *et al.* Hexane extraction coupled with methane : ammonia chemical ionization gave a detection limit of 0.5 ng/ml for the drug (Detrick and Foltz, 1976). Simultaneous measurement of delta-9-THC and its -11-oic acid metabolite was achieved in 1979 (Foltz *et al.*, 1979) with separation employing a 25 m SE-30 capillary column. The use of positive ammonia CI improved the detection limit for delta-9-THC to 0.2 ng/ml (Foltz and Hidy, 1982) in an assay for the drug and its two major metabolites and the method was reported to have been in use for 4 years at Batelle Columbus Laboratories. Over 2000 samples had been processed and one person could complete the analysis of 36 samples in two 8-hour days. In their latest method, Foltz *et al.* (1983) have used electron capture negative chemical ionization for the measurement of delta-9-THC, 11-hydroxy-delta-9-THC and delta-9-THC-11-oic acid to 0.2, 0.5 and 0.1 ng/ml respectively. Protein was first precipitated with acetonitrile and the cannabinoids were extracted with a 9 : 1 mixture of hexane and acetic acid. THC and the alcohol metabolite were extracted at high pH and the pH was then lowered for extraction of the acid.

Electron capture MS appears to be the method of choice for several drug assays at low concentration and has also been employed by Karlsson *et al.* (1983) for measurements of delta-9-THC-11-oic acid in urine. The metabolite was extracted with a mixture of hexane and diethyl ether after hydrolysis and converted, without further purification, into its pentafluoropropyl-pentafluoropropionyl derivative in a single-step reaction. Samples containing the deuterated internal standard were introduced into the mass spectrometer with a fused silica capillary column and the detection limit was below 1 ng/ml. As a demonstration of the sensitivity of negative electron-capture CI, the most sensitive detection of delta-9-THC reported to date has employed this technique on the pentafluorobenzoyl ester where as little as 10 fg of the pure compound could be detected (Hunt and Crow, 1978). Unfortunately, the method was not developed to measurements of the drug in biological fluids.

The great disparity between detection limits of this drug and, indeed of many others, when measured as the pure compound or when extracted from biological media, is due to the presence in the latter samples of co-extracted material which produces ions that interfere with those from the drug being measured. This produces noise, termed 'biological noise' which frequently reduces the level of detection, in the case of delta-9-THC, by a factor of 1000 or more. THC assays are particularly prone to interference of this type as the compound, which is unusual for a drug in that it contains no nitrogen, is chemically similar to many of the coextracted lipids contributing to the noise. With electron impact mass spectrometry, detection limits in the low picogram or upper femtogram range are attainable with the pure drug or, preferably, a derivative, but as can be seen from the above discussion, the best techniques for the measurement of the drug in biological media are limited to the 0.1–1.0 ng/ml range.

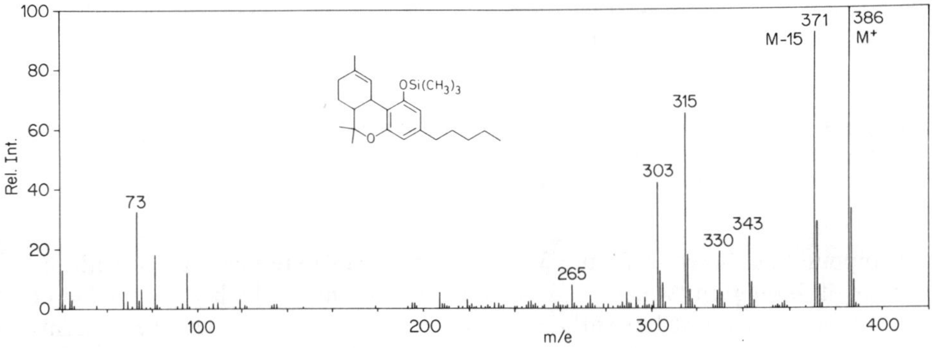

Figure 2: Electron impact (25 eV) mass spectrum of the TMS derivative of delta-9-THC.

In our original attempts to improve the sensitivity of detection of delta-9-THC we employed conventional methods with electron-impact mass spectrometry without much success. The mass spectrum of delta-9-THC and its TMS derivative are similar (Figure 2 shows the spectrum of the TMS ether) and contain several ions suitable for selected ion monitoring. The molecular ion possesses many of the properties necessary for high sensitivity detection; it is of relatively high mass, high relative abundance and at even mass. Single-ion recording of this ion enabled the pure drug as its TMS derivative to be detected to 500 fg using a packed SE-30 column interfaced to a VG 70/70F mass spectrometer. This was an improvement over detection of the underivatized

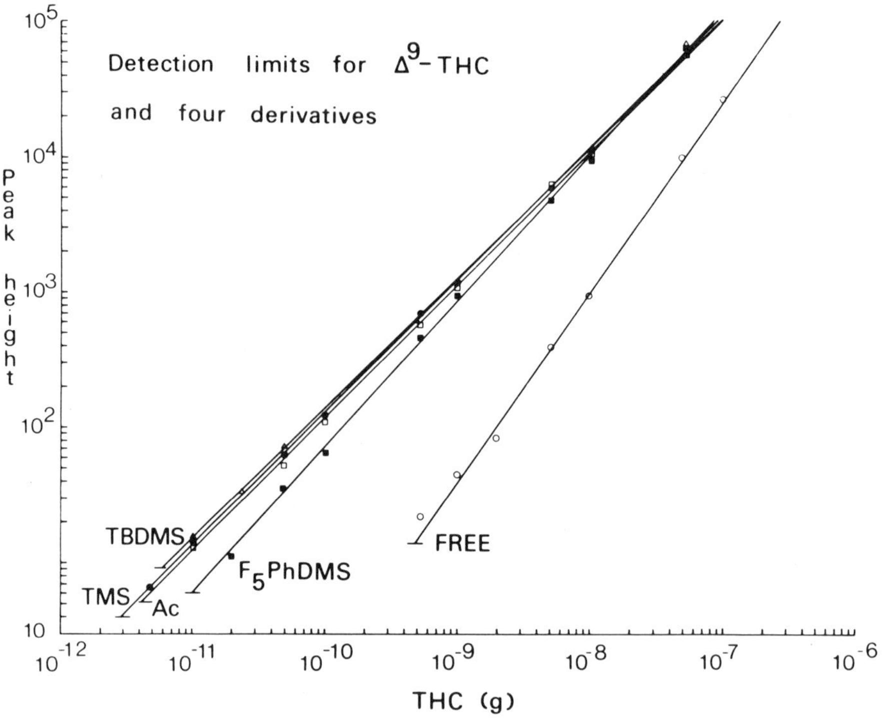

Figure 3: Plot of response against concentration for delta-9-THC and four of its derivatives. The horizontal lines at the bottom of each plot represent the noise level. *Reproduced from Harvey* et al. (*1979*), *by permission of Pergamon Press.*

compound as shown in Figure 3. Other derivatives were investigated with the hope of finding one which produced an ion containing a higher proportion of the total ion current but without much success. Figure 3 shows the concentration versus response curves for the TMS, *tert*-butyldimethylsilyl (TBDMS), pentafluorophenyldimethylsilyl and acetyl derivatives and it is apparent that similar detection limits are achieved with all four derivatives with the TMS derivative being marginally the best (Harvey *et al.*, 1979). (Since the data in Figure 3 were recorded, the sensitivity of the mass spectrometer has been increased by a factor of about 5.) Other derivatives gave similar results (Harvey, 1978) and this was attributed to the fragmentation of the molecule being directed by charge localization on the heterocyclic oxygen atom (Harvey, 1981). Thus the spectra of most derivatives were largely independent of the nature of the derivative. In fact, those derivatives such as the TBDMS ether which did produce fragment ions from the derivatized phenol moiety tended to do so in competition with ions whose production was initiated by charge localization on the heterocyclic oxygen atom, thus further reducing the percentage of ion current carried by a particular ion. Attempts to improve sensitivity by monitoring ions other than the molecular ion in these derivatives did not produce any increase in sensitivity. Clearly a different approach was needed.

20.2. THE METASTABLE ION MONITORING METHOD

Two methods were being investigated at the time of this early work, in addition to selective ionization, for trace analysis by selected ion monitoring, high resolution selected ion monitoring and the technique of metastable ion monitoring just introduced by Gaskell and Millington (1978) for steroid measurements. The latter technique appeared more appropriate as our instrument did not have the peak locking facility necessary for high resolution monitoring. Both methods depend for their sensitivity on improving the signal-to-noise ratio by selective rejection of ions produced from fragmentation of endogenous compounds co-extracted with the drug. High resolution monitoring selects ions of a given composition and has been used for delta-9-THC measurements by Valentine *et al.* (1977) at a resolution of 2500. The detection limit reported was, however not as good as later methods employing negative ion CI. Metastable ion monitoring selects fragment ions formed in metastable transitions in the first field-free region of a double focusing mass spectrometer and thus imposes the constraint that the ion must have arisen from a given precursor. In this way, many of the unwanted, endogenously-derived, fragment ions can be rejected. This method was adopted in our technique for delta-9-THC measurement (Harvey *et al.*, 1980).

An examination of the mass spectrum of the TMS derivative of delta-9-THC showed that the $M^{+\cdot}$ (m/z 386) to $[M - CH_3]^+$ (m/z 371) transition gave the most abundant metastable ion and even though the loss of a methyl group was not very specific to THC, it was felt that the abundance of the ion would result in the best sensitivity. Focusing of the VG 70/70F mass spectrometer was achieved by tuning the magnet to m/z 371 with normal electric sector and acceleration potentials and then raising the accelerating voltage from its normal setting of 4.0 kV to 4.16 kV to bring the metastable ion into focus. The ion-source controls were then adjusted for maximum sensitivity. Ions of mass 371 which were formed in other transitions possessed the wrong energy to be transmitted by the electric sector and were thus filtered out. Delta-9-THC, as its TMS derivative, was injected *via* a 2 m SE-30 packed column at 230°C and the signal was recorded with a chart recorder connected to the main amplifier which was operated with a response of 1 s to reduce noise. The detection limit was determined at 500 fg for the pure drug, comparable to that obtained by conventional ion recording but with the amplifier operating at a gain 10 times greater.

The highest gain and precision in GC–MS assays is obtained using single ion recording as this maximizes the signal and eliminates errors introduced by field switching between the peak from the analyte and that from its internal standard (Lee and Millard, 1975). In order to focus a stable-isotope labelled internal standard in metastable ion monitoring it is generally necessary to change two fields. Therefore, to maximize precision, the technique of single ion monitoring was adopted. This required an internal standard having a metastable ion in common with that of delta-9-THC but which separated on the GLC column. At the same time, a structure similar to that of delta-9-THC was preferable in order to minimize extraction and derivatization errors. Isomeric THCs were investigated but found to be insufficiently separated on the packed column employed. Among the other cannabinoids which could be separated from delta-9-THC by GLC was CBN. This cannabinoid showed a similar metastable transition, $M^{+\cdot} \rightarrow [M - CH_3]^+$, following electron impact but had a mass which was four mass units too low. This mass discrepancy was rectified by the introduction of four deuterium atoms to give [1′,1′,2′,2′-$^2H_4$]CBN which had the same molecular weight as delta-9-THC and yielded a similar metastable ion. Its response was also similar to that of delta-9-THC as shown in Figure 4a which shows a metastable ion chromatogram of delta-9-THC and CBN corresponding to 0.01 ng of delta-9-THC injected onto the column.

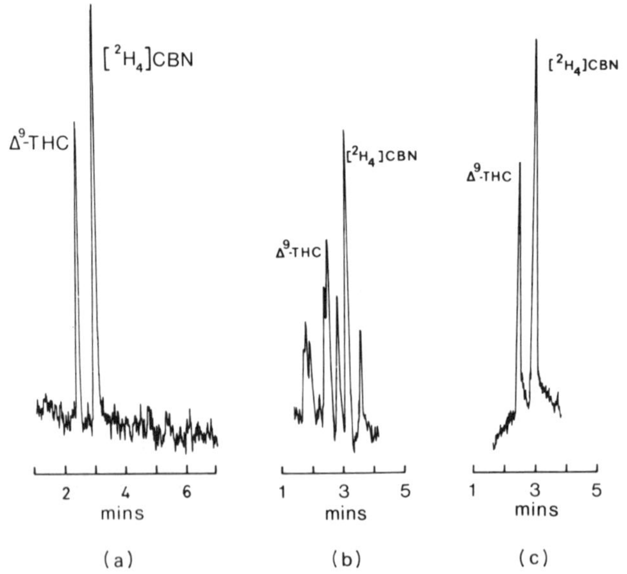

Figure 4: Metastable ion chromatograms for GC–MS of: (a) 0.01 ng each of the TMS derivatives of delta-9-THC and [1′, 1′, 2′, 2′ – $^2H_4$]CBN, (b) the same compounds extracted from plasma without methylation; (c) the same with methylation. *Reproduced from Harvey* et al. (*1984*) *by permission of Academic Press.*

Hexane appeared to be the best solvent for extraction of delta-9-THC from plasma as its low polarity should minimize the concentration of lipid material co-extracted with the drug and avoid the need for further purification with its attendant loss of sample. Delta-9-THC and the internal standard were thus added to plasma, extracted with hexane and examined directly as TMS derivatives by metastable ion monitoring. The result, shown in Figure 4b, was disappointing as a considerable amount of background remained. Nevertheless, the peak produced by the drug was clearly visible at the 0.01 ng level. No signficant reduction in the noise level was produced by changing the GLC phase. However, by working on the assumption, based on their GLC behaviour, that the contaminating compounds were fatty acids, it was found that by treating the extract with diazomethane prior to TMS ether formation, it was possible to derivatize these compounds selectively and to remove them from the chromatogram. This left a clean background against which the drug could be detected in plasma to 0.001 ng/ml. Figure 4c shows the chromatogram obtained by this technique corresponding to 0.01 ng of delta-9-THC added to plasma and extracted with an efficiency of 70%. No reaction of THC or CBN with diazomethane was observed even after prolonged standing.

Plasma

↓

Add internal standard ([$^2H_4$] CBN), leave for 30 min.

↓

Extract (× 3) with hexane, centrifuge

↓

Methylate with diazomethane

↓

TMS ether formation (BSTFA)

↓

GC–MS, metastable ion monitoring of m/z 386 to m/z 371
(VG Micromass 70/70 F, 2 m × 2 mm SE-30, 70 eV)

Scheme 1
Metastable ion monitoring of delta-9-THC in plasma.

The final technique (Scheme 1) thus involved adding the internal standard, dissolved in ethanol, to the plasma sample, extracting three times with redistilled hexane with centrifugation at 1500 g to break the resulting emulsion, treatment with ethereal diazomethane and conversion of the cannabinoids to TMS ethers with BSTFA. No clean-up stage was required and the procedure could be completed in about 1 hour. The calibration was linear from 1000 ng/ml to 0.005 ng/ml and at the 0.03 ng range delta-9-THC could be measured with a precision of 12.1% (CV, $n = 5$). With human plasma, however, there is still room for improvement as some samples contain very much more background than is found in animal plasma and this is not removed by reaction with diazomethane. Even so, the method is more sensitive than even the negative CI-based techniques discussed above.

20.3. APPLICATIONS AND FURTHER TECHNIQUE DEVELOPMENT

The technique has been used to study the pharmacokinetics of delta-9-THC in rabbits following intravenous injection into the marginal ear vein. After a single dose of 1 mg/kg the drug could be measured for 7–10 days, depending on the rabbit, with terminal half-lives in the range of 30–66 hours, and after a single dose of 0.1 mg/kg the drug could be measured for 3.5 days (Figure 5). After

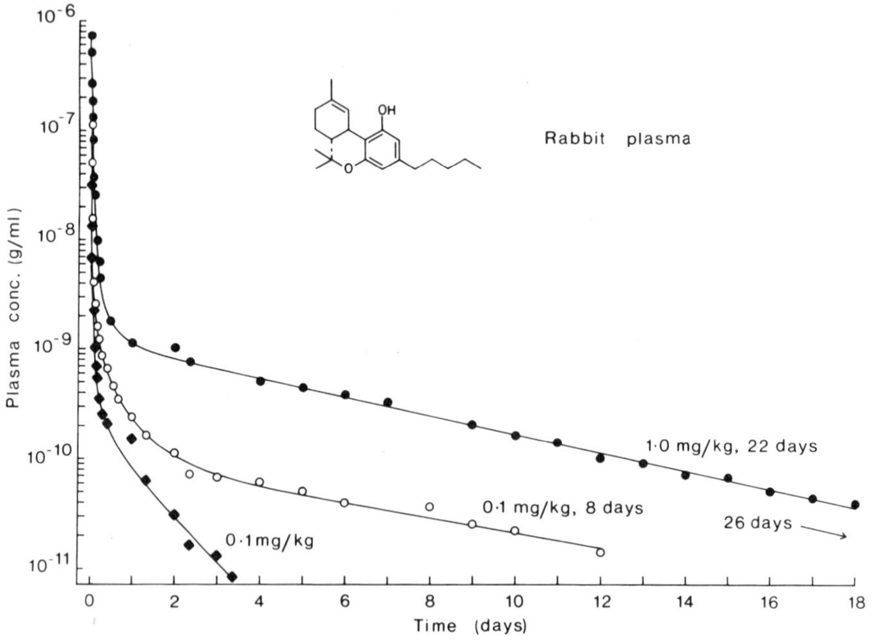

Figure 5: Concentration–time plots of the concentration of delta-9-THC in rabbit plasma after a single 0.1 mg intravenous dose (◆), after 8 daily injections of 0.1 mg delta-9-THC (O) and after 22 doeses of the drug at 1.0 mg/kg (●). *Reproduced from Harvey* et al. *(1982) by permission of Elsevier Science Publishers.*

repeated administration, considerable accumulation of the drug occurred, and longer half-lives were recorded (Figure 5). Thus after daily injections of 0.1 mg for 8 days, the drug could be measured for 12 days, and after 1 mg/kg for 22 days it was possible to follow the decline in concentration of the drug in plasma for 26 days when it showed a half-life of 83.0 hours. Use of the method for measurements of the drug in tissues required the introduction of a clean-up stage based on Sephadex LH-20 column chromatography to remove the excess of lipid and gave a detection limit of 50 ng/g wet weight. This method was used to show accumulation of the drug in fat (Harvey *et al.*, 1981, 1982, 1984b) and to determine its distribution in the eye (Leuschner *et al.*, 1985). The method was sensitive enough to enable the drug to be measured in mouse erythrocyte membranes after intravenous administration (Leuschner *et al.*, 1984).

Extension of the method to the measurement of metabolites required field switching as different metastables were present for each compound (Harvey *et al.*, 1984a). This was achieved by keeping the accelerating voltage constant at 4.0 kV, setting the electric sector voltage by means of the eight channel peak-selector to the value determined by M_2/M_1, where M_1 and M_2 are the masses of the parent and daughter ions respectively, and setting the magnetic field manually to the value given by M_2^2/M_1. These values are given in Table 1 for delta-9-THC and 11-hydroxy-delta-9-THC as their TMS derivatives and for delta-9-THC-11-oic acid as its methyl ester, TMS ether. In a typical experiment the instrument was set up as follows. The fields were adjusted to the values given in Table 1 to record delta-9-THC and the internal standard. After these compounds had appeared, the electric sector voltage was set to the value for the hydroxy-metabolite by switching the eight channel peak-selector to channel 2 and the magnetic field was set manually with the aid of the mass marker. Finally the fields were switched again to record the signal from the acid metabolite. A

TABLE 1. Derivatives and focusing conditions for metastable ion monitoring of delta-9-THC and its two major metabolites.

Compound	Derivative	Fragmentation	Metastable	Electric sector voltage[a]
THC	TMS	386 to 371	356.6	961.1
[$^2H_4$] CBN	TMS	386 to 371	356.6	961.1
11-OH-THC	TMS	474 to 371	290.4	782.7
THC-11-COOH	Me : TMS	430 to 371	320.1	862.8

[a] Normal setting = 1000.0

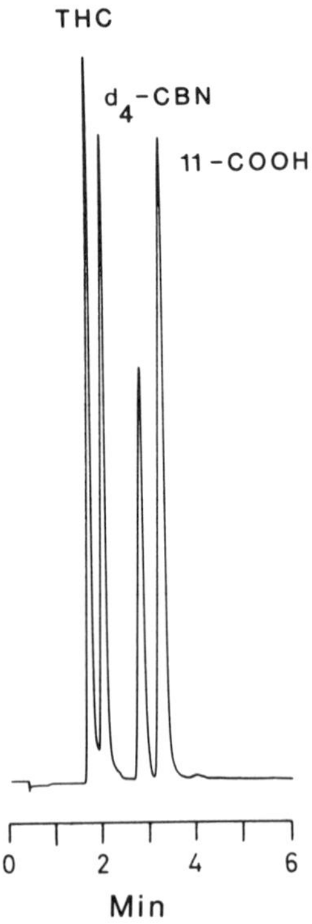

Figure 6: Metastable ion chromatogram for GC–MS 1.0 ng each of delta-9-THC, [$^2H_4$]CBN, 11-hydroxy-delta-9-THC and delta-9-THC-11-oic acid as TMS and methyl ester, TMS derivatives.

typical chromatogram, recorded from standard compounds, is shown in Figure 6. Detection limits were: delta-9-THC, 0.005 ng; 11-hydroxy-delta-9-THC, 0.01 ng; delta-9-THC-11-oic acid, 0.005 ng.

For extraction of the cannabinoids from plasma in a single fraction, it was necessary to employ a solvent more polar than hexane. None of the acid metabolite could be recovered with hexane although the hydroxy-compound could be extracted to the extent of about 35%. Ethyl acetate was adopted as the extraction solvent but this required a clean-up stage to be introduced. A small column of Sephadex LH-20 packed into a 50 × 5 mm Pasteur pipette was found to be suitable; the cannabinoids were collected in the fraction eluted between 1.5 and 7.5 ml with 10% methanol in chloroform. The method is summarized in

Plasma

↓

Add internal standard ([$^2H_4$] CBN), leave for 30 min.

↓

Extract ($\times$ 3) with ethyl acetate, centrifuge

↓

Chromatograph on Sephadex LH–20 in chloroform

↓

Methylate with diazomethane

↓

TMS ether formation (BSTFA)

↓

GC–MS, metastable ion monitoring on 3 channels according to Table 1
(VG Micromass 70/70 F, 2 m $\times$ 2 mm SE-30, 70 eV)

Scheme 2

Metastable ion monitoring of delta-9-THC and its two major metabolites in plasma.

Scheme 2. It was initially feared that monoglycerides, particularly mono-palmitin which has a retention time on SE-30 and a mass spectrum very similar to that of the TMS derivative of the 11-hydroxy-metabolite (Harvey and Paton, 1980), would interfere with measurement of this metabolite. However, because the molecular ion is very unstable, so little survives to the first field-free region that the metastable ion abundance is negligible. Detection limits for the cannabinoids were 0.1 ng/ml for delta-9-THC and the hydroxy-metabolite and 0.05 ng/ml for the acid. A typical chromatogram is shown in Figure 7. The method is still not fully developed but shows promise, particularly for the acid metabolite which, as stated above, is of most interest to forensic scientists.

Mass spectrometry has thus played a considerable part in the trace analysis of the cannabinoids but there is some way yet to go before the methods are sufficiently sensitive to define the pharmacokinetics of the drug fully in humans. A combination of some of the above techniques such as the use of a capillary column in the metastable assay should lead to some improvement. The most useful approach for improving the sensitivity of this assay will probably be the use of a multi-analyzer instrument in order to improve the resolution of the metastable ions. However, the cost of these instruments will probably restrict their use to a few specialist laboratories even though their wider availability could have a considerable impact on trace analysis in general.

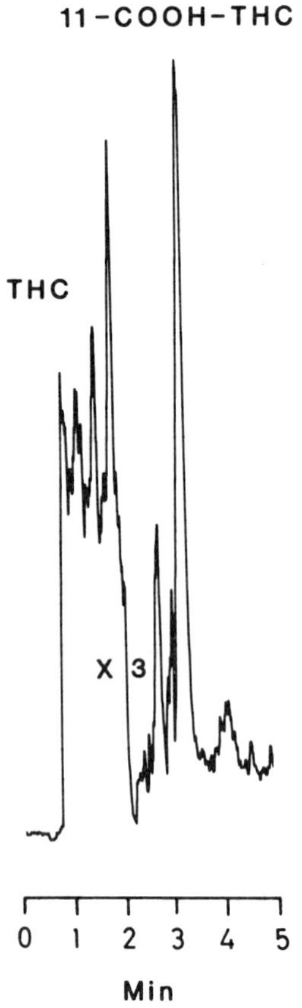

Figure 7: Metastable ion chromatogram for GC–MS of the compounds shown in Figure 6 extracted from plasma.

REFERENCES

Agurell, S., Gustafsson, B., Holmstedt, B., Leander, K., Lindgren, J.-E., Nilsson, I., Sandberg, F., and Asberg, M. (1973). Quantitation of delta-1-THC in plasma from cannabis smokers, *J. Pharm. Pharmacol.*, **25**, 554–558.

Agurell, S., Gillespie, H., Halldin, M., Hollister, L. E., Johansson, E., Lindgren, J. E., Ohlsson, A., Szirmai, M., and Widman, M. (1985). A review of recent studies on the pharmacokinetics and metabolism of delta-1-tetrahydrocannabinol, cannabinol and cannabidiol in man, in *Marihuana '84; Proceedings of the Oxford Symposium on Cannabis* (Ed. D. J. Harvey), pp. 49–62, IRL Press, Oxford.

Detrick, R., and Foltz, R. (1976). Quantitation of delta-9-tetrahydrocannabinol in body fluids by gas chromatography/chemical ionization mass spectrometry, in *Cannabinoid Assays in Humans, Res. Monogr. Ser.* No. 7. (Ed. R. E. Willette), pp. 88–95, NIDA, Rockville.

Foltz, R. (1984). Analysis of cannabinoids in physiological specimens by gas chromatography/mass spectrometry, *Adv. Anal. Toxicol.*, **1**, 125–157.

Foltz, R. L., Clark, P. A., Hidy, B. J., Lin, D. C. K., Graffeo, A. P., and Petersen, B. A. (1979). Quantitation of delta-9-tetrahydrocannabinol and 11-nor-delta-9-tetrahydrocannabinol-9-carboxylic acid in body fluids by GC/CI–MS, in *Cannabinoid Analysis in Physiological Fluids, ACS Symp. Ser.* No. 98 (Ed. J. A. Vinson), pp. 59–71, American Chemical Society, Washington, D.C.

Foltz, R. L., and Hidy, B. J. (1982). Quantitative analysis for delta-9-THC, 11-hydroxy-delta-9-THC and 9-carboxy-delta-9-THC in plasma using GC/CI–MS, in *The Analysis of Cannabinoids in Biological Fluids*, Res. Monogr. Ser. No. 42 (Ed. R. L. Hawks), pp. 99–118, NIDA, Rockville.

Foltz, R. L., McGinnis, K. M., and Chin, D. M. (1983). Quantitative measurement of delta-9-tetrahydrocannabinol and two major metabolites in physiological specimens using capillary column gas chromatography negative chemical ionization mass spectrometry, *Biomed. Mass Spectrom.*, **10**, 316–323.

Garrett, E. R., and Hunt, C. A. (1977). Pharmacokinetics of delta-9-tetrahydrocannabinol in dogs, *J. Pharm. Sci.*, **66**, 395–407.

Gaskell, S. J., and Millington, D. S. (1978). Selected metastable peak monitoring: a new specific technique in quantitative gas chromatography mass spectrometry, *Biomed. Mass Spectrom.*, **5**, 557–558.

Harvey, D. J. (1978). Comparison of fourteen substituted silyl derivatives for the characterization of alcohols, steroids and cannabinoids by combined gas-liquid chromatography and mass spectrometry, *J. Chromatogr.*, **147**, 291–298.

Harvey, D. J. (1981). The mass spectra of the trimethylsilyl derivatives of delta-1- and delta-6-tetrahydrocannabinol, *Biomed. Mass Spectrom.*, **8**, 575–578.

Harvey, D. J. (1984). Analysis of the cannabinoids, in *Analytical Methods in Human Toxicology*, Part 1 (Ed. A. S. Curry), pp. 257–310, Macmillan, London.

Harvey, D. J. (1985). Advances in methods for detection and measurement of the cannabinoids, in *Marihuana '84; Proceedings of the Oxford Symposium on Cannabis* (Ed. D. J. Harvey), pp. 121–136, IRL Press, Oxford.

Harvey, D. J., and Paton, W. D. M. (1980). The use of deuterium labelling in structural and quantitative studies of tetrahydrocannabinol metabolites by mass spectrometry, *Adv. Mass Spectrom.*, **8**, 1194–1203.

Harvey, D. J., and Paton, W. D. M. (1984). Metabolism of the cannabinoids, *Rev. Biochem. Toxicol.*, **6**, 221–264.

Harvey, D. J., Martin, B. R., and Paton, W. D. M. (1979). Identification and measurement of cannabinoids and their in vivo metabolites in liver by gas chromatography–mass spectrometry, in *Marihuana; Biological Effects* (Eds. G. G. Nahas and W. D. M. Paton), pp. 45–62, Pergamon, Oxford.

Harvey, D. J., Leuschner, J. T. A., and Paton, W. D. M. (1980). Measurement of delta-1-tetrahydrocannabinol in plasma to the low picogram range by gas chromatography–mass spectrometry using metastable ion detection, *J. Chromatogr.*, **202**, 83–92.

Harvey, D. J., Leuschner, J. T. A., and Paton, W. D. M. (1981). Pharmacokinetics and metabolism of delta-1-tetrahydrocannabinol in rabbits and mice, *Br. J. Pharmacol.*, **74**, 771p.

Harvey, D. J., Leuschner, J. T. A., and Paton, W. D. M. (1982). Gas chromatographic and mass spectrometric studies on the metabolism and pharmacokinetics of delta-1-tetrahydrocannabinol in the rabbit, *J. Chromatogr.*, **239**, 243–250.

Harvey, D. J., Leuschner, J. T. A., and Paton, W. D. M. (1984a). Pharmacokinetic studies on delta-1-tetrahydrocannabinol and its major metabolites using metastable ion monitoring, *Spectroscopy Int. J.*, **3**, 195–209.

Harvey, D. J., Leuschner, J. T. A., Wing, D. R., and Paton, W. D. M. (1984b). Pharmacokinetics of delta-9-THC in rabbits and mice with measurements to the low picogram range by combined gas chromatography/mass spectrometry using metastable ion detection, in *The Cannabinoids: Chemical, Pharmacologic and Therapeutic Aspects* (Eds. S. Agurell, W. L. Dewey and R. E. Willette), pp. 291–308, Academic Press, Orlando, Florida.

Hunt, D. F., and Crow, F. W. (1978). Electron capture negative ion chemical ionization mass spectrometry, *Anal. Chem.*, **50**, 1781–1784.

Karlsson, L., Jonsson, J., Aberg, K., and Roos, C. (1983). Determination of delta-9-tetrahydrocannabinol-11-oic acid in urine as its pentafluoropropyl-pentafluoropropionyl derivative by GC/MS utilizing negative chemical ionization, *J. Anal. Toxicol.*, **7**, 198–202.

Law, B., and Moffat, A. C. (1985). The influence of the metabolism and elimination of cannabinoids on forensic analysis and interpretation, in *Marihuana '84; Proceedings of the Oxford Symposium on Cannabis* (Ed. D. J. Harvey), pp. 197–204, IRL Press, Oxford.

Lee, M. G., and Millard, B. J. (1975). A comparison of labelled and unlabelled internal standards for quantification by single and multiple ion monitoring, *Biomed. Mass Spectrom.*, **2**, 78–81.

Leuschner, J. T. A., Wing, D. R., Harvey, D. J., Brent, G. A., Dempsey, C. E., Watts, A., and Paton, W. D. M. (1984). The partitioning of delta-1-tetrahydrocannabinol into erythrocyte membranes *in vivo* and its effect on membrane fluidity. *Experientia*, **40**, 866–868.

Leuschner, J. T. A., Harvey, D. J., and Paton, W. D. M. (1985). Distribution of delta-1-tetrahydrocannabinol in the eye and cerebrospinal fluid of rabbits, in *Marihuana '84; Proceedings of the Oxford Symposium on Cannabis* (Ed. D. J. Harvey), pp. 93–100, IRL Press, Oxford.

Rosenthal, D., Harvey, T. M., Bursey, J. T., Brine, D. R., and Wall, M. E. (1978). Comparison of gas chromatography mass spectrometry methods for the determination of delta-9-tetrahydrocannabinol in plasma, *Biomed. Mass Spectrom.*, **5**, 312–316.

Turner, C. E., ElSohly, M. A., and Boeren, E. G. (1980). Constituents of *Cannabis sativa* L. XVII. A review of the natural constituents, *J. Nat. Prods.*, **43**, 169–234.

Valentine, J. L., Bryant, P. J., Gutshall, P. L., Gan, O. H. M., Lovegreen, P. D., Thompson, E. D., and Niu, H. C. (1977). High pressure liquid chromatographic mass spectrometric determination of delta-9-tetrahydrocannabinol in human plasma following marijuana smoking. *J. Pharm. Sci.*, **66**, 1263–1266.

Williams, P. L., and Moffat, A. C. (1980). Identification in human urine of delta-9-tetrahydrocannabinol-11-oic acid glucuronide; a tetrahydrocannabinol metabolite, *Biochem. Pharmacol.*, **32**, 445–448.

Mass Spectrometry in Biomedical Research
Edited by S. J. Gaskell
© 1986 John Wiley & Sons Ltd.

Chapter 21

ANALYSIS OF DRUGS IN BIOLOGICAL MATRICES USING GAS CHROMATOGRAPHY– ELECTRON CAPTURE NEGATIVE CHEMICAL IONIZATION MASS SPECTROMETY

William A. Garland and Michael P. Barbalas

21.1. INTRODUCTION

In pharmaceutical research, sensitive and specific analytical methods are required to, first, measure the concentrations of a drug and active metabolites of a drug in blood or tissue from either man or a pharmacological test animal and, second, determine the structures of metabolites of a drug. The first requirement is significant because of the current practice of using pharmacokinetic parameters derived from blood concentration measurements as the experimental standard upon which to base decisions concerning a drug's bioavailability, interaction with other drugs, dose proportionality, optimum dosing regimen, etc. (Cabana, 1984). The second requirement is significant because data on the identity and formation of metabolites are needed both to understand the mechanism of action or toxicity of a drug, and to assist in the design of improved drugs (Nelson, 1982).

This chapter is a personal evaluation of the utility of gas chromatography–electron capture negative chemical ionization mass spectrometry (GC–EC/NCI MS) to meet the two requirements stated above.

EC/NCI MS uses the low energy electrons generated when the high energy

(> 200 eV) electrons emitted from the filament in a chemical ionization source are consumed in producing primary ions. For example, the following reactions are observed in a typical CI ion source with methane (used for all the work reported here) as the reagent gas:

$$\left.\begin{array}{l} e^- \text{ (high energy)} + CH_4 \rightarrow 2e^- \text{ (low energy)} + CH_4^{+\cdot} \\ e^- \text{ (high energy)} + CH_4 \rightarrow 2e^- \text{ (low energy)} + CH_3^+ + H^\cdot \end{array}\right\} \text{Primary ions}$$

These primary ions are consumed in reactions with un-ionized methane to form the well-known positive chemical ionization (PCI) reagent ions, CH_5^+ (*m*/*z* 17) and $C_2H_5^+$ (*m*/*z* 29)

$$\left.\begin{array}{l} CH_4^{+\cdot} + CH_4 \rightarrow CH_5^+ + CH_3^\cdot \\ CH_3^+ + CH_4 \rightarrow C_2H_5^+ + H_2 \end{array}\right\} \text{Secondary ions}$$

The energy of the generated low energy electrons is further reduced to near thermal energy (0 – 0.1 eV) by non-reacting collisions with un-ionized methane. These non-reacting collisions are important because electron capture reactions ($AB + e^- \rightarrow AB^-$) have significant cross sections only at thermal or near thermal electron energies; dissociative resonance electron capture reactions ($AB + e^- \rightarrow A^\cdot + B^-$) and ion pair production reactions ($AB + e^- \rightarrow A^+ + B^- + e^-$), two potential loss mechanisms for molecular weight information, occur at 0–15 eV and > 10 eV, respectively,

Additional discussion of reactions in the chemical ionization source is found in Chapter 22 (sections 22.2–22.5).

21.2. QUANTIFICATION

For quantitative purposes, the ideal ionization technique would generate an intense ion characteristic of the molecular weight of the drug, produce no fragment ions of the drug, and not generate any ions from the biological matrix. Formation of a molecular anion is favoured by a high electron affinity of the analyte, a low energy for the ionizing electron, and frequent stabilizing collisions of the anion with neutral reagent gas molecules to remove excess internal energy. This latter consideration is important since negative ions can easily eject an electron to lose excess internal energy, a neutralization mechanism not available to the more stable positive ions. A high electron affinity is shown by compounds with (Poole and Zlatkis, 1980):

(a) either a halogen or nitro group attached to a π-bonded carbon system;
(b) a highly conjugated carbon system;
(c) a conjugated system attached to a weak electron capturing group such as carbonyl.

Since a low electron affinity is shown by alcohols, amines, phenols, aliphatic saturated aldehydes, thioethers, ethers, fatty acid esters, hydrocarbons and halogen atoms attached to a vinyl carbon (Poole and Zlatkis, 1980), EC/NCI MS also has the promise of having a low response towards most potential interfering compounds in a biological matrix.

21.2.1. DRUGS WITH INTRINSIC EC RESPONSE

The class of compound most studied by EC/NCI MS is the 1,3-dihydro-5-phenyl-1,4-benzodiazepines (Garland and Miwa, 1980, 1983; Garland and Min, 1979; Miwa *et al.*, 1981; Rubio *et al.*, 1982a,b). Chemicals in this class are used extensively in medicine as anxiolytic (Valium ®, Librium ®, Serax ®, Ativan ®, Verstran ®, Tranxene ®, Xanax ®), anticonvulsant (Clonopin ®) and hypnotic (Dalmane ®, Mogadon ®, Rohypnol ®, Halcion ® and Versed ®) agents. Because of the high potency of these compounds and their large volume of distribution, they are present, following dosing, at relatively low concentrations in blood.

The structures of a common 1,3-dihydro-5-phenyl-1,4-benzodiazepine, diazepam (Valium ®), and one of its principal metabolites, nordiazepam, are given below. Using the criteria previously stated, both compounds should have a high

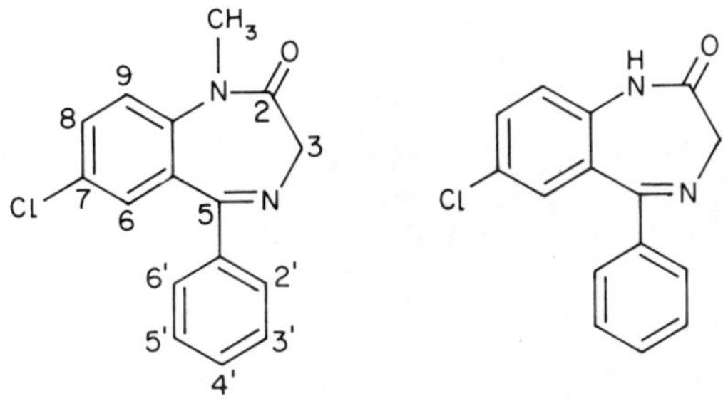

diazepam nordiazepam

electron affinity because a halogen atom (chlorine) is attached to a conjugated carbon system and this prediction is realized in practice. Typical EC/NCI mass spectra of diazepam and nordiazepam are given in Figures 1a and 2a. It is significant that the signal for the $M^{-\cdot}$ ions of diazepam and nordiazepam in these spectra are approximately 50 times more intense than the corresponding $(M + H)^+$ ions generated by methane positive chemical ionization (PCI).

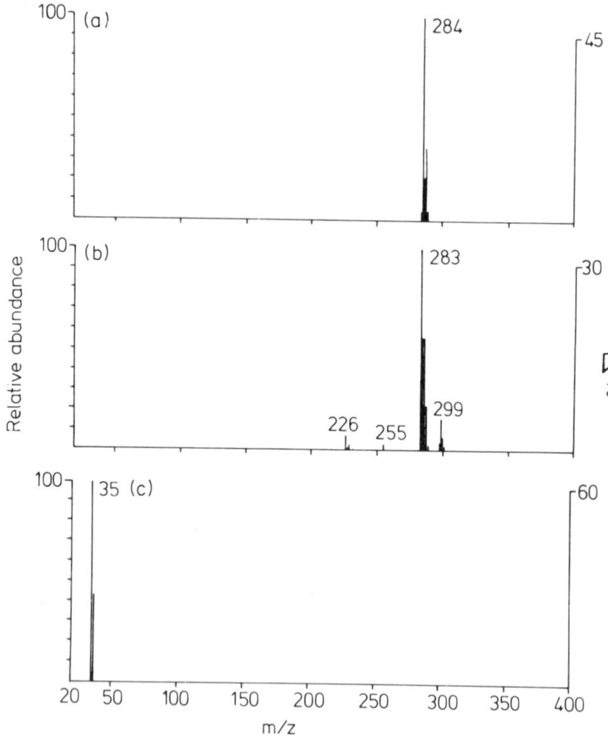

Figure 1: NCI mass spectra of diazepam ($M_r = 284$) under EC/NCI conditions, using methane (a) or methane/trace of O_2 (b). c: spectrum recorded with high ion source temperature (200°).

The reason that these spectra are only 'typical' is that for a period of several years the spectra in Figures 1a and 2b showing $(M-1)^-$ rather than $M^{-\cdot}$ ions, and not the spectra in Figures 1a and 2a, were periodically observed for diazepam and nordiazepam. The reason for this was investigated. The methane EC/NCI mass spectra of diazepam and nordiazepam were evaluated (Garland and Miwa, 1983) as a function of the temperature of the ion source, electron energy and emission of the electron filament (McEwen and Rudat, 1981), the nature of the hydrocarbon reagent gas (C_2H_6, iso-C_4H_{10} and C_3H_8), reagent gas pressure, and the concentration of air and water in the ion source. Of these variables, only the presence of a trace of air in the ion source and the temperature of the ion source (Figures 3 and 4) had any effect on the appearance of the spectra. An increase in the ion source temperature did not significantly promote $(M-1)^-$ formation, but increased formation of the Cl^- ion at the expense of the $M^{-\cdot}$ ion (Figures 1c and 2c). This illustrates the thermal nature

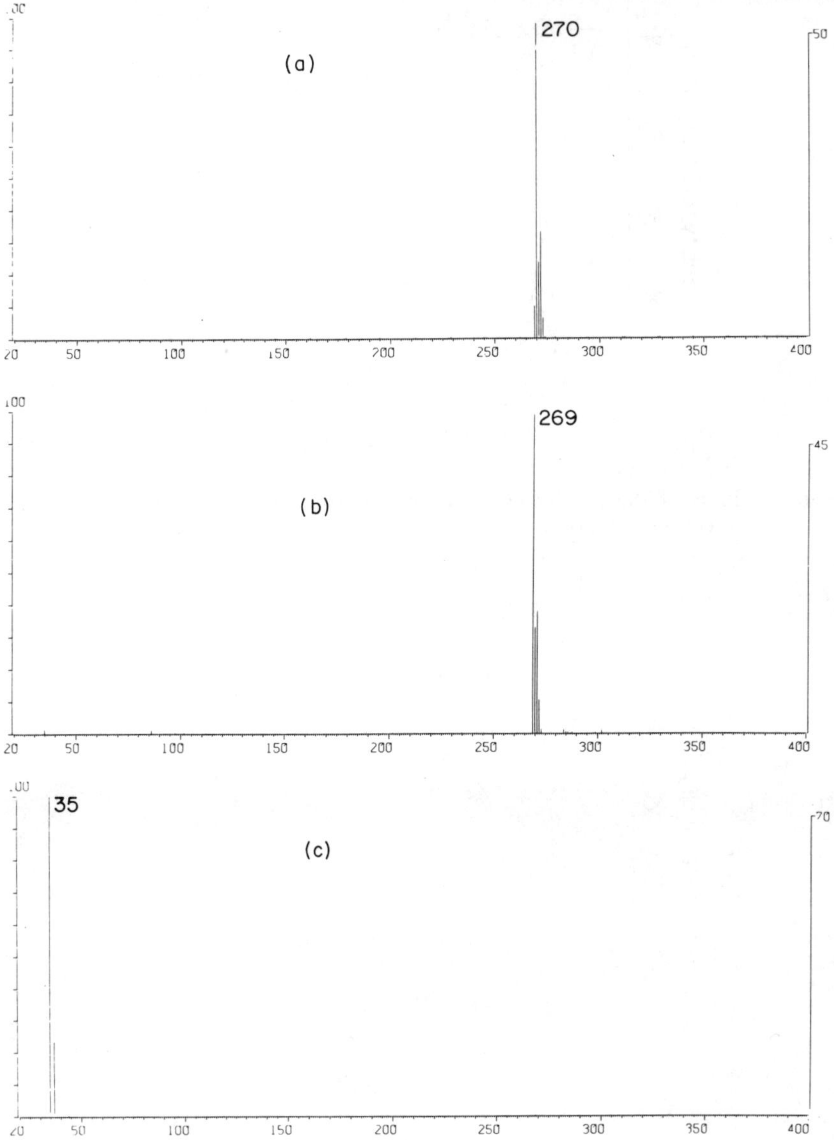

Figure 2: NCI mass spectra of nordiazepam (M_r = 270) under EC/NCI conditions, using methane (a) or methane/trace of O_2 (b). c : spectrum recorded with high ion source temperature (200°).

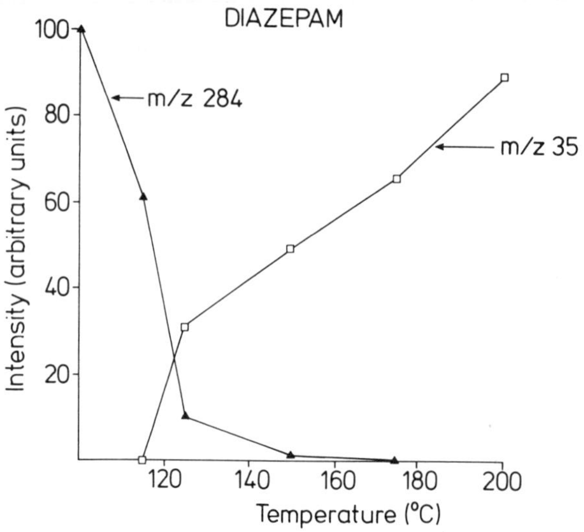

Figure 3: Plots of the ion intensities of *m*/*z* 35 and *m*/*z* 284 in the EC/NCI mass spectrum of diazepam as a function of ion source temperature.

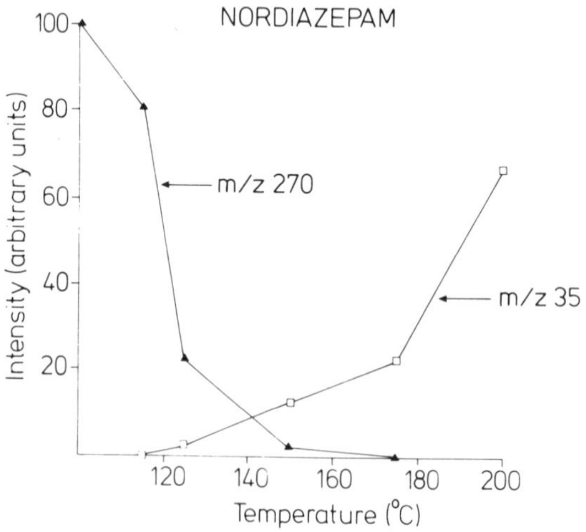

Figure 4: Plots of the ion intensities of *m*/*z* 35 and *m*/*z* 270 in the EC/NCI mass spectrum of nordiazepam as a function of ion source temperature.

of the reacting electron and the relative ease with which an electron capture mechanism can turn into a dissociative capture mechanism. The amount of air needed to cause the disappearance of the $M^{-\cdot}$ ion and the appearance of the $(M - 1)^-$ ion increased the analyzer ion gauge pressure reading by only 5×10^{-7} Torr, corresponding to an ion source pressure of approximately 10^{-3} Torr. This amount of air is too small either to be measured on the thermocouple pressure gauge adjacent to the ion source, or to change the methane NCI background mass spectrum, i.e. H^-, $O^{-\cdot}$, $O_2^{-\cdot}$, OH^-, etc. are not observed. By bleeding small amounts of either N_2 or O_2 into the ion source, it was determined that the active agent was O_2, not N_2. Consistent with this observation, a precise mass determination of the $(M + 15)^-$ addition ions in the methane/(trace of air) NCI spectra of diazepam showed the ion to be $(M - H + O)^-$ and not $(M + CH_3)^-$. A precise mass determination of the $(M - 1)^-$ ion in the methane/(trace of air) NCI mass spectrum of nordiazepam indicated that the $(M - 1)^-$ ion was $(M - H)^-$.

Based on the fact that there are no reagent ions in the methane/(trace of air) NCI mass spectra, it was proposed that the $(M - H)^-$ ions in the methane/(trace of air) mass spectra of diazepam and nordiazepam are caused by a reaction of O_2 with the $(M)^{-\cdot}$ ions generated by reaction with the thermal electrons in the methane plasma. If this is true, the reaction must be very efficient since the $M^{-\cdot}$ ion when generated would be expected to undergo only up to 2000 collisions with neutral gas molecules before exiting the ion source (Richter and Schwarz, 1978). Given an ion source pressure of 0.4 Torr of methane and an estimated 10^{-3} Torr of air (2×10^{-4} Torr of oxygen) only one collision of the exiting $M^{-\cdot}$ ion would be expected with residual oxygen (oxygen concentration of 1 part per 2000). Thus, the conversion of $M^{-\cdot}$ to $(M - H)^-$ must occur at every collision, i.e. it is a diffusion controlled reaction. In this regard, the reaction of O_2 with the $M^{-\cdot}$ ions would be expected to be favoured by the fact that the electrons in both O_2 and the radical ions are in an unpaired 'triplet state'. Decomposition of the resulting ion complex generates the $(M - H)^-$ ion and the relatively stable HO_2 radical (Howard, 1980). Previously, Levonowich *et al.* (1975) postulated that residual O_2 in the ion source was responsible for the generation of the $(M - 19)^-$ ion $(M^{-\cdot} - Cl + O)$ in the methane NCI mass spectrum of certain polychlorobiphenyls.

Further support for the suggestion of an efficient reaction between O_2 and $M^{-\cdot}$ as the cause of $(M - H)^-$ formation comes from a consideration of the relative sensitivities of analogues of either diazepam or nordiazepam substituted with NH_2, H, F or acetyl at C7 when analyzed by either methane NCI, methane/(trace of air) NCI, or N_2O/methane NCI (Table 1). When these analogues are ionized by either methane or methane/(trace of air) NCI, their sensitivities substantially increase as the substituent at C7 becomes more electronegative. If ionization by methane/(trace of air) NCI occurs via a reactant ion, all the analogues of diazepam or nordiazepam would be expected to have similar sensitivities, which is what is observed when the analogues are ionized by $OH^-(N_2O/CH_4)$ reactant ion NCI.

There is an almost perfect correlation between the logarithms of the relative sensitivities of analogues of nordiazepam (Table 1) and the values of the analogues' Hammett sigma meta constant (Shorter, 1973) for the substituent at C7 (correlation coefficient is 0.996 for ionization with methane, and 0.997 for ionization with methane/(trace of air)). The same correlation with each analogue's sigma para constant is not as good (correlation coefficient is 0.878 for ionization with methane and 0.889 for ionization with methane/(trace of air)). This suggests that the initial site of ionization is the 4,5-imine bond, the functional group meta to C7. Oelschlager *et al.* (1966, 1967) have suggested that this same bond is reduced in the polarographic analysis of diazepam and nordiazepam.

Table 1. Relative sensitivities of various 1,4-benzodiazepin-2-ones toward negative chemical ionization in a plasma of methane (0.8 Torr) or methane (0.8 Torr) + N_2O (0.2 Torr). The total ion currents of the M^{-}. and $(M\text{–}H)^-$ ions in the molecular cluster were determined. Except where noted these were the principal ions (i.e. intensity greater than 10% of base peak ion) in the spectrum between *m/z* 35 and *m/z* 435. Samples were introduced through the gas chromatograph inlet. Similar results were observed with the direct insertion probe inlet. Analogues with $R_2 = NO_2$ were not analysed because of the tendency of these compounds to undergo nitro-reduction in the ion source (Min and Garland, 1977).

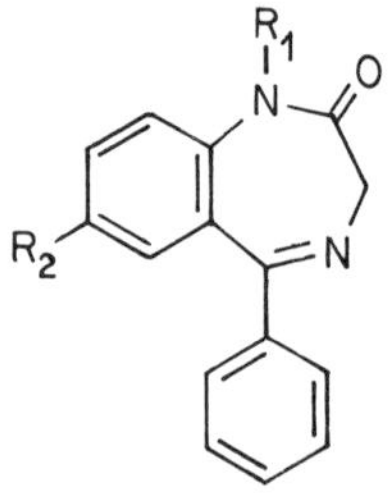

R_1	R_2	Methane	Methane + trace of air	Methane + N_2O
H	NH_2	0.3	0.2	0.5
H	H	1	1	1
H	F	25	27	1
H	Cl	48	60	1
H	$O{=}C{-}CH_3$	53	68	1
CH_3	NH_2	2	7	1
CH_3	H	1	1	1[b]
CH_3	F	31	59	3
CH_3	Cl	91	174[a]	7
CH_3	$O{=}C{-}CH_3$	214	370[a]	3

[a] Also includes $[M - 58]^-$. [b] Also includes $[M - 27]^-$.

Excluding the analogue where $R_2 = NH_2$, the relative sensitivities in Table 1 also suggest that the 4,5-imine bond is the site of ionization of analogues of diazepam (correlation coefficient of the logarithm of the relative sensitivities with sigma meta values is 0.981 for ionization with methane and 0.966 for ionization with methane/(trace of air); the correlation coefficients with sigma para values are 0.826 and 0.784, respectively). Including the analogue where $R_2 = NH_2$, the correlations are poorer (correlation coefficient with sigma meta values is 0.915 for ionization with methane and 0.852 for ionization with methane/(trace of air); the respective correlation coefficients with sigma para values are 0.768 and 0.643). The reason for the anomalous behaviour of the analogues of diazepam where $R_2 = NH_2$ is not known.

The methane/(trace of air) NCI mass spectra of various deuterated analogues of nordiazepam were obtained in an attempt to locate the $H^{\cdot}$ lost to generate the $(M - H)^-$ ion. Since the methane/(trace of air) NCI mass spectra of diazepam and diazepam trideuterated in the *N*-methyl group showed $(M - 1)^-$ ions, it was initially assumed that the $H^{\cdot}$ lost in the mass spectrum of nordiazepam was not that on the amide nitrogen. However, the methane/(trace of air) NCI mass spectrum of one nordiazepam analogue, perdeuterated at all positions except the amide nitrogen, showed $(M - 1)^-$ ions with no $(M - 2)^-$ ions. The methane/(trace of air) NCI mass spectrum of nordiazepam deuterated on the amide nitrogen was carefully obtained with just enough O_2 to give both $(M - 1)^-$ and $M^{-\cdot}$ ions for unlabelled nordiazepam, and the spectrum did show an $(M - 2)^-$ ion in addition to the $M^{-\cdot}$ ion. The methane/(trace of air) NCI mass spectra of [2′,3′,4′,5′,6′-$^2H_5$]diazepam, [6,8,9-$^2H_3$]diazepam and diazepam trideuterated in the methyl group all showed only $(M - 1)^-$ ions. Therefore, the $(M - 1)^-$ ion in the methane/(trace of air) NCI mass spectrum of diazepam is formed solely by the loss of the hydrogen atom on C3, a fact confirmed by the EC/NCI mass spectrum of [3,3-$^2H_2$]diazepam. In contrast to these results, the hydrogen atom lost to generate the $(M - H)^+$ ion in the positive electron ionization mass spectra of diazepam and nordiazepam is on the 2′ (or the equivalent 6′) carbon (Benz *et al.*, 1979).

The methane/(trace of air) NCI mass spectra of diazepam and nordiazepam differ considerably from one another. The mass spectrum of diazepam shows fragment ions at *m/z* 226 and *m/z* 255, in addition to the $(M - H)^-$ and $(M - H + O)^-$ ions (Figure 1b); however, the mass spectrum of nordiazepam shows only an $(M - H)^-$ ion (Figure 2b). In the methane/(trace of air) NCI spectrum of diazepam trideuterated in the methyl group, the *m/z* 226 of diazepam remains at *m/z* 226; in the methane/(trace of air) NCI spectrum of [3,3-$^2H_2$]diazepam, the *m/z* 226 ion shifts to *m/z* 227. This suggests that the ion at *m/z* 226 results from the loss of CH_3–N=C=O from the $(M - 1)^-$ ion. In the methane/(trace of air) NCI mass spectrum of diazepam trideuterated in the methyl group, the *m/z* 255 ion of diazepam shifts to *m/z* 258, while in the methane/(trace of air) NCI mass spectrum of [3,3-$^2H_2$]diazepam the *m/z* 255 ion shifts to m/z 256. This suggests that the m/z 255 ion results from the loss of carbon monoxide from the $(M - 1)^-$ ion.

Thus, compared to positive CI or EI, the conditions and structural features needed to reliably produce intense molecular anions for quantification of this class of compounds are complex. The rewards for this complexity are shown in Figure 5. This Figure shows selected ion current profiles from the GC–EC/NCI analyses of a rat brain tissue homogenate and rat blood fortified with 1 ng of nordiazepam. For both determinations, 1 of 100 μl was injected of an acetone/hexane solution (10 : 90) prepared by reconstituting the residue from evaporation of a dichloroethane/benzene (20 : 80) extract of the matrix. The selected ion current profiles are remarkable in that they illustrate that the sensitivity and specificity of the EC/NCI technique permit the determination of low concentrations of a compound with intrinsic electron capturing properties in a complex matrix like brain tissue with relatively simple sample work-up.

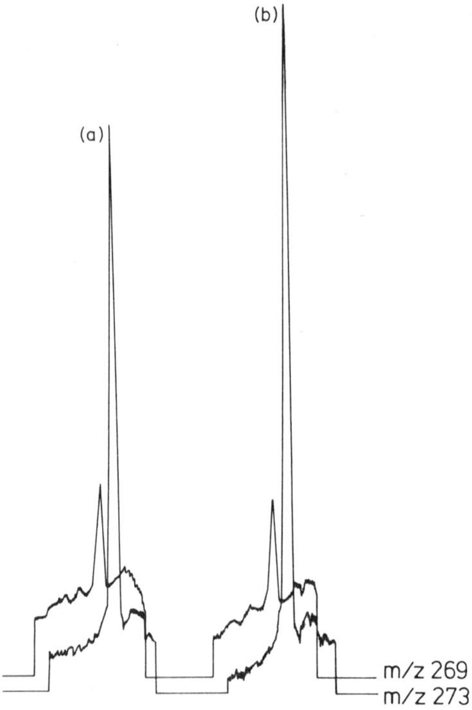

Figure 5: Selected ion current profiles from the analysis of extracts of (a) 1 ml of rat brain homogenate and (b) 1 ml of rat blood, each spiked with 1 ng of nordiazepam (m/z 269) and 10 ng of dideuterated nordiazepam (m/z 271). The analyses are under the 'trace of oxygen' conditions. The ^{37}Cl ion (m/z 273) for the dideuterated nordiazepam internal standard was monitored.

A perhaps even more dramatic example of the low limits of quantification possible using GC–EC/NCI to analyze for benzodiazepines is provided by the assay for flurazepam·2 HCl (Dalmane ®) in plasma (Miwa *et al.*, 1981). Prior to development of this assay there was no procedure sensitive enough to determine the pharmacokinetic profile of flurazepam in man. Selected ion current profiles of flurazepam and the deuterated flurazepam analogue used as internal standard in the assay are given in Figure 6. The profiles in Figure 6a are from the analysis by low resolution MS (with a back-extraction step in the sample work-up) of a plasma sample containing 12.5 pg/ml of flurazepam, while the profiles in Figure 6b are from the analysis by high resolution MS (with no back-extraction step in the sample work-up) of a plasma sample containing 12.5 pg/ml of flurazepam. The profiles in both Figures represent less than 0.5 pg injected on column.

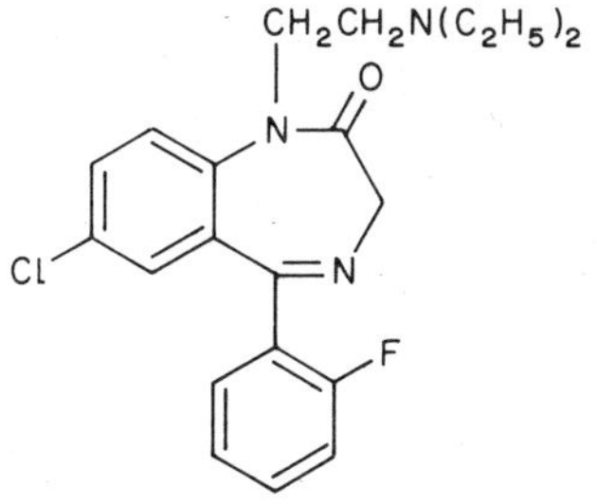

flurazepam

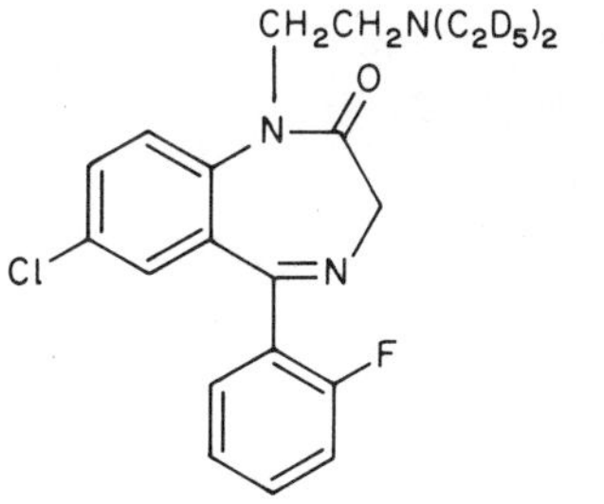

$[^2H_{10}]$flurazepam

Other GC–EC/NCI MS assays have been reported for the following benzodiazepines: midazolam (Rubio *et al.*, 1982a), clonazepam (Garland and Min, 1979) and imidazobenzodiazepine-3-carboxamide (Rubio *et al.*, 1982b). All three assays have excellent limits of quantification (less than 200 pg/ml), although for the last compound the principal ion formed was a $(M - HCl)^{-\cdot}$ fragment ion. Besides the benzodiazepines, drugs with nitro groups and/or extended conjugated systems should also be susceptible to ionization by electron capture and initial reports support this notion. Thus, GC–EC/NCI MS has been used for the quantification of nitroglycerin in plasma (Settlage *et al.*, 1983; Ottoila *et al.*, 1982; Gaudry *et al.*, 1984), ipronidazole in turkey skin and muscle (Garland *et al.*, 1980) and retinoic acid in plasma (Napoli *et al.*, 1985). This last assay is ten times more sensitive than a PCI assay for the same compound. For retinoic acid, $M^{-\cdot}$ ions are produced, while from nitroglycerin only the much less desirable NO_3^- is observed. Ipronidazole, having a nitro-group attached to a pseudo-aromatic ring, generates a $M^{-\cdot}$ ion.

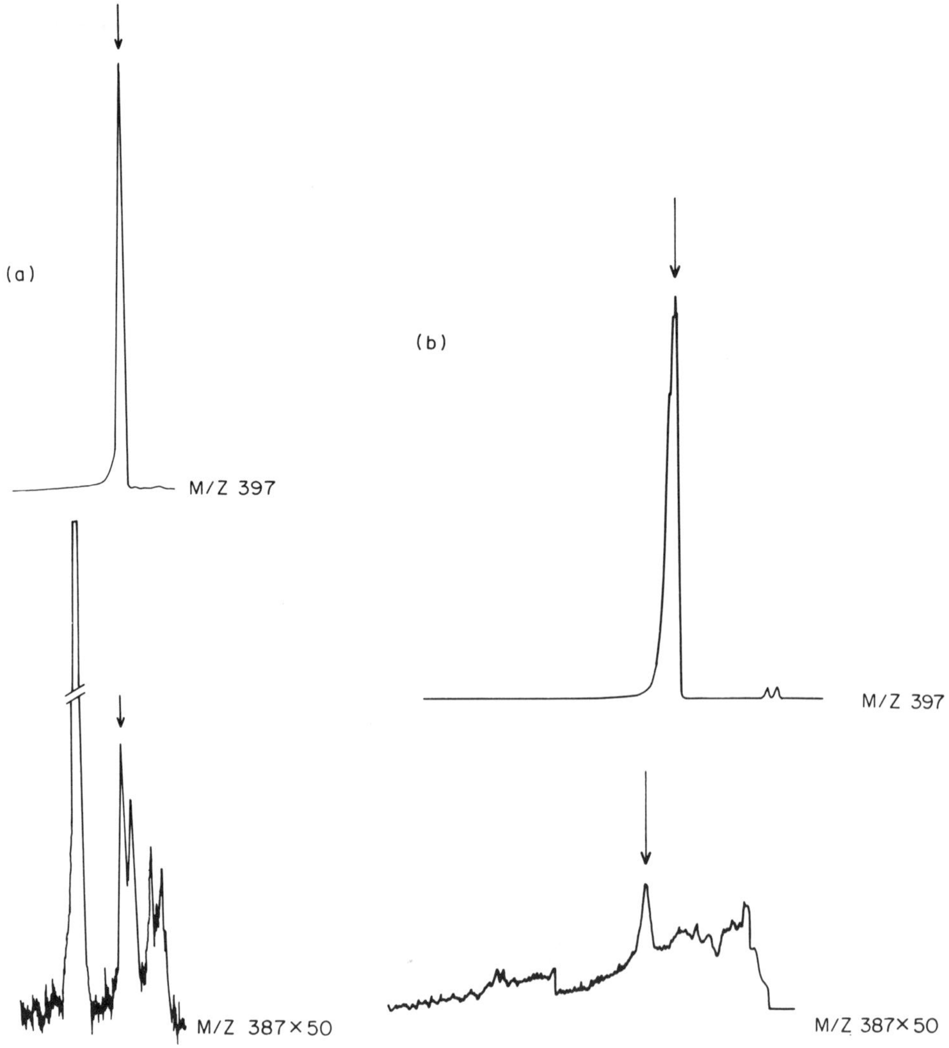

Figure 6: Selected ion current profiles from the analysis of an extract of 2 ml of drug-free plasma fortified with 25 pg of flurazepam (12.5 pg/ml) by either low resolution (a) or high resolution (b) GC–EC/NCI. The [$^{2}H_{10}$]-analogue was used as internal standard.

21.2.2. DRUGS REQUIRING DERIVATIZATION TO ACHIEVE ELECTRON CAPTURE

Drugs without an intrinsic EC response must be derivatized before analysis by EC/NCI. Unfortunately, some specificity is then lost since other components of the matrix will also be derivatized. Nevertheless, increased sensitivity can be obtained by derivatization. To date, the most important derivatization reaction has been the pentafluorobenzylation of carboxylic acids such as trimoprostil, a prostaglandin E_2 analogue (Min *et al.*, 1980; Rubio and Garland, 1985). The EC/NCI mass spectrum of pentafluorobenzylated trimoprostil is shown in Figure 7. No $M^{-\cdot}$ ion is observed, but the base peak ion, $(M - C_6F_5CH_2)^-$, still provides molecular weight data for trimoprostil. The EC/NCI response of the derivative is approximately 100 times more sensitive than the methane PCI generated $(M + H)^+$ ion of the methyl ester of trimoprostil, or the $(M - C_3H_6O_2)^{+\cdot}$ base peak ion in the positive electron ionization mass spectrum of methylated trimoprostil. An injection of 4 pg of pentafluorobenzylated trimoprostil into the GC–MS gave a selected ion current profile with a signal-to-noise ratio of 5 : 1. The limit of quantification of the assay for trimoprostil was 200 pg/ml. Even lower limits of quantification for drug assays using pentafluorobenzyl ester derivatives have been reported (Cosulich *et al.*, 1985).

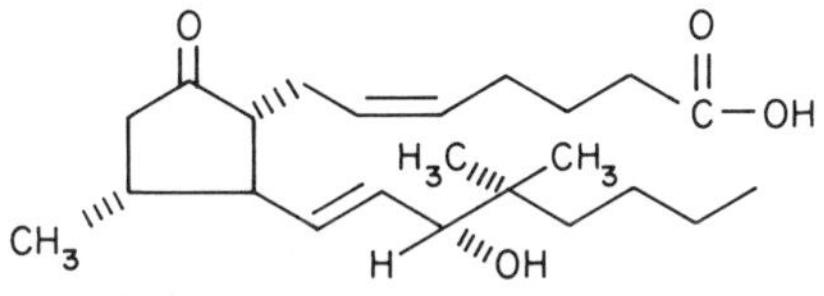

trimoprostil

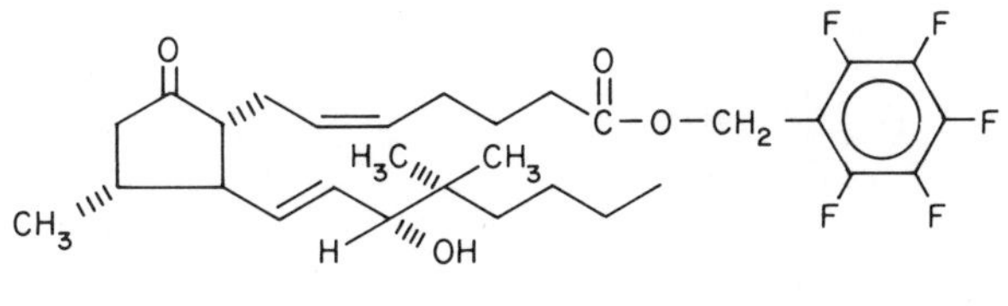

pentafluorobenzylated trimoprostil

Amines (or alcohols) can be perfluoroacylated to give amide (or ester) derivatives. For example, cibenzoline is a cyclopropane derivative with a quieting effect on cardiac arrhythmias. The EC/NCI mass spectrum of the pentafluoropropionyl derivative of cibenzoline is shown in Figure 8a. No $M^{-\cdot}$

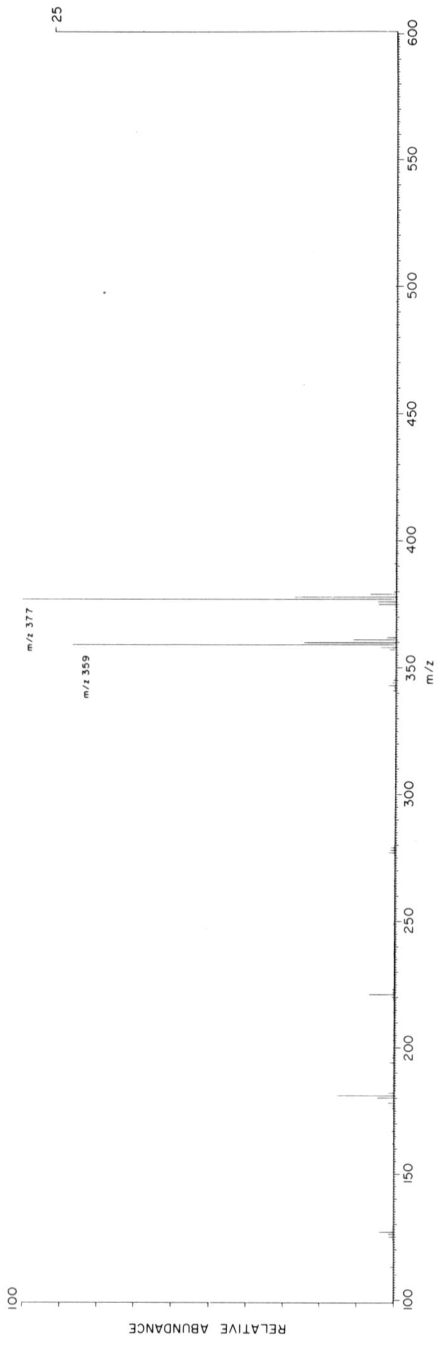

Figure 7: EC/NCI mass spectrum of the pentafluorobenzyl ester derivative of trimoprostil.

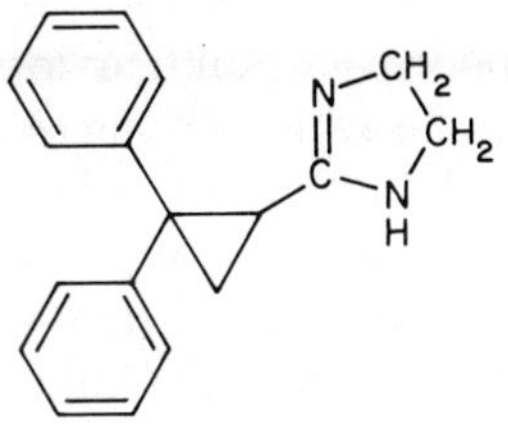

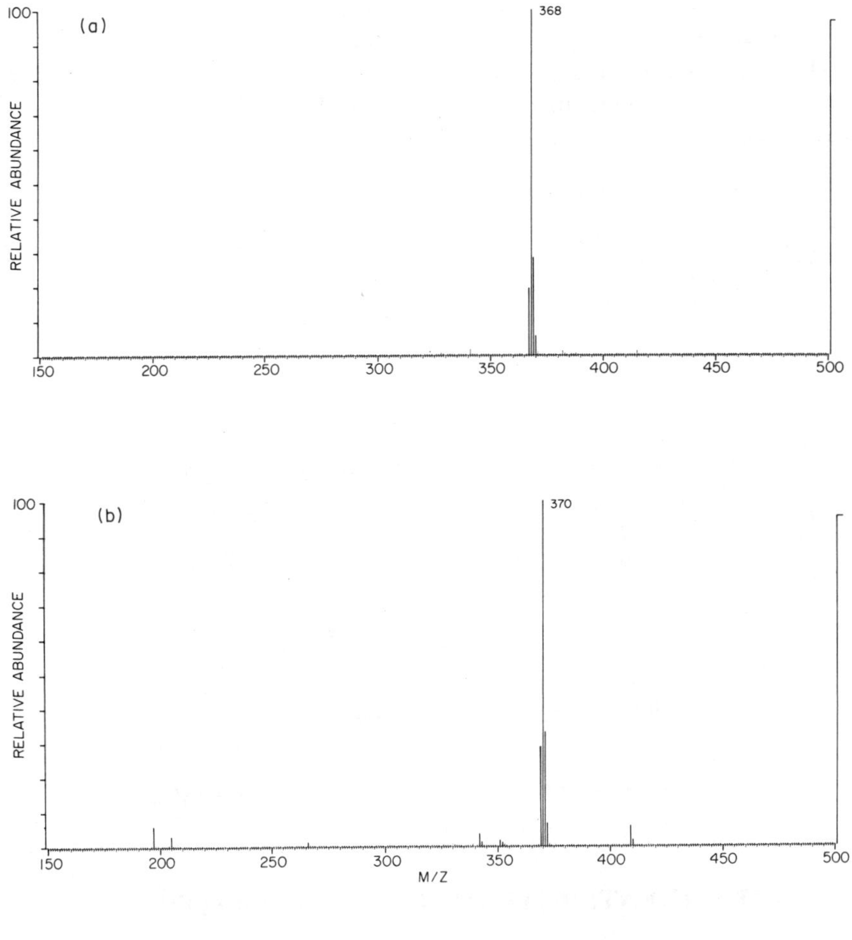

Figure 8: EC/NCI mass spectra of pentafluoropropionated derivatives of cibenzoline (a) and [$^{15}N_2$] cibenzoline (b), used as the internal standard in the assay.

ion is observed at m/z 408 and the base peak ion at m/z 368 corresponds to the loss of 2HF. From a metabolic viewpoint, a loss of 40 dalton could also represent, for example, loss of one aromatic ring (minus 76 dalton), addition of two hydroxy groups (plus 32 dalton) and partial reduction of the remaining aromatic ring (plus 4 dalton). Thus, some mass specificity is lost, although a GC–EC/NCI MS assay for cibenzoline in plasma, with a limit of quantification of 1 ng/ml, has been developed (Min and Garland, 1984). The loss of one molecule of HF from perfluoroacylated amines and alcohols has often been reported (Wood, 1982; Lewy and Markey, 1978), although useful $M^{-\cdot}$ ions have also been observed with the perfluoroacylated derivatives (Ibrahim *et al.*, 1984; Foltz *et al.*, 1983). The sensitivity increase over PCI analysis reported for these derivatives tends to be less than a factor of ten, unlike the larger sensitivity increase typically found with pentafluorobenzyl derivatives.

The derivatization of drugs containing a guanidine moiety with hexafluoroacetone provides a derivative with an extremely intense $M^{-\cdot}$ ion (Julien-La Rose *et al.*, 1983).

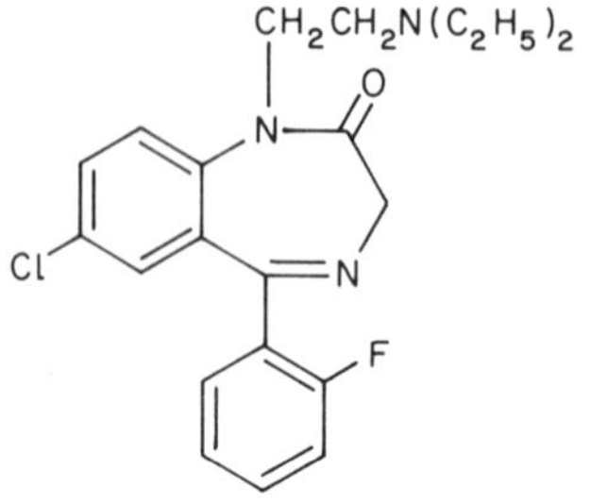

flurazepam

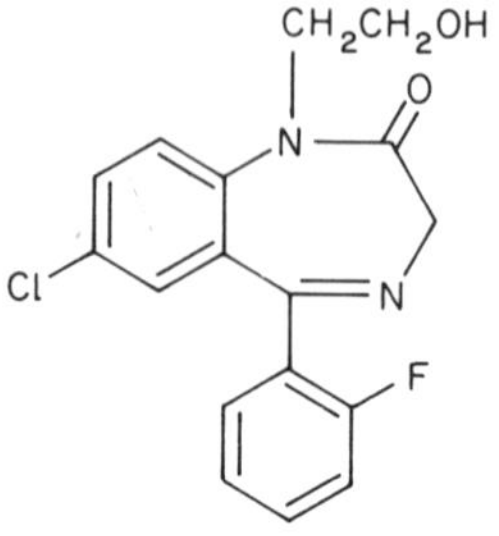

hydroxyethyl flurazepam

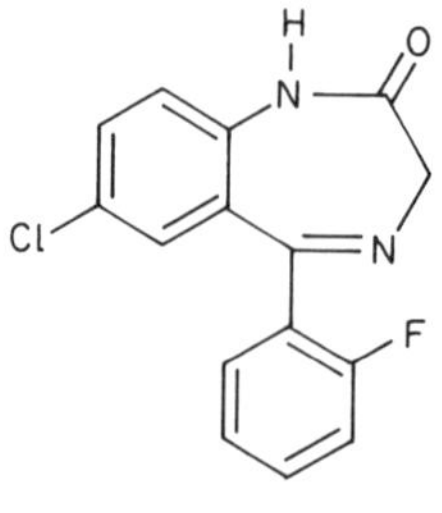

desalkyl flurazepam

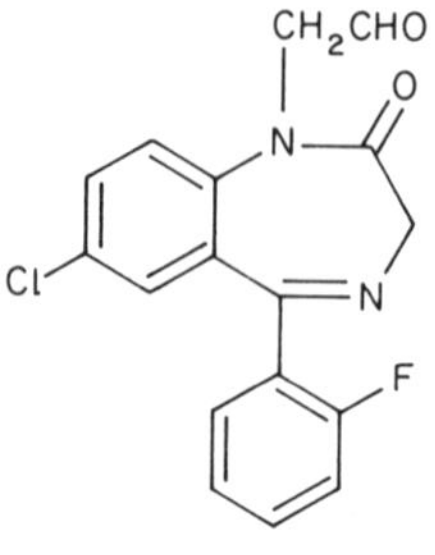

N-(2-acetaldehyde)-desalkyl flurazepam

21.3. IDENTIFICATION OF METABOLITES

For structure identification, the ideal ionization technique would generate both an ion characteristic of the molecular weight of the metabolite and structurally

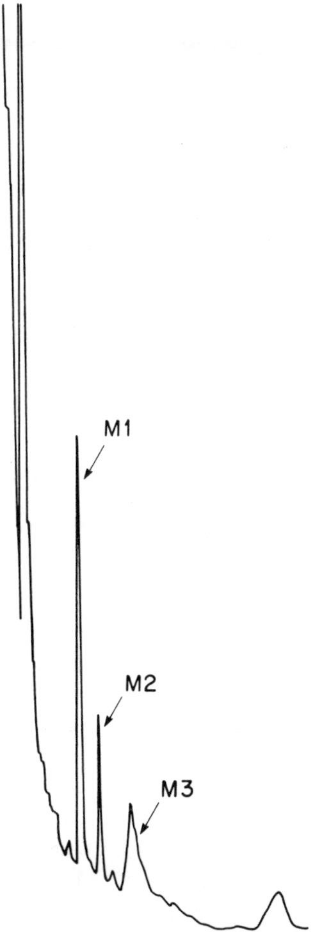

Figure 9: EC–GC trace from the analysis of a reconstituted residue of plasma extracts from a human given a 30 mg dose of flurazepam·2HCl.

significant fragment ions. In addition, the spectra of the metabolites would be predictable from the spectra of the drug and other structurally similar reference compounds. Based on these criteria, the limited results to date using EC/NCI MS to identify metabolites of the 1,3-dihydro-5-phenyl-1,4-benzodiazepines only partially support the use of EC/NCI MS for metabolite identification. For example (Garland *et al.*, 1983), Figure 9 shows the electron capture gas chromatography (EC–GC) trace following the injection of an aliquot of the organic solvent extract of plasma from a human volunteer given a 3 mg dose of flurazepam·2HCl (Dalmane ®). Desalkyl flurazepam (M1) and hydroxyethyl

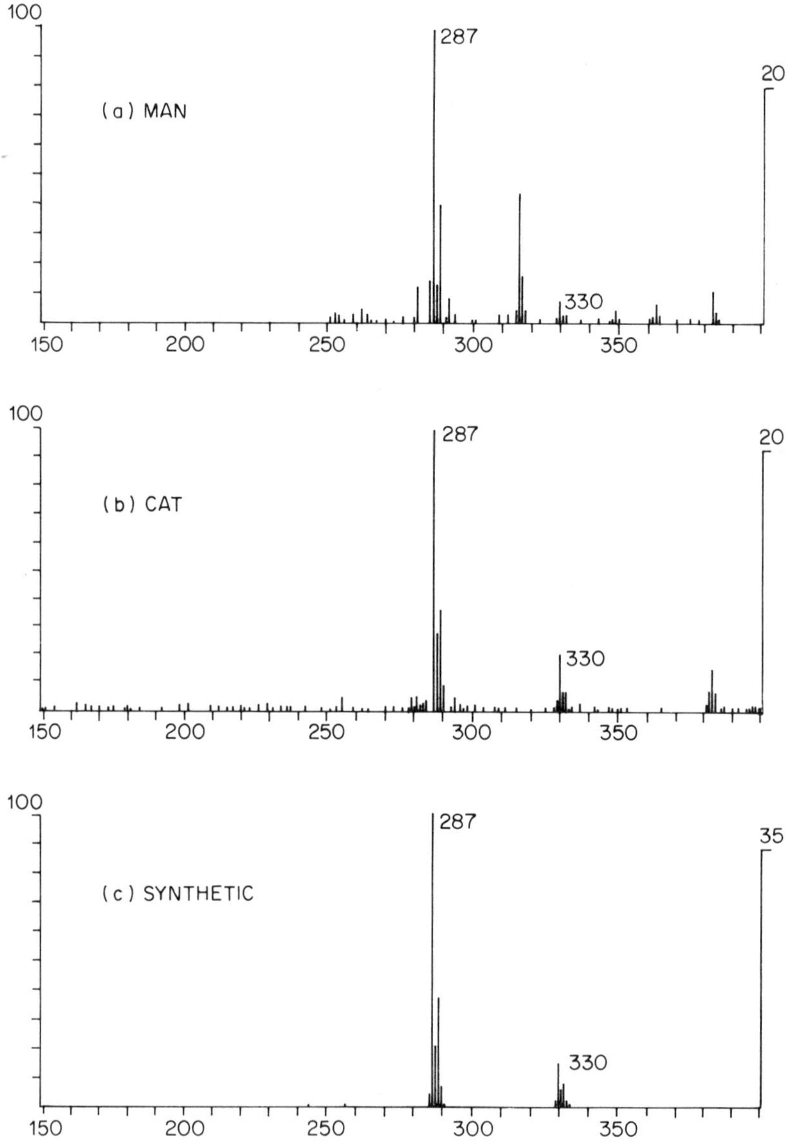

Figure 10: EC/NCI mass spectra of: (a) unknown metabolite of flurazepam in human plasma, (b) unknown metabolite of flurazepam in cat plasma, (c) synthesized 2-acetaldehyde)-desalkyl flurazepam, (d) authentic flurazepam, (e) authentic desalkyl flurazepam, (f) authentic hydroxyethyl flurazepam.

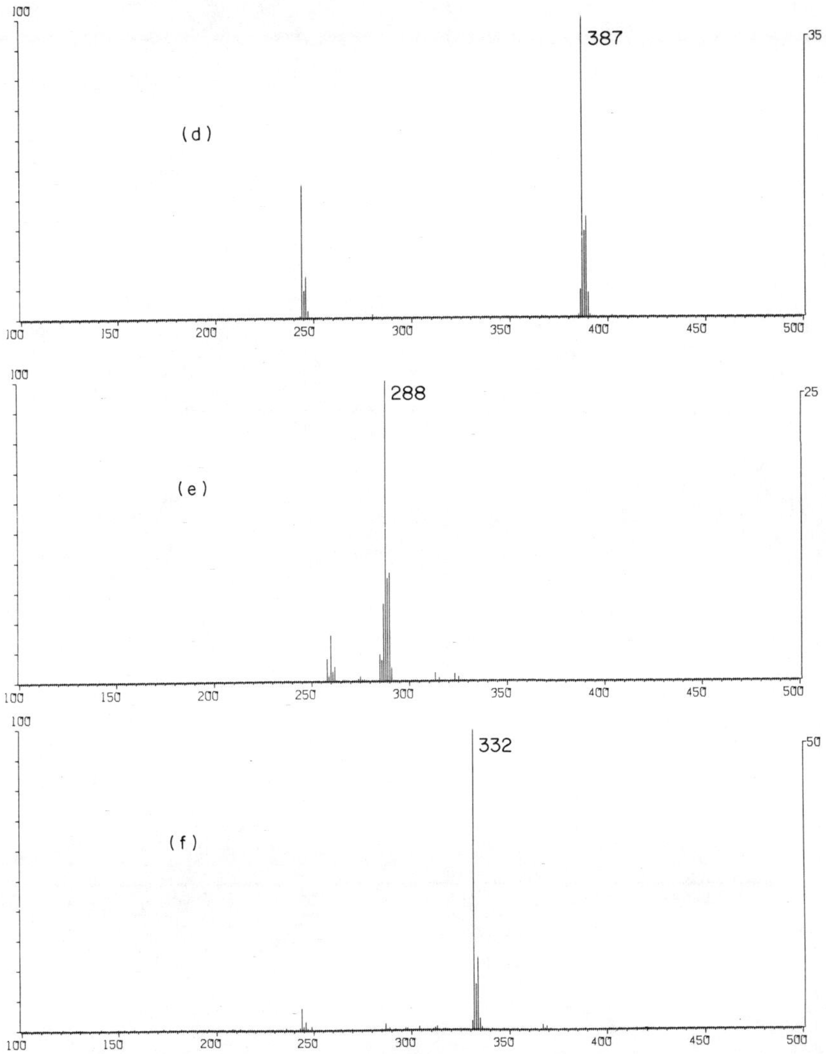

Figure 10 (cont.)

flurazepam (M3) are two known metabolites of flurazepam·2HCl, while the identity of the peak marked M2 is unknown. GC–EC/NCI mass spectra of M2 are given in Figures 10a and 10b. What is the $M^{-\cdot}$ ion in these spectra? The GC–EC/NCI mass spectra of flurazepam (M_r = 387), desalkyl flurazepam (M_r = 288) and hydroxyethyl flurazepam (M_r = 332) (Figures 10d-f) are all dominated by intense $M^{-\cdot}$ ions, and this would suggest that *m/z* 287 is the $M^{-\cdot}$ ion in the mass spectrum of M2, and that *m/z* 330 is from a co-eluting

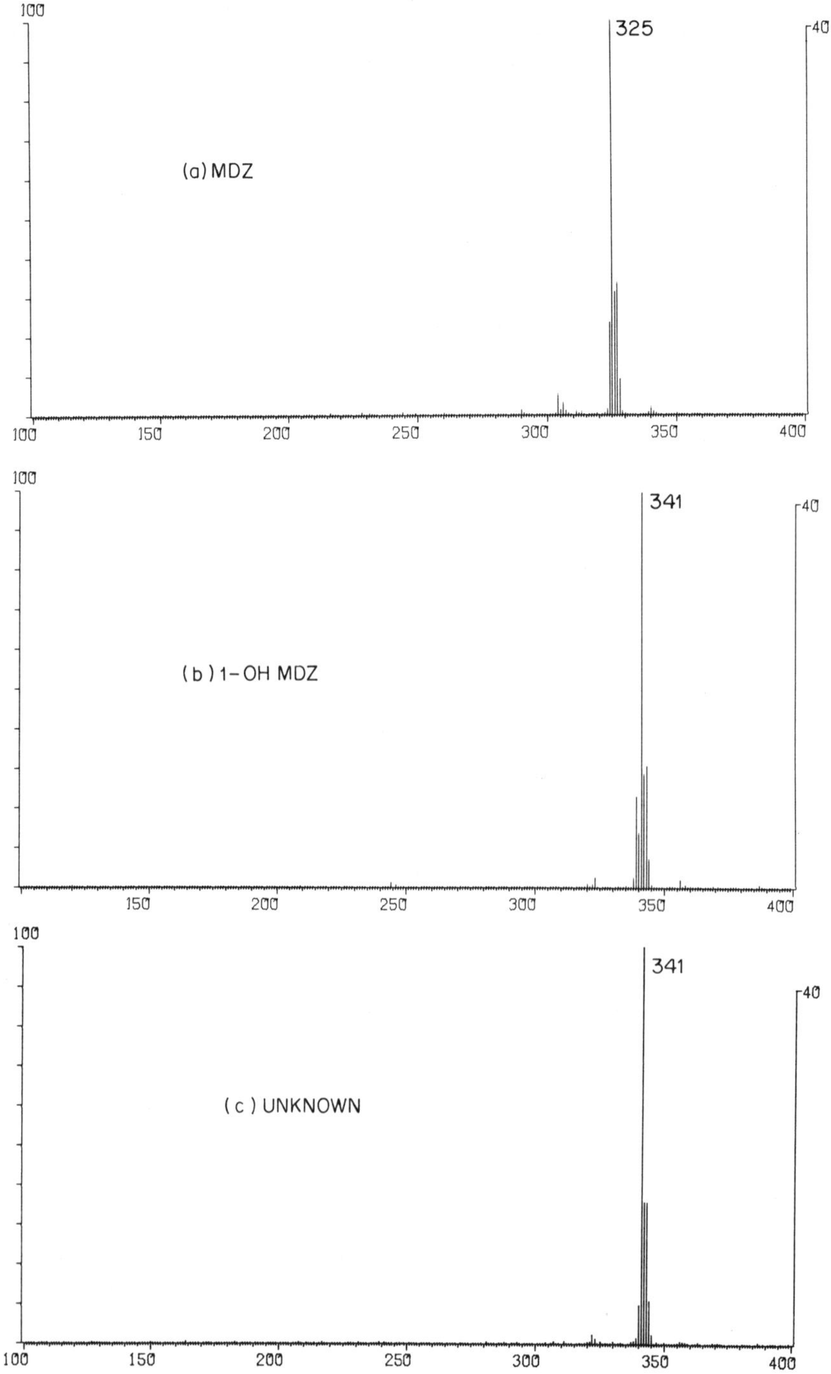
100
325
40
(a) MDZ
100
150
200
250
300
350
400
100
341
40
(b) 1-OH MDZ
150
200
250
300
350
400
100
341
40
(c) UNKNOWN
100
150
200
250
300
350
400

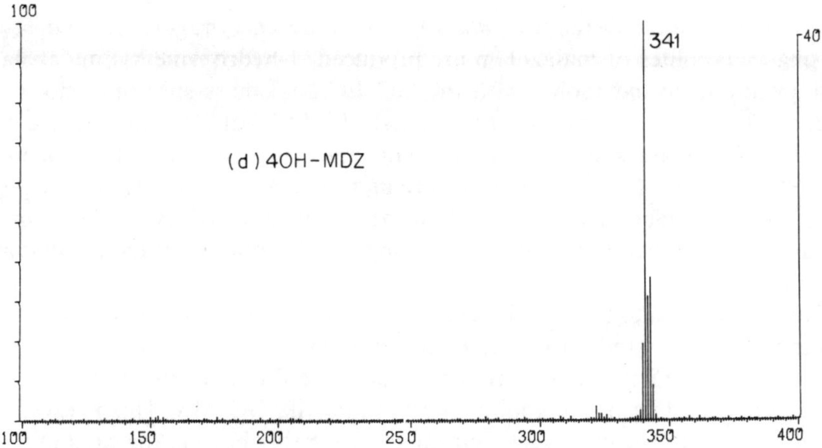

Figure 11: EC/NCI mass spectra of (a) midazolam, (b) 1-hydroxymethyl midazolam, (c) the unidentified metabolite of midazolam, (d) authentic 4-hydroxy midazolam.

interfering compound. In fact, this is not true, the molecular weight of M2 is 330, and the metabolite is *N*-(2-acetaldehyde)-desalkyl flurazepam. Thus, in this example a relatively small structural change (relative to desalkyl flurazepam, flurazepam and hydroxyethyl flurazepam) has caused almost the complete disappearance of the $M^{-\cdot}$ ion.

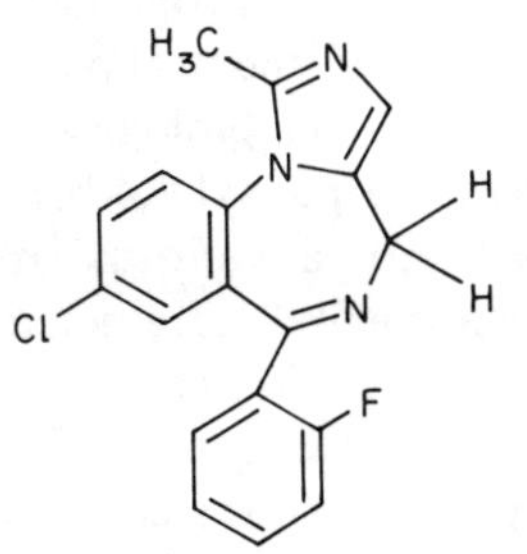

midazolam

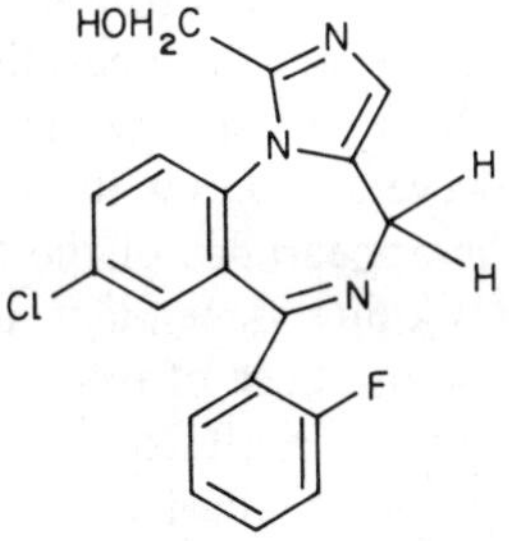

1-hydroxymethyl midazolam

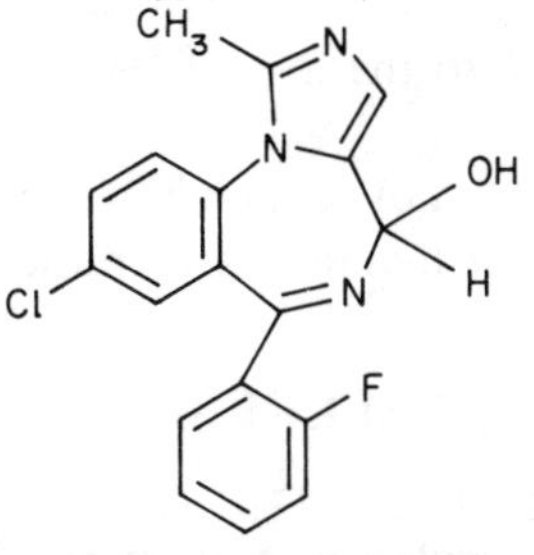

4-hydroxy midazolam

Another example (Arendt *et al.*, 1984) is more encouraging. In man, two plasma metabolites of midazolam are produced, 1-hydroxymethyl midazolam and an unknown metabolite with the GC–EC/NCI mass spectrum shown in Figure 11c. Is the intense ion at *m/z* 341 the $M^{-\cdot}$ ion? In this regard, the GC–EC/NCI mass spectra of midazolam and 1-hydroxymethyl midazolam both show intense $M^{-\cdot}$ ions (Figures 11a and 11b), and in this example the ion at *m/z* 341 is, in fact, an $M^{-\cdot}$ ion and the metabolite is 4-hydroxy midazolam. It should be noted, however, that in this example no informative fragment ions are generated.

Thus, compared to PCI, NCI seems a less reliable method for determining the molecular weights and structures of metabolites. However, two observations are worthwhile. First, the sensitivity is remarkable. The spectrum in Figure 11a represents 1 to 3 ng of metabolite injected into the GC–MS. This is excellent sensitivity for 'full-scan' mass spectrometry, probably beyond the capability of PCI for this class of compounds. Second, although not shown in Figures 10 or 11, the mass spectra of both metabolites are remarkably simple for masses below *m/z* 100. This is quite different from the spectra commonly generated by PCI or positive electron ionization, and this fact may be useful in identifying metabolites of certain classes of drugs.

21.4. SUMMARY

GC–EC/NCI MS is the method of choice for the analysis of drugs and drug metabolites having both intrinsic electron capturing properties and spectra showing molecular anion generation, because of the technique's excellent sensitivity and the low response of most chemicals in a biological matrix to the technique. However, studies with the 1,3-dihydro-5-phenyl-1,4-benzodiazepines suggest that the appearance of the NCI spectra, relative to the appearance of PCI spectra, is quite dependent upon instrumental parameters, structural features, and the presence of traces of O_2 in the ion source.

Formation of the pentaflurobenzyl esters of drugs which contain a carboxylic acid function provides a derivative whose spectrum shows complete molecular weight data and gives a very significant increase in sensitivity. Formation of perfluroacyl derivatives gives NCI spectra which are often dominated by fragment ions corresponding to the loss of HF or 2HF with a modest gain in sensitivity.

Compared to PCI, NCI may not be as reliable in providing molecular weight and structural information for metabolite identification.

REFERENCES

Arendt, R. M., Greenblatt, D. J., and Garland, W. A. (1984). Quantitation by gas chromatography of the 1- and 4-hydroxy metabolites of midazolam in human plasma, *Pharmacology*, **29**, 158–164.

Benz, W., Vane, F. M., and Rapp, U. (1979). The fragmentation of 5-phenyl-1,4-benzodiazepin-2-ones. Mechanism for formation of $(M + H)^+$, *Org. Mass Spectrom.*, **14**, 154–159.

Cabana, B. E. (1984). Bioavailability and pharmacokinetics in drug development, in *Pharmacokinetics: A Modern View* (Eds. Benet, L. E., Levy, G. and Ferraiolo, B. L.), pp. 113–132, Plenum, New York.

Cosulich, D. B., Perkinson, N. A., and Batra, V. K. (1985). Capillary gas chromatographic–mass spectrometric analysis of 15-deoxy-16-hydroxy-16-vinylprostaglandin E_2, *J. Pharm. Sci.*, **74**, 76.

Foltz, R. L., McGinnis, K. M., and Chinn, D. M. (1983). Quantitative measurement of Δ^9-tetrahydrocannabinol and two major metabolites in physiological specimens using capillary column negative ion chemical ionization mass spectrometry, *Biomed. Mass Spectrom.*, **10**, 316–323.

Garland, W. A., and Min, B. H. (1979). Determination of clonazepam in human plasma by gas chromatography–negative ion chemical ionization mass spectrometry, *J. Chromatogr.*, **172**, 279–286.

Garland, W. A., and Miwa, B. J. (1980). Methane negative chemical ionization analysis of 1,3-dihydro-5-phenyl-1,4-benzodiazepin-2-ones, *Environmental Health Perspectives*, **36**, 69–76.

Garland, W. A., and Miwa, B. J. (1983). The $[M - 1]^-$ ion in the negative chemical ionization mass spectra of diazepam and nordiazepam, *Biomed. Mass Spectrom.*, **10**, 126–129.

Garland, W. A., Hodshon, B. J., Chen, G., Weiss, G., Felicito, N. R., and MacDonald, A. (1980). Determination of ipronidazole and its principal metabolite in turkey skin and muscle by combined gas chromatography–negative chemical ionization mass spectrometry–stable isotope dilution, *J. Agric. Food Chem.*, **28**, 273–277.

Garland, W. A., Miwa, B. J., Dairman, W., Krappell, B., Chiueh, M. C. C., Divoll, M., and Greenblatt, D. J. (1983). Identification of 7-chloro-5-(2-fluorophenyl)-2,3-dihydro-2-oxo-1H-1,4-benzodiazepine-1-acetaldehyde, a new metabolite of flurazepam in man, *Drug Metab. and Dispos.*, **11**, 70–72.

Gaudrey, D., Gerardin, A., Briand, C., and Wantiez, D. (1984). Quantitative determination of nitroglycerol in human plasma by gas chromatography negative ion mass spectrometry using ^{15}N labelled nitroglycerol as internal standard, *Biomed. Mass Spectrom.*, **11**, 276–277.

Howard, C. J. (1980). Kinetic study of the equilibrium $HO_2 + NO \rightleftharpoons OH + NO_2$ and the thermochemistry of HO_2, *J. Am. Chem. Soc.*, **102**, 6937–6941.

Ibrahim, K. E., Couch, M. W., Williams, G. M., Budd, M. B., Yost, R. A., and Midgley, J. M. (1984). Quantitative measurement of octopamines and synephrines in urine using capillary column gas chromatography negative ion chemical ionization mass spectrometry, *Anal. Chem.*, **56**, 1695–1699.

Julien-La Rose, A., Lange, C., Lavene, D., Kiechel, J. R., and Basselier, J. J. (1983). Determination of guanfacine in human plasma by GC–MS with electron capture negative chemical ionization, *Int. J. Mass Spectrom. Ion Proc.*, **48**, 221–224.

Levonowich, P. F., Tannenbaum, H. P., and Dougherty, R. C. (1975). Negative chemical ionization as a model for reactions in solution: new nucleophilic reactions of superoxide, *J. Chem. Soc. Chem. Comm.*, **1975**, 597–598.

Lewy, A. J., and Markey, S. P. (1978). Analysis of melatonin in human plasma by gas chromatography negative chemical ionization mass spectrometry, *Science*, **201**, 741–743.

McEwen, C. N., and Rudat, M. A. (1981). Radical trapping in a mass spectrometer ion source, *J. Am. Chem. Soc.*, **103**, 4343–4349.

Min, B. H., and Garland, W. A. (1977). Determination of clonazepam and its 7-amino metabolite in plasma and blood by gas chromatography–chemical ionization mass spectrometry, *J. Chromatogr.*, **139**, 121–133.

Min, B. H., and Garland, W. A. (1984). Quantitation of cibenzoline in human plasma by gas chromatography–negative ion chemical-ionization mass spectrometry, *J. Chromatogr.*, **336**, 403–409.

Min, B. H., Pao, J., Garland, W. A., de Silva, J. A. F., and Parsonnet, M. (1980). Determination of an antisecretory trimethyl prostaglandin E_2 analog in human plasma by combined capillary column gas chromatography–negative chemical ionization mass spectrometry, *J. Chromatogr.*, **183**, 411–419.

Miwa, B. J., Garland, W. A., and Blumenthal, P. (1981). Determination of flurazepam in human plasma by gas chromatography–electron capture negative chemical ionization mass spectrometry, *Anal. Chem.*, **53**, 793–797.

Napoli, J. L., Pramanik, B. C., William, J. B., Dawson, M. I., and Hobbs, P. D. (1985). Quantification of retinoic acid by gas–liquid chromatography–mass spectrometry; total versus all-trans-retinoic acid in human plasma, *J. Lipid Res.*, **26**, 387–392.

Nelson, S. D. (1982). Metabolic activation and drug toxicity, *J. Med. Chem.*, **25**, 753–765.

Oelschlager, H., Volke, J., and Hoffman, H. (1966). Arzneimittelanalysen mittels polargraphic oder oszillopolarographic, *Collect. Czech. Chem. Commun.*, **31**, 1264–1272.

Oelschlager, H., Volke, J., Hoffman, H., and Kurek, E. (1967). Mechanisms der polarographischen reduktion des chlordiazepoxide, *Arch. Pharm. Weinheim.*, **300**, 250–257.

Ottoila, P., Taskinen, J., and Sothmann, A. (1982). Quantitative determination of nitroglycerin in human plasma by capillary gas chromatography negative ion chemical ionization mass spectrometry, *Biomed. Mass Spectrom.*, **9**, 108–110.

Poole, C. F., and Zlatkis, A. (1980). Derivatization techniques for the electron capture detector, *Anal. Chem.*, **52**, 1002A–1016A.

Richter, W. J., and Schwarz, H. (1978). Chemical ionization: a mass spectrometric analytical procedure of rapidly increasing importance, *Angew. Chem. Int. Ed. Engl.*, **17**, 424–439.

Rubio, F., and Garland, W. A. (1985). Determination of acetyltrimoprostil and its metabolite trimoprostil in human or dog plasma by gas chromatography–negative ion chemical ionization mass spectrometry, *J. Chromatogr.*, **339**, 313–320.

Rubio, F., Miwa, B. J., and Garland, W. A. (1982a). Determination of midazolam and two metabolites of midazolam in human plasma by gas chromatography–negative chemical-ionization mass spectrometry, *J. Chromatogr.*, **233**, 157–165.

Rubio, F., Miwa, B. J., and Garland, W. A. (1982b). Determination of imidazobenzodiazepine-3-carboxamide, a new anxiolytic agent, in human plasma by gas chromatography–negative chemical ionization mass spectrometry, *J. Chromatogr.*, **233**, 167–173.

Settlage, J. A., Gielsdorf, W., and Jaeger, H. (1983). Femtogram level quantitative determination of nitroglycerin and metabolites in human plasma by GC–MS negative ion chemical ionization, *J. High Res. Chromatogr. Commun.*, **6**, 68–71.

Shorter, J. (1973). *Correlation Analysis in Organic Chemistry: An Introduction to Linear Free-Energy Relationships*, p. 17, Clarendon Press, Oxford.

Wood, P. L. (1982). A selected ion monitoring assay for dopamine and its metabolites using negative chemical ionization, *Biomed. Mass Spectrom.*, **9**, 302–306.

Mass Spectrometry in Biomedical Research
Edited by S. J. Gaskell
© 1986 John Wiley & Sons Ltd.

Chapter 22

QUANTIFICATION OF NEUROTRANSMITTERS AND RELATED COMPOUNDS BY ELECTRON CAPTURE CHEMICAL IONIZATION GC–MS

Kym F. Faull and Olof Beck

22.1. INTRODUCTION

One goal of analytical neurochemistry is to develop methods for quantifying neurotransmitters and related compounds in tissues and body fluids of experimental animals and human subjects. The trend in the methodology employed in this work has been toward assays of improved specificity and sensitivity, such that the precision of the quantitative measurements has increased simultaneously with a decrease in the limits of detection. This trend has allowed the measurements to be made on smaller samples, which is important in anatomical studies as it permits more precise localization of compounds within substructures of the brain and other tissues, and in behavioural studies as it provides for reliable measurements from small samples of body fluids (e.g. blood plasma and cerebrospinal fluid).

Gas chromatography (GC) has made an important contribution to the development of this field. The use of a mass spectrometer as the GC detector was a major advance offering increased specificity and sensitivity. The subsequent refinements of selected ion monitoring (SIM) and fused silica capillary gas chromatography have provided a combination of sensitivity and selectivity virtually unequalled by other methods which have found application in the field.

Electron capture negative ion chemical ionization has increased the sensitivity obtainable with GC–MS by up to an order of magnitude. Developed primarily by Hunt and colleagues (Hunt *et al.*, 1976), the most notable feature of this technique is that it can be used to detect and quantify electron capturing species at great dilution. To describe this ionization technique, we have chosen to use the term electron capture chemical ionization (ECCI). It should be remembered, however, that this term is a compromise between what is strictly correct and what is in common usage. As will be demonstrated in the following discussion, the processes involved here are not strictly chemical ionization but simply involve capture of low energy electrons by sample molecules with electron affinity.

In this chapter we will outline the salient features of the ECCI process and discuss some recent applications. The examples chosen have been confined to GC–MS applications, although the ECCI technique has been used in GC–MS/MS (Chapter 25).

22.2. REACTIONS IN THE CHEMICAL IONIZATION SOURCE

The ionization chamber is filled with a reagent gas at relatively high pressure (often around 1 Torr). An electron beam, usually emitted from an incandescent filament, is fired into the ionization chamber. Because of the relative abundance of reagent gas molecules, almost all the initial ionization by the electron beam is of the reagent gas. Each ionizing event is accompanied by the production of a low energy secondary electron and the loss of 20 to 30 eV of energy from the primary electron (Klots, 1968). Additional ionizing and non-ionizing collisions between these electrons and neutral reagent gas molecules reduce the kinetic energy of the electrons to a mean level of $2–5 \times 10^{-2}$ eV within 0.01 μs (Pellizzari, 1974).

Reagent gases which have zero or negative electron affinities act by producing a large but finite population of low energy or near-thermal electrons and a large population of positive reagent gas ions. Methane, isobutane, argon, nitrogen and ammonia are reagent gases of this type. These reagent gases do not produce any significant abundance of negative reagent gas ions. Therefore, in the negative ion mode, the major reaction which is observed is capture of the low energy electrons by sample molecules possessing electron affinity. Of course, in the positive ion mode, the reagent gas ions can be used to ionize sample molecules via ion–molecule reactions and this is the basis of the widely used chemical ionization technique.

The range of possible events is more complicated with reagent gases which have positive electron affinities. Because of the relative preponderance of reagent gas molecules, sample molecules with electron affinity face severe competition for the available low energy electrons. For this reason reagent gases with positive electron affinities have not been used for work in the purely ECCI mode, although gases such as N_2O, CH_2Cl_2 and NF_3 among others, react with a wide array of organic molecules to produce useful negative ion mass spectra.

22.3. ELECTRON CAPTURE CHEMICAL IONIZATION

The formation of negative ions by the interaction between electrons and sample molecules is generally described in terms of three different mechanisms:

$AB + e^- \rightarrow [AB]^{-\cdot}$ Resonance capture (1)

$AB + e^- \rightarrow A + B^-$ Dissociative resonance capture (2)

$AB + e^- \rightarrow A^+ + B^- + e^-$ Ion-pair production (3)

In mechanisms (1) and (2) the electron must have an energy level comparable to one of the available energy levels of the capturing molecule because there are no product electrons to carry excess kinetic energy. The resonance capture mechanism (1) uses electrons with close to 0 eV of energy (near-thermal electrons) and yields molecular anions. The dissociative resonance capture mechanism uses electrons with energies in the range 0–15 eV and yields fragment anions. The third mechanism, ion-pair production (3), usually requires electron energies above 10 eV and often yields low mass and structurally insignificant fragment anions. Consequently, this mechanism is generally not important for the production of informative negative ion mass spectra.

22.4. RELATIVE SENSITIVITY WITH ELECTRON CAPTURE CHEMICAL IONIZATION

The most notable feature of ECCI is that sample molecules possessing strong electron affinity can be detected at relative sensitivities of up to three orders of magnitude greater than in other ionization modes. It is not surprising, therefore, that in the area of trace level work ECCI has had a major impact.

Several factors contribute to this large sensitivity increase. The low energy mechanisms involved in ECCI usually favour limited fragmentation of the parent molecule. Because only a few ions of different mass carry the bulk of the negative ion current, a significant signal enhancement is available for SIM (e.g., compare the EI and the ECCI spectra for the three different derivatives of GABA shown in Figure 1). The low background noise that occurs in the negative ion mode contributes to the sensitivity enhancement. Also, the higher reaction rate of electron capture reactions compared to other ionization reactions contributes to the sensitivity enhancement (Harrison, 1982). The

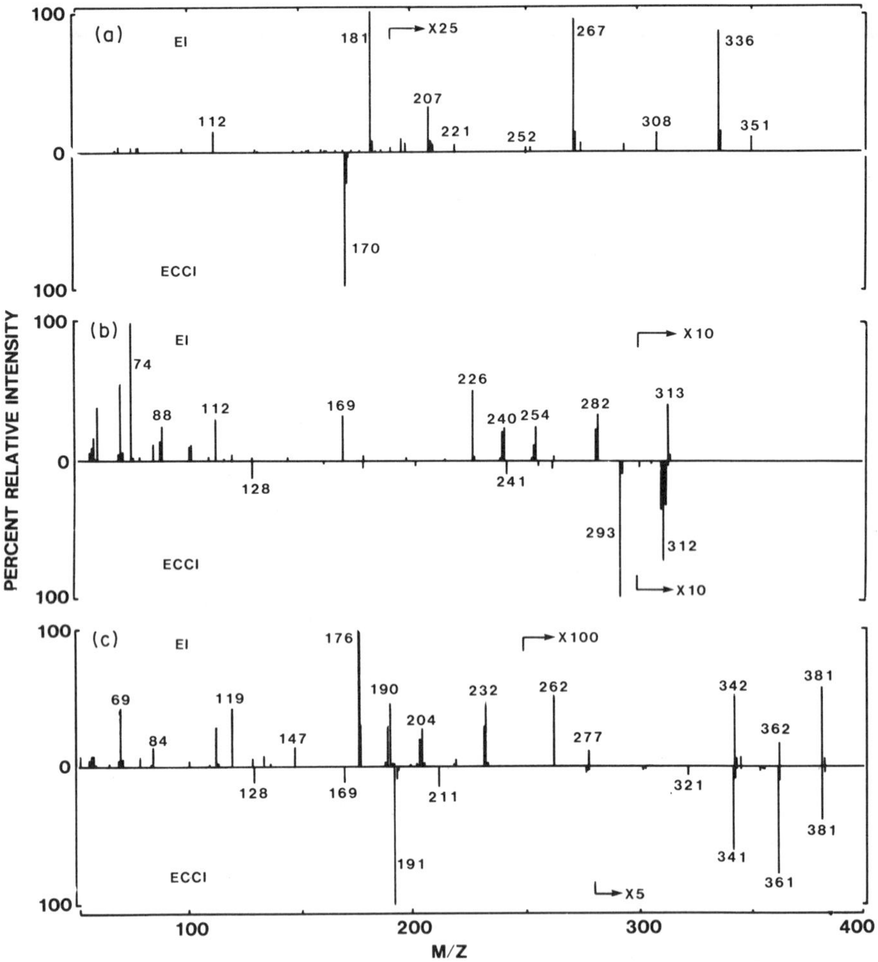

Figure 1: EI (top and ECCI (bottom) mass spectra of three derivatives of γ-aminobutyric acid (GABA).

a) the neopentylidene Schiff base pentafluorobenzyl ester ($(CH_3)_3CCH{=}N(CH_2)_3COOCH_2C_6F_5$). The probable structures of the ions at *m/z* 351, 336, 267, 181 and 170 are $M^{+\cdot}$, $(M-15)^+$, $(C_6F_5CH_2COO(CH_2)_3)^+$, $(C_6F_5CH_2)^+$ and $((CH_3)_3CCH{=}N(CH_2)_3COO)^-$, respectively.

b) the *N*-heptafluorobutyryl methyl ester ($C_3F_7CONH(CH_2)_3COOCH_3$). The probable structures of the ions at *m/z* 313, 282, 254, 240, 226, 169, 312 and 293 are $M^{+\cdot}$, $(C_3F_7CONH(CH_2)_3CO)^+$, $(C_3F_7CONH(CH_2)_3)^+$, $(C_3F_7CONH(CH_2)_2)^+$, $(C_3F_7NHCOCH_2)^+$, $(C_3F_7)^+$, $(M-H)^-$ and $(M-HF)^-$, respectively.

c) the *N*-pentafluoropropionyl pentafluoropropyl ester ($C_2F_5CONH(CH_2)_3COOCH_2C_2F_5$). The probable structures of the ions at *m/z* 381, 362, 262, 232, 204, 190, 176, 119, 69, 361, 341, 321 and 191 are $M^{\pm}$, $(M-F)^+$, $(M-C_2F_5)^+$, $(C_2F_5CONH(CH_2)_3CO)^+$, $(C_2F_5CONH(CH_2)_3)^+$, $(C_2F_5CONH(CH_2)_2)^+$, $(C_2F_5CONHCH_2)^+$, $(C_2F_5)^+$, $(CF_3)^+$, $(M-HF)^{+\cdot}$, $(M-2HF)^{+\cdot}$, $(M-3HF)^{-\cdot}$ and $(C_2F_5CH_2OCOCH_2)^-$, respectively.

relative importance of reduced fragmentation, reduced background noise, and higher reaction rates presumably varies between applications.

22.5. APPLICATION OF ELECTRON CAPTURE CHEMICAL IONIZATION TO BIOMEDICAL PROBLEMS

22.5.1. GENERAL COMMENTS

To attain the enhanced sensitivity available in the ECCI mode, a molecule must have positive electron affinity. Furthermore, to be applicable to gas chromatography, a molecule must be sufficiently non-polar, volatile and thermally stable so that it may traverse the chromatographic column. Few biological molecules meet these two requirements in and of themselves. Chemical derivatization is therefore usually necessary. The successful application of the ECCI technique to biomedical problems requires a method which efficiently extracts small amounts of biogenic compounds from complex biological samples and produces in reasonable yield a thermally stable chemical derivative suitable for GC and the ECCI technique.

This is not a trivial task. Although limits of detection and ultimate levels of sensitivity can be judged from experiments in which milligram or microgram quantities of pure compounds are first derivatized and then diluted in a suitable solvent, these results are frequently unrealistic for biological experiments involving nanogram or picogram quantities of material in crude extracts. Extraction recoveries and the yield of chemical derivatives frequently deteriorate when working at the nanogram and picogram level, and losses due to adsorption can escalate. These types of problems are frequently encountered in other trace level GC and GC–MS procedures.

Derivatizing reactions conferring both volatility and electron capturing ability have been extensively explored and widely applied to GC–electron capture detection (Blau and King, 1978: Knapp, 1979). Those derivatives suitable for GC/ECD which also produce anions characteristic of the target analyte, are also suitable for GC/ECCI. However, there remains considerable opportunity for the design of derivatives which have favourable fragmentation patterns under ECCI conditions, particularly in the production of high mass ions suitable for SIM.

In other respects it appears that the ECCI process is not as robust as the conventional electron impact and positive ion chemical ionization processes, and can be affected in a major way by small changes in some operating parameters. For example, the relative intensities of ions in an ECCI spectrum can be drastically altered by small changes in source temperature (Foltz, 1983; Garland and Miwa, 1983; Johnson *et al.*, 1984), with lower temperatures favouring the high mass ions. A compromise must therefore be reached between high temperatures to minimize source contamination and low temperatures to enhance high mass ion intensity. Experience has shown, however, that premature source contamination leading to a rapid deterioration in sensitivity is not a major problem. We have been able to complete large numbers of repetitive analyses with a source temperature of 100°C without a significant loss in sensitivity. However, for certain applications which demand maximal sensitivity (e.g., plasma melatonin quantification (Lewy and Markey, 1978)), avoidance of source contamination can be essential for the maintenance of performance.

Reports have been made of variability between different laboratories in ECCI fragmentation patterns for the same compounds ostensibly run under the same conditions. These types of differences are presumably attributable to small changes in parameters which are difficult to control, such as low levels of impurities in the CI reagent gas and/or GC carrier gas, particularly oxygen and water (for example, see Garland and Miwa, 1983). However, given the same operating conditions, we and others have found the ECCI source to yield reliable results with little variation in relative intensities of ions over long periods of time.

Attention must be drawn to the potential problem of depletion of the thermal electron supply in the ECCI source. With concentrated extracts this problem can result in a loss of sensitivity and alteration in fragmentation patterns (Williams *et al.*, 1981; Garland and Min, 1979). Complete quenching of the response precludes the use of the ECCI technique with direct insertion probe introduction of derivatized extracts (Johnson *et al.*, 1984). Because of the finite population of thermal electrons, the dynamic range of the ECCI source is probably less than that of other types of ion sources. However, it has been our experience that this is not a major problem for most applications, provided attention is given to the quality of chromatography and care is taken to avoid burdening the ion source with excessive sample.

22.5.2. EQUIPMENT

One hardware modification which we have found advantageous is the replacement of the conventional Grob-type split/splitless GC injector with a so-called dropping needle injector for use with capillary columns (Van den Berg and Cox, 1972, Figure 2). Although not in widespread use, one advantage of this solvent-free injector is that it avoids placing on the column residual trace amounts of derivatizing reagents which are frequently present in derivatized extracts. In the ECCI mode, this is important because it serves to lower the background of electron capturing species eluting from the column and results in a more stable baseline. Other advantages include an extension of column life because the injection procedure avoids depositing large volumes of solvent on the column. Furthermore, excellent transfer efficiencies from the dropping needle to the column and the opportunity to deposit relatively large volumes on the needle by repeated application and evaporation are advantageous when working with dilute solutions. We have also observed that with compounds which have a tendency for adsorption to chromatographic surfaces the dropping needle injector produces fewer memory effects. Adsorption can be further reduced by silanization of the injector which is made entirely of glass.

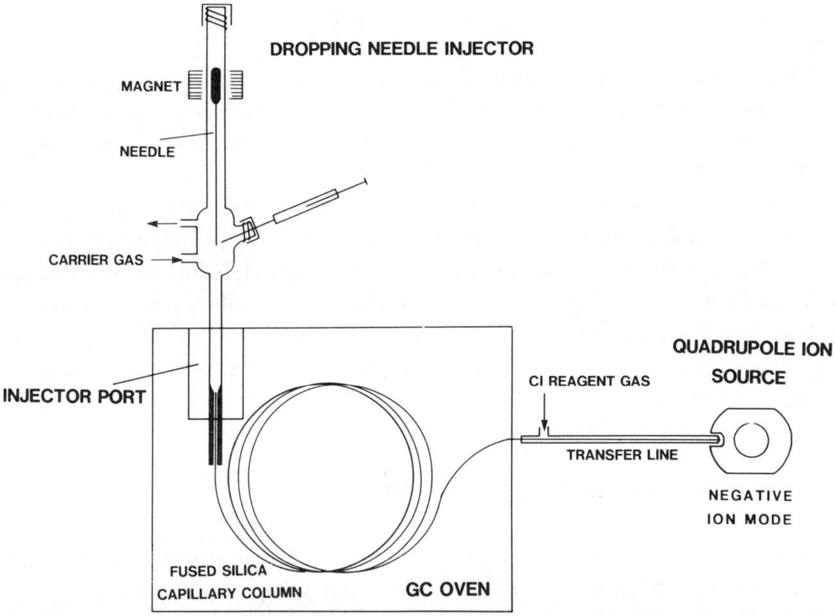

Figure 2: Schematic representation of the dropping-needle GC injector and GC–MS interface used for routine ECCI work.

Rarely is mention made of the CI source pressure necessary for optimal performance, or the degree of control of the source presssure necessary for reproducibility and the maintenance of performance. Some workers favour a relatively low source pressure for electron capture, in the range 10^{-3} to 10^{-2} Torr (Brandenberger, 1983). However, casual observations in our laboratory suggest that for most applications indicated source pressures from 0.3 to 1.0 Torr produce maximal sensitivities. Significant negative ion generation attributed to electron capture can frequently be observed in the CI source in the complete absence of reagent gas. In this case, the low energy secondary electrons produced in the source at electrode and metal surfaces and from gas phase ionization reactions are probably responsible for the observed reactions, but as might be expected, a considerable enhancement of signal is produced immediately upon the introduction of the reagent gas.

A variety of routes have been successfully adopted for the introduction of the reagent gas into the source. The physical constraints imposed by instrument and source design seem in many instances to have dictated the approach adopted. Coaxial introduction of the reagent gas with the column effluent into the ion source may have an advantage because the reagent gas would sweep the end of the column and effect rapid transport of the effluent into the ion source. This approach serves to maintain chromatographic fidelity and avoids the problems which can arise when the GC effluent and reagent gas enter the ion source from opposing directions.

Any discussion on trace level analyses should include at least a brief mention of the value of direct GC–MS coupling. Made possible primarily through the availability of flexible fused silica capillary GC columns, direct GC/MS coupling in which the end of the GC column is brought in close proximity to the ion source of the mass spectrometer eliminates the need for separators and transfer lines and consequently avoids many of the adsorption problems which can otherwise harass trace level analyses. This approach has some disadvantages, however, including some loss of chromatographic fidelity since the resulting velocity of carrier gas flow through the column is usually higher than that appropriate for optimal chromatographic resolution with even the lowest of head pressures.

22.5.3 TRACE LEVEL QUANTIFICATION

The procedure for quantitative analysis by GC–MS is essentially the same, irrespective of the ionization process. To each sample a known amount of internal standard is added. Ions characteristic of the internal standard and of the analyte are monitored. The ratio of peak intensities for the analyte and internal standard provides a quantitative measurement, relatable by inverse regression analysis to a standard curve. The standard curve is constructed by determining the ratio of the peak intensities in a series of samples containing the internal standard and the analyte at various dilutions. It is generally considered good laboratory practice to generate a standard curve with each batch of unknown samples and to carry the samples comprising the standard curve through the entire preparatory procedure. For each experiment it is then possible to check on the linearity of the response.

When working in the nanogram and picogram concentration range, however, this approach can produce some problems. We have repeatedly observed that the peak area for an internal standard, whether it be a stable isotope analogue or a chemical homologue, is much greater in a biological sample than in a pure standard which has been carried through the entire work-up procedure. For example, in the analysis of 3-methoxytyramine (3-MT) in brain tissue, the peak area for the trideuterated-internal standard in the biological samples is approximately 33 per cent greater than in the pure standards (Table 1). A likely explanation for this difference is that the tissue extract contains many other compounds which tend to decrease losses of the metabolite during sample work-up, derivatization and chromatography. Despite the lower peak area with pure standards, the ratio of metabolite peak area/internal standard peak area is as expected, provided the peaks for the internal standard and metabolite are sufficiently strong for their accurate measurement.

TABLE 1. Peak areas for trideuterated 3-MT in extracted and derivatized samples.

Standards		Biological Samples	
Tube No.	Peak Area	Tube No.	Peak Area
10S	14 723	401A	16 786
7S	13 078	401B	16 729
6S	6880	401C	17 908
5S	10 607	403A	10 916
4S	15 180	403C	12 176
3S	12 952	406A	7610
2S	6010	406B	22 324
1S	16 630	406C	20 287
Mean ± standard error	12 007 ± 1370		15 592 ± 1756

For this example, a random selection was made from a large body of available data (K. A. Greene, O. Beck and K. F. Faull, unpublished observations). The measured peak was derived from the $(M^{-}C_2F_5COH)^{-}$ fragment ion of the di-pentafluoropropionyl derivative of $[^2H_3]$-3-methoxytyramine at *m/z* 314. The internal standard (10 ng) was added in equal amounts to a set of standard tubes (1S–10S) and a set of tubes each containing mouse brain tissue (401A–406C, approximately 400 mg wet weight of tissue per tube). The samples were analyzed as described in Section 22.5.4.

In some instances, particularly when working with compounds which show a strong tendency for adsorption to glassware, we have observed more than a ten-fold loss of the signal from the standards which are carried through the entire work-up and derivatization procedure, although the biological extracts prepared at the same time and under the same conditions yield chromatograms with reasonably intense peaks for the analyte and internal standard. The amine–aldehyde condensation product tryptoline (tetrahydro-β-carboline) represents a good example of this problem. One approach we have used to circumvent the problem is to add increasing amounts of the unlabelled compound with the same amount of internal standard to a series of samples and record the relative intensities of ions from the analyte and internal standard. Illustrated graphically in Figure 3, the intercept on the ordinate (b) of the plot of peak area ratio against weight ratio is determined by the amount of analyte in the original extract plus the proportion of ion current in the analyte trace contributed by the internal standard (b′). The latter component (b′) can be measured accurately from a relatively large amount of the internal standard. A theoretical line can then be constructed using the slope of the observed line (a) and the intercept (b′). The interpolated value from the intercept (b) using the theoretical line yields the amount of metabolite in the original extract. While this approach allows for visual inspection of the linearity of the response under the complete conditions of the assay, it of course requires a larger amount of sample.

COMPOUNDS

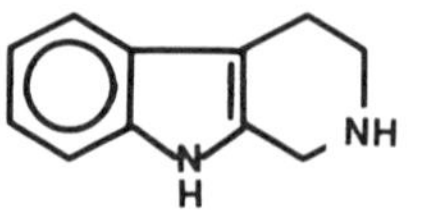

TRYPTOLINE I
(TETRAHYDRO-β-CARBOLINE)

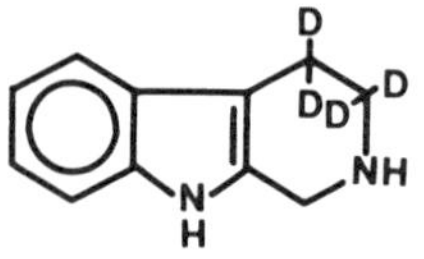

$^2H_4$-TRYPTOLINE II

EXPERIMENT

SAMPLE	PICOGRAMS ADDED I	II[2]
1. 1ml supernatant[1]	0 (0)	1000
2. 1ml supernatant	82.5 (1)	1000
3. 1ml supernatant	82.5 (1)	1000
4. 1ml supernatant	165 (2)	1000
5. 1ml supernatant	165 (2)	1000
6. 1ml supernatant	412.5 (3)	1000

RESULTS

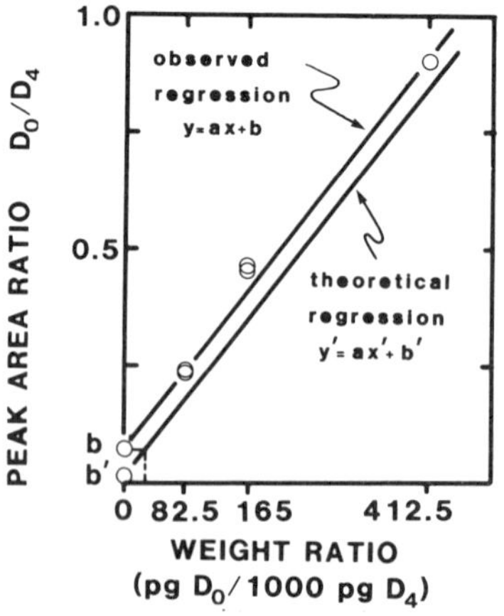

Figure 3: Quantification of tryptoline in brain tissue. [1]Corresponding to the extract from 156 mg of tissue. [2]Approximately 50 per cent pure. b′ calculated from the internal standard as the quotient of the peaks observed in the [$^2H_0$] and [$^2H_4$] ion traces. *Reproduced from Faull* et al. (1982) *by permission of Alan R. Liss, Inc.*

22.5.4. EXTRACTIVE ACYLATION OF *O*-METHYLATED CATECHOLAMINES

We have developed a novel approach for the isolation of the *O*-methylated catecholamine metabolites from aqueous media and their simultaneous conversion to chemical derivatives amenable to the ECCI process. This method could be applicable to other classes of compounds with similar functional groups.

The chemical analysis of the catecholamine metabolites 3-methoxytyramine (3-MT), normetanephrine (NMN) and metanephrine (MN) (Figure 4) is complicated by the low levels in which they occur and the amphoteric nature of the functional groups, which hampers their isolation from aqueous media. In developing an extraction and derivatization method which maximally exploits the potential sensitivity of the ECCI technique, we attempted to make use of the procedure of acylation in aqueous media which had already been developed for the work-up of these compounds from dilute solutions and biological samples (Baker *et al.*, 1981). However, the previously used reagents, acetic and propionic anhydride, yield products that are not suitable for the ECCI technique. Efforts to convert the products of 3-MT, NMN, and MN obtained by acetic anhydride extraction into suitable derivatives by transacylation or by using different hydrolysis procedures were unsuccessful. An effort to use pentafluoropropionic anhydride (PFPA) instead of acetic anhydride demonstrated that the desirable derivative could be recovered in low yields. By incorporating an organic solvent into the mixture and by vigorously mixing the two layers, the yield of product was dramatically improved. The best solvent was found to be dichloromethane, probably because it formed a semi-stable gel which gave intimate mixing of the two layers. The procedure is best described as an extractive acylation rather than acylation in aqueous media.

The developed method involves cooling the samples to 0°C, followed by addition of solid $NaHCO_3$, dichloromethane and treatment with pentafluoropropionic anhydride. The mixture is vortexed vigorously for 10 minutes during which time it warms to room temperature. Additional solid $NaHCO_3$ is added to maintain a small excess of crystals in the bottom of the tube. After standing for an additional 10 minutes at room temperature and then centrifuging, the water layer is discarded and the organic phase washed with water. The organic phase, excluding the residual $NaHCO_3$, is evaporated under a stream of nitrogen and the residue treated with additional PFPA at 75°C for 15 minutes. Finally, the excess reagent is evaporated in a stream of nitrogen, and the residue redissolved in ethyl acetate or heptane prior to injection onto the GC column.

Fused silica GC (helium carrier gas)-selected ion monitoring MS in the negative ion CI (methane 1 Torr) mode has been used to determine the relative intensities of the pairs of diagnostic ions at *m/z* 621 and 626 ($M^{-\cdot}$), 635 and 640 ($M^{-\cdot}$) and 311 and 315 ($(M - C_2F_5COH)^-$) derived, respectively, from NMN, MN and 3-MT and their deuterated internal standards.

Figure 4: Structures of the catecholamine *O*-methylated metabolites and their chemical derivatives.

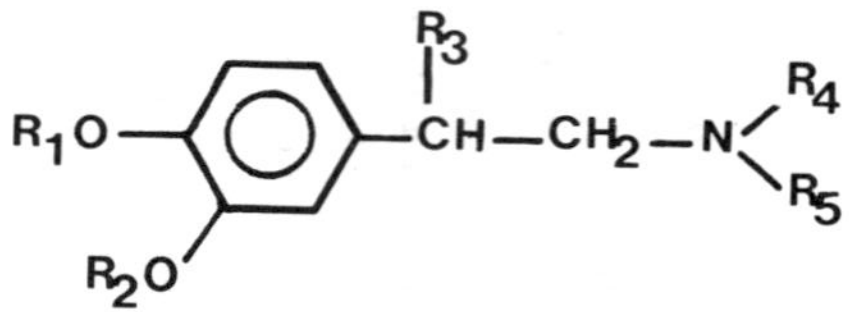

CATECHOLAMINE	O-METHYLATED METABOLITE	CHEMICAL DERIVATIVE	R_1	R_2	R_3	R_4	R_5
I Dopamine	—	—	H	H	H	H	H
II	3-Methoxytyramine	—	H	CH_3	H	H	H
III		A	COC_2F_5	CH_3	H	COC_2F_5	H
IV Norepinephrine	—	—	H	H	OH	H	H
V	Normetanephrine	—	H	CH_3	OH	H	H
VI		A	COC_2F_5	CH_3	OH	COC_2F_5	H
VII		B	COC_2F_5	CH_3	COC_2F_5	COC_2F_5	H
VIII Epinephrine	—	—	H	H	OH	H	CH_3
IX	Metanephrine	—	H	CH_3	OH	H	CH_3
X		A	COC_2F_5	CH_3	OH	COC_2F_5	CH_3
XI		B	COC_2F_5	CH_3	COC_2F_5	COC_2F_5	CH_3

A. Formed during the extractive acylation step.
B. Formed by treatment with pentafluoropropionic anhydride under anhydrous conditions.
(*Taken from Beck and Faull (1986) with permission of Academic Press*)

The yield of the initial derivatization/extraction step was estimated by analyzing for underivatized 3-MT, NMN and MN remaining in the aqueous phase. The analysis was accomplished using HPLC on a reversed phase octadecylsilane column with amperometric detection of the *O*-methylated catecholamines in the eluate. The result indicated that less than 5 per cent of the parent compounds remained in the aqueous phase following treatment with PFPA and dichloromethane extraction. This result was in agreement with recovery estimates of the complete work-up procedure made using GC–MS by comparing peak areas of standard samples that were directly treated with PFPA with those of samples run through the entire procedure. The results suggested a recovery of 90–100 per cent for each compound.

TABLE 2. Concentrations of 3-methoxytyramine, normetanephrine and metanephrine in biological fluids.

Biological Sample	3-MT		NMN		MN	
	Free[c]	Total[d]	Free	Total	Free	Total
CSF[a]						
Squirrel monkey	1.18	13.80	4.69	9.78	0.37	0.59
Canine	0.49	1.28	0.36	1.37	0.15	0.59
Human	1.13	11.60	0.86	11.10	0.13	1.90
Mouse Brain Tissue[b]						
Sacrifice by decapitation	54.5	56.0	28.6	34.7	—	—
Sacrifice by microwave irradiation	14.6	15.6	6.3	9.1	—	—
Sacrifice by freezing in liquid nitrogen	28.9	30.2	17.2	17.3	—	—

[a] Units pmol/ml.
[b] Units ng/g.
[c] Unconjugated form of the metabolites.
[d] Concentration of unconjugated plus conjugated forms of the metabolite measured after enzymatic hydrolysis with a mixture of sulphatase and β-glucuronidase. (*Taken from Beck* et al., *1985*).

The products formed in the initial derivatization/extraction step were identified as being the dipentafluoropropionyl (PFP) derivatives of 3-MT, NMN and MN. The EI mass spectra of the MN and NMN derivatives contained significant $(M\text{-}18)^{+\cdot}$ fragments indicating that the PFP groups were located on the amine and phenol groups of the molecules, thus leaving the β-hydroxyl groups unchanged. The alcoholic hydroxyl groups of NMN and MN were acylated only during the second treatment with PFPA under anhydrous conditions.

This work-up procedure and subsequent analysis using capillary column gas chromatography–ECCI mass spectrometry has been successfully applied to the identification and quantification of trace levels of 3-MT, NMN, and MN in human CSF and mouse brain tissue extracts. A typical selected ion chromatogram from human CSF is shown in Figure 5. A summary of our results, shown

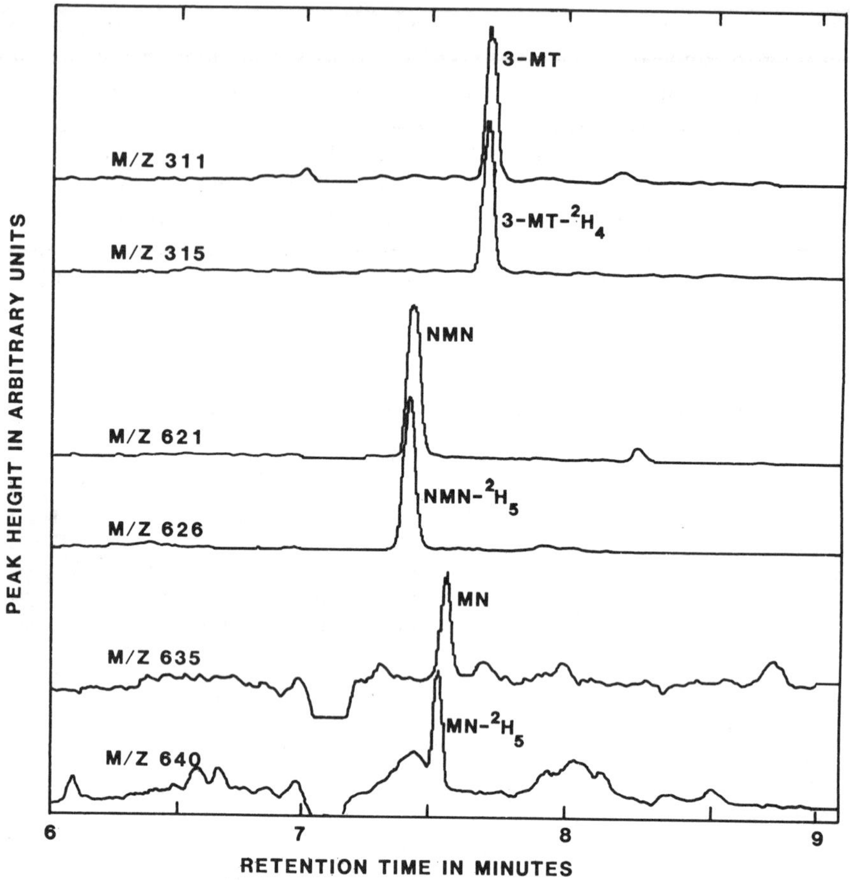

Figure 5: Selected ion traces from the assay of a human cerebrospinal fluid extract showing peaks attributable to 3-methoxytyramine, normetanephrine and metanephrine, and their respective deuterated analogues. *Reproduced from Beck and Faull (1986) by permission of Academic Press.*

in Table 2, indicates significant quantities of conjugated forms of these metabolites in squirrel monkey, canine and human CSF but smaller quantities of the conjugates in mouse brain tissue. The results also indicate a big effect of the method of sacrifice on mouse brain tissue levels of the unconjugated metabolites.

22.5.5. GAMMA-AMINOBUTYRIC ACID (GABA)

A number of procedures have been published which use conventional EI and CI GC–MS for quantification of GABA in blood, cerebrospinal fluid and brain tissue homogenates. In general, the level of sensitivity demonstrated by these procedures extends to the low nanogram and high picogram range. In an attempt to extend the simplicity and sensitivity of these methods, we have developed a method for the extraction of GABA from aqueous samples and methods for converting GABA to a number of different derivatives suitable for GC and ECCI (C. Borrego, C. J. Evans and K. F. Faull, unpublished observations).

Organic solvent extraction of amino acids from aqueous solutions is hampered by the zwitterionic character of the functional groups. Reaction of the amino group with a water soluble carbodiimide can yield an *N*-protected species which is readily extractable into an organic solvent. We have developed a scheme for the extraction of GABA from aqueous solutions by reaction with di-*tert*-butylpyrocarbonate (Figure 6). The reaction is achieved by treating a solution of GABA in sodium carbonate buffer (pH 10) with 100-fold molar excess of the reagent dissolved in *t*-butanol. The solution is then acidified with acetic acid to pH 4.0 and extracted with ethyl acetate. The organic phase is blown to dryness in a stream of nitrogen, and the residue is treated with mineral acid (6 M HCl) which cleaves the amino protecting group. Removal of the by-products and water is then achieved by lyophilization and the dried residue is ready for chemical derivatization.

Direct examination of the reaction mixture using fast atom bombardment MS has been used to confirm the reaction sequence. This has shown appropriate signals for the mono-, di- and tri-sodiated species of the protonated *N*-protected molecule in the aqueous reaction mixture (Figure 7a) and a strong signal for the protonated *N*-protected molecule in the ethyl acetate extract (Figure 7b). Recovery of GABA after treatment with mineral acid is demonstrated by the signal at *m/z* 104 in the FAB spectrum of the hydrochloric acid solution of the final product (Figure 7c). Measurements made using GC–MS have indicated this method yields recoveries of GABA of between 52 and 78 per cent. The

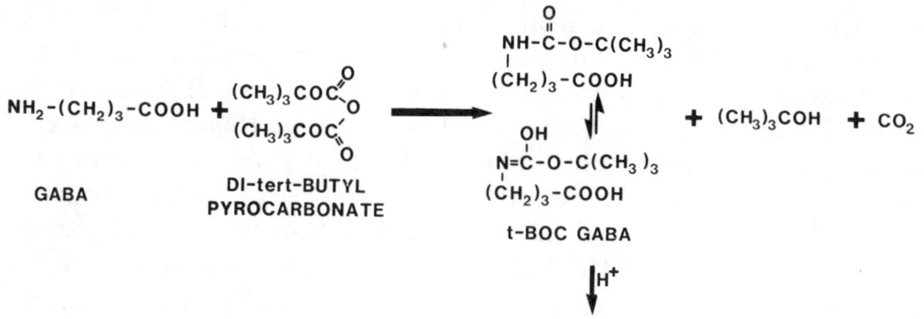

Figure 6: Reaction sequence for the derivatization and extraction of GABA. (*Taken from unpublished observations of C. Borrego, C. J. Evans and K. F. Faull.*)

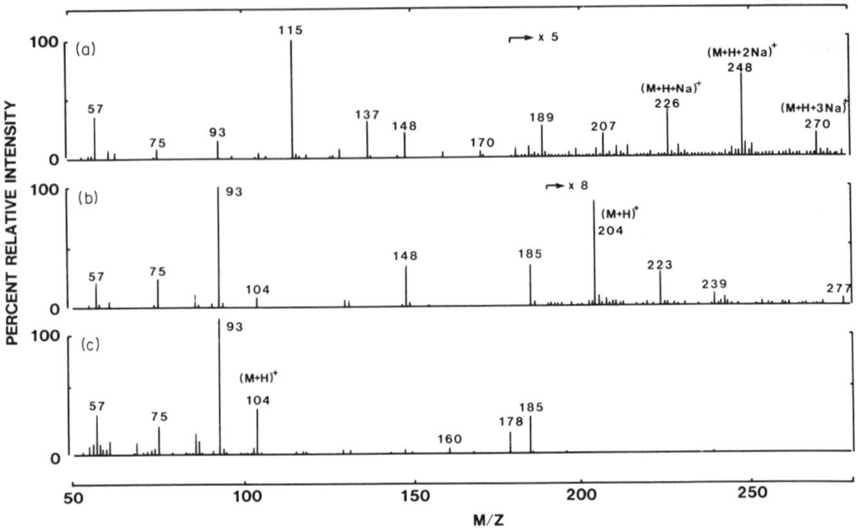

Figure 7: Fast atom bombardment mass spectra of aliquots of the reaction mixture containing GABA and di-*tert*-butylpyrocarbonate taken at various stages in the reaction and extraction sequence (see Figure 6). a) the aqueous reaction mixture containing the reactants in sodium carbonate buffer (pH 10) showing ions at m/z 226, 248 and 270 corresponding to sodiated adducts of protonated t-BOC GABA. b) the ethyl acetate extract obtained following acidification to pH 4.0, showing ions at m/z 204 and 104 corresponding to protonated t-BOC GABA and protonated GABA respectively. c), after 6M HCl treatment of the dried residue remaining after evaporation of the ethyl acetate extract, showing an intense ion at m/z 104 corresponding to protonated GABA.

simplicity of the approach and avoidance of time consuming ion exchange procedures suggests the method may have merit for the organic solvent extraction of other amino acids with similar functional groups.

We have also extensively explored a variety of chemical derivatization procedures for converting GABA to a form suitable for ECCI GC–MS. The EI and ECCI mass spectra of three promising derivatives are shown in Figure 1. A variety of fragmentations are observed under ECCI conditions when the electron capturing group is positioned on the carboxyl end (Figure 1a), the amino end (Figure 1b) and both ends of the molecule (Figure 1c). Careful comparison of the responses obtained for each derivative relative to benzophenone have shown the carboxylate anion derived from the neopentylidene Schiff base pentafluorobenzyl ester provides the greatest potential for sensitivity under ECCI conditions (Table 3). This finding was not unexpected because carboxylate anions derived from pentafluorbenzyl esters of a number of other organic molecules, including phenylacetic acid (Faull and Barchas, 1983), tetrahydrocannabinol (Foltz and Chinn, 1981), peptides (Morris *et al.*, 1981) and prostaglandins and other arachidonic acid derivatives (Min *et al.*, 1980; Blair *et al.*, 1982; Waddell *et al.*, 1983, 1984), have been previously shown to produce extraordinarily strong responses under ECCI conditions.

TABLE 3. Responses of GABA derivatives relative to that of benzophenone (BP) under EI and ECCI conditions.

	EI			ECCI		
Derivative	Neopentylidene Schiff base pentafluorobenzyl ester					
Ion Ratio GABA/BP	TIC[a] / TIC	$(C_6F_5CH_2)^+$ / $M^{+\cdot}$	$(M–CH_3)^+$ / $M^{+\cdot}$	TIC / TIC	170[b] / $M^{-\cdot}$	
Response	0.4	1.2	0.1	17.8	74.3	
Derivative	*N*-heptafluorobutyryl methyl ester					
Ion Ratio GABA/BP	TIC / TIC	$(M–OCH_3)^+$ / $M^{+\cdot}$	$M^{+\cdot}$ / $M^{+\cdot}$	TIC / TIC	$(M–HF)^{-\cdot}$ / $M^{-\cdot}$	
Response	8.7	2.4	0.3	18.7	6.0	
Derivative	*N*-pentafluoropropionyl pentafluoropropyl ester					
Ion Ratio GABA/BP	TIC / TIC	176[c] / $M^{+\cdot}$	232[d] / $M^{+\cdot}$	TIC / TIC	191[e] / $M^{-\cdot}$	$(M–HF)^{-\cdot}$ / $M^{-\cdot}$
Response	6.3	5.7	2.6	21.5	19.0	3.8

Data taken from C. Borrego and K. F. Faull (unpublished observations). Experiments were carried out by derivatizing 400 μg of GABA, adding 700 μg of benzophenone and injecting 0.001 per cent of the final mixture. The values presented represent the peak area obtained from the GABA derivatives expressed as a ratio of the peak area of benzophenone. Chromatography was carried out on a 15 m medium polarity bonded phase fused silica capillary column, programmed from 80°C to 250°C at 30° per minute. An indicated source pressure of 1 Torr was maintained with methane. Source temperatures of 200°C and 100°C were used in the EI and ECCI modes, respectively. Identity of ions: [a]total ion chromatogram; [b]$((CH_3)_3CCH = N(CH_2)_3COO(^-$; [c]$(_2F_5CONHCH_2)^+$; [d]$(C_2F_5CONH(CH_2)_3CO)^+$; [e]$(C_2F_5CH_2OCOCH_2)^-$. The complete EI and ECCI spectra of each derivative are shown in Figure 1.

A breakthrough in this work has come with the development of a derivatization scheme which reliably produces a stable Schiff base PFB ester derivative in high yield. This has now been achieved by first forming the Schiff base using freshly redistilled pivaldehyde and triethylamine, then rapidly forming the PFB ester using pentafluorobenzylbromide, and finally redissolving the derivative in an anhydrous solvent containing a small amount of pivaldehyde. By using the solvent-free injector any adverse effects of the aldehyde on the GC column are avoided and excellent levels of sensitivity down to 5–10 picograms can be produced.

22.6. CONCLUSIONS

The potential of electron capture negative ion chemical ionization GC–MS for providing increased sensitivity without sacrificing specificity can be exploited for the analysis of electron capturing compounds in chemically derivatized crude biological extracts provided attention is given to a number of operating parameters. These include the quality of chromatography, the need to keep to a minimum the level of residual electron capturing reagents in the sample, the importance of contaminants in the GC carrier gas and reagent gas, and the sensitivity of the high mass, negatively charged ions to high source temperatures. Undoubtedly the most perplexing problem encountered by all high sensitivity GC–MS analyses, including ECCI procedures, is the loss of analyte by adsorption to surfaces. Despite these problems, the literature contains an ever increasing array of examples which exploit the ECCI process for high sensitivity work.

ACKNOWLEDGMENTS

The authors gratefully acknowledge the support of Dr. Jack D. Barchas for his encouragement in preparing this chapter. Work supported by grants to J. D. Barchas, M.D. from the Scottish Rite Schizophrenia Research Program, the National Institute on Alcohol Abuse and Alcoholism (AA 05972), the National Institute of Mental Health (MH 23861), and the MacArthur Foundation Network on the Psychobiology of Depression.

REFERENCES

Baker, G. B., Coutts, R. T., and Martin, I. L. (1981). Analysis of amines in the central nervous system by gas chromatography with electron-capture detection, *Progr. Neurobiol.*, **17**, 1–24.

Beck, O., and Faull, K. F. (1985). Extractive acylation and mass spectrometric assay of 3-methoxytyramine, normetanephrine and metanephrine in cerebrospinal fluid, *Anal. Biochem.*, **149**, 492–500.

Beck, O., Greene, K. A., Faull, K. F., and Barchas, J. D. (1985). Convenient extraction and mass spectrometric assay of 3-methoxytyramine, normetanephrine and metanephrine in biological fluids, in *Proc. 33rd Ann. Conf. on Mass Spectrometry and Allied Topics, San Diego;* American Society for Mass Spectrometry; pp. 216–217.

Blair, I. A., Barrow, S. E., Waddell, K. A., Lewis, P. J., and Dollery, C. T. (1982). Prostacyclin is not a circulating hormone in man, *Prostaglandins*, **23**, 579–589.

Blau, K., and King, G. S. Eds., (1978). *Handbook of Derivatives for Chromatography*, Heyden and Son, London.

Brandenberger, H. (1983). Negative ion mass spectrometry by low pressure chemical ionization: inherent possibilities for structural analysis, *Int. J. Mass Spectrom. Ion Phys.* **47**, 213–216.

Faull, K. F., and Barchas, J. D. (1983). Negative ion mass spectrometry fused-silica capillary gas chromatography of neurotransmitters and related compounds, in *Methods of Biochemical Analysis*, Vol. 29 (Ed. D. Glick), pp. 325–383, Wiley, New York.

Faull, K. F., Holman, R. B., Elliott, G. R., and Barchas, J. D. (1982). Tryptolines: artifact or reality? A new method of analysis using GC/MS, in *Beta-Carbolines and Tetrahydroisoquinolines* (Eds. E. Usdin, F. Bloom and J. D. Barchas), pp. 135–154, Alan R. Liss, Inc., New York.

Foltz, R. L. (1983). High sensitivity quantitative analysis of drugs and metabolites by GC/MS. in *Proc. 31st Ann. Conf. on Mass Spectrometry and Allied Topics, Boston*; American Society for Mass Spectrometry; pp. 454–455.

Foltz, R. L., and Chinn, D. M. (1981). Quantitative analysis of Δ^9-tetrahydrocannabinol and metabolites using negative ion chemical ionisation, in *Proc. 29th Ann. Conf. on Mass Spectrometry and Allied Topics, Minneapolis*; American Society for Mass Spectrometry; pp. 321–322.
Garland, W. A., and Min, B. H. (1979). Determination of clonazepam in human plasma by gas chromatography–negative ion chemical ionization mass spectrometry, *J. Chromatogr.*, **172**, 279–286.
Garland, W. A., and Miwa, B. J. (1983). The $(M-1)^-$ ion in the negative chemical ionization mass spectra of diazepam and nordiazepam, *Biomed. Mass Spectrom.*, **10**, 126–129.
Harrison, A. G. (1982). *Chemical Ionization Mass Spectrometry*, CRC Press, Boca Raton, Florida.
Hunt, D. F., Stafford, G. C. Jr., Crow, F., and Russell, J. W. (1976). Electron capture negative ion chemical ionization mass spectrometry, *Anal. Chem.*, **48**, 2098–2105.
Johnson, J. V., Yost, R. A., and Faull, K. F. (1984). Tandem mass spectrometry for the trace determination of tryptolines in crude brain extracts, *Anal. Chem.*, **56**, 1655–1661.
Klots, C. E. (1968). Energy deposition mechanisms, in *Fundamental Processes in Radiation Chemistry*, (Ed. P. Ausloos), pp. 1–57, Wiley, New York.
Knapp, D. R. (1979). *Handbook of Analytical Derivatization Reactions*, Wiley-Interscience, New York.
Lewy, A. J., and Markey, S. P. (1978). Analysis of melatonin in human plasma by gas chromatography negative chemical ionization mass spectrometry, *Science*, **201**, 741–743.
Min, B. H., Pao, J., Garland, W. A., de Silva, J. A. F., and Parsonnet, M. (1980). Determination of an antisecretory trimethyl prostaglandin E_2 analog in human plasma by combined capillary column gas chromatography–negative chemical ionization mass spectrometry, *J. Chromatogr.*, **183**, 411–419.
Morris, H. R., Dell, A., Judkins, M., McDowell, R. A., Panico, M., and Taylor, G. W. (1981). Fast atom bombardment and other new developments in mass spectrometric peptide sequencing, in *Proc. 7th American Peptide Symposium*, (Eds. D. H. Rich and E. Gross), pp. 745–761, Pierce Chemical Co., Rockford, IL.
Pellizzari, E. D. (1974). Electron capture detection in gas chromatography, *J. Chromatogr.*, **98**, 323–361.
Van den Berg, P. M. J., and Cox, Th. P. H. (1972). An all-glass solid sampling device for open tubular columns in gas chromatography, *Chromatographia*, **5**, 301–305.
Waddell, K. A., Blair, I. A., and Wellby, J. (1983). Combined capillary column gas chromatography negative ion chemical ionization mass spectrometry of prostanoids, *Biomed. Mass Spectrom.*, **10**, 83–88.
Waddell, K. A., Barrow, S. E., Robinson, C., Orchard, M. A., Dollery, C. T., and Blair, I. A. (1984). Quantitative analysis of prostanoids in biological fluids by combined capillary column gas chromatography negative ion chemical ionization mass spectrometry, *Biomed. Mass Spectrom.*, **11**, 68–74.
Williams, C. M., Crowley, J. R., and Couch, M. W. (1981). Limitations of NICI–GCMS for analysis of biogenic amines, in *Proc. 29th Ann. Conf. on Mass Spectrometry and Allied Topics, Minneapolis*; American Society for Mass Spectrometry; pp. 685–686.

Mass Spectrometry in Biomedical Research
Edited by S. J. Gaskell
© 1986 John Wiley & Sons Ltd.

Chapter 23

MICRODETERMINATION OF PROSTAGLANDINS AND THROMBOXANE B_2 BY GAS CHROMATOGRAPHY–HIGH RESOLUTION SELECTED ION MONITORING

Masataka Ishibashi, Kouwa Yamashita, Keiko Watanabe, and Hiroshi Miyazaki

23.1. INTRODUCTION

With advances in the elucidation of the physiological roles of various prostaglandins (PGs) and thromboxane A_2 (TXA_2) microanalytical methods with high sensitivity and high specificity have been developed in order to investigate the distribution and/or occurrence of these compounds in biological fluids and tissues. For this purpose, gas chromatography–selected ion monitoring (GC–SIM) has been widely used as the most specific method in microanalysis of PGs and TXB_2 as it provides both structural information and GC retention times. In the analysis of biochemical extracts, however, there is a risk that GC–SIM may be affected by unexpected impurities having quite similar physical and/or chemical properties to the compound of interest. In order to enhance the reliability of the analytical results, a number of intensive investigations have been performed in purification, derivatization, GC separation and instrumentation.

Derivatization plays an important role in microanalysis by GC–SIM. It is employed to enhance the GC performance by improving volatility and thermal stability. Trimethylsilyl (TMS) ethers, methyl esters and methyloximes have generally been used as derivatives for GC and GC–mass spectrometric analyses of prostaglandins and thromboxane (Hamberg and Samuelsson, 1967; Vane and Horning, 1969; Rosello *et al.*, 1977). As the sensitivity in microanalysis using GC–SIM is expected to be improved by choosing a derivative with a mass spectrum in which the ion current is concentrated in characteristic ions in the high mass region, a number of new derivatives, such as dimethylethylsilyl (DMES) (Miyazaki *et al.*, 1978), dimethyl-n-propylsilyl (DMNPS) (Miyazaki *et al.*, 1978), dimethylisopropylsilyl (DMIPS) (Miyazaki *et al.*, 1981) and tert-butyldimethylsilyl (TBDMS) (Kelly and Taylor, 1976; Smith *et al.*, 1977; Bazan and Knapp, 1982; Mai *et al.*, 1983) ether derivatives and cyclic alkaneboronates (Pace-Asciak and Wolf, 1971; Smith *et al.*, 1977) have been extensively examined for quantification of PGs and TXB_2 by means of GC–SIM.

In general, electron impact–mass spectrometry (EI–MS) gives spectra which contain many ions useful for identification of the compound, but characteristic ions in the high mass region, suitable for quantification by GC–SIM, are frequently of low abundance. Chemical ionization (CI) MS has been a powerful technique for GC–SIM because the spectra give abundant ions retaining the intact molecular structure. The concentration of ion current in these ions provides higher sensitivity during GC–SIM. In particular, GC–electron capture chemical ionization/SIM, one of the most sensitive techniques for compounds having high electron affinity, is advantageous for the analysis of low levels of compounds containing halogen atoms. This technique has been applied to the microanalysis of PGs and TXB_2 (Min *et al.*, 1980; Barrow *et al.*, 1982; Waddell *et al.*, 1983; Miyazaki *et al.*, 1984; Pace-Asciak and Micallef, 1984). Pentafluorobenzyl esters of PGs and TXB_2 provide particularly favourable properties for analysis by GC–electron capture MS/SIM, resulting in sensitivities that may be 10 to 100 times higher than those of other techniques (Waddell *et al.*, 1983).

GC–high resolution (HR) SIM, developed by Millington (Millington, 1975; Millington *et al.*, 1975) and made popular by Gaskell *et al.* (Gaskell and Pike, 1978; Gaskell *et al.*, 1980; Finlay and Gaskell, 1981; Gaskell and Brownsey, 1983), has been applied to the microanalysis of hormonal steroids in human tissues. As this technique enables monitoring of characteristic ions of selected exact mass or elemental composition, it is expected to give a high degree of reliability in trace analyses of PGs and TXB_2. Indeed, chemical specificity and sensitivity have been augmented by development of high resolution SIM, as well as by positive CI and electron capture CI/SIM. Another approach to obtaining higher specificity and sensitivity is to utilize a more efficient chromatographic technique. HR SIM coupled with high resolution GC may be recognized as the most specific and reliable method available for the microanalysis of PGs and TXB_2 in biological fluids and tissues. This chapter describes briefly our studies on the development of the microanalysis of PGs and TXB_2 by capillary GC–HR SIM and its biological applications, in collaboration with Professor Hayaishi's group (Osaka Medical college).

23.2. GAS CHROMATOGRAPHIC PROPERTIES OF DMES, DMNPS AND DMIPS ETHER DERIVATIVES (Miyazaki *et al.*, 1978; 1981)

Prostaglandin methyl esters and methyloxime-methyl esters, usually prepared by treatment with diazomethane and then with methoxylamine hydrochloride, have generally been converted to their TMS ether derivatives. As TMS ether derivatives give diagnostically useful ions, they have been widely used for GC–MS. However, the disadvantage of these derivatives for GC–SIM is that characteristic ions in the high mass region may be of extremely low intensity when many hydroxyl groups are present in the parent molecule because of successive eliminations of trimethylsilanol. In response to this problem, other silylation reactions, especially tert-butyldimethylsilylation, have been extensively examined for PG methyl esters and methyloxime-methyl esters in GC and GC–MS.

The DMIPS ether derivatives of PGs and TXB_2 methyl esters and methyloxime-methyl esters were prepared and used for comparison of their GC and GC–MS properties with the corresponding TMS, DMES and DMNPS ether derivatives. Scheme 1 shows the derivatization procedures for the PGs and TXB_2, affording DMIPS ethers of methyl esters and methyloxime-methyl esters. Other silyl ethers were prepared using the appropriate alkylsilylimidazole. The DMES, DMNPS and DMIPS ether derivatives of PGs and TXB_2 methyl esters and methyloxime-methyl esters gave high methylene unit (MU) values in comparison with the corresponding TMS ether derivatives, and this was multiplied in proportion to the number of hydroxyl groups in the molecule.

Figure 1 shows the increment of MU values of the TMS, DMES, DMNPS and DMIPS ether derivatives. As shown in this figure, the retention times of TMS ether derivatives increased in the order of $PGF_{2\alpha}$, PGD_2, $PGF_{1\alpha}$, PGE_2, PGE_1, TXB_2 and 6-keto-$PGF_{1\alpha}$, and those of DMES, DMNPS and DMIPS ether derivatives increased in the order of PGD_2, PGE_2, PGE_1, $PGF_{2\alpha}$, $PGF_{1\alpha}$, TXB_2 and 6-keto-$PGF_{1\alpha}$. The separation of PGE_1, $PGF_{2\alpha}$ and $PGF_{1\alpha}$ was enhanced with an increase in the number of carbon atoms in the silyl groups. Figure 1 also indicates that the DMES, DMNPS and DMIPS ether derivatives allowed classification of PGs and TXB_2 into two distinct groups of di- and trihydroxylated compounds and greatly enhanced the separation of individual PGs and TXB_2 in comparison with the corresponding TMS ether derivatives. The DMES, DMNPS and DMIPS ether derivatives made it possible to perform better separation of PGs and TXB_2 which were not as well separated as the TMS ether derivatives.

Figure 2 shows the capillary GC separation of a standard mixture of PGs and TXB_2 as the DMIPS ether derivatives of methyl esters and methyloxime-methyl esters. All of these derivatives showed well-defined and resolved peaks. Separation of the DMIPS ether derivatives could be achieved within 20 min using a

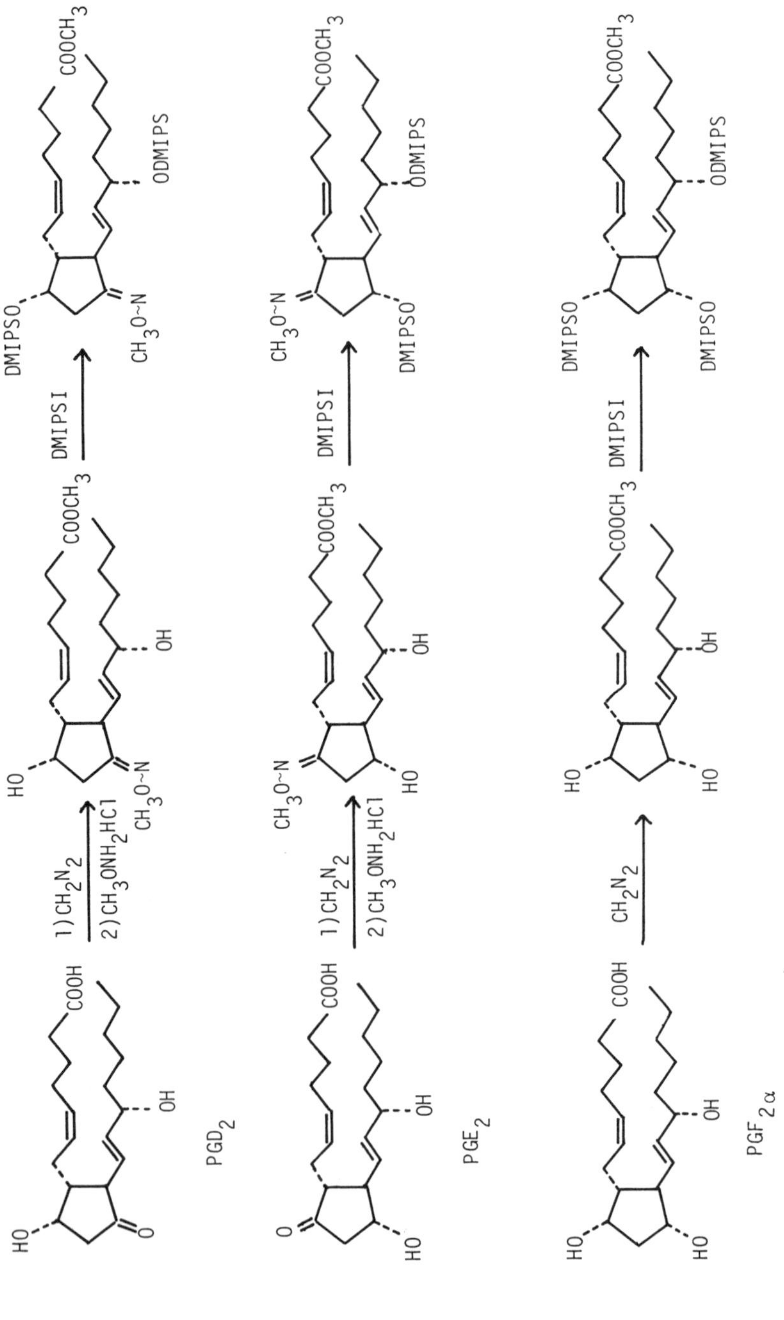
PGD2
PGE2
PGF2α
1) CH2N2
2) CH3ONH2HCl
CH2N2
DMIPSI
COOH
COOCH3
OH
HO
O
CH3O~N
DMIPSO
ODMIPS

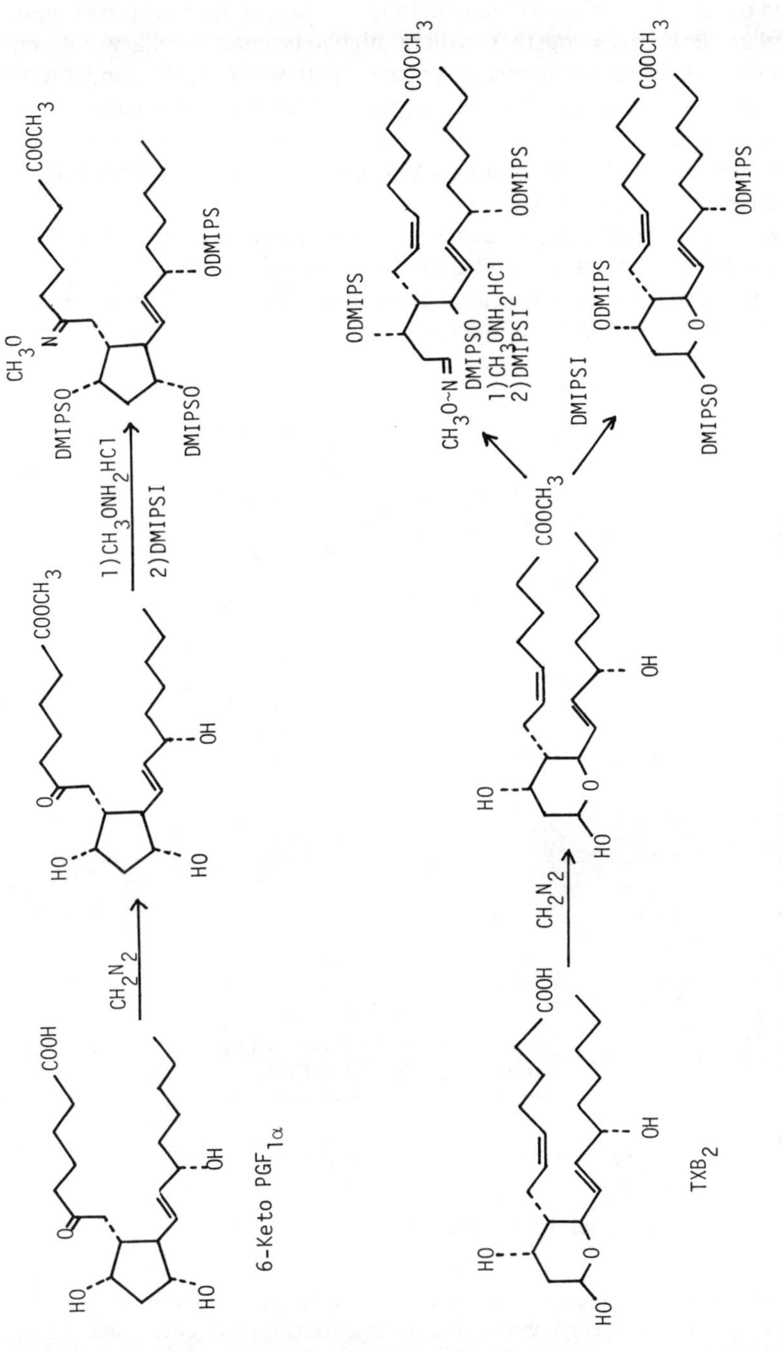

Scheme 1

Derivatization procedures for PGs and TXB_2

wall-coated open tubular glass capillary column, indicating that the use of these DMIPS ether derivatives together with a high efficiency capillary column provided sufficient resolution for the concurrent analysis of the PGs and TXB_2 in a single run. The *syn* and *anti* isomers of the DMIPS ether of 6-keto-$PGF_{1\alpha}$ methyloxime-methyl ester were observed as two well-resolved peaks with equal peak height, whereas the corresponding TMS ether derivatives provided poorly resolved peaks on a capillary column.

The operating temperature of 260 °C for the analysis of DMIPS ether derivatives of PGs and TXB_2 is well below the thermal limit of the cross-linked (or chemically bonded) fused silica capillary column. Therefore, the column can be very well conditioned and will maintain resolution for long periods of use.

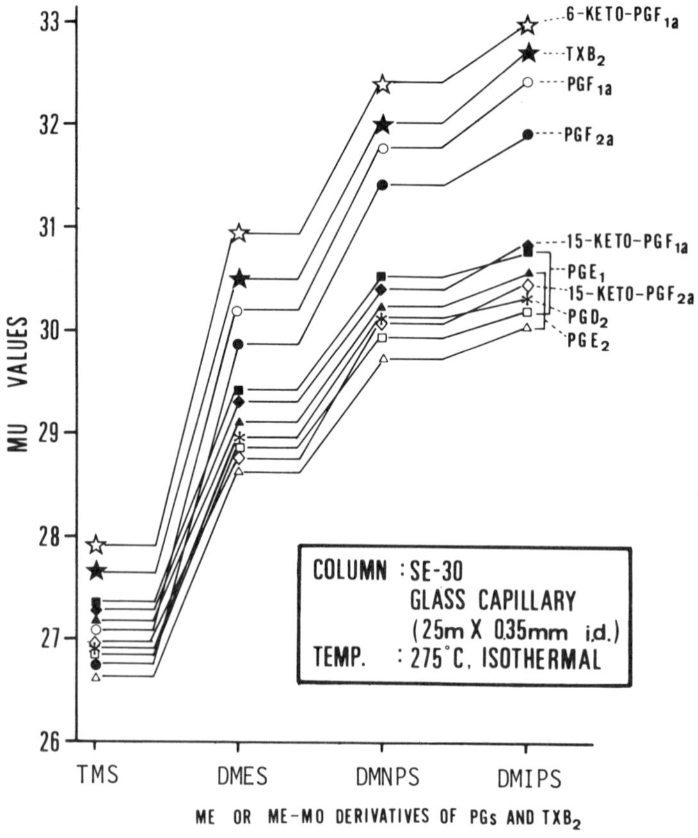

Figure 1: Increment of methylene unit values in TMS, DMES, DMNPS and DMIPS ether derivatives of PG methyl esters, PG methyloxime-methyl esters and TXB_2 methyloxime-methyl ester.

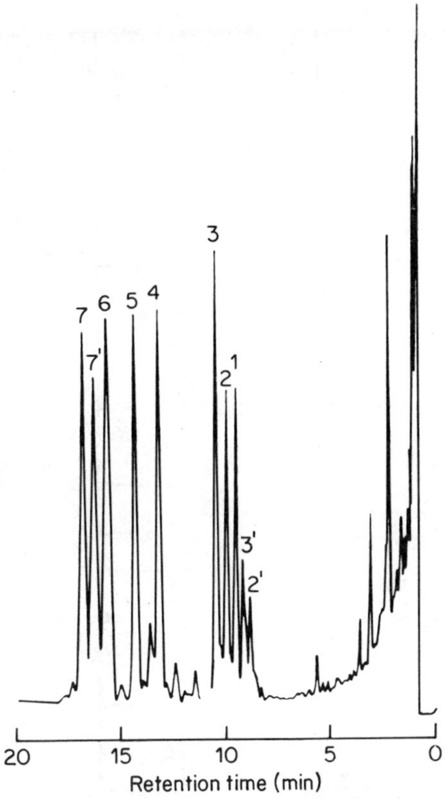

Figure 2: Gas chromatographic separation of DMIPS ether derivatives of PG methyl esters, PG methyloxime-methyl esters and TXB_2 methyloxime-methyl ester. PGD_2(1), PGE_2(2,2′), PGE_1(3,3′), $PGF_{2\alpha}$(4), $PGF_{1\alpha}$(5), TXB_2(6), 6-keto-$PGF_{1\alpha}$(7,7′).

On the other hand, the TBDMS ethers of PGs and TXB_2 need high GC temperatures because TBDMS ethers of hydroxylated compounds in general have retention times 2 to 3 MU higher than TMS ether derivatives per hydroxyl group.

23.3. MASS SPECTROMETRIC PROPERTIES OF DMES, DMNPS AND DMIPS ETHER DERIVATIVES (Miyazaki *et al.*, 1978, 1981)

TBDMS ethers have been used for the GC–MS analysis of steroids (Phillipou *et al.*, 1975; Gaskell and Brooks, 1976) and PGs and TXB_2 (Kelly and Taylor, 1976; Smith *et al.*, 1977; Bazan and Knapp, 1982). The advantage of these

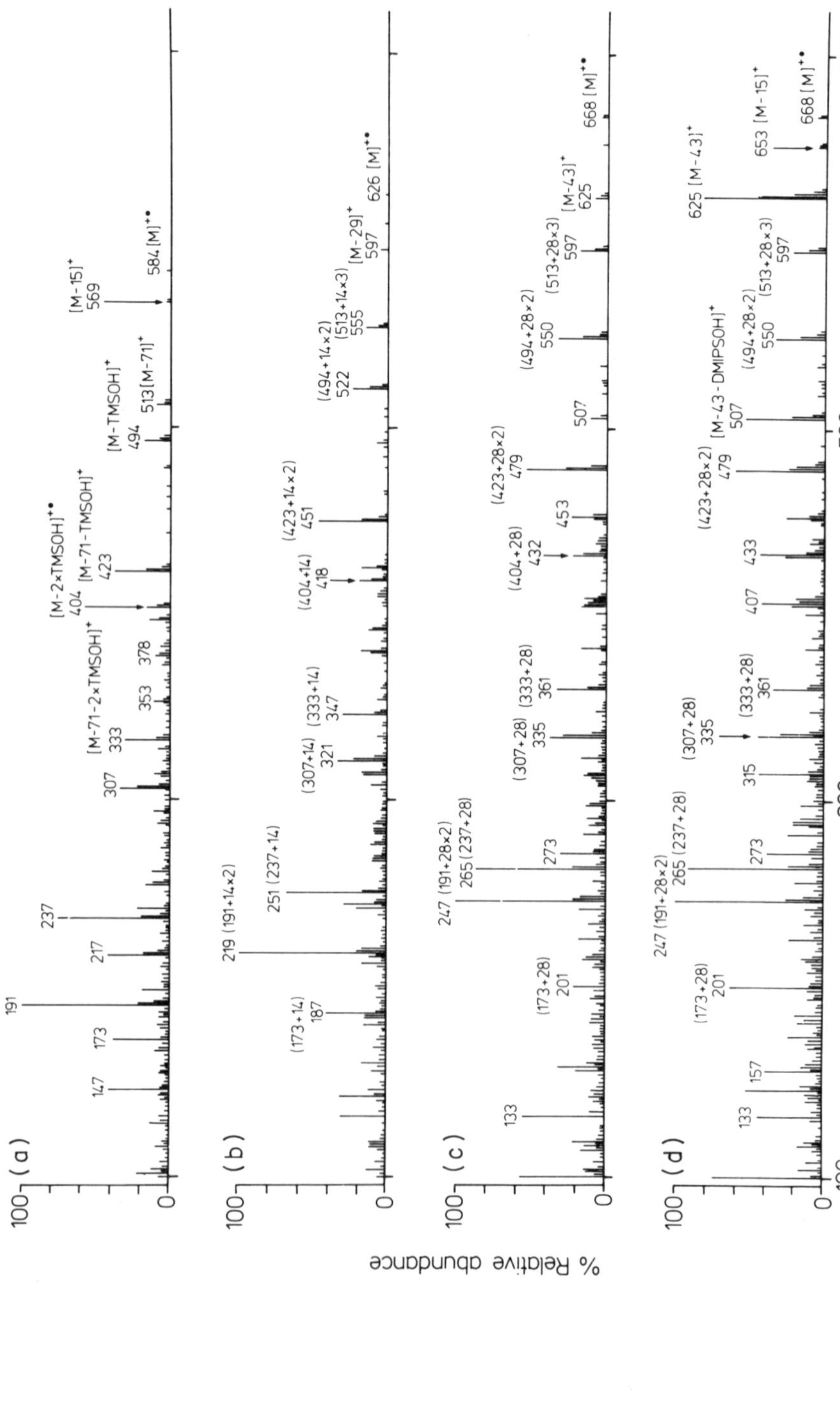

Figure 3: Electron impact mass spectra of TMS(a), DMES(b), DMNPS(c) and DMIPS(d) ether derivatives of $PGF_{2\alpha}$ methyl ester.

TABLE 1. Mass spectral data of DMIPS either derivatives of PGs methyl esters, PGs methyloxime-methyl esters and TXB_2 methlyoxime-methyl ester.

Compound	Mol·wt	$[M]^{+\cdot}$	$[M-15]^+$	$[M-31]^+$	$[M-43]^+$	Other ions (rel. int. (%))							
PGD_2[c]	595	2.9	3.2	24.4	79.5	524(100)[d]	477(7.9)[e]	446(7.9)[f]	434(10.9)[g]	406(52.3)[h]	328(11.2)[i]	227(28.6)	
PGE_2	595	a 2.8	3.5	17.1	100.0	524(43.3)[d]	477(5.7)[e]	446(19.7)[f]	434(6.3)[g]	406(35.8)[h]	393(19.7)	323(8.3)	253(52.3)
		b 2.7	3.9	19.8	25.4	524(2.1)[d]	477(7.8)[e]	446(8.3)[f]	434(1.0)[g]	406(2.0)[h]	394(16.1)	323(78.1)	253(100)
PGE_1	597	a 0.3	4.9	18.8	100.0	526(68.0)[d]	479(3.1)[e]	448(19.6)[f]	436(5.7)[g]	408(31.3)[h]	396(4.0)	325(6.9)	253(3.6)
		b 0.2	1.2	3.1	18.4	526(1.4)[d]	479(2.8)[e]	448(5.5)[f]	436(2.8)[g]	408(4.0)[h]	396(43.9)	325(100)	253(16.6)
$PGF_{2\alpha}$	668	2.8	2.6	0.1	63.8	597(18.4)[d]	550(25.7)[e]	519(4.3)[f]	507(43.2)[g]	479(58.4)[h]	361(33.8)[i]	247(100)	
$PGF_{1\alpha}$	670	0.5	3.2	0.2	87.1	599(38.1)[d]	552(36.8)[e]	521(5.1)[f]	509(50.0)[g]	481(100)[h]	363(19.7)[i]	247(87.4)	
TXB_2	713	0.2	1.0	5.5	7.5	642(0.2)[d]	595(1.0)[e]	564(12.7)[f]	552(8.1)[g]	524(39.1)[h]	446(15.6)[i]	357(100)	239(71.3)
6-Keto $PGF_{1\alpha}$[c]	713	6.6	5.9	52.0	100.0	642(17.0)[d]	595(14.2)[e]	564(72.9)[f]	552(18.1)[g]	524(4.2)[h]	446(65.5)[i]	406(95.8)[j]	201(61.5)

[a] The minor isomer of the methyloxime derivative.
[b] The major isomer of the methyloxime derivative.
[c] Data of the major isomers of methyloxime derivatives are cited.
[d] $[M - 71(C_5H_{11})]^+$
[e] $[M - 118(DMiPSOH)]^{+\cdot}$
[f] $[M - (31 + 118)]^+$
[g] $[M - (43 + 118)]^+$
[h] $[M - (71 + 118)]^+$
[i] $[M - (31 + (2 \times 118))]^+$
[j] $[M - (71 + (2 \times 118))]^+$

derivatives is that they generally afford simple EI mass spectra with intense $[M\text{-}C_4H_9]^+$ ions, suitable for highly specific and sensitive assays by GC–SIM. DMIPS ether derivatives of hydroxylated steroids also provided characteristic mass spectra which were comparable to the corresponding TBDMS ether derivatives (Miyazaki *et al.*, 1979).

Figure 3 shows the mass spectra of the TMS, DMES, DMNPS and DMIPS ether derivatives of $PGF_{2\alpha}$ methyl ester. The mass spectral fragmentation patterns were closely related to those of the TMS ether derivative except for 14 or 28 u shifts for each hydroxyl group. The relative abundances of the $[M\text{-alkyl}]^+$ ions in the DMES and DMNPS ether derivatives of PGs and TXB_2 were not enhanced in comparison with those of $[M\text{-}CH_3]^+$ ions in the corresponding TMS ether derivatives but the $[M\text{-}C_3H_7]^+$ ion for the DMIPS ether was significantly more abundant.

Table 1 summarizes the mass spectral data of DMIPS ether derivatives of PGs and TXB_2 methyl esters and methyloxime-methyl esters. $[M\text{-}C_3H_7]^+$ was prominent except in the case of the TXB_2 derivative. The elimination of the alkylsilanol and production of the subsequent characteristic fragment ions were found in all PGs and TXB_2. All DMIPS ether derivatives except for TXB_2, as shown in Table 1, provided abundant ions in the high mass region, suggesting that these were useful for profile analyses of PGs by SIM.

23.4. STABILITY OF THE DMES, DMNPS AND DMIPS ETHER DERIVATIVES (Miyazaki *et al.*, 1978, 1981, 1984)

To achieve successful trace analyses of PGs and TXB_2, it is essential to carry out purification by use of extensive chromatographic separations during clean-up and derivatization procedures. The TMS ether derivatives of PGs and TXB_2 methyl esters and methyloxime-methyl esters have the serious disadvantage of susceptibility to hydrolysis by trace amounts of moisture. In response to this problem, attention has been focused on identifying derivatives which possess good chemical stability in addition to mass spectral characteristics favourable for SIM.

DMIPS ether derivatives of PGs and TXB_2 are relatively insensitive to moisture and were stable to purification using Sephadex LH-20 and silica gel column chromatography. Thus it is possible to achieve a high degree of purification, permitting the injection of relatively large amounts of sample without overloading a capillary column in GC–SIM. Consequently, the ability of GC–SIM to detect low concentrations of these compounds is enhanced. In contrast, the removal of excess reagent and impurities from TMS ethers is difficult while avoiding hydrolysis of the derivatives.

In addition, the storage stability of the DMIPS ether derivative of $PGF_{2\alpha}$ methyl ester in n-hexane was compared with that of the corresponding TMS ether derivative. The DMIPS ether derivative was stable for at least 7 days at room temperature, whereas the TMS ether derivative was decomposed to the extent of more than 30%.

23.5. SENSITIVITY DURING HIGH RESOLUTION SELECTED ION MONITORING

(Yamashita *et al.*, 1984; **Miyazaki** *et al.*, 1986)

The major factor limiting the sensitivity is interference by endogenous substances having similar extractive and analytical properties to the compound of interest. Sensitive and selective detection by GC–low resolution (LR) SIM is compromised by the presence of ions with the same nominal mass which are derived from other endogenous substances. Significant improvement has been achieved by increasing the mass spectrometic resolution to permit detection only of the ion of exact mass corresponding to a specific elemental composition. This method reduces interference by co-eluting impurities.

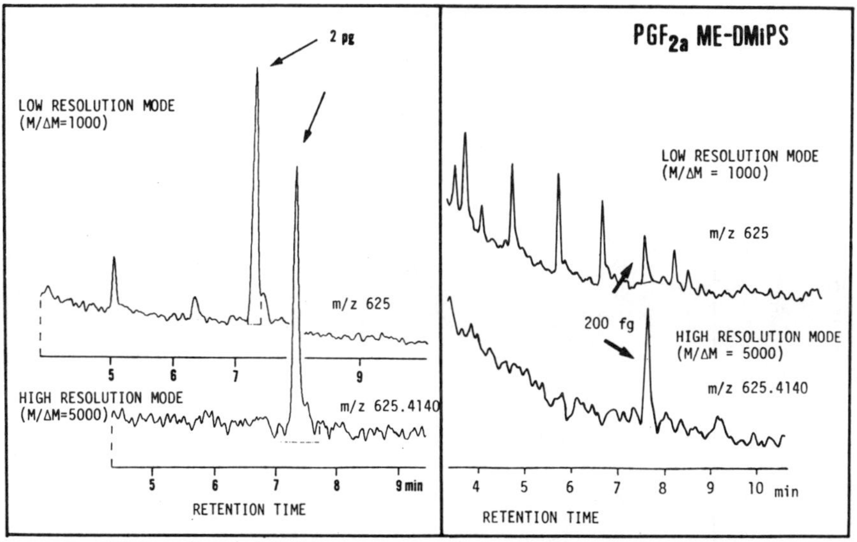

Figure 4: SIM analyses of $PGF_{2\alpha}$ corresponding to injection of 2 pg (left) and 0.2 pg (right) of DMIPS ether derivatives of its methyl ester. For SIM at high resolution ($M/\Delta M = 5000$), ions of 625.4140 were focused by reference to the ion of *m/z* 604.9632 derived from perfluorokerosene, which was independently introduced into the ion source. A VG Analytical MM ZAB–HF GC–MS system (Manchester, UK) equipped with a VandenBerg solventless injector was employed. Column: crosslinked Ultra ♯1 fused silica capillary column (25 m × 0.32 mm, Hewlett Packard, PA, USA); column temperature: 260°C; injection port temperature: 320°C; trap current: 200 μA; ionization energy: 70 eV; ion source temperature: 270°C; accelerating voltage: 8 KV; carrier gas: He, linear velocity at 260°C, μ = 25 cm/s.

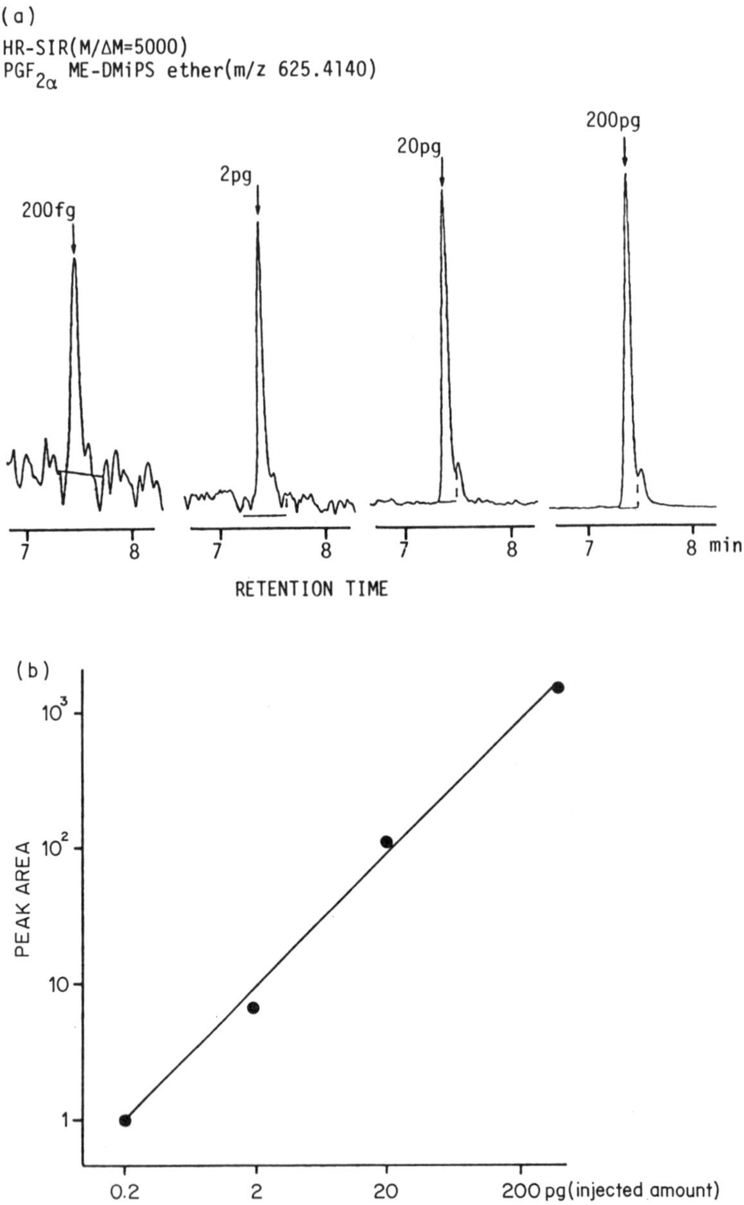

Figure 5: a) SIM responses for DMIPS ether derivative of $PGF_{2\alpha}$ methyl ester obtained by monitoring its characteristic ion of $[M\text{–}C_3H_7]^+$ (m/z 625.4140) at a mass resolution of 5000. b) Absolute calibration graph for $PGF_{2\alpha}$ methyl ester-DMIPS ether derivative prepared by using the data of Figure 5a.

Figure 4 shows analyses of the DMIPS ether derivatives of $PGF_{2\alpha}$ methyl ester obtained by monitoring the characteristic fragment ion $[M\text{-}C_3H_7]^+$ ($C_{33}H_{65}O_5Si_3$, *m/z* 625.4140) with two different resolutions of 1000 and 5000 (10% valley). When 2 pg of $PGF_{2\alpha}$ was injected, as shown in Figure 4a, both traces showed practically the same S/N, of more than 25, although the absolute ion intensity was reduced to about 20% by increasing the MS resolution from 1000 to 5000. This loss could be recovered by increasing the gain of the final amplifier, which did not affect the baseline noise level because the baseline signal was reduced at high resolution.

Figure 4b shows the comparison of the results obtained with different resolutions by injection of 0.2 pg of $PGF_{2\alpha}$. The existence of $PGF_{2\alpha}$ might be recognized from its retention time, but the many unidentified peaks observed in the low resolution analysis cast doubt on the identification. The analysis at higher resolution (Figure 4b) gave a well-shaped peak corresponding to $PGF_{2\alpha}$, with S/N of 5, demonstrating a greater reliability of the result at higher resolution.

These results lead us to conclude that the loss of signal intensity which is found at higher resolution is more than compensated by enhancement of the S/N, because of the reduction of background noise. Use of high-resolution SIM, in this case, yielded a high degree of confidence in the ability to detect 10^{-13} g levels of PGs and TXB_2.

Figure 5a illustrates the response in the high resolution (5000) mode to injection of several trace amounts of $PGF_{2\alpha}$. When the peak area of $PGF_{2\alpha}$ was plotted against the amount injected, the detector response (peak area) was found to be proportional to the amounts of $PGF_{2\alpha}$ injected within the range of 0.2 to 200 pg (Figure 5b). Operating conditions in this experiment are indicated in the Figure legend.

23.6. APPLICATION TO TRACE ANALYSES OF PROSTAGLANDINS IN HUMAN BRAIN (Ogorochi *et al.*, 1984)

Figure 6 shows the extraction, purification and derivatization procedure for PGs in the human brain. Tissues were homogenized with a Polytron homogenizer in 10 ml of distilled ethanol precooled to −20°C. Fifty ng of $[^2H_7]PGD_2$ and 40 ng each of $[^2H_7]PGE_2$ and $[^2H_5]PGF_{2\alpha}$ were added to the homogenate as internal standards. After centrifugation (1000 g; 10 min), supernatant solutions were applied to Sep-Pak C18 cartridges and a PG fraction collected.

PGs were derivatized to the DMIPS ether derivatives of methyl esters or methyloxime-methyl esters, as follows. Methyl esters were formed by treatment of a sample with ethereal diazomethane in methanol. After purification by silica

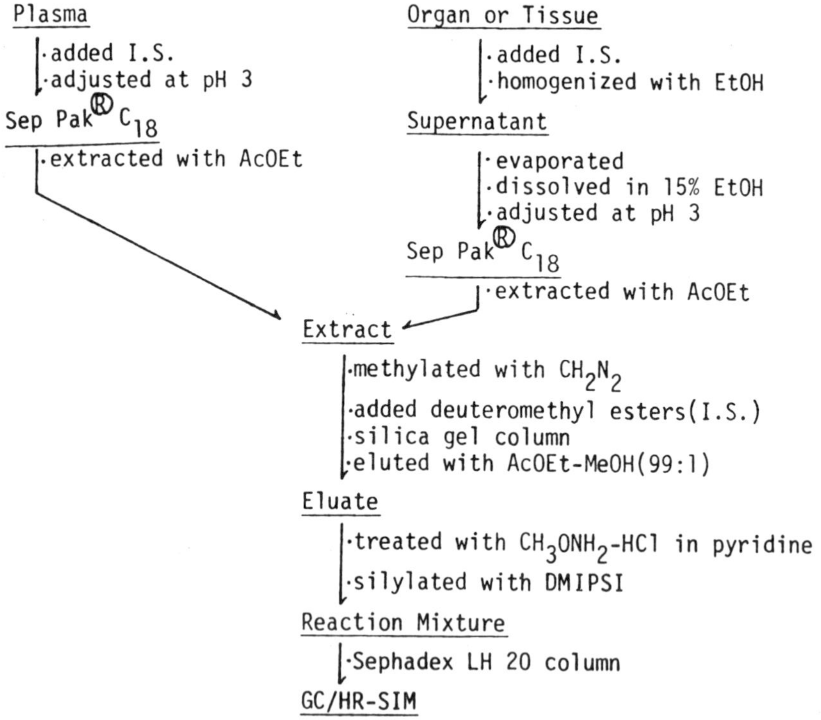

Figure 6: Flow diagram for the analysis of PGs and TXB_2 in biological fluids and tissues.

gel column chromatography using ethyl acetate–methanol (99 : 1) as solvent, the [^{13}C, $^{2}H_3$]methyl esters of PGE_1, $PGF_{1\alpha}$ and 6-keto $PGF_{1\alpha}$ were added to the sample as internal standards. The methyl esters were treated with a saturated solution of methoxylamine hydrochloride in pyridine at 60°C for 1 h. DMIPS imidazole was then added to the reaction mixture, and the solution was allowed to stand for another hour at room temperature. The reaction mixture was applied to a Sephadex LH-20 column (5 × 0.8 cm) and the final derivative was eluted with 2.8 ml of $CHCl_3$–n-hexane–methanol (10 : 10 : 1). After evaporation, the residue was dissolved in n-hexane containing 1% by vol. pyridine and used for GC–HR SIM.

For the extraction and purification of PGs and TXB_2 from biological samples, reverse-phase octadecylsilyl silica cartridges such as Sep-Pak C18 and Bond-Elut C18 have been successfully used. The problem in extracting these analytes from brain homogenates arises from their binding to tissue components. In the case of samples with relatively low concentrations of fatty lipids such as plasma, pleural fluids and cell culture medium, extraction of PGs and

TXB_2 can be performed by transferring directly onto the cartridge following the adjustment of pH to 3. The brain, however, is recognized to be one of the organs most rich in fatty lipids, so that a modified procedure was required. Prior to extraction of PGs and TXB_2 with the C18 cartridge, brain homogenates were diluted with 15% aqueous ethanol and the pH was adjusted to 3 in order to reduce interference in the extraction step by fatty lipids which are derived mainly from cell membranes. In order to assess losses through the extraction, a recovery test was carried out: 10 ng each of PGE_2, $PGF_{2\alpha}$ and their [^{14}C] labelled analogues as markers were added to the brain homogenates and the pH was adjusted to 3 with a few drops of 1M HCl. After application of the sample to the cartridge and washing with distilled water, the PGs were eluted with ethyl acetate. The recoveries of radioactivity at this step were better than 85%.

As mentioned above, the DMIPS ether derivatives exhibited excellent hydrolytic stability on column chromatography with Sephadex LH-20 or silica gel. When the DMIPS ether derivative of 10 ng of [^{14}C]$PGF_{2\alpha}$ was chromatographed on Sephadex LH-20, the recovery was more than 95%. This purification step is enough to keep to a maximum the resolution power of the GC capillary column by avoiding overloading.

The recovery of added [^{14}C]$PGF_{2\alpha}$ through the entire procedure was 39–41% in four experiments. Calibration graphs were prepared for PGD_2, PGE_2, $PGF_{2\alpha}$ and 6-keto-$PGF_{1\alpha}$; in the range 0.25–10 ng, linear relationships were observed between quantity of analyte and response ratio of analyte/internal standard. From these data it was determined that the minimum quantity of each PG in brain homogenates that may be measured conveniently is approximately 200 pg/g tissue if it is assumed that a S/N of 5 is required.

Figure 7 shows typical SIM analyses for $PGF_{2\alpha}$ in an extract from arachnoid. $PGF_{2\alpha}$ could not be determined by GC–LR SIM because of the very low concentration and the large amounts of interfering substances giving an ion with the same nominal mass. As shown in the right-hand panel of Figure 7, response to interfering ions from biological compounds was drastically reduced by monitoring at 5000 resolution the characteristic ions $[M\text{-}C_3H_7]^+$ with exact masses of 625.4140 for $PGF_{2\alpha}$ and 630.4452 for the internal standard. The peaks detected corresponded to about 40 pg of $PGF_{2\alpha}$ and 250 pg of the internal standard.

Figure 8 shows SIM analyses for PGD_2 and PGE_2 in the same extract of arachnoid, performed at a MS resolution of 5000. The ion at m/z 552.3540 was used for monitoring the DMIPS ether derivatives of methyloxime-methyl esters of PGD_2 and PGE_2 and the ion of m/z 559.3980 was followed for [$^2H_7$]PGD_2 and [$^2H_7$]PGE_2, added as internal standards. Peaks corresponding to m/z 552.3540 were clearly observed at the same retention times as those of the corresponding deuterated internal standards, suggesting the presence of PGD_2 and PGE_2 in the extract of arachnoid. As is obvious from Figure 8, interfering peaks of unknown origin were almost negligible.

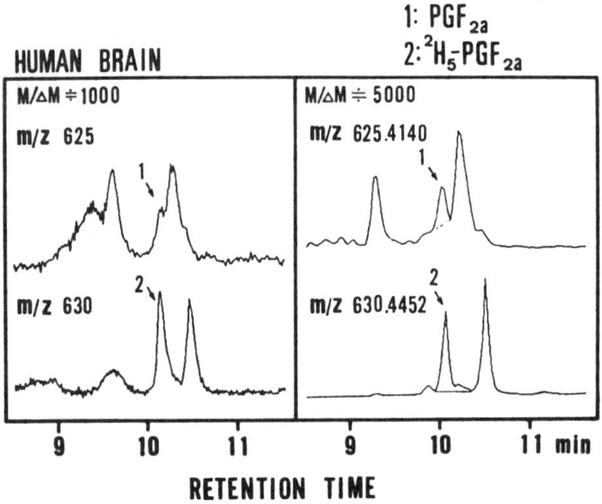

Figure 7: SIM analyses of DMIPS ether derivatives of $PGF_{2\alpha}$(1) and $[^2H_5]PGF_{2\alpha}$(2) methyl esters in the extract from human brain (arachnoid) obtained by monitoring at low (left) and high (right) resolution.

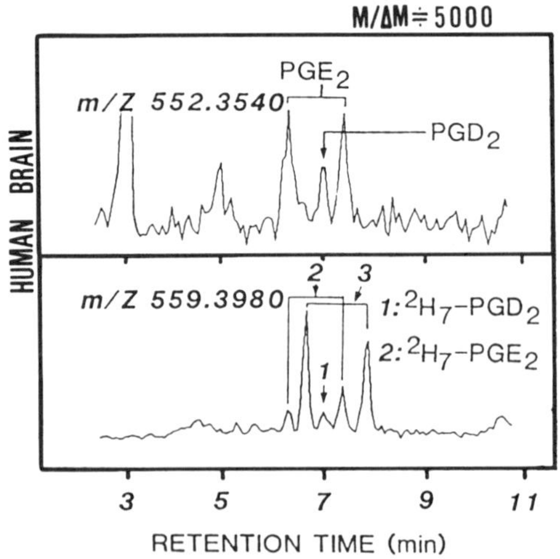

Figure 8: SIM analyses of DMIPS ether derivatives of PGD_2 and PGE_2 methyloxime-methyl esters in an extract from human brain (arachnoid). Traces were obtained by monitoring at *m/z* 552.3450 for PGD_2 and PGE_2 (top) and at *m/z* 559.3980 for $[^2H_7]PGD_2$(peak 1, bottom) and $[^2H_7]PGE_2$(peak 2, bottom). Peak 3 in the bottom of panel represents the isotope peak of PGE_1 $[^{13}C_1,^2H_3]$methyl ester.

The GC–HR SIM procedure revealed that PGD_2, PGE_2, $PGF_{2\alpha}$ and 6-keto-$PGF_{1\alpha}$ were present in almost every region of human brain tested. There were considerable differences between PG levels in different regions of the brain. Most values lay between 0.5 and 10 ng/g wet weight, but a wide range of several hundred pg/g to 40 ng/g wet weight was observed. PGD_2 and PGE_2 showed consistent regional distributions. High concentrations were observed in the pineal body, pituitary, olfactory bulb, hypothalamus and preoptic area. Recently, investigations on the distribution of the binding proteins specific for these PGs in post-mortem human brain were carried out. These demonstrated that binding proteins for PGD_2 and PGE_2 are also concentrated in the hypothalamus and basal forebrain regions. These results suggest possible physicochemical roles for PGD_2 and PGE_2 in these areas. On the other hand, neither $PGF_{2\alpha}$ nor its binding protein showed marked regional abundance. Thus, GC–HR SIM enabled the determination of the profile, levels and distributions of several PGs in post-mortem human brain. Details of these investigations have been given by Ogorochi *et al.* (1984).

23.7. CONCLUSIONS

On the basis of the above mentioned fundamental investigations, it may be concluded that the combination of the DMIPS ether derivatives and capillary GC–HR SIM should be a very powerful technique in the trace analysis of PGs and TXB_2 in biological fluids and tissues. The GC–MS properties of the DMIPS ether derivatives are superior to other silyl ether derivatives in enhancement of GC separation, improvement of chemical stability for purification with column chromatography and enhancement of the intensity of characteristic ions for quantification by GC–SIM. In addition, it is clear from the application reported here that the present method, including the use of DMIPS ether derivatives, purification procedures and selection of exact masses during GC–SIM, makes possible the sensitive and selective detection of PGs in human brain without serious interference from other endogenous substances. The improvement in sensitivity in the high resolution mode may be attributed to enhancement of the S/N ratio in the detection of selected ions.

REFERENCES

Barrow, S. E., Waddell, K. A., Ennis, M., Dollery, C. T., and Blair, I. A. (1982). Analysis of picomolar concentrations of 6-keto-prostaglandin $F_{1\alpha}$ in biological fluids, *J. Chromatogr.*, **239**, 71–80.

Bazan, A. C., and Knapp, D. R. (1982). Improved derivative of 6-keto-prostaglandin $F_{1\alpha}$ for gas chromatographic–mass spectrometric analysis, *J. Chromatogr.*, **236**, 201–207.

Finlay, E. M. H., and Gaskell, S. J. (1981). Determination of testosterone in plasma from men by gas chromatography/mass spectrometry, with high-resolution selected-ion monitoring and metastable peak monitoring, *Clin. Chem.*, **27**, 1165–1170.

Gaskell, S. J., and Brooks, C. J. W. (1976). t-Butyldimethylsilyl derivatives in the gas chromatography–mass spectrometry of steroids, *Biochem. Soc. Trans.*, **4**, 111–113.

Gaskell, S. J., and Brownsey, B. G. (1983). Immunoadsorption to improve gas chromatography/high resolution mass spectrometry of estradiol-17β in plasma, *Clin. Chem.*, **29**, 677–680.

Gaskell, S. J., and Pike, A. W. (1978). Gas chromatography–high resolution mass spectrometry in the analysis of steroids in physiological fluids, in *Quantitative Mass Spectrometry II* (Eds. A. P. De Leenheer, R. R. Roncucci and C. Van Peteghem), pp. 181–189, Elsevier, Amsterdam.

Gaskell, S. J., Finlay, E. M. H., and Pike, A. W. (1980). Analyses of steroids in saliva using highly selective mass spectrometric techniques, *Biomed. Mass Spectrom.*, **7**, 500–504.

Hamberg, M., and Samuelsson, B. (1967). On the mechanism of the biosynthesis of prostaglandins E_1 and $F_{1\alpha}$, *J. Biol. Chem.*, **242**, 5336–5343.

Kelly, R. W., and Taylor, P. L. (1976). tert-Butyldimethylsilyl ethers as derivatives for qualitative analysis of steroids and prostaglandins by gas phase methods, *Anal. Chem.*, **48**, 465–467.

Mai, J., German, B., and Kinsella, J. E. (1983). Prostaglandins and thromboxane: *tert*-butyldimethylsilyl ethers as derivatives for their analyses by gas chromatography–mass spectrometry–selected ion monitoring, *J. Chromatogr.*, **254**, 91–107.

Millington, D. S. (1975). Determination of hormonal steroid concentrations in biological extracts by high resolution mass fragmentography, *J. Steroid Biochem.*, **6**, 239–245.

Millington, D. S., Buoy, M. E., Brooks, G., Harper, M. E., and Griffiths, K. (1975). Thin-layer chromatography and high resolution selected ion monitoring for the analysis of C_{19} steroids in human hyperplastic prostate tissue, *Biomed. Mass Spectrom.*, **2**, 219–224.

Min, B. H., Pao, J., Garland, W. A., DeSilva, J. A. F., and Parsonnet, M. (1980). Determination of an antisecretory trimethyl prostaglandin E_2 analog in human plasma by combined capillary column gas chromatography–negative chemical ionization mass spectrometry, *J. Chromatogr.*, **183**, 411–419.

Miyazaki, H., Ishibashi, M., Yamashita, K., and Katori, M. (1978). Use of new silylating agents for separation and identification of prostaglandins by gas chromatography and gas chromatography–mass spectrometry, *J. Chromatogr.*, **153**, 83–90.

Miyazaki, H., Ishibashi, M., and Yamashita, K. (1979). Dimethylisopropylsilyl ether derivatives in gas chromatography mass spectrometry of hydroxysteroids, *Biomed. Mass Spectrom.*, **6**, 57–62.

Miyazaki, H., Ishibashi, M., Yamashita, K., Nishikawa, Y., and Katori, M. (1981). Dimethylisopropylsilyl ether derivatives in gas chromatography mass spectrometry of prostaglandins and thromboxane B_2, *Biomed. Mass Spectrom.*, **8**, 521–526.

Miyazaki, H., Ishibashi, M., Takayama, H., Yamashita, K., Suwa, I., and Katori, M. (1984). Microdetermination of prostaglandins and thromboxane B_2 by capillary gas chromatography and negative ion chemical ionization mass spectrometry, *J. Chromatogr.*, **289**, 249–258.

Miyazaki, H., Yamashita, K., and Ishibashi, M. (1986). Microdetermination of prostaglandins and thromboxane B_2 by gas chromatography/high resolution-selected ion monitoring, in *Advances in Prostaglandin, Thromboxane and Leukotriene Research, vol. 15* (Eds. O. Hayaishi and S. Yamamoto), pp. 44–46, Raven Press, New York.

Ogorochi, T., Narumiya, S., Mizuno, N., Yamashita, K., Miyazaki, H., and Hayaishi, O. (1984). Regional distribution of prostaglandins D_2, E_2 and $F_{2\alpha}$ and related enzymes in postmortem human brain, *J. Neurochem.*, **43**, 71–82.
Pace-Asciak, C. R., and Wolf, L. S. (1971). N-Butylboronate derivatives of the F prostaglandins. Resolution of prostaglandins of the E and F series by gas–liquid chromatography, *J. Chromatogr.*, **56**, 129–133.
Pace-Asciak, C. R., and Micallef, S. (1984). Gas chromatographic–mass spectrometric profiling with negative-ion chemical ionization detection of prostaglandins and their 15-keto and 15-keto-13,14-dihydro catabolites in rat blood, *J. Chromatogr.*, **310**, 233–242.
Phillipou, G., Bigham, D. A., and Seamark, R. F. (1975). Steroid t-butyldimethylsilyl ethers as derivatives for mass fragmentography, *Steroids*, **26**, 516–524.
Rosello, J., Tusell, J., and Gelpi, E. (1977). Profiles of prostaglandins A, B, E and F (series I and II) obtained by gas chromatography with multiple-ion detection, *J. Chromatogr.*, **130**, 65–76.
Smith, A. G., Harland, W. A., and Brooks, C. J. W. (1977). Gas–liquid chromatography–mass spectrometry of thromboxane B_2 and its detection in semen and human aorta by selected ion monitoring, *J. Chromatogr.*, **142**, 533–547.
Vane, F., and Horning, M. G. (1969). Separation and characterization of the prostaglandins by gas chromatography and mass spectrometry, *Anal. Letters*, **2**, 357–371.
Waddell, K. A., Blair, I. A., and Wellby, J. (1983). Combined capillary column gas chromatography negative ion chemical ionization mass spectrometry of prostanoids, *Biomed. Mass Spectrom.*, **10**, 83–88 (1983).
Yamashita, K., Ishibashi, M., Miyazaki, H., Narumiya, S., Ogorochi, T., and Hayaishi, O. (1984). Microdetermination of prostaglandins and thromboxane B_2 by gas chromatography/high resolution-selected ion monitoring, *J. Pharm. Dyn.*, **7**, 12.

Mass Spectrometry in Biomedical Research
Edited by S. J. Gaskell
© 1986 John Wiley & Sons Ltd.,

Chapter 24

MASS SPECTROMETRIC MEASUREMENT OF NEUROPEPTIDES

Dominic M. Desiderio and Genevieve H. Fridland

24.1. INTRODUCTION

This chapter reviews the basic principles of novel mass spectrometric (MS) techniques useful for the fast and facile measurement of endogenous neuropeptides in biological extracts (Desiderio, 1984). MS offers the highest level of molecular specificity compared with that obtained using other analytical techniques such as radio-immunoassay (RIA) and radioreceptor assay (RRA), where the latter two methods are commonly used for the measurement of endogenous peptides.

24.1.1. PEPTIDE FAMILIES

Mass spectra and accurate mass determinations were used to elucidate the amino acid sequence of the first hypothalamic releasing peptide, TRF (pyroglutamyl-histidyl-proline-amide) (Burgus *et al.*, 1970). That study led to an increased level of activity in brain neuropeptide research (Krieger *et al.*, 1983), and the amino acid sequences of many other endogenous peptides were determined. Two pentapeptides, methionine enkephalin (ME = Tyr-Gly-Gly-Phe-Met) and leucine enkephalin (LE = Tyr-Gly-Gly-Phe-Leu) were found to be endogenous ligands that bind to the morphine receptor (mu) and their amino acid sequences were determined with the help of MS (Hughes *et al.*, 1975). Larger peptides containing ME at their N-termini were also identified; for example β-endorphin is composed of 31 amino acids and contains the pentapeptide ME as its N-Terminus.

Another group of peptides that exhibit extraordinarily high opiate activity is the dynorphins ('dynamic morphine') (Goldstein *et al.*, 1979). These peptides interact preferentially with the kappa receptor and contain the LE sequence at their N-termini (Chavkin *et al.*, 1982). Another peptide suggested to be implicated in an organism's modulation of pain stimuli (Jessell, 1982) is the undecapeptide amide substance P (SP) (Chang *et al.*, 1971).

Each one of the peptide families mentioned above (enkephalins, endorphins, dynorphins, and substance P) derives from gene product protein precursors that are approximately 300 amino acids long and contain several copies of the individual target peptides (Krieger, 1983; Hollt, 1983). Figure 1 is a schematic representation of the three opioid peptidergic families, bovine pre-proenkephalin A, porcine pre-proenkephalin B, and bovine pre-proopiomelanocortin (Hollt, 1983). An individual working peptide derives from its precursor in a tissue-specific manner that reflects the current metabolic demand in that organism. The pituitary and hypothalamus generally contain high levels of the larger peptides; internal brain segments such as amygdala, mid-brain, and

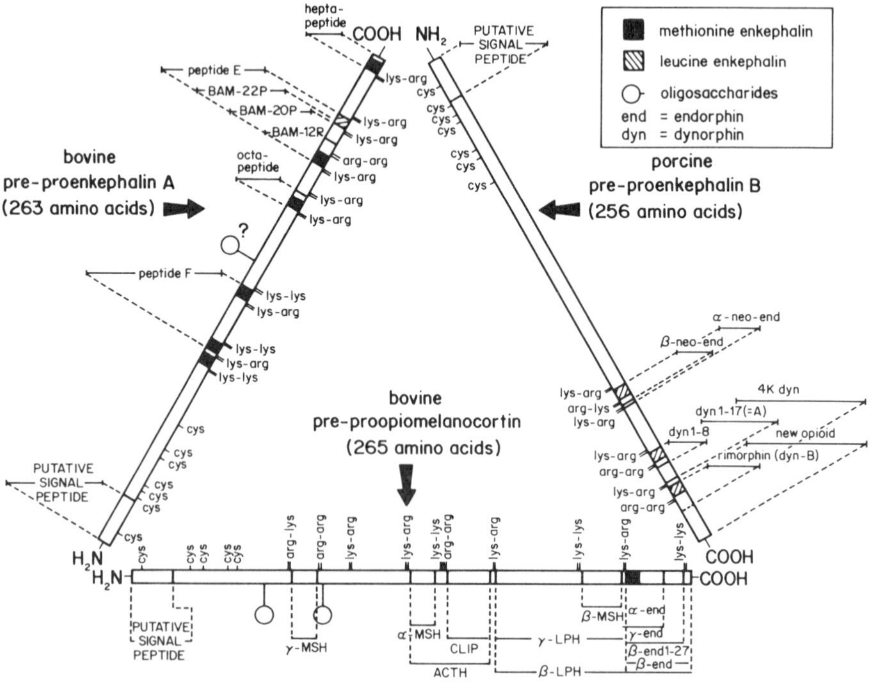

Figure 1: Schematic representation of the three gene product opioid peptidergic precursors: bovine pre-proopiomelanocortin, bovine pre-proenkephalin A, and porcine pre-proenkephalin B. *Reproduced from Hollt (1983) by permission of Elsevier Publications.*

pons–medulla contain mostly shorter peptides; and end-target organs such as tooth pulp contain smaller peptides.

24.1.2. SAMPLE PREPARATION

The major aim of our research programme is to identify and quantify with maximum molecular specificity the individual opioid peptides present in several biological tissues. We have correlated data for these peptides from several neuropeptidergic precursors obtained using several analytical techniques (reversed phase (RP) high performance liquid chromatography (HPLC), RIA, RRA, and MS). Figure 2 illustrates the comprehensive analytical scheme and the methodologies used towards that aim. We use tissue of canine and human origin from several areas of the brain (pituitary, hypothalamus, thalamus, caudate nucleus, amygdala, pons–medulla, etc.), tooth pulp, and CSF—all obtained under normal and/or stressed conditions.

The animal is anaesthetized, and the tissue surgically removed as quickly as possible and immersed in liquid nitrogen to minimize enzymatic degradation and metabolic interconversion. Human tissues are obtained from surgery or the dental clinic and rapidly frozen in liquid nitrogen. The tissues are stored at $-70°C$ until processed.

The samples are weighed and immediately homogenized in cold acetic acid upon defrosting (Desiderio *et al.*, 1982a). Internal standard (IS) is added at this point to those samples to be used for MS quantification (Lawson *et al.*, 1980; Desiderio and Kai, 1983a,b). Acidified acetone is added to precipitate protein. After centrifugation, acetone is evaporated in a stream of N_2. The supernatants are lyophilized and resuspended in trifluoroacetic acid (TFA) and applied to a pre-wetted C_{18} Sep-Pak cartridge; the cartridge is washed with TFA, and the peptides eluted with acetonitrile (CH_3CN) in TFA. This step removes salts and other interfering substances and produces a 'peptide-rich fraction'.

This peptide-rich fraction is then injected onto an octadecylsilyl silica RP–HPLC analytical column, which is then subjected to gradient elution. Although peptides are generally quite polar due to their zwitterionic termini, polar side chains, and interchain hydrogen bonding, they elute quite well from an RP column when a ion-pairing reagent is added to form a peptide : buffer ion-pair that possesses reduced polarity and interacts with the C_{18} side chains on the column (Desiderio, 1983). Because we detect and quantify peptides of interest with various forms of post-HPLC detection (see Figure 2), it is important that the HPLC buffer be volatile to avoid interference with subsequent modes of peptide detection. We have developed a volatile triethylamonium formate buffer (TEAF) as the ion-pairing reagent (Desiderio and Cunningham, 1981). Acetonitrile is used as the organic modifier. The use of this buffer system, combined with an increasing concentration of organic modifier, permits efficient and rapid

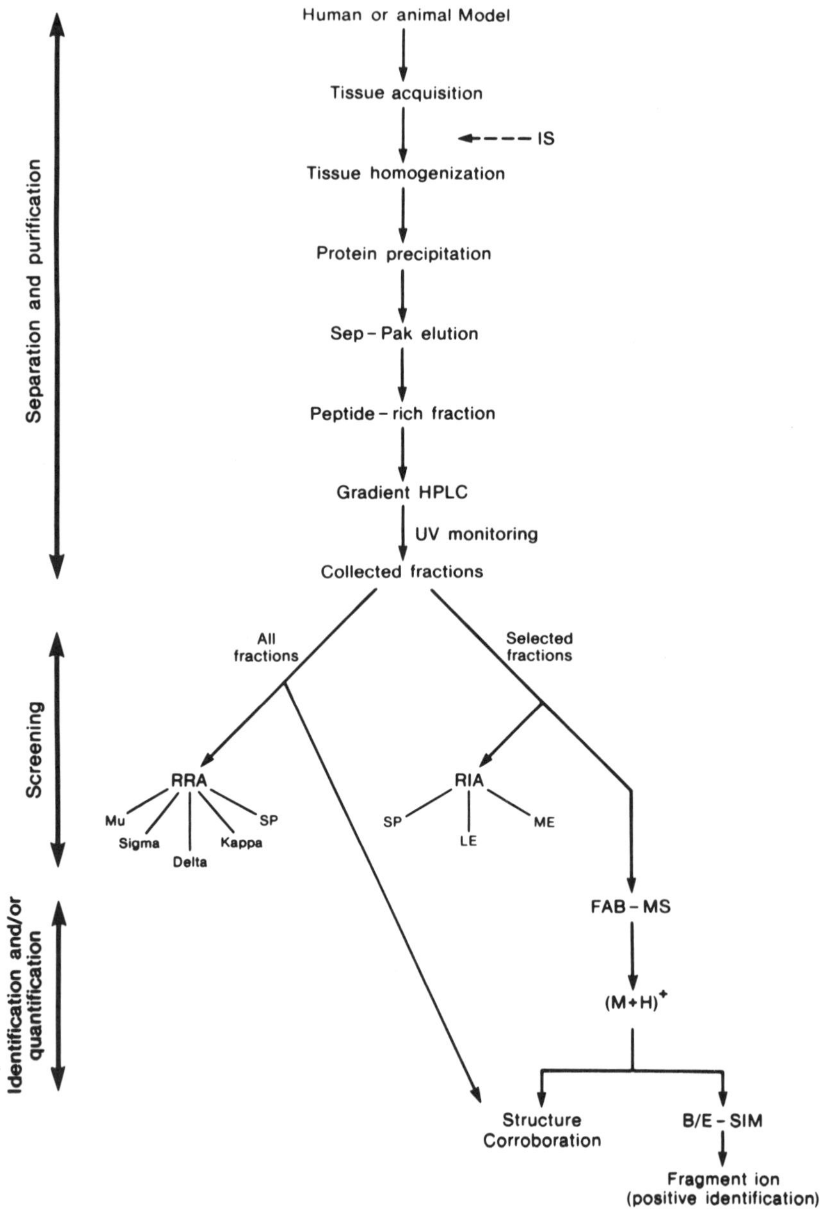

Separation and purification
Screening
Identification and/or quantification
Human or animal Model
Tissue acquisition
IS
Tissue homogenization
Protein precipitation
Sep-Pak elution
Peptide-rich fraction
Gradient HPLC
UV monitoring
Collected fractions
All fractions
Selected fractions
RRA
Mu
Sigma
Delta
Kappa
SP
RIA
SP
LE
ME
FAB-MS
(M+H)+
Structure Corroboration
B/E-SIM
Fragment ion (positive identification)

separation of peptides of a wide range of sizes and hydrophobicities in the course of a single HPLC gradient. The eluant is monitored by UV at 200 nm, taking advantage of the absorbance of the peptide bond at this wavelength. We have shown that, with the tooth pulp as model, most UV-absorbing material eluting under these conditions is peptidic in nature (May *et al.*, 1982).

Figure 3 shows an HPLC chromatogram of a peptide-rich fraction from canine hypothalamus. The gradient profile is shown as the solid line corresponding to 10 to 30% CH_3CN on the right-hand axis. The baseline shift occurring as the chromatogram develops is due to the UV absorbance of the TEAF buffer. The elution times of selected synthetic peptides are shown along the top of the figure. It must be stressed that, without further structural elucidation, a molecular structure cannot be assigned to a peak eluting at the same retention time as a standard. Furthermore, a discrete chromatographic peak is not expected to be observed with UV detection at the elution time of the peptides of interest, because an endogenous peptide is generally present at the pmol level in the tissues studied. RP–HPLC is used as a separation step, and subsequent detection of the endogenous peptides is achieved by more sensitive means (RRA, RIA, MS); positive identification is made by MS.

24.2. POST-HPLC DETECTION

24.2.1. RADIO-IMMUNOASSAY (RIA)

Immunoassays are based on the binding affinity of an antibody (Ab) for its (putatively unique) antigen (Ag), and may be used analytically to measure either the endogenous Ag or the Ab (Yalow, 1980). A radioactive isotope generally labels the Ag, and the term ‘radio-immunoassay’ is used. The measurement depends on the ability of the unlabelled Ag (the unknown) to inhibit or compete with the binding of the radioactive antigen (Ag*) to an Ab. Femtomoles of a

---

Figure 2: Flow chart showing purification and separation steps leading to screening, identification, and/or quantification of neuropeptides with maximum molecular specificity. IS: internal standard; HPLC: high performance liquid chromatography; UV: ultraviolet detection; RIA: radio-immunoassay; ME: methionine enkephalin; SP: substance P; RRA: radioreceptor assay; FAB-MS: fast atom bombardment mass spectrometry; $(M + H)^+$: protonated molecular ion; SIM: selected ion monitoring; *B*/*E*: linked-field scanning. FAB–MS is used in three modes. In some studies, FAB–MS is used to screen all fractions for the presence of $(M + H)^+$ ions. In other studies, FAB–MS produces an $(M + H)^+$ to corroborate the molecular weight of a peptide or *B*/*E*-SIM techniques are used to produce an amino acid sequence-determining ion, which can be used either for positive identification or quantification of a peptide.

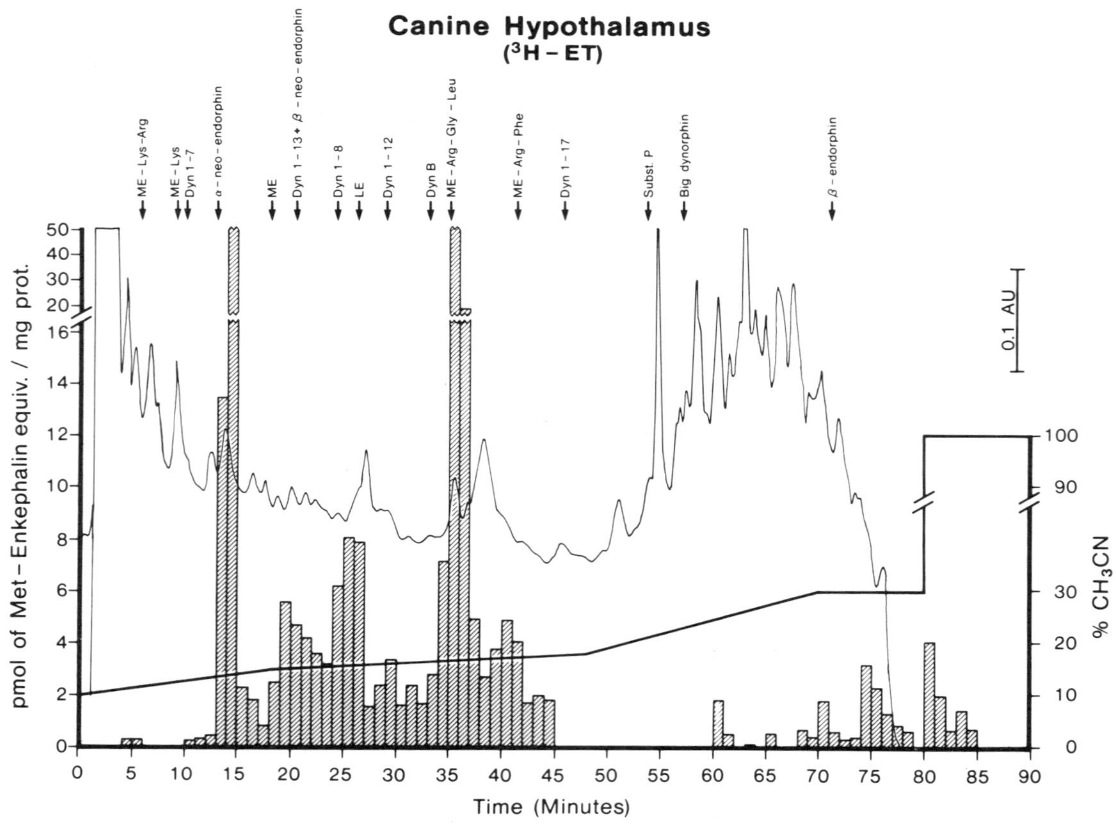

Figure 3: Gradient RP–HPLC chromatographic separation of a peptide-rich fraction derived from canine hypothalamus. The *x* axis represents time (minutes) and corresponds to fractions. The left-hand axis indicates the receptoractive-opiate activity as picomoles of ME equivalents per fraction (determined using [^{3}H] etorphine as radioligand). The right-hand axis indicates the gradient profile of the organic modifier acetonitrile. The chromatographic profile shown is obtained from UV absorption at 200 nm; the heights of the bars correlate with the amount of receptoractive-peptide within each fraction. The retention times of synthetic peptide standards are indicated by the arrows along the top of the chromatogram (Desiderio, 1984).

compound can be analyzed by RIA. Whenever the size of the original Ag molecule to be analyzed is too small (less than approximately 4000 daltons), it can be chemically conjugated to a larger carrier molecule such as ovalbumin, thyroglobulin, or bovine serum albumin to increase its immunogenicity. For example, up to 30 molecules of ME are chemically conjugated to ovalbumin (Gros *et al.*, 1978). The synthetic complex (small peptide : large carrier protein) is injected into an animal and antibodies are raised to the antigenic complex, but, it must be realized, not necessarily only to the original oligopeptide ME.

Although the sensitivity of detection of the RIA method is excellent and many authors believe that the molecular specificity is also quite high, more recent analytical and structural data indicate that the molecular specificity of RIA is not as unambiguous as generally presumed. Nevertheless, RIA is probably the assay method most widely utilized around the world.

Commercial RIA kits are routinely used in our laboratory for quantification of immunoreactive ME, LE, SP, and occasionally other peptides of interest in the fractions eluted from the gradient HPLC (Figure 3). Because we utilize HPLC-separated fractions eluting at the retention times of the synthetic peptides, we minimize the problem of cross-reactivity of the Ab with related peptides in our RIA screening.

Table 1 collects some of the RIA data obtained in our laboratory and shows the amount of immunoreactive ME found in the peptide-rich fraction of human tooth pulp extracts. These data are part of our study to determine if stress mobilizes an opioid peptidergic pathway (Tanzer *et al.*, 1986), and are presented here to illustrate the endogenous levels of immunoreactive ME.

24.2.2. RADIORECEPTOR ASSAY (RRA)

Because radiolabelled ligands with high specific activity are easily synthesized, it is a simple matter to utilize these radiolabelled ligands to determine the subcellular localization of individual peptide receptors. Receptors are thought to be large protein–lipid complexes that assume a particular conformation and interact specifically with appropriate ligands (Childers, 1983). Receptor-enriched preparations are prepared readily for detection and/or semi-quantification of endogenous peptides. The sensitivity of RRA is lower than RIA, but in many cases it is sensitive at the part per billion or ng/g tissue level and can be used for semi-quantitative purposes. The molecular specificity of RRA is also lower than RIA, yet RRA provides an excellent analytical procedure to efficiently screen a biological extract and to yield a metabolic profile of opioid activity in diverse tissues and extracts (Desiderio, 1984).

The RRA is based on the competition between an endogenous peptide and a labelled ligand for binding to a receptor population present in a brain membrane preparation. A receptor-rich synaptosomal preparation (Gray and

TABLE 1. Radio-immunoassay and radioreceptor assay of endogenous methionine enkephalin (ME) concentration in HPLC-purified human tooth pulp extracts (pg peptide/mg protein).

Subject	Tooth No.[a]	Tooth pulp wet weight (g)	Protein (mg)[c]	Immunoreactive-ME	Receptoractive-ME
DD	5[b]	0.1181	6.8	1181	[d]
	12	0.0576	3.3	86	9094
	28	0.1181	6.8	87	816
JG	1	0.0394	2.3	272	115 219
	16[b]	0.0252	1.5	8400	53 000
	17	0.0543	3.1	126	98 231
	32	0.0649	3.8	284	25 533
JW	[b]	0.0213	0.7	20 000	107 151
		0.0273	1.6	121	7183
		0.0478	2.8	22	22 859
CB	5	0.0710	4.1	3048	5057
	12[b]	0.0843	4.9	2546	30 124
	21	0.0496	2.9	872	1830
	28	0.0556	3.2	2693	1373
TH (treated)	5[b]	0.0448	2.6	327	1568
	12	0.0456	2.6	70	912
	21	0.0396	2.3	n.d.	1154
	28	0.0401	2.3	262	1293
SG (treated)	5[b]	0.0486	2.8	208	n.d.
	12	0.0489	2.8	576	384
	21	0.1107	6.4	294	502
	28	0.0925	5.4	17	n.d.
SS (treated)	5[b]	0.1303	7.6	34	1146
	12	0.1506	6.1	35	367
	21	0.0773	4.5	52	432
	28	0.0732	4.2	n.d.	n.d.

[a] Universal numbering system. JW—tooth number data not recorded.
[b] First tooth extracted.
[c] Human tooth pulp contains 5.8% protein.
[d] This one tooth was also tested by MS, and $(M + H)^+$ for ME at m/z 574 was observed.
n.d. = not detected.

Whittaker, 1962) is obtained from the canine limbic system (Childers, 1983) and incubated (37°C) to release endogenous receptor ligands. Both labelled ligand and unlabelled HPLC fraction are added to the synaptosomal membrane preparation and incubated (2.5 h, 0°C). Unbound label is rapidly removed by filtration through glass fibre filters. After washing the filters (4 ×) and drying them (infrared), the radioactivity remaining on the filter is measured. Non-specific binding (NS) is determined in the presence of an excess of unlabelled peptide, while total binding (T) is determined in the absence of competing cold ligand. Specific binding (SB) = T − NS of each fraction is used to quantify the receptoractive-opioid activity of individual fractions when compared to the binding of known amounts of the opioid peptide ME with the same synaptosomal membrane preparation. Hence, the receptoractive-peptide activity is expressed as pmol of ME equivalents rather than actual peptide amounts.

Figure 3 presents the RRA data superimposed on the gradient RP–HPLC chromatogram. The receptoractive-opiate activity of each fraction is indicated as a bar relating to the scale on the left-hand side. The radioligand used is tritiated etorphin, which is broad-based and interacts with mu, delta, and sigma opioid receptors. The information available from the RRA is that a given fraction such as fraction 14 in Figure 3 contains a receptoractive-peptide that interacts with the opioid receptor present in the synaptosomal preparation at a level equivalent to the interaction of 50 pmol of ME, under these experimental conditions. The RRA provides an efficient and rapid screen to identify the presence of receptoractive-opioid species in the HPLC fractions from a biological tissue, and permits one to selectively pursue the identification and/or quantification of peptides of interest in those fractions that show opioid receptor activity.

24.3. MASS SPECTROMETRY

It is the authors' experience that MS must be incorporated in the analytical scheme for the quantification and identification of biologically important peptides because MS provides the researcher with the most important analytical parameter–structure (Desiderio, 1984). This conclusion reflects the fact that so much subsequent importance is placed on single analytical measurements of biological peptides. For example, clinical diagnosis and prognosis, pharmacological treatment, surgical intervention, and research money and time are some of the sequelae to analytical measurement. It is our opinion that, if such an enormous amount of time, money, effort, and quality of human life are dependent on a measurement, then that analytical measurement should be made with state-of-the-art techniques that maximize sensitivity without compromising molecular specificity.

24.3.1. FIELD DESORPTION (FD)

The first use of MS for measurement of endogenous peptides was provided by FD MS (Desiderio *et al.*, 1981a, 1981b; Desiderio and Yamada, 1982; Desiderio and Kai, 1983a,b; Tanzer *et al.*, 1981; Yamada and Desiderio, 1982), where the $(M + H)^+$ ion was produced readily for peptides such as the enkephalins, and an $[^{18}O]$-labelled enkephalin IS yielded another $(M + H)^+$ ion located four mass units higher (Desiderio and Kai, 1983a,b). The IS is added to the tissue before homogenization (Figure 2). A microcomputer (COM) calculates the ratio of the two $(M + H)^+$ integrated ion currents (Desiderio *et al.*, 1984) and, when the amount of $[^{18}O]$peptide IS originally added to the extract is taken into account, the amount of endogenous peptide is calculated. This measurement process worked well for several years, but a significant fraction of the FD emitters would break, no matter how high the level of technical expertise of the laboratory personnel.

24.3.2. FAST ATOM BOMBARDMENT (FAB)

FAB MS (Barber *et al.*, 1981) rapidly superseded the use of FD MS for the analytical measurement of endogenous peptides (Desiderio *et al.*, 1983a,b) because a high sensitivity is attained with a long-lasting ion current in FAB, a liquid matrix such as glycerol prevents or at least minimizes analyte damage during FAB irradiation, and fragile emitters are not needed for FAB. In general, FAB of a peptide produces an abundant ion current corresponding to the $(M + H)^+$ of the peptide. Amino acid sequence-determining fragment ions may also occur, but at a lower (generally 1%) relative intensity.

24.3.3. LINKED-FIELD MASS SPECTROMETRY

An important experimental consideration for the unambiguous measurement of any endogenous biological peptide is structural resolution (versus mass resolution) to address the problem of differentiation of isobaric oligopeptides. For example, for a hypothetical tripeptide composed of three amino acids (A,B,C), 3! (six) different amino acid sequences (ABC, ACB, BAC, BCA, CAB, CBA) are possible; therefore, even if the accurate mass of the $(M + H)^+$ ion of a peptide were determined, crucial amino acid sequence-determining information may not be present. Increased mass resolution will not solve this polymer problem, but increased 'structural resolution' would.

For peptides in particular, the oligopeptide problem is minimized by obtaining the mass spectrum of a significant ion in the mass spectrum of the peptide. The most important ion is the $(M + H)^+$ ion, and it is a relatively simple matter

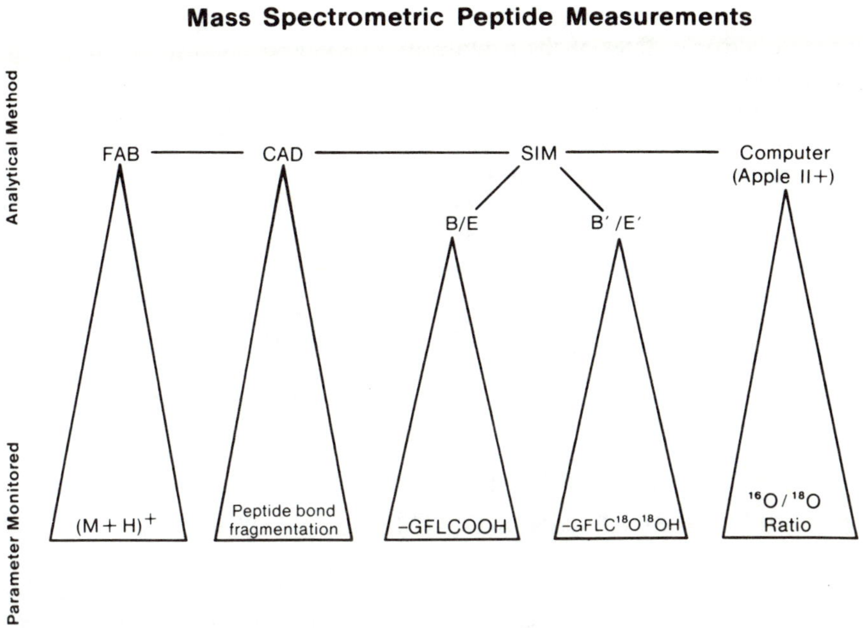

Figure 4: Schematic representation of the FAB-CAD-*B*/*E*,*B*′/*E*′-COM MS procedure for measurement of endogenous peptides showing the parameter monitored by each individual analytical method, where −GFLCOOH and $-GFLC^{18}O^{18}OH$ are the *C*-terminal containing tripeptide fragment ions of endogenous and exogenous LE, respectively.

to obtain the mass spectrum of the $(M + H)^+$ ion by linked-field scanning mass spectrometry (Jennings and Mason, 1983). The *B*/*E* linked-field scan method maintains the accelerating voltage (*V*) constant, uncouples *V* from the electric sector voltage (*E*), and scans *B* and *E*, keeping the ratio of *B*/*E* (*B* = magnetic field) constant. In this manner, all product ions that arise from one selected precursor ion are readily collected. This structurally-fixed relationship between product and precursor ions is a significant analytical feature because, as mentioned above, even if HPLC is utilized to purify a peptide from a biological source, there is no guarantee that the HPLC fraction is a pure compound. Nonetheless, if different molecular ions are present in that one HPLC fraction, then a mass spectrum may be taken of that particular $(M + H)^+$ ion produced by FAB. The product ion spectra will include ions specific to a particular sequence and will therefore permit the distinction of isobaric oligopeptides.

In some cases, the target peptide will not spontaneously fragment during transit in an MS field-free region, and amino acid sequence-determining fragment ions will not be formed. In some of those cases, it is possible to induce peptide bond fragmentation by collision-activated dissociation (CAD) processes. For example, when an 8 kV-accelerated $(M + H)^+$ ion collides with an inert gas, some of the energy arising from the collision may be converted into excited electronic states of the peptide, and peptide bond cleavage may result (Todd and McLafferty, 1983). A unique amino acid sequence-determining fragment ion is then monitored in the selected ion monitoring (SIM) mode.

Our laboratory uses a FAB-CAD-*B*/*E*,*B*′/*E*′-SIM-COM combination of instrumental techniques to quantify HPLC-purified peptides that derive from the peptide-rich fraction of a biological tissue. The instrumental combination of FAB-CAD-*B*/*E*,*B*′/*E*′-SIM-COM is rather easy to achieve experimentally and, when utilized as described above, stable isotope-incorporated internal standards are used in the *B*/*E*, *B*′/*E*′-SIM mode to permit quantification of the peptide (Tanzer *et al.*, 1981; Yamada and Desiderio, 1982; Desiderio *et al.*, 1983a,b; Desiderio, 1984). Figure 4 is a schematic representation of this technique and correlates the parameters experimentally monitored with the analytical methods.

24.3.4. MS MEASUREMENTS OF ENDOGENOUS PEPTIDES

Table 2 contains representative MS measurements of RP–HPLC-purified endogenous peptides in canine tissue extracts and in CSF. LE and ME are measured, either by FD MS-SIM-COM (method 1) or by FAB-CAD-*B*/*E*, *B*′/*E*′-SIM-COM (method 2). Amounts range from 2 ng LE/g to 3760 ng ME/g in posterior pituitary tissue.

TABLE 2. Mass spectrometric measurement of enkephalins in canine tissue extracts (ng/g) and CSF (ng/ml).

Source	Enkephalin	Amount	Method[a]
Hypothalamus	LE	170	1
Cerebrospinal fluid	LE	44	1
Pituitary			
Anterior	LE	70	2
	ME	2,950	2
Posterior	LE	2	2
	ME	3,760	2
Caudate Nucleus	LE	1,500	2
Tooth pulp[b]	LE	30	2
	ME	179	2
Tooth Pulp			
Control	LE	20	2
	ME	487	2
Electrostimulated	LE	45	2
	ME	390	2

[a] Method 1: FD MS-SIM-COM
2: FAB-CAD-*B*/*E*-*B*′/*E*′-SIM-COM.

[b] Unstimulated pooled tissue from five dogs.

24.4. COMPARATIVE SPECIFICITY AND SENSITIVITY OF MEASUREMENTS

The authors feel that FAB-CAD-*B*/*E*,*B*′/*E*′-SIM-COM methodology produces the optimum molecular specificity that can be obtained to date. Currently, the MS procedures described can be used to quantify endogenous peptides at the picogram (generally the femtomole) level, a sensitivity level directly competitive with RIA. However, the MS procedures offer an incomparably higher level of molecular specificity. For many biological tissue extracts such as several brain regions and tooth pulp (Table 2), the level of the MS detection sensitivity is quite adequate for quantification of endogenous peptides. Furthermore, computer techniques and negative ion MS techniques are currently under investigation and may offer higher levels of detection sensitivity.

Up to now, many authors have stated the opinion that RIA is less expensive to perform than MS. However, one should consider that, with the purchase of a $500,000 MS instrument, one obtains state-of-the-art molecular specificity and detection sensitivity and readily measures an endogenous peptide every 10 minutes. Therefore, 48 analyses can be done in an 8 hour work-day, 96 in two shifts, and 144 in three shifts. An estimate for an RIA analysis is approximately $50 (a wide range exists here); thus one can then readily calculate that, in a few months, an entire mass spectrometer system is amortized.

24.5. CONCLUSIONS

The analytical methodology described in this chapter focuses exclusively on the measurement of endogenous neuropeptides. On one hand, RRA was demonstrated to be an excellent metabolic profiling method in conjunction with gradient RP–HPLC. On the other hand, RIA can be used for high sensitivity and relatively high molecular specificity measurement of peptides. The unique feature demonstrated in this chapter is the development of the combination methodology that utilizes FAB, CAD, *B*/*E* linked-field scanning, SIM, computer methodologies, and stable isotope-incorporated peptide internal standards. This MS method offers the highest level of molecular specificity that can be currently obtained. Experience to date indicates that the picogram (femtomole) level of sensitivity is obtained; newer developments suggest that increased sensitivity will be obtained for peptide measurement.

ACKNOWLEDGMENTS

The authors gratefully acknowledge the financial assistance provided to this research program by NIH (GM-26666, GM-28611, and RR-01651) and the typing assistance of Dianne Cubbins.

REFERENCES

Barber, M., Bordoli, R. S., Sedgwick, R. D., Tyler, A. N., and Whalley, E. T. (1981). FAB MS of bradykinin and related oligopeptides, *Biomed. Mass Spectrom.*, **8**, 337–342.

Burgus, R., Dunn, T. E., Desiderio, D. M., Ward, D. N., Vale, W., and Guillemin, R. (1970). Characterization of ovine hypothalamic hypophysiotropic TSH-releasing factor, *Nature*, **226**, 321–325.

Chang, M. M., Leeman, S. E., and Niall, H. D. (1971). Amino-acid sequence of substance P, *Nature*, **232**, 86–87.

Chavkin, C., James, I. F., and Goldstein, A. (1982). Dynorphin is a specific endogenous ligand of the kappa opioid receptor, *Science*, **215**, 413–415.

Childers, S. R. (1983). Receptors for opiates and opioid peptides, in *Principles of Receptorology*, (Ed. M. K. Agarwal), pp. 629–668, de Gruyter & Co., New York.

Desiderio, D. M. (1983). High performance liquid chromatography and mass spectrometry of biologically important neuropeptides, in *Advances in Chromatography*, vol. 22 (Eds. J. C. Giddings, E. Grushka, J. Cazes and P. Brown), pp. 1–36, Marcel Dekker, New York.

Desiderio, D. M. (1984). *Analysis of Neuropeptides by Liquid Chromatography and Mass Spectrometry*, Elsevier Scientific Publishers, Amsterdam.

Desiderio, D. M., and Cunningham, M. D. (1981). Triethylamine formate buffer for HPLC–FDMS of oligopeptides, *J. Liq. Chromatogr.*, **4**, 721–722.

Desiderio, D. M., and Kai, M. (1983a). Field desorption mass spectral measurement of enkephalins in canine brain with ^{18}O peptide internal standards, *Int. J. Mass Spectrom. Ion Phys.*, **48**, 261–264.

Desiderio, D. M., and Kai, M. (1983b). Preparation of stable isotope-incorporated peptide internal standards for FDMS quantification of peptides in biologic tissue, *Biomed. Mass Spectrom.*, **10**, 471–479.

Desiderio, D. M., and Yamada, S. (1982). Measurement of endogenous leucine enkephalin in canine thalamus by HPLC and FD–MS, *J. Chromatogr.*, **239**, 87–95.

Desiderio, D. M., Yamada, S., Tanzer, E. S., Horton, J., and Trimble, J. (1981a). High performance liquid chromatography and field desorption mass spectrometric measurement of picomole amounts of endogenous neuropeptides in biologic tissue, *J. Chromatogr.*, **217**, 437–452.

Desiderio, D. M., Yamada, S., Sabbatini, J. Z., and Tanzer, F. S. (1981b). Quantification of picomole amounts of underivatized leu-enkephalin by means of field desorption mass spectrometry, *Biomed. Mass Spectrom.*, **8**, 10–12.

Desiderio, D. M., Tanzer, F. S., Kai, M., Wakelyn, C., Fridland, G., and Trimble, J. (1983a). Measurement with optimal molecular specificity of endogenous peptides in nerve tissue by means of FAB-CAD-B/E-B′/E′-SIM methodology, in *Peptides: Structure and Function* (Eds. V. J. Hruby and D. H. Rich), pp. 715–718, Pierce Chemical Co., Rockford, IL.

Desiderio, D. M., Katakuse, I., and Kai, M. (1983b). Measurement of leucine enkephalin in caudate nucleus tissue with fast atom bombardment–collision activated dissociation-linked field scanning mass spectrometry, *Biomed. Mass Spectrom.*, **10**, 426–429.

Desiderio, D. M., Laughter, J. S., Katakuse, I., Kai, M., and Trimble, J. (1984). A microcomputer data acquisition system for mass spectrometry, *Comp. Enhan. Spectros.*, **2**, 21–31.

Goldstein, A., Tachibana, S., Lowney, L. I., Hunkapiller, M., and Hood, L. (1979). Dynorphin-(1–13), an extraordinarily potent opioid peptide, *Proc. Nat. Acad. Sci. USA*, **76**, 6666–6670.

Gray, E. G., and Whittaker, V. P. (1962). The isolation of nerve endings from brain: an electron-microscopic study of cell fragments derived by homogenization and centrifugation, *J. Anat.*, **96**, 79–88.

Gros, C., Pradelles, P., Rouget, C., Bepoldin, O., Dray, F., Fournie-Zaluski, M. C., Roques, B. P., Polard, H., Llorens-Cortes, C., and Schwartz, J. C. (1978). Radioimmunoassay of methionine- and leucine-enkephalin in regions of rat brain and comparison with endorphins estimated by a radioreceptor assay, *J. Neurochem.*, **31**, 29–39.

Hollt, V. (1983). Multiple endogenous opioid peptides, *Trends Neurosci.*, **6**, 24–26.

Hughes, J., Smith, T. W., Kosterlitz, H. W., Fothergill L. H., Morgan B. A., and Morris, H. (1975). Identification of two related pentapeptides from the brain with potent opioid activity, *Nature*, **255**, 577–579.

Jennings, K. R., and Mason, R. S. (1983). Tandem mass spectrometry utilizing linked scanning of double focusing instruments, in *Tandem Mass Spectrometry* (Ed. F. W. McLafferty), pp. 197–222, Wiley, New York.

Jessell, T. M. (1982). Substance P in nociceptive sensory neurons, in *Substance P in the Nervous System* (Eds. R. Porter and M. O'Conner), pp. 225–248, Pitman, London.

Krieger, D. T. (1983) Brain Peptides, in *Vitamins and Hormones*, vol. 40 (Ed. D. D. McCormick), pp. 1–50, Academic Press, Orlando, FL.

Krieger, D. T., Brownstein, M. J., and Martin, J. B. (1983). *Brain Peptides*, Wiley, New York.

Lawson, A. M., Lim, C. K., Richmond, W., Samson, D. M., Setchell, K. D. R., and Thomas, A. C. S. (1980). Isotope dilution mass spectrometry as a basis for accuracy in clinical chemistry, in *Current Developments in the Clinical Applications of HPLC, GC and MS* (Eds. A. M. Lawson, C. K. Lim, and W. Richmond) pp. 135–153, Academic Press, London.

May, H. E., Tanzer, F. S., Fridland, G., Wakelyn, C., and Desiderio, D. M. (1982). High performance liquid chromatography and proteolytic enzyme characterization of peptides in tooth pulp extracts, *J. Liq. Chromatogr.*, **5**, 2135–2154.

Tanzer, F. S., Desiderio, D. M., and Yamada, S. (1981). HPLC isolation and FD–MS quantification of picomole amounts of met-enkephalin in canine tooth pulp, in *Peptides: Synthesis–Structure–Function* (Eds. D. H. Rich, and E. Gross) pp. 761–764, Pierce Chemical Co., Rockford, IL.

Tanzer, F. S., Desiderio, D. M., Walker, J., Harris, E. F., and Wakelyn, C. (1986). Stress decreases endogenous methionine enkephalin in human teeth, *Am. J. Orthodon.*, in press.

Todd, P. J., and McLafferty, F. W. (1983). Collisionally activated dissociation of high kinetic energy ions, in *Tandem Mass Spectrometry* (Ed. F. W. McLafferty) pp. 149–174, Wiley, New York.

Yalow, R. S. (1980). Radioimmunoassay, *Ann. Rev. Biophys. Bioeng.*, **9**, 327–345.

Yamada, S., and Desiderio, D. M. (1982). Measurement of endogenous neuropeptides in canine caudate nuclei and hypothalami with high performance liquid chromatography and field desorption mass spectrometry, *Anal. Biochem.*, **127**, 213–221.

Mass Spectrometry in Biomedical Research
Edited by S. J. Gaskell
© 1986 John Wiley & Sons Ltd.

Chapter 25

TRIPLE QUADRUPOLE MS/MS IN BIOMEDICAL RESEARCH

Jodie V. Johnson, Mike S. Lee, Maw Rong Lee,
Harry O. Brotherton and Richard A. Yost

25.1. INTRODUCTION

Tandem mass spectrometry (MS/MS), although a relatively new analytical technique, has gained rapid acceptance with the analytical community since its introduction in the 1970s. Based upon mass spectrometers capable of providing two (or more) sequential stages of mass separation, MS/MS integrates the two basic operations of chemical analysis, separation and identification, in one instrumental technique. Thus, the acceptance of MS/MS has been due largely to its ability to provide rapid, sensitive, and selective analyses of complex mixtures, often with minimal, if any, sample clean-up. A recent book (McLafferty, 1983) and several recent reviews (Johnson and Yost, 1985; Yost and Fetterolf, 1983; Cooks and Glish, 1981; McLafferty, 1981; Kondrat and Cooks, 1978) contain extended explanations of the theory, instrumentation, and applications of MS/MS. This chapter will describe specific applications from research performed in our laboratory to illustrate how triple quadrupole MS/MS can be used to solve difficult problems encountered in biomedical research.

25.2. MS/MS FOR TRACE ANALYSIS

In biomedical research, often one is trying to determine the presence and quantity of trace components in highly complex biological matrices. To perform such trace analyses successfully, one must consider carefully the figures of merit for analytical techniques: sensitivity, selectivity, speed of analysis, and cost effectiveness. Mass spectrometry is unique among trace instrumental techniques in being able to detect trace components at the part-per-trillion level and below (Cairns and Rogers, 1983). The high *sensitivity*, defined by the slope of a calibration curve (Table 1), of mass spectrometric analysis provides detectable signals for minute quantities of analyte. However, the high sensitivity achieved for 'pure' analytes may have little bearing on the ability to determine trace analytes in increasingly complex mixtures. This is due to the interferences which can arise from other chemical constituents of the mixture or from the mass spectrometer's background; these interferences can be referred to as 'chemical noise' (Kondrat and Cooks, 1978). The limiting factor in mass spectrometry is rarely the response of the detector, but rather the mass spectrometer's ability to discern the signal of the analyte from the chemical noise, that is its *selectivity*. Thus, a more descriptive figure of merit for a trace analytical technique is the limit of detection, defined as the amount of analyte which yields a signal-to-noise ratio (S/N) adequate to provide the desired confidence (typically S/N = 3) (Table 1). Thus, the performance of a technique for trace analysis is a function of both sensitivity and selectivity (inversely related to the chemical noise).

TABLE 1. Figures of merit for analytical techniques.

$$\text{Sensitivity} = [\text{slope of the calibration curve}] = \frac{\text{d(signal)}}{\text{d(amount)}}$$

$$\text{Limit of Detection} = [\text{amount for S/N} = 3] = 3\left(\frac{\text{noise}}{\text{sensitivity}}\right)$$

Cooks and Busch (1982) have shown that the selectivity of a technique can be improved by increasing the number of stages of analysis. As the number of stages of analysis is increased, the absolute levels of the signal and noise decrease; however, if the additional stages increase the selectivity of the technique, the chemical noise will decrease more rapidly than the signal. If the chemical noise dominates over the electrical noise (as is almost always the case in MS), an overall improvement in the S/N ratio will result. Thus, as long as there is a detectable signal above the electrical noise level, increasing the number of analytical stages will result in improved limits of detection. These added stages of analysis can include the chromatographic step in GC–MS, the second stage of mass analysis in MS/MS, or even both in GC–MS/MS. The increase in selectivity provided by the two stages of mass analysis in MS/MS reduces the chemical noise and thus, can lead to improved (lower) limits of detection compared to traditional MS, even though some sensitivity is inevitably lost in the second stage.

A further benefit of the sensitivity and selectivity of MS/MS is its potential for increased *speed of analysis*. This results from the ability either to eliminate the chromatographic separation step (Brotherton and Yost, 1984) or to use rapid, limited chromatographic separation with short packed and capillary columns (Trehy *et al.*, 1985; Yost *et al.*, 1984). The loss in chromatographic selectivity is balanced by the gain in the mass spectrometric (MS/MS) selectivity. Speed is also gained since it is often possible to analyze complex mixtures with little or no sample preparation. This increased speed of analysis is often reflected in an increased *cost effectiveness* of analysis, despite the higher initial instrumental cost.

25.3. MS/MS INSTRUMENTATION

A conceptual diagram of a tandem mass spectrometer is illustrated in Figure 1 and compared to the, perhaps more familiar, GC–MS. A tandem mass spectrometer consists of an ion source, two mass analyzers separated by a fragmentation region, and an ion detector. Although several instrumental approaches have been employed for mass analysis in MS/MS and GC–MS, the quadrupole mass filter shown is by far the most common and is the one in use in our laboratory. Further information on instrumentation for MS/MS can be found in a recent review (Yost and Fetterolf, 1983). Regardless of the type of

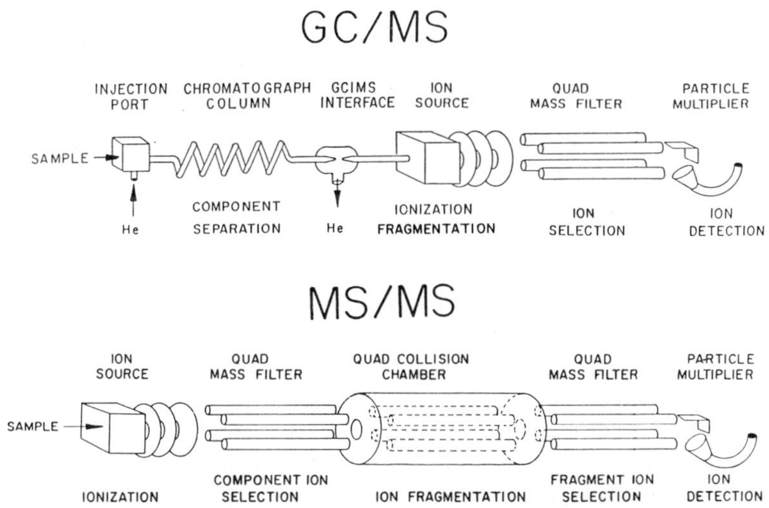

Figure 1: Conceptual diagram of the triple quadrupole tandem mass spectrometer (MS/MS) and a traditional gas chromatograph/quadrupole mass spectrometer (GC–MS). Note that the same analytical operations are provided by both instruments.

mass analyzer, the principles of MS/MS are straightforward, and can be compared to conventional GC–MS. A mixture is introduced into the ion source of the tandem mass spectrometer, where ionization of the mixture produces ions characteristic of the individual components. The separation of the analyte from the other mixture components (the chromatographic step of GC–MS) is then achieved by the mass selection of the ion characteristic of the analyte by the first mass analyzer. This parent ion undergoes collisionally-activated dissociation (CAD) through collisions with neutral gas molecules in the fragmentation region to yield various daughter ions, analogous to the fragmentation occurring immediately following ionization in GC–MS. Subsequent mass analysis of the daughter ions by the second mass analyzer permits identification and quantification of the separated components.

The four most common MS/MS operating modes of the tandem mass spectrometer are daughter scan, parent scan, neutral loss scan, and selected reaction monitoring (Figure 2). The specific operational mode chosen for a particular analysis will depend on the information desired. A *daughter scan*

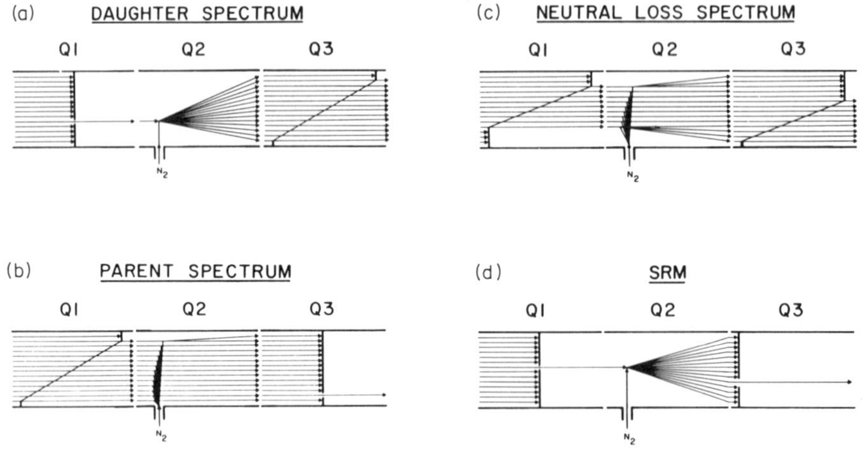

Figure 2: The four most common MS/MS operational modes of a triple quadrupole tandem mass spectrometer. In the MS/MS modes, quadrupoles 1 and 3 (Q1 and Q3, respectively) are mass filters, while quadrupole 2 (Q2) serves as an efficient collision and focusing device.

(Figure 2a) consists of selecting a parent ion characteristic of the analyte with the first mass analyzer, fragmenting it by CAD in the fragmentation region, and scanning the second mass analyzer to obtain a daughter mass spectrum. Analogous to a normal mass spectrum in GC–MS, the daughter mass spectrum can be used for identification of an analyte (separated by the first mass analyzer) by interpretation or by matching the spectrum with that of an authentic sample.

In a *parent scan* (Figure 2b), a specific daughter ion is selected with the second mass analyzer, and the first mass analyzer is scanned over a specific mass range, selecting parent ions of different m/z that fragment to yield the specific daughter ion. This mode can be used to screen for a class of compounds that fragment to yield a common substructure. In a *neutral loss scan* (Figure 2c), both mass analyzers are scanned with a constant difference in mass. The resulting neutral loss spectrum contains the daughter ions which arise from the loss of a specific neutral fragment from the parent ions. This mode is useful for screening for a class of compounds characterized by a common fragmentation pathway. The parent and neutral loss scans can be used for rapidly screening mixtures for classes of compounds. The necessary confirmation can then be accomplished by obtaining a complete daughter spectrum of each parent ion detected in the mixture.

Although the three operational modes just described are very selective, as in full scan normal mass spectrometry they may lack the sensitivity necessary for the determination of trace components. Thus, for trace analysis, *selected reaction monitoring* (SRM) (Figure 2d) is often used, wherein a limited number of parent ion–daughter ion pairs are monitored for each analyte. Thus, an enhancement in the sensitivity is obtained, albeit at the expense of some selectivity. This selected reaction monitoring is analogous to selected ion monitoring (SIM) used to obtain maximum sensitivity in conventional GC–MS. However, with SRM, an increased confidence in the experimental results is obtained due to the selection of both parent and daughter ions.

Because of its potential for rapid, sensitive, and selective trace mixture analysis, MS/MS is finding increased use in biomedical applications. In the sections below, we present some of these applications, taken from work performed in our laboratory.

25.4. RAPID SCREENING AND CONFIRMATION OF EXOGENOUS COMPOUNDS

The determination of exogenous compounds (e.g. drugs, pesticides) in biological matrices is often accomplished by a two-step screening/confirmation procedure. The rapid screening step is used to identify samples containing the analyte(s) (positives) and to eliminate those lacking the analyte(s) (negatives). The screening method should be sensitive enough to prevent false negatives, but will generally lack the selectivity to prevent false positives. All positives are then subjected to a second, more selective, confirmation step in order to identify positively those samples which do contain the analyte(s). If the rapid screening step effectively reduces the number of samples which must be subjected to the more time-consuming confirmation step, then this two-step procedure saves time and expense. Typically, the two steps employ two different analytical techniques. MS/MS offers the advantage of being able to perform both of these steps on the same instrument.

As an example of such an analytical problem, consider the testing of racing animals for illicit drugs. Presently, screening for illicit drugs is most commonly performed by thin-layer chromatography (TLC), with confirmation performed on any positives by GC–MS. These methods require extensive sample work-up prior to analysis, limiting the number of animals that can be tested. However, we have developed an MS/MS technique which permits both rapid screening and confirmation with minimal sample work-up (Brotherton and Yost, 1984). With sample introduction via a heated solids probe, screening by SRM permits the detection of most of the illicit drugs in the ppm range (ng on probe) in raw blood serum and urine and in the low ppb range (pg on probe) in simple extracts of these fluids. The necessary confirmation of positives is accomplished by comparing a complete daughter spectrum of each detected parent ion in the mixture with that obtained from an authentic standard. This allows confirmation of most of the drugs at the ppm level in the simple extracts. With this

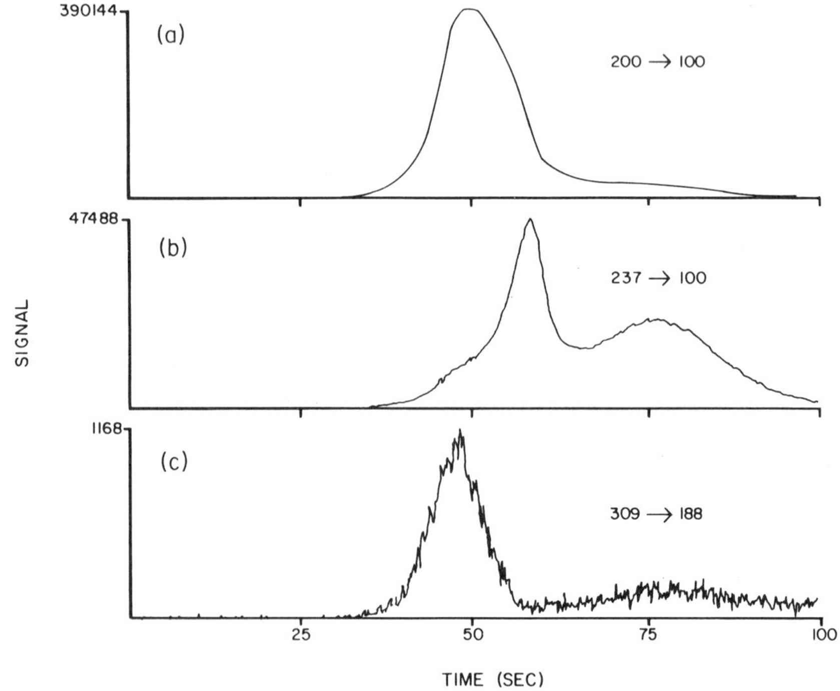

Figure 3: Methane positive CI–SRM screening for three drugs in the urine of a greyhound dosed with therapeutic doses of diethylcarbamazine (a), procaine (b), and phenylbutazone (c). Sample introduction is by direct insertion probe. Results are positive for all three drugs.

procedure, it is possible to screen for as many as 50 drugs in a single 1 μl serum or urine sample in less than 5 minutes at a cost estimated to be equivalent to that of current TLC screening. This procedure is illustrated more fully below with the analysis of greyhound urine.

Urine samples were obtained from two greyhounds which had been dosed with a daily therapeutic dose of diethylcarbamazine for 4 days, and, on the fourth day, with a single therapeutic dose of procaine and phenylbutazone. With the introduction of a 1 μl urine sample into the chemical ionization (CI) source via a heated solids probe, methane positive CI-SRM screening indicated a positive for each of the three drugs (Figure 3). A second 1 μl urine sample was then heated and a complete daughter spectrum was obtained of the protonated molecule, $(M + H)^+$, of each drug. A good match with the library daughter spectrum of the authentic drug confirmed the presence of each drug (shown for phenylbutazone, present at the lowest concentration, in Figure 4). This MS/MS procedure permits the sensitive and selective screening and confirmation of illicit drugs rapidly and with minimal sample preparation.

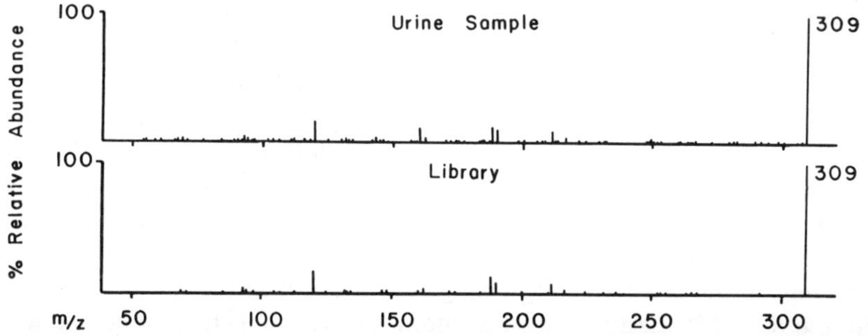

Figure 4: Comparison of the daughter spectrum of *m/z* 309 in greyhound urine with a reference library daughter spectrum of the $(M + H)^+$, *m/z* 309, of phenylbutazone (in chloroform).

We are currently developing an MS/MS technique to screen for lysergic acid diethylamide (LSD) in human urine (Johnson *et al.*, 1986). LSD has a very low psychoactive dose (1 μg/kg) and is readily metabolized by the body, necessitating a very low limit of detection. Preliminary studies have established a limit of detection of 80 pg of LSD in 5 μl of spiked urine introduced on a solids probe with methane positive CI-SRM.

The enhanced selectivity of MS/MS over traditional MS makes possible trace analysis with extremely low limits of detection. For example, we have demonstrated the ability to detect hexachlorobenzene and trichlorophenol in simple extracts of human serum and urine with detection limits of 0.05 pg and 0.25 pg, respectively, at a rate of 100 sample injections per hour (Yost *et al.*, 1984). These low detection limits were made possible by the use of a short packed GC column for sample introduction, electron capture negative CI, and SRM MS/MS. The enhanced selectivity of MS/MS permitted the use of the 50 cm × 0.1 cm (i.d.) packed GC column to provide short retention times (< 30 s) for rapid analysis.

25.5. QUANTIFICATION OF ENDOGENOUS COMPOUNDS

The determination of endogenous compounds in biological matrices is often a more difficult problem than that of exogenous compounds. This is due to the lack of true blanks, the presence of other, very similar chemical compounds which can lead to interferences, and the effects (possibly unknown) of health, diet, and other environmental factors upon the analytes' concentrations. As these compounds often are biologically active in very minute amounts, some type of preconcentration and sample clean-up may be necessary prior to analysis. Such sample manipulations, however, increase the possibilities for sample contamination, sample loss through adsorption onto glassware, oxidation, etc., and artifactual formation of the analytes. Quantification of these endogenous compounds is best accomplished with the use of isotopically-labelled analogues of the analytes as internal standards to correct for some of these effects. The sensitivity and selectivity of MS/MS allows a reduction in the amount of required sample work-up and thus, a concomitant reduction in the magnitude of the associated problems. As an example of this type of application, consider the MS/MS quantification of tryptolines (tetrahydro-β-carbolines) in the low to sub-ppb range in rat brain extracts.

In the GC–MS analysis of rat brain extracts for tryptolines, extensive clean-up is required, raising doubts as to whether the tryptolines are artifactually produced during this clean-up (Faull *et al.*, 1982). The initial goal of this work was to use the additional selectivity of MS/MS to analyze brain tissue directly, with a minimum of sample purification and without chemical derivatization, and thus to reduce the possibility of artifactual formation of the tryptolines. However, upon investigating the solids probe CI-SRM technique (used for the determination of illicit drugs above), the limits of detection (*ca.* 15 ppm) obtained for standards were not sufficient for determining the ppb-levels of tryptolines (Johnson *et al.*, 1985a). To increase the sensitivity and make the compounds amenable to GC, the heptafluorobutyryl (HFB) derivatives of the tryptolines were investigated (Johnson *et al.*, 1984). With the short-column GC–MS/MS approach (detailed above), the electron capture negative CI-SRM limit of detection for tryptoline-HFB standards was 0.11 pg injected. However, in the short-column GC–MS/MS analysis of the HFB-derivatized crude extracts of rat brain, there was a severe reduction of the tryptoline signal. This was due to the fact that during the derivatization of brain tissue, large numbers of compounds, many present at levels significantly higher than those of the tryptolines, were also derivatized. The short GC column was not adequate to resolve these compounds from the analytes, leading to severe quenching in the electron capture negative CI process and difficulty in detecting even 1 ng of the analytes. Indeed, all the selectivity possible in the MS/MS analysis cannot correct for competitive ionization and quenching from interferences in the negative CI source.

One possible solution to this problem which was investigated was the use of positive CI. Although less prone to ionization interferences than electron capture negative CI, it lacked the required sensitivity to detect the trace tryptolines. Another possible solution, then investigated, was the use of a longer capillary GC column to separate the analytes from the other derivatized sample components. With chromatographic separation on an 11 m fused-silica capillary column, the quenching problem was indeed eliminated, making it possible to achieve detection limits similar to those achieved with the short column, but with longer retention times (*ca.* 10 min) and therefore, less rapid analysis. The need for the added selectivity of MS/MS, even with high resolution chromatographic separation, is illustrated in a comparison of the GC–MS and GC–MS/MS chromatograms for this application (Figure 5) (Johnson *et al.*,

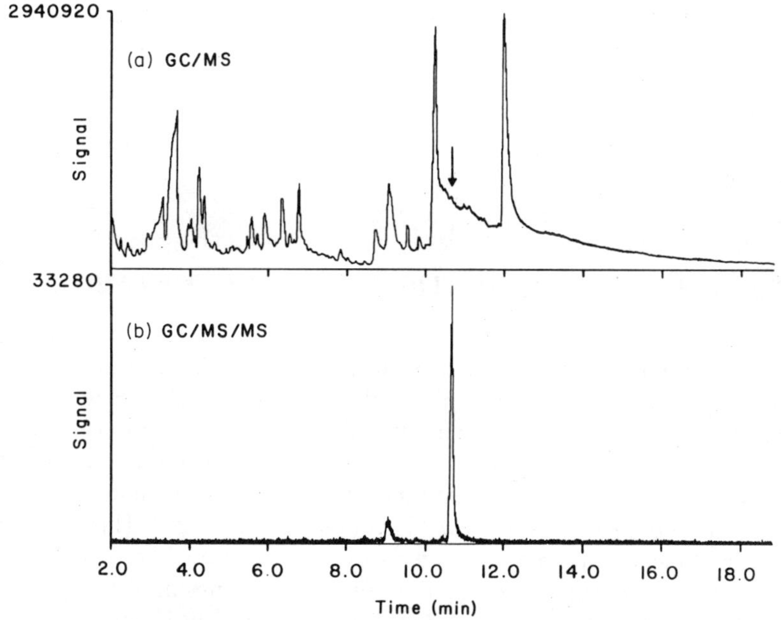

Figure 5: Chromatograms for the determination of methtryptoline (indicated by the arrow) in heptafluorobutyryl-derivatized rat brain extracts by (a) GC/negative CI–SIM (m/z 362) and (b) GC/negative CI–SRM (m/z 362 → m/z 179) *Reproduction from Johnson* et al., *(1985) by permission of Alan R. Liss, Inc.*

1985a). Although monitoring only a single ion (corresponding to $(M - HF)^-$ of methtryptoline HFB), the GC–MS/SIM chromatogram shows numerous compounds, but no discernible GC peak for methtryptoline HFB (Figure 5a). However, GC–MS/MS/SRM is transparent to the co-eluting and interfering compounds, producing a clean chromatogram with an intense GC peak for methtryptoline HFB (estimated to be in the low pg range injected) (Figure 5b). Therefore, it is apparent that the chromatographic separation was necessary only to prevent the quenching of the analyte's NCI signal and not to increase the selectivity of the short-column GC–MS/MS technique.

This GC–MS/MS/SRM technique was then applied to the quantification of tryptoline and methtryptoline. With deuterated tryptolines as internal standards, the levels of tryptoline and methtryptoline were found to range from 1.1 to 25 ng/g brain tissue and from 0.5 to 3.1 ng/g brain tissue, respectively (Johnson *et al.*, 1985b). Although replicate injections of the same sample showed that the GC peak areas had poor reproducibility (60% CV), the use of deuterium-labelled internal standards permitted precise (5% CV) quantification in the HFB-derivatized crude extracts of rat brains. Experiments which monitored for artifactual formation during sample preparation showed that little, if any, occurred to form tryptoline and methtryptoline.

These studies have illustrated the potential of MS/MS for providing rapid, sensitive, and selective analyses of biological matrices for trace exogenous and endogenous compounds. Rapid screening and confirmation is possible for large numbers of targeted compounds at the ppm to ppb level in complex mixtures with minimal sample preparation. Trace analysis at the sub-ppb levels is possible by a combination of the sensitivity of electron capture negative CI with the selectivity of capillary GC–MS/MS.

25.6. RAPID IDENTIFICATION OF DRUG METABOLITES

The identification of drug metabolites rapidly (in less than one day) is desirable but is not often possible. Most methods for drug metabolite identification require the use of various chromatographic techniques for the isolation and separation of each metabolite present in the sample prior to identification. Although chromatography is an extremely powerful and valuable technique for the separation of drug metabolites, it is not particularly rapid. This is because a collection of metabolites may contain a variety of functional groups resulting in a wide range of polarities, and not all of these metabolites may be separated or detected under the same chromatographic conditions. This is especially true for identification techniques which employ gas chromatography. Therefore, multiple chromatographic conditions may have to be developed, requiring a considerable amount of time.

The technique of MS/MS, in which samples can be rapidly introduced into the instrument with minimal sample preparation, and all metabolites vaporized, ionized, and separated under the same set of conditions, is clearly advantageous. Not only is the time required for chromatographic elution eliminated, but also the time spent on method development is avoided. The result is a much faster method of drug metabolite identification.

Consider the structures of drug metabolites, their relationship to the original parent drug structure, and the functions of the neutral loss and parent scan modes of a tandem mass spectrometer. Since drug metabolites usually retain a large portion of the parent drug molecule, both would be expected to undergo similar CAD fragmentations. Specifically, both should lose some of the same neutral fragments and/or form some of the same daughter ions. If this does occur, both the neutral loss and parent modes of the MS/MS instrument could be effectively used to selectively screen the biological sample for those species which contain the characteristic functionalities or substructures of the pure drug, without separation of the components in the mixture. The ions contained in the neutral loss and parent spectra would correspond to possible metabolites, and their daughter spectra could be obtained. With knowledge of both the fundamentals of drug metabolism and mass spectral interpretation, metabolite structures could thus be rapidly identified.

The MS/MS procedure for metabolite identification can be summarized as follows:

(1) Obtain a *normal CI mass spectrum* of the *pure drug*.

(2) Obtain *daughter spectra* of the *pure drug*, selecting the protonated molecule, adduct ions, and abundant fragment ions for CAD. Note the major neutral losses and abundant daughter ions.

(3) Obtain *neutral loss spectra* of the *biological sample*, selecting the major neutral losses of the pure drug observed in step 2. The ions contained in these spectra represent possible metabolites.

(4) Obtain *parent spectra* of the *biological sample*, selecting the abundant daughter ions observed in step 2. The ions contained in these spectra also represent possible metabolites.

(5) Obtain *daughter spectra* of all the ions (*possible metabolites*) observed in steps 3 and 4 in the *biological sample*. Metabolite structures can then be identified either by direct spectral interpretation or by comparison with spectra of pure reference standards.

(6) Repeat steps 3 and 4 on the biological sample, selecting any new neutral losses and daughter ions observed from the daughter spectra of metabolites identified in step 5. This will permit the identification of further metabolites with similar substructures present in the previously identified metabolites, but not necessarily present in the parent drug.

With this MS/MS method, major metabolites of primidone, cinromide, and phenytoin have been identified in urine and plasma (Perchalski *et al*., 1982). By application of a knowledge of typical metabolic pathways, the major metabolites of these drugs were identified in only a few hours, with minimal sample preparation. Sample preparation involved the mild deproteination of plasma followed by extraction, and the hydrolysis and extraction of urine to yield free, volatile species in a volatile solvent. This simple work-up was performed in less time than that required for the evaporation of 3 µl of plasma or urine in a solids probe vial at room temperature. Samples were deposited in probe vials in 2 µl aliquots and introduced into the instrument on a direct insertion probe, where they were heated ballistically from 25 to 350°C in 2 minutes.

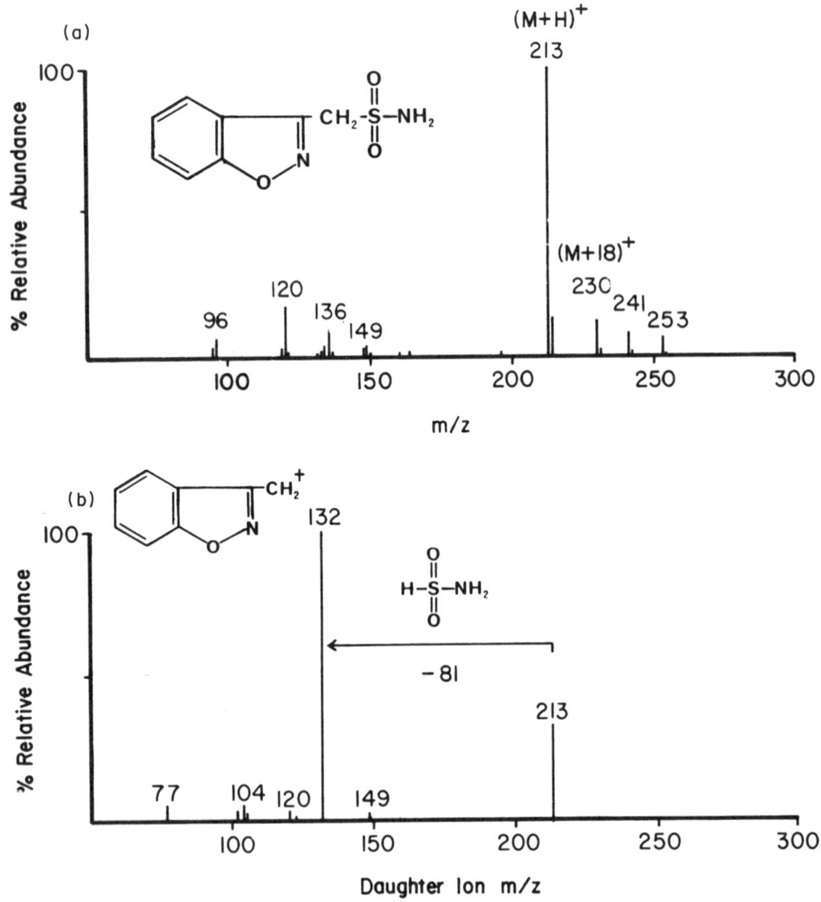

Figure 6: a) Positive CI normal mass spectrum of pure zonisamide, showing an intense $(M + H)^+$ ion and characteristic methane adduct ions at $(M + 29)^+$ and $(M + 41)^+$. The $(M + 18)^+$ ion, m/z 230, is due to a trace impurity of NH_3 in the reagent gas. b) Positive CI–CAD daughter mass spectrum of $(M + H)^+$ of pure zonisamide.

An example which demonstrates the utility of MS/MS involves the rapid identification of the metabolites of a new anti-epileptic drug, zonisamide (Warner-Lambert/Parke-Davis Pharmaceutical, Ann Arbor, MI), from an enzyme-hydrolyzed urine extract. In this example, the method was used to identify the simple and unique fragmentations of the pure drug and its metabolites. Typically, all of the fragmentation information obtained from the pure drug is used to identify metabolite structures with this method. However, to demonstrate more clearly the use and applicability of this MS/MS method, only one major neutral loss and one abundant fragment ion are chosen for discussion below.

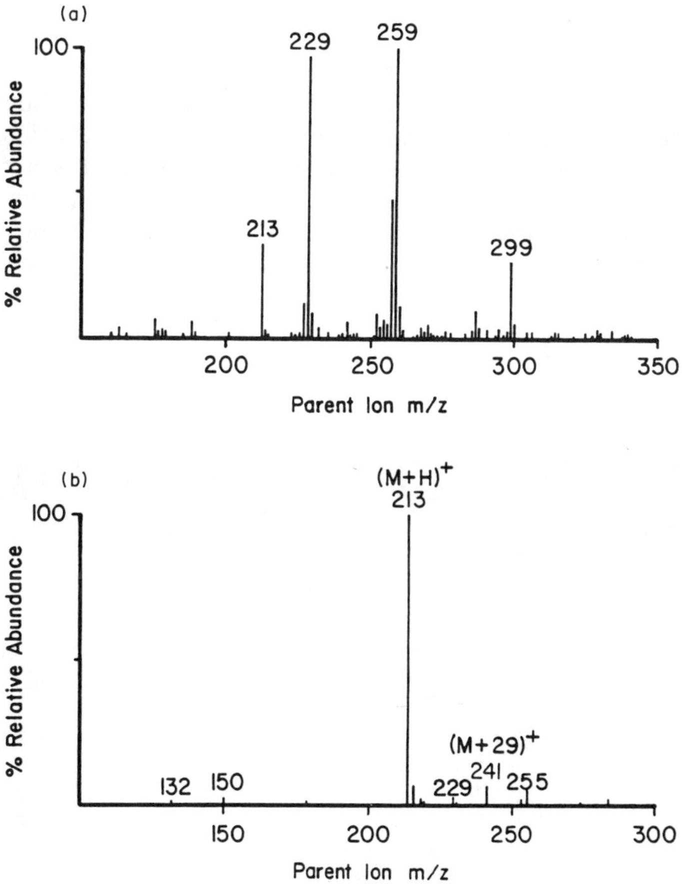

Figure 7: a) Neutral loss (of 81 u) spectrum of an extract of enzyme-hydrolyzed urine of a subject administered the antiepileptic drug, zonisamide. b) Parent spectrum (of *m/z* 132) of the enzyme-hydrolyzed urine extract.

An intense protonated molecule, $(M + H)^+$ at *m/z* 213, and abundant $(M + 29)^+$ and $(M + 41)^+$ adduct ions were observed in the methane positive CI mass spectrum of pure zonisamide (step 1) with very little fragmentation (Figure 6a). The daughter spectrum of the protonated molecule of pure zonisamide (step 2) indicated a major neutral loss of 81 u (due to the loss of the sulphonamide portion of the molecule), and the formation of an abundant daughter ion at *m/z* 132 (representing the ring substructure of the molecule) (Figure 6b). Thus, a characteristic fragmentation of zonisamide is the loss of the sulphonamide group to form the ring substructure daughter ion. Therefore, if steps 3 and 4 are performed on the urine extract sample, selecting the loss of 81 u in the neutral loss scan mode and the parents of *m/z* 132 in the parent scan mode, the resulting spectra should yield ions which correspond to the metabolites of zonisamide. The neutral loss (of 81 u) spectrum of the urine extract (step 3) contained abundant ions at *m/z* 229, 259 and 299, as well as 213 ($(M + H)^+$ of zonisamide) (Figure 7a). The parent spectrum (of *m/z* 132) of the urine extract (step 4) yielded, in addition to the ions characteristic of the original drug, abundant ions at *m/z* 229 and 255 (Figure 7b). The abundant ions in these two spectra were then chosen for the daughter ion experiment (step 5).

The four metabolites of zonisamide identified in urine with this method are shown in Figure 8, along with their major CAD fragmentations observed from

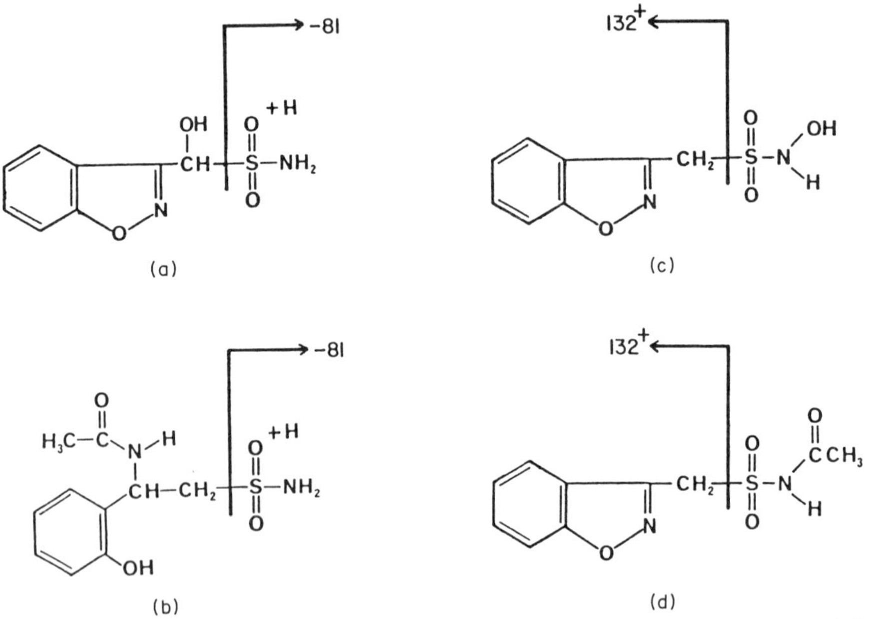

Figure 8: Metabolites of zonisamide identified by the MS/MS technique and their characteristic fragmentations observed in the daughter spectra of their $(M + H)^+$ ions.

their daughter spectra. The metabolites a and b were identified by use of the neutral loss–daughter spectra sequence, steps 3 and 5. This was possible because the metabolite structures retained the intact, unmetabolized sulphonamide portion of the parent drug, which was easily lost under CAD conditions (neutral loss of 81 u). Thus, metabolism occurred on the ring portion of the drug. The metabolites c and d are the zonisamide metabolites which retained the ring substructure of the parent drug. These metabolites were the result of metabolism occurring on the sulphonamide portion of the drug. Under CAD conditions, metabolites c and d formed an abundant daughter ion at m/z 132, corresponding to the ring substructure unit of the parent drug, and thus were detected with the parent scan of m/z 132, step 4. An interesting note here is that the two isomeric hydroxy-zonisamide metabolites, a and c, were identified, since they distilled off the probe at different times and yielded different CAD spectra (Figure 8). It would not normally be expected to be possible to distinguish two isomers from an MS/MS experiment.

The use of this MS/MS technique offers several advantages. One advantage is speed of analysis. Since all species are vaporized off the probe under the same conditions, the identification of a wide range of metabolites with different polarities is possible. Furthermore, the time required for chromatographic elution is eliminated, and the need for development of multiple chromatographic conditions is avoided. The speed with which a metabolite search can be performed makes this method applicable for rapid metabolite screening, for monitoring of patients, and for pharmacokinetic studies. A second advantage of this technique is that it is based only on the mass spectrometric behaviour of the pure drug. Therefore, the probability of detecting active and bioavailable metabolites is high. Finally, there is no need either for radioactively-labelled compounds or for metabolite standards. Standards may be used but are not required.

A disadvantage of this technique is the inability to determine the position of substituents. Therefore, the various isomers of a metabolite cannot generally be distinguished. Chromatography in combination with MS/MS (GC–MS/MS and LC–MS/MS) could be used for further identification of isomeric compounds.

TABLE 2. MS/MS for biomedical research: characteristics and capabilities.

1. *Improved detection limits* through reduction in chemical noise.
2. *Increased confidence* in experimental results due to selection of both parent and daughter ion masses.
3. *Decreased sample clean-up* needed, often permitting direct analysis of crude samples.
4. *Minimization or elimination of chromatographic separation.*
5. *Extremely rapid analysis* due to decreased time for sample preparation and chromatographic separation.
6. *Unique analytical schemes* possible based upon a versatile selection of operational modes.

25.7. CONCLUSIONS

These studies have demonstrated that triple quadrupole tandem mass spectrometry is a powerful tool for solving difficult problems encountered in biomedical research. Due to its unique characteristics (Table 2), MS/MS is capable of providing rapid screening and quantification of targeted compounds in very complex biological matrices at the part-per-billion level and below. In addition, the unique capabilities of MS/MS offer new analytical techniques, such as the metabolite identification method presented here.

ACKNOWLEDGMENTS

The reported work at the University of Florida has been supported by the National Science Foundation, the Petroleum Research Fund of the American Chemical Society, the National Institutes of Health Biomedical Research Support program, the Naval Research Laboratory, the Medical Research Service of the Veterans Administration, a Procter and Gamble Fellowship, the Lederle Laboratories, and the Epilepsy Research Foundation of Florida. We are grateful for the collaborative assistance of Clyde Williams and B. J. Wilder of the Veterans Administration Medical Center (Gainesville), Ron Gronwall of the UF School of Veterinary Medicine, and Kym F. Faull, Olof Beck, and Jack D. Barchas at the Stanford University School of Medicine.

REFERENCES

Brotherton, H. O., and Yost, R. A. (1984). Rapid screening and confirmation for drugs and metabolites in racing animals by tandem mass spectrometry, *Am. J. Vet. Res.*, **45**, 2436–2440.

Cairns, T., and Rogers, W. M. (1983). Acceptable analytical data for trace analysis, *Anal. Chem.*, **55**, 54A–57A.

Cooks, R. G., and Busch, K. L. (1982). Counting molecules by desorption ionization and mass spectrometry/mass spectrometry, *J. Chem. Educ.*, **59**, 926–933.

Cooks, R. G., and Glish, G. L. (1981). Mass spectrometry/mass spectrometry, *Chem. Eng. News*, **1981**, (Nov. 30), 40–52.

Faull, K. F., Holman, R. B., Elliott, G. R., and Barchas, J. D. (1982). Tryptolines: artifact or reality? A new method of analysis using GC/MS, in *Beta-Carbolines and Tetrahydroisoquinolines* (Ed. F. E. Bloom), pp. 135–154, Alan R. Liss, Inc., New York.

Johnson, J. V., and Yost, R. A. (1985). Tandem mass spectrometry for trace analysis, *Anal. Chem.*, **57**, 758A–768A.

Johnson, J. V., Yost, R. A., and Faull, K. F. (1984). Tandem mass spectrometry for the trace determination of tryptolines in crude brain extracts, *Anal. Chem.*, **56**, 1655–1661.

Johnson, J. V., Yost, R. A., Beck, O., and Faull, K. F. (1985a). The use of tandem mass spectrometry for the identification and quantitation of tryptolines (tetrahydro-β-carbolines) in tissue extracts, in *Aldehyde Adducts in Alcoholism* (Ed. M. A. Collins), pp. 161–177, Alan R. Liss, New York.

Johnson, J. V., Yost, R. A., Beck, O., and Faull, K. F. (1985b). GC/MS/MS for the simultaneous determination of indoleamines and tryptolines in crude extracts of rat brains, in *Proc. 33rd. Ann. Conf. Mass Spectrometry and Allied Topics, San Diego,* American Society for Mass Spectrometry, pp. 498–499.

Johnson, J. V., Lee, M. R., Lee, M. S., and Yost, R. A. (1986). Tandem MS for the determination of trace levels of LSD in human urine, in *Proc. 34th. Ann. Conf. Mass Spectrometry and Allied Topics, Cincinnati,* American Society for Mass Spectrometry, paper RPC 8.

Kondrat, R. W., and Cooks, R. G. (1978). Direct analysis of mixtures by mass spectrometry, *Anal. Chem.*, **50**, 81A–92A.

McLafferty, F. W. (1981). Tandem mass spectrometry, *Science*, **214**, 280–287.

McLafferty, F. W., Ed. (1983). *Tandem Mass Spectrometry*, John Wiley & Sons, New York.

Perchalski, R. J., Yost, R. A., and Wilder, B. J. (1982). Structural elucidation of drug metabolites by triple quadrupole mass spectrometry, *Anal. Chem.*, **54**, 1466–1471.

Trehy, M. L., Yost, R. A., and Dorsey, J. G. (1985). Short open tubular columns in gas chromatography/mass spectrometry, *Anal. Chem.*, **58**, 14–19.

Yost, R. A., and Fetterolf, D. D. (1983). Tandem mass spectrometry (MS/MS) instrumentation, *Mass Spectrom. Rev.*, **2**, 1–45.

Yost, R. A., Fetterolf, D. D., Hass, J. R., Harvan, D. J., Weston, A. F., Skotnicki, P. A., and Simon, N. A. (1984). Comparison of mass spectrometric methods for trace level screening of hexachlorobenzene and trichlorophenol in human blood serum and urine, *Anal. Chem.*, **56**, 2223–2228.

Mass Spectrometry in Biomedical Research
Edited by S. J. Gaskell
© 1986 John Wiley & Sons Ltd.

Chapter 26

THE DETERMINATION OF DRUGS IN URINE BY LC–MS AND LC–MS/MS

Jack Henion and Thomas Covey

26.1. INTRODUCTION

The identification of trace levels of organic compounds in complex biological matrices represents a challenging problem for analytical chemists. The 'search for zero' confronts analysts with the problem of both detection and identification of analytes at ultra-trace levels that seemed impossible just a few years ago.

Due to the increasing importance of regulatory constraints for many drugs or toxic compounds, it has become essential that analytes be unequivocally identified when they are 'detected' in biological or environmental samples. This is no small task since it requires modern analytical equipment and highly competent personnel. It also increasingly requires refined sample work-up methodologies which may deviate from classical techniques. These and the intrinsic problems of ultra-trace analysis create a considerable challenge for the analyst.

The three areas of importance for the ultra-trace analysis of biological samples include sample work-up, screening, and confirmation of the analyte. The preferred sample work-up method should utilize limited sample and allow high sample throughput such that up to 50–100 samples per day may be analyzed. The desirable screening method must be both as sensitive and specific as possible to preclude unnecessary efforts spent trying to confirm a false positive. Finally, the confirmatory technique should be both ultra-sensitive (sub part per billion, ppb) and unequivocal. These analytical attributes are possible today, but they are more easily said than done.

The confirmatory step is the most important stage in the analytical scheme for those samples which are deemed positive in the screening step and may be subject to legal scrutiny. There is little disagreement that mass spectrometry is required for unequivocal confirmation of organic analytes. Mass spectral confirmation requires a clean, unambiguous mass spectrum which agrees in all respects with an authentic sample of the suspected analyte. Unfortunately, the analysis of complex biological samples often can result in mass spectra which contain interfering ions due to endogenous materials from the sample matrix. To minimize this possibility one often uses high resolution capillary chromatographic columns coupled to the mass spectrometer (GC–MS). Under optimized conditions, such coupling offers remarkable separation and resolution of the analyte from the sample interfering components. The analysis times, however, often require in excess of 30 min per sample, and compromise high sample throughput capability.

Recently, the unique capabilities of the mass spectrometer as an HPLC detector have offered new analytical opportunities for the analysis of biological samples (Lee and Henion, 1985). The technique of HPLC is extremely useful for separating biological matrix interferents from analytes of interest (Snyder and Kirkland, 1979). In addition, this technique requires no elevated temperatures, as does the GC, so there is less possibility of thermal degradation of trace analytes and HPLC analysis times can be much shorter than capillary GC. Thus combined HPLC–MS (LC–MS) could offer significant opportunities for the ultra-trace analysis of important analytes in biological samples (Skrabalak *et al.*, 1984), albeit with reduced chromatographic efficiency compared to capillary GC columns.

Finally, a recent advance in mass spectrometry has provided the technique of tandem mass spectrometry (MS/MS) (McLafferty, 1983). This modern analytical technique reduces the problem of the mixed mass spectra noted above. The MS/MS technique allows one to specifically focus on the ions particular to the analyte of interest even when it is in the presence of interfering material (Henion *et al.*, 1983; Johnson and Yost, 1985). Thus MS/MS provides a final 'separation' after either GC or HPLC separation of materials introduced into the mass spectrometer. We have recently investigated the merits of both solid probe MS/MS and LC–MS/MS applied to the characterization of drugs in complex biological matrices (Henion *et al.*, 1983; Covey and Henion, 1985).

26.2. EXPERIMENTAL

In order to utilize conventional HPLC flow rates and achieve very short run times during LC–MS/MS, we have investigated the use of short, three micron HPLC columns for ultra-trace analysis of drugs and their metabolites in biological samples. The system used included two model 6000A pumps, a M-660 solvent programmer, and a model 440 UV detector (Waters Assoc., Milford, MA) and was equipped with a Rheodyne model 7510 injector driven by

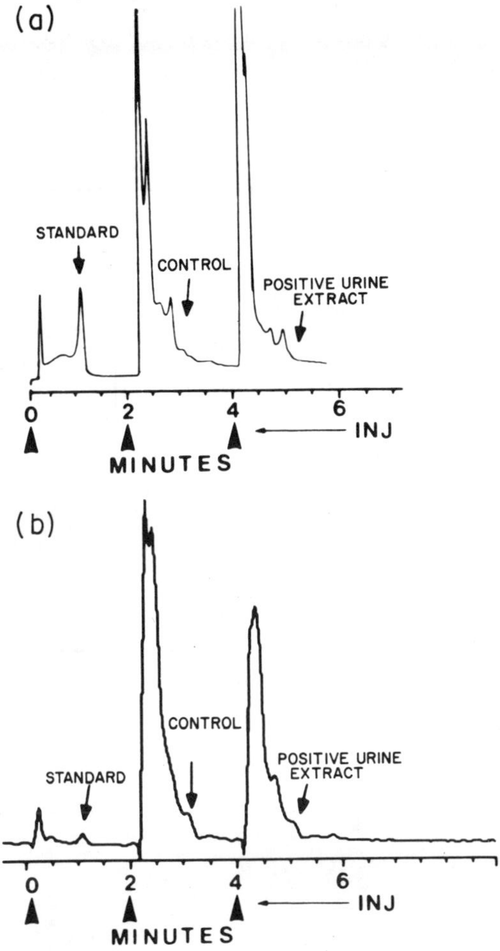

Figure 1: a) HPLC UV chromatogram at 254 nm for the successive injection at 2 minute intervals of furosemide, a control equine urine extract, and an extract of a racehorse urine extract thought to contain furosemide. b) Total ion current chromatogram under negative ion APCI LC–MS conditions of the same sample sequence shown in (a).

a Model ISS-100 automatic sampler, and a three micron C18 'fast' HPLC column (4.6 mm i.d. × 3 cm) from Perkin Elmer (Perkin Elmer Corp., Norwalk, CN). The mobile phase was 67/33 methanol/0.5% acetic acid maintained at 1 ml/min. The exit of the UV detector was connected to a heated nebulizer LC–MS probe which was inserted into the atmospheric pressure chemical ionization (APCI) source of a TAGA 6000E triple quadrupole mass spectrometer (Sciex, Inc., Thornhill, Ontario, Canada). HPLC flow rates of 1 ml/min could be continuously introduced into the APCI source under either LC–MS or LC–MS/MS conditions. The mass spectrometer could be operated in either the full scan or selected reaction monitoring (SRM) mode in the LC–MS/MS experiment (McLafferty, 1983).

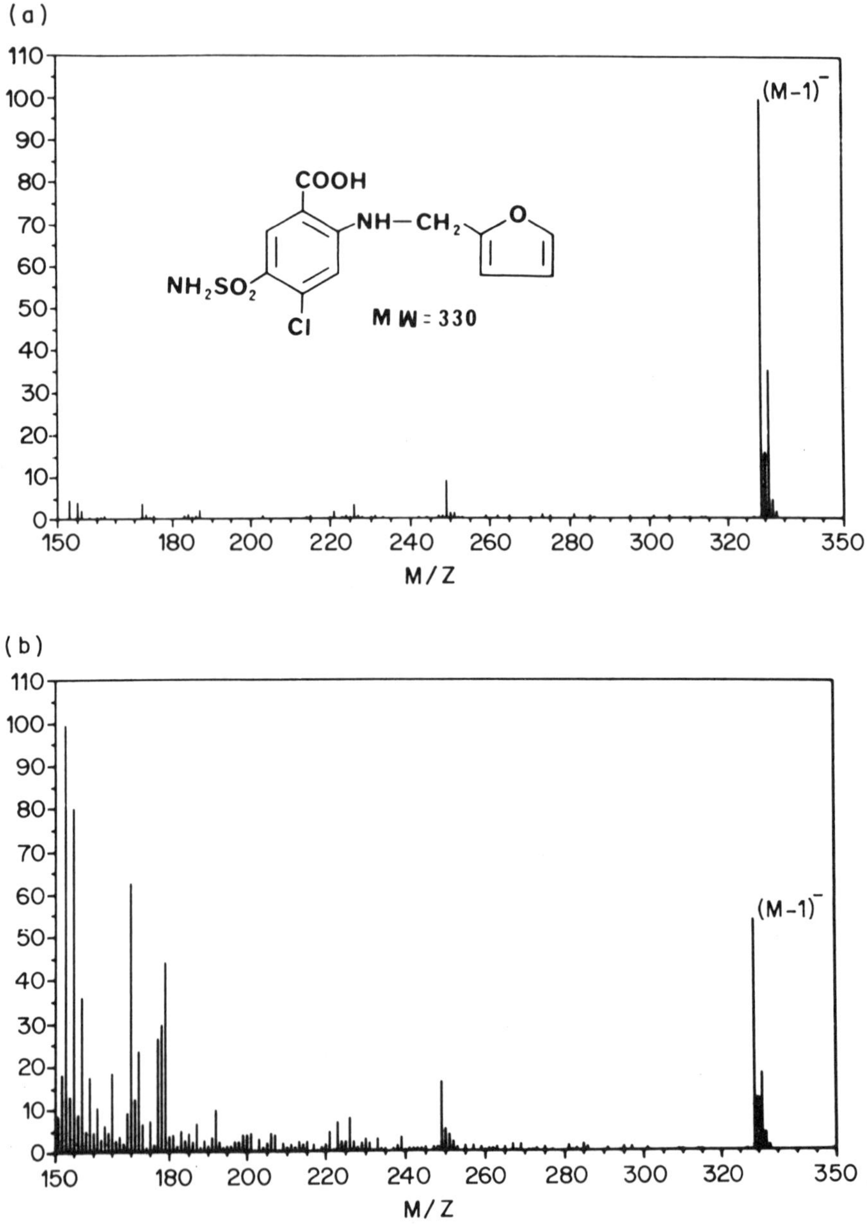

(c)

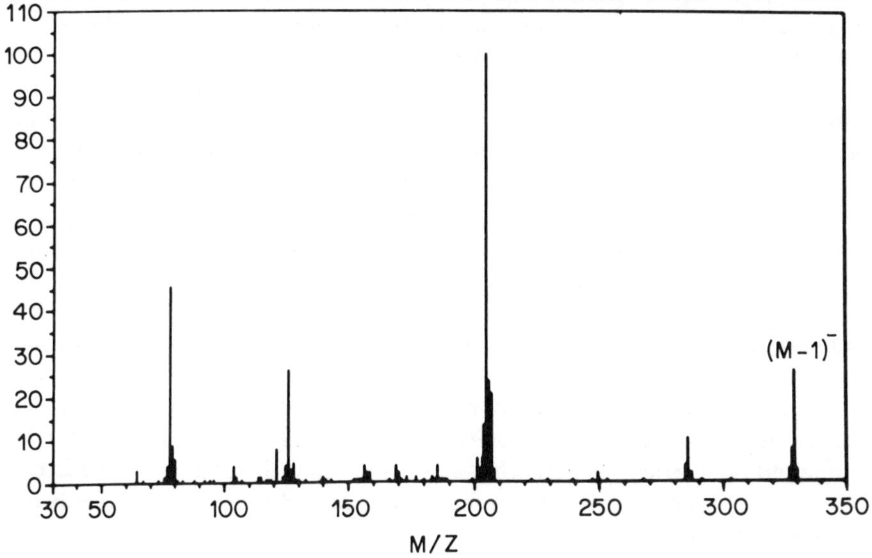

Figure 2: a) LC–MS APCI negative ion mass spectrum of the 1.1 min component shown in Figure 1b. b) LC–MS APCI negative ion background-subtracted mass spectrum of the 5.1 min component of the racehorse urine shown in Figure 1b. c) LC–MS/MS APCI negative ion daughter mass spectrum of *m/z* 329 from furosemide.

In a continuing effort to achieve rapid sample analysis with high sensitivity and specificity, we have evaluated so-called 'fast' HPLC as an inlet system for tandem mass spectrometry. The focus of this approach is to combine abbreviated sample clean-up techniques with LC–MS/MS. In this way the complex biological extract may be further chromatographed by a short, high efficiency HPLC column. Components which co-elute with the analyte of interest may be separated in a final stage by the tandem mass spectrometer. The approach taken is demonstrated in the examples which follow.

26.3. RESULTS AND DISCUSSION

An example of the capabilities of LC–MS and LC–MS/MS will be shown by means of the determination of the potent diuretic drug, furosemide, in equine urine. Since the analysis of the crude organic extract was accomplished by fast HPLC in a very short time period, the only useful separation accomplished by the HPLC column was that of the analyte of interest from the high levels of matrix interference.

Figure 1a shows the HPLC UV chromatogram at 254 nm for the successive injection at 2 minute intervals of an authentic standard of furosemide followed by a control urine extract and finally an extract of racehorse urine thought to contain furosemide as indicated by preliminary TLC screening results. The arrows indicate the expected retention time for the elution of underivatized furosemide under the HPLC conditions used and the injection points are indicated under the time axis. It is clear from Figure 1a that there is no evidence for furosemide in the UV chromatogram of the racehorse positive shown. Similarly, the total ion current chromatogram (TIC) shown in Figure 1b has little evidence for furosemide when compared with the TIC for the control urine extract.

After determining that negative ion detection with atmospheric pressure chemical ionization (APCI) is the ionization mode of choice for furosemide under these conditions, the LC–MS mass spectrum shown in Figure 2a was obtained from the 1.1 min retention time component observed in Figure 1b. Because APCI is a mild ionization technique, most of the ion current for the compound is carried by the $(M-1)^-$ ion and there is little structural information. Note that the mass range does not include fragment ions below m/z 150 due to the presence of abundant solvent adduct ions in the low mass region of the mass spectrum under LC–MS conditions. When the background-subtracted APCI mass spectrum of the sample eluting at 5.1 min in Figure 1b was recorded we obtained the 'mixed' mass spectrum shown in Figure 2b. The 'additional' ions seen in Figure 2b are presumably due to co-eluting interfering components which are to be expected from the unresolved 'bump' and peak tail seen in the TIC of the sample shown in Figure 1b. The mass spectrum shown in Figure 2b is not acceptable for forensic purposes, and requires either better chromatography efficiency by the HPLC column or the mixture analysis capability provided by tandem mass spectrometry (Finkle, 1981).

The triple quadrupole mass spectrometer used in this work can readily accomplish LC–MS/MS in addition to the LC–MS experiments described above. The differences between the mass spectral data available from these two types of experiment are evident from inspection of figures 2a and 2c. The collisionally activated decomposition (CAD) mass spectrum of furosemide (Figure 2c) obtained from the LC–MS/MS experiment includes increased structural information which spans the entire mass range. The technique also provides mixture analysis capability which is needed in the racehorse positive urine extract described above.

Although full scan LC–MS/MS analysis of crude biological samples can provide the most specificity, as shown in the CAD mass spectrum in Figure 2c, the technique of selected reaction monitoring (SRM) nicely sorts out the analyte from the interferents and provides improved detection limits. SRM is analogous to selected ion monitoring (SIM) in that it monitors selected ions of the CAD mass spectrum which result from a selected precursor. As an example let us look

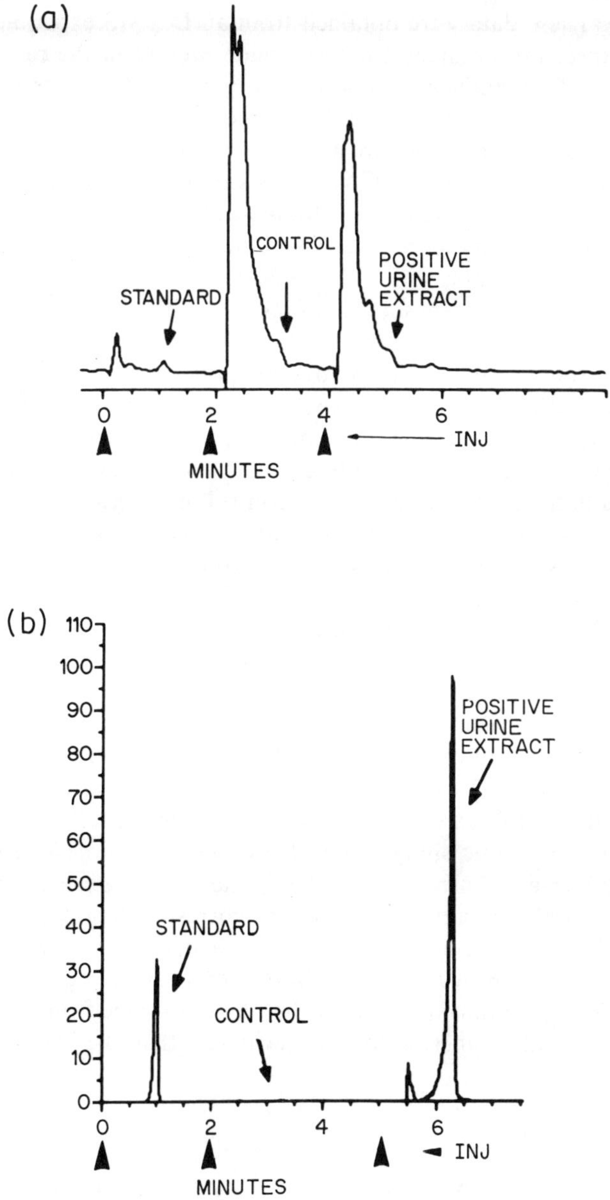

Figure 3: a) Total ion current chromatogram under negative ion APCI LC–MS conditions of the sample sequence shown in Figure 1a. b) LC–MS/MS APCI negative ion selected reaction monitoring (SRM) chromatogram of the same sample sequence.

at the data shown in Figure 3a,b. Figure 3a shows the same TIC that was shown in Figure 1b. These data were obtained from an LC–MS experiment, but the APCI mass spectrum obtained for furosemide present in the racehorse urine was not suitable for identification purposes (Figure 2b). However, if we conduct an LC–MS/MS experiment in the SRM mode we obtain the ion current chromatogram shown in Figure 3b. The trace represents the signal for the five combined selected ions from the CAD mass spectrum of furosemide (*m/z* 329, 285, 205, 125, and 78). These data dramatically show the mixture analysis capability of tandem mass spectrometry and the fast (1 minute analyte retention time) analysis times possible with LC–MS/MS. If we compare the complexity of the HPLC UV chromatogram shown in Figure 1a with the simplicity of the data shown in Figure 3b we see how readily the analyte of interest may be distinguished from any interfering matrix components. As expected, no analyte was detected in the control extract while analysis of the sample of authentic furosemide produces the SRM ion current profile observed from the first injection shown in Figure 3b. It is interesting to note that very little ion current was observed in the SRM analysis of the control urine extract even though the UV chromatogram clearly indicates a complex mixture. This obtains because the tandem mass spectrometer was set to focus only the parent *m/z* 329 negative ions through the first quadrupole, followed by CAD of these ions. The facts that the biological matrix components are not particularly sensitive to negative ionization and that those parent *m/z* 329 ions which are observed do not produce the selected daughter ions common to furosemide explain why the control urine extract yields the paucity of ion current observed in Figure 3b.

Finally, for forensic purposes the partial CAD spectra abstracted from SRM data (Figure 4a,b) may be presented along with the rest of the relevant analytical data for this sample (e.g. extraction chemistry, TLC $R_f$values and response to chromogenic sprays, and HPLC retention times) to identify the analyte as furosemide. Inspection of these data for the positive urine extract (Figure 4b) shows good agreement with the corresponding data for authentic furosemide shown in Figure 4a. Again, it is particularly noteworthy that the CAD daughter ions provide specific information spanning the entire mass range. These data, combined with the other data available from the analysis of this sample, provide sufficient forensic evidence that the sample contained furosemide.

26.4. SUMMARY AND CONCLUSIONS

An early claim for tandem mass spectrometry was that it would preclude any sample clean-up whatsoever. We find that MS/MS indeed can provide impressive mixture analysis for trace analytes, but not without some preliminary sample purification. In the case cited here the sample was subjected to a brief, simple liquid–liquid extraction and very crude (fast) HPLC purification on-line with tandem mass spectrometry. If HPLC analysis is eliminated by simply placing the extract on a solid probe, the matrix and other effects inhibit satisfactory results. However, by using the HPLC column to provide simple separation of the unretained, high levels of endogenous biological matrix components, the MS/MS technique can become a very fast, effective means of both screening and confirming the presence of targeted compounds.

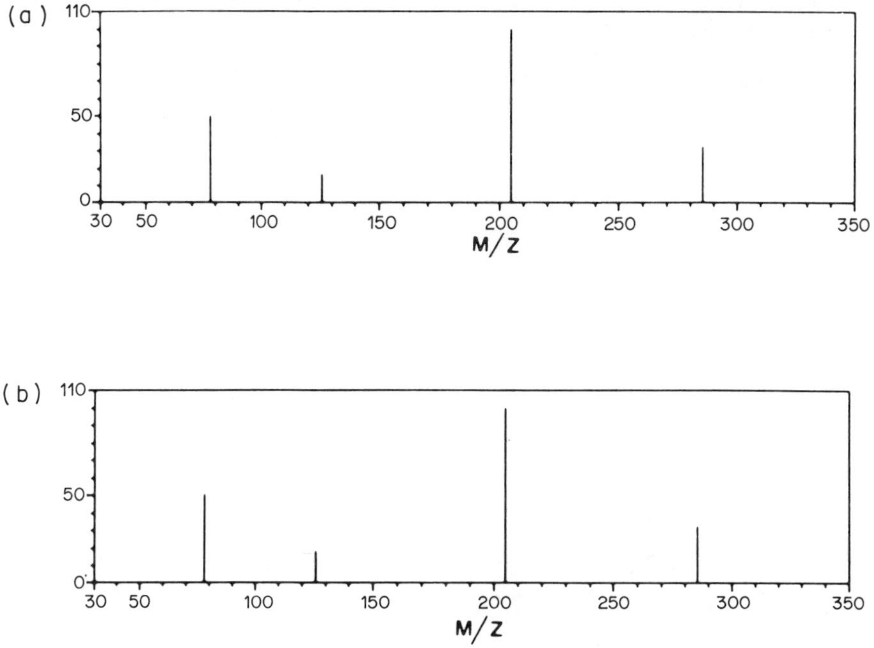

Figure 4: a) LC–MS/MS APCI negative ion bargraph plot of the SRM daughter ions *m/z* 285, 205, 125 and 78 of *m/z* 329 from furosemide. b) LC–MS/MS APCI negative ion plot of the SRM daughter ions *m/z* 285, 205, 125, and 78 of *m/z* 329 from the racehorse urine extract described in Figure 1a,b.

REFERENCES

Covey, T. R., and Henion, J. D. (1985). Quantitative selected ion monitoring gas chromatography/mass spectrometry of diethylstilbestrol in bovine liver, *Biomed. Mass Spectrom.*, **12**, 274–287.

Finkle, B. S. (1981). Forensic toxicology in the 1980's—the role of the analytical chemist, *Anal. Chem.* **54**, 433A–447A.

Henion, J. D., Maylin, G. A., and Thomson, B. A. (1983). Determination of drugs in biological samples by thin layer chromatography–tandem mass spectrometry, *J. Chromatogr.*, **271**, 107–124.

Johnson, J. V., and Yost, R. A. (1985). Tandem mass spectrometry for trace analysis, *Anal. Chem.*, **57**, 758A–768A.

Lee, L. D., and Henion, J. D. (1985). Micro-liquid chromatography mass spectrometry with direct liquid introduction, *J. Chromatogr. Sci.*, **23**, 253–264.

McLafferty, F. W., Ed., (1983). *Tandem Mass Spectrometry*, John Wiley and Sons, New York.

Skrabalak, D. S., Covey, T. R., and Henion, J. D. (1984). Qualitative detection of corticosteroids in equine biological fluids and the comparison of relative dexamethasone metabolite/dexamethasone concentration in equine urine by micro-liquid chromatography-mass spectrometry, *J. Chromatogr.*, **315**, 359–372.

Snyder, L. R., and Kirkland, J. J. (1979). *Introduction to Modern Liquid Chromatography*, John Wiley and Sons, New York.

INDEX

For abbreviations, see page xv.